YASI

The Danny Weil and Colvin Radio Expeditions

James D. Cain

Sweet dreams are made of this
Who am I to disagree?
I travel the world and the Seven Seas
Everybody's looking for something.

- "Sweet Dreams," The Eurythmics, 1983, (Dave Stewart and Annie Lenox).

"Hamming is everything in our lives"
Lloyd Colvin, to Leon Fletcher,
October, 1987.

"I never wanted to be put on a pedestal." -
Danny Weil, to the author, October, 2001.

"Every time YASME sails we die a little down here." - Dick Spenceley, KV4AA, in *The YASME Newsletter*.

Long you live and high you fly
And smiles you'll give and tears you'll cry
And all you touch and all you see
Is all your life will ever be.

- "The Darkside of the Moon,"
Pink Floyd, 1973, (Roger Waters, David Gilmour, Richard Wright).

Among the many boxes of amateur radio archives saved by Lloyd Colvin, W6KG, and Iris Colvin, W6QL, over more than 60 years, was *QST* for March 1930, just a few months after 14-year-old Lloyd received his first operator's license. Perhaps he gazed at the world map on the cover while he exchanged Morse code with other operators barely a hundred miles away from his boyhood home in Boise, Idaho, thinking of what it would be like to have his call sign on that cover, or to visit those far-away places on the map. In the years to come he, along with his wife, Iris, would achieve those goals, and much more.

To the memory of my father,
Kenneth Earl Cain, 1914-2000,
a member of "The Greatest Generation"

Printed in USA

ISBN: 0-87259-893-4

First Edition

Contents

Foreward

In April 1997 an international group of amateur radio fans gathered on an uninhabited reef in the South China Sea called Scarborough Reef - among them Chinese, Filipinos, and Americans. Mean-while, thousands of "hams" around the world waited eagerly for the team - which also included members of a Chinese scientific research group - to come on the air. Whoever managed to make a radio contact with this "expedition," whose call sign was BS7H, would be assured a spot at the top of the amateur radio world.

Unfortunately, a political dispute between China and the Philippines cut the BS7H operation short - to three days instead of the planned seven. The Philippine government said it had spotted three Chinese military vessels in the vicinity of the Spratly Islands, which are some 300 miles south of Scarborough Reef and were claimed by both the Philippines and China. (In the early 1980s two amateur radio operators attempting to land and operate from the Spratlys had died as the result of hostilities there.) The Philippines ordered an increased military presence in the entire area, and on the first day of the BS7H operation two reconnaissance jets flew over the reef; the next day a Philippine Navy ship landed at the reef and two Philippine Navy officers visited the three ham radio operating sites set up on Scarborough. The next day the operation was aborted.

This brief amateur radio operation from Scarbor-ough Reef was only the second from there in history, and it was funded in part by a monetary gift left by Lloyd Colvin, who had died in 1993 and is one of the three subjects of this book. During his lifetime Colvin and his wife Iris Colvin (who died in 1998) had visited nearly every country on earth, and had operated amateur radio from well over a hundred of them. The exciting expedition to Scarborough Reef was in the very spirit in which the Colvins had lived their lives and was a fitting tribute to them.

That spirit is the subject of this book.

Introduction - An Amateur Radio Primer

The terms "Amateur Radio" and "Ham Radio" are interchangeable. They refer to people in nearly every country on earth who hold government licenses to transmit on frequency bands set aside for personal, non-commercial communication by radio. Hams today communicate by Morse code, voice, television, and a number of computerized, digital modes - direct station-to-station, and through both land-based repeater stations and through orbiting earth satellites.

In the early days of the 20th Century hams were radio experimenters who were not even required to be licensed. Government licensing came in the nineteen-teens. From the beginning, the greatest lure of the hobby was to extend the geographic distance over which communication could be established. The very first radio communicators were thrilled to span a few miles; by the 1920s, amateur radio operators using the shortwave bands could contact other hams on the other side of the globe.

By the mid-1930s, hams were keeping track of how many countries of the world they had "worked." They would exchange written confirmations of their contacts, called "QSL cards," to prove their accomplishments. An award was established, for hams able to prove contact with 100 different countries. This award, called the DX Century Club (DXCC), recognized such entities as islands and political anomalies to count for the award - examples being Puerto Rico (a territory) and Vatican City (a political entity).

"DX" means "distance." Your car radio may even have a button labeled "DX."

After World War 2, during which amateur radio was suspended in most countries, the DXCC award program was restarted, and it once again became the most popular aspect of the hobby. Many of the country entities on the new DXCC list were uninhabited islands, while others were real countries which did not have any licensed hams. When an individual ham or group made a special visit to one of these places in order to put it on the air, the operation was called a "DXpedition," short for "DX expedition."

The subjects of this book are arguably the three most famous of all hams in the world of DXpeditions. The first, Danny Weil, left his home in England in 1954 and sailed, alone, across the Atlantic, through the Caribbean, then to islands in the Pacific, over the course of eight years and three boats. He landed in the U.S. Virgin Islands, on his first voyage, knowing nothing of Amateur Radio. Needing a reliable means of communication while sailing, he learned Morse code and radio theory, then obtained a British amateur radio license, issued in the British Virgin Islands. He then landed on islands devoid of ham radio activity and put them on the air. He is remembered by most hams as "the first DXpeditioner."

In the mid-1960s, on the heels of Danny Weil's adventures, Iris and Lloyd Colvin, a wealthy married couple from the San Francisco Bay area, began traveling the world. They were seasoned hams; he had been first licensed in 1929, she in 1945. They also were experienced world travelers; Lloyd had retired in 1961 as a U.S. Army lieutenant colonel, and his Army assignments had already taken them through many countries. From 1965 until Lloyd's death, in Turkey in 1993, they visited most of the countries on earth and operated ham radio from as many of them as they could, well over 100.

On most of these visits they stayed several weeks and made thousands of radio contacts. In the course of their combined lives they made more than a million ham radio contacts and had the largest collection of QSL cards known. It seems impossible that the sheer numeric measure of their accomplishments will ever be equaled in the amateur radio world.

More importantly, they served as unofficial ambassadors, as U.S. citizens and as amateur radio celebrities.

This book is about three people with serious wanderlust, ham radio being the common bond between Danny Weil and Lloyd and Iris Colvin. As people, they could not have been more different. This thread extends to the entire worldwide community of ham radio aficionados, people of different cultures, backgrounds, and status, all of whom are driven to "Work DX."

1 · The Birth of YASME

"... and so it came that one fine day I purchased for myself an eight foot sailing dinghy." - Danny Weil

Danny Weil was born on 14 December 1918, in a suburb of London. He was an only child. His father, Jack, died during World War 2, a victim of the Blitz - German air raids on London and other cities in 1940 and 1941. His mother, Christine Eckles, remarried and lived until 1994. Danny Weil lives today, in 2002, at age 84, in San Antonio, Texas.

He began school at age three, which was not unusual in England in those days, eventually earning a university degree in mechanical engineering and navigation. Danny came from a line of watchmakers going back hundreds of years and he apprenticed in the trade.

After working as a truck driver and for Dehaviland Aircraft, Danny joined the Royal Air Force in 1935, at age 17. He spent some ten years in the RAF, through World War 2, serving in Ceylon (now Sri Lanka), Norway, and Stalingrad (now Volgograd), before finishing his military career in Southampton. He taught workshop practices for the RAF and was also trained as an aircraft rigger.

After the war, Danny opened a watch shop in Southampton.

> I am a Londoner, generally known as 'Cockney,' being born within the sound of Bow Bells, and it was here that I trained as first a watchmaker, this being a family trade, and secondly, by choice, a mechanical engineer in which I graduated. With this knowledge I joined the Air Force and spent some ten years amusing myself, some in peace, others in World War 2, eventually escaping unscathed.

Danny opened a watch making business in Bournemouth in the south of England, where "It was away from all the smog of London, I had a fine business location, the climate was fine, and most important of all, so were the women ... it was a seaside resort.

"My business was a great success, but sitting at a bench all day, delving into the insides of defunct watches, listening to old ladies give me their life's history drove me almost nuts. I tried many ways to divert my mind from its deep dismal rut -- dancing, skating, motor cycle and car racing; they all took their course, but to no avail; I had run out of legitimate sports, or so it seemed then. Now, the ways of the sea were unknown to me, although in my leisure hours I had often watched small sail boats cruise up and down the beautiful River Avon. I think the urge to attempt this form of pastime came to me because it seemed so very different from any other form of sport, and so it came that one fine day I purchased for myself an eight foot sailing dinghy."

This was 1948, and Danny was 30 years old. He joined the local yacht club "to make things legitimate" and began to learn how to handle a sailboat, albeit a very small one.

In 1947, the world had been captivated by the voyage of Thor Heyerdahl, a 32-year-old Norwegian who sailed a 40-square-foot raft 4,300 miles from the west coast of South America to Tahiti. Heyderdahl, who died in 2002, set out with five others to demonstrate his theory that Polynesia's first settlers had come from South America. Although anthropologists now doubt his theory, Heyerdahl became an instant popular hero in 1947 and his book "Kon Tiki" became a best-seller when it was published on 5 September 1950. It was translated into 65 languages and a movie based on the book won an Oscar.

Danny Weil read the book "Kon Tiki" as soon as it came out in 1950, while he was preparing his first boat, and the primitive radio equipment he took with him was very similar to that aboard Kon Tiki.

Heyderdahl described the Kon-Tiki's equipment as a transmitter especially designed for the voyage, a back-up "secret sabotage set used during the war," a National NC-173 receiver (popular among amateurs at the time), all run

Danny Weil at age 6

on dry batteries. Two of Heyderdahl's five companions were experienced radio operators, and daily location and weather reports from Kon Tiki were transmitted on schedule, on 13,990 kc, just below the amateur 20-meter band, with just six watts of power.

The expedition's most critical use of the radios followed the landing of Kon-Tiki on an uninhabited reef in French Polynesia. After some drying out the receiver began to work, then, finally, the transmitter, just in time to transmit a report to a radio amateur on Rarotonga. There were no batteries left, so the four non-radio men on the crew took turns cranking a hand generator to power the equipment.

As a boy in the 1920s Danny read Jules Verne's 20,000 Leagues Under the Sea, which had first been published in English in 1873. The book was not so much science fiction as science fact; the human-powered Confederate submarine Hunley already had sunk a Union frigate during the Civil War. 20,000 Leagues was really a travelogue-by-boat, and it mentioned not only the Marquesas Islands (Danny's favorite) but even the most recent (in 2002) island to be anointed DXCC status: Ducie Island.

Danny's descriptions of his early lessons in sailing are filled with great good humor and self-deprecation.

> Very quickly the wind filled that tiny scrap of canvas and I leaned back into the stern of the dinghy, holding the tiller and feeling very proud of the whole set up. The boat heeled at an alarming angle, and I put my entire weight on the side to prevent a spill, but for some unknown reason we didn't move. I think it must have taken me a full five minutes to realize the fact that I had failed to cast off the mooring line, and in my endeavors to appear quite normal to the very amused crowd, I cast off the line holding the boat, but failed to release the rope at the same time which in turn, decided to pull me into the water. For some unknown reason I endeavoured to hang onto the rope and the dinghy, and it became a tug of war between myself and the wind, which was doing its level best to sail the dinghy into the quay wall. It finally penetrated my mind that I had better release the rope and board the dinghy, but the dinghy had other ideas. The sail jibed, the boat took off on another tack, and with the sudden motion, I was forced to release the side of the boat and drop into the water.

Danny Weil in a British WW 2 plane

At the time, Danny hardly knew the meaning of such words as "jib" and "tack" but later he would buy, and read, books on sailing and navigation, books he carried with him on his voyages across the Atlantic and Pacific oceans, alone. On this first river "voyage" he managed to, shall we say, become separated from his tiny vessel, while

> The dinghy meantime, had sailed itself to another part of the harbour and after gouging out a series of holes in other people's boats, finally ran aground in a big mud patch. With the crowd screaming with laughter and the irate owners of the other boats screaming with anger, I think you have a real good idea how I felt at that time.

Danny's second attempt to sail "this confounded boat" suffered from a lack of wind and he found his attempts at rowing "so bad the dinghy made better headway without my interference." This time, he got his dinghy into the river proper with the sail filling and the tide in his favor. Despite numerous groundings in the mud - he'd neglected to bring a can to bail out the boat - Danny was "beginning to learn that the buoys had a purpose, and for the last two miles, managed to sail without any further mishap.

> It was a perfect day for the sailor, the wind still blew steadily, the water was calm, and I was sailing in the right direction without a worry in the world. As I neared the entrance to the sea, so the current increased its speed, and the faster I went, so grew my enthusiasm for this sailing racket. The entrance, know locally as "The Run" came closer, and I could see the water bubbling and seething through this narrow entrance, and to my uninitiated mind, it looked like real good fun ... that's what I thought...

> Now up to this point, apart from the earlier incident, I couldn't see why everyone made such a fuss about the technicalities of sailing. There was I, the first time at sea, sailing along without any bother at all... it was just too easy. The mere fact I had slipped up at the beginning was simple to understand, but to actually sail the boat ... well at the

time, I couldn't see any point in it.

My knowledge of tides and currents was exactly nil, but I did know that at sometime it did change, but then, once again, I hadn't the vaguest idea which way I would go when the tide turned.

Danny persevered through several groundings and overturnings of his 150-pound dinghy, which "seemed to me to be immovable. As I pushed it out, so the sea would pick it up and plunk it back, knocking me over at the same time. Tears came to my eyes in exasperation as I struggled against the elements to win. Somehow, the beach became completely flooded and without any assistance from me, the boat floated off and drifted out to sea with me sifting, bowed down with the weight of the responsibility of my life."

Danny sold this first dinghy and "went back to ice skating." He also went back to "plugging away at the shop daily, back in the same old groove," but he was not to be defeated. He read his books and, more important, walked along the shore watching experienced sailors, "noticing how skilled helmsmen swung the tiny craft around in enclosed spaces, how they took every advantage of patchy wind. All these little points I tried to get stuck into my head for future use."

Danny began the search for a larger sailboat and found one in Sandbanks, a tiny place in Dorset, a "particular type of dinghy known as the 12-ft. "International" class."

Not daring to let the last owner know my lack of knowledge, I talked for some considerable time on other people's exploits, trying to give him the idea that I was an experienced sailor. Quite frankly, I don't think I took him in one little bit, but I will admit he was a good listener and never once disputed any of my very vague facts. After all, he was trying to sell me the boat, but at the time I thought I had really conveyed my 'deep knowledge' to him. I know now I didn't fool him one little bit."

Always a questioner of authority, Danny said "The trouble was, I was so intent on giving him my ideas that I didn't listen too closely to his, and that is where I made my big mistake."

Danny bought the boat and took it back to Christchurch, the docks closest to home. I was "anxious to get the boat into the water as quickly as possible," he said, and "the mistakes I made in rigging it were numerous, but impatience had always been a habit of mine and the job took three times longer than really necessary."

Danny was finding yachting both expensive and dangerous and decided once again to give it up until "I should either be able to buy a new boat or get more knowledge. Falling back into the old rut again gave me a big pain, and I found that on those off days I would wander down to the boat yards gazing enviously at all the yachts moored up, yachts capable of traveling long distances without any worry, but my pocket book was far too small to cope with my big ideas."

Among the books Danny read were several by individuals who had built their own boats capable of ocean voyages,. but they had been experienced mariners. Danny faced a double whammy of having little money and no sailing experience. He was a watchmaker and mechanical engineer, "hardly allied trades to the sea. I had the feeling that without the knowledge so very necessary, I would still know the boat when I found it, but to find it was another question."

In January 1951 Danny Weil began the "biggest project of my life," rebuilding the boat that would be named "YASME," a take on the Japanese yasume, meaning

The hulk that would become YASME I

"freedom." He had found a 40-foot, 21-ton "yacht." Danny fitted it with a 10-HP gasoline engine. It was too small for the job but a better, diesel engine, was financially unattainable. By the end of 1953, working in his spare time, Danny found that completion was still at least a year away. "There was only one thing to do: give up work." Danny found someone to take over the watch shop and "threw the whole mess into the manager's hands with a warning that whatever happened I wasn't to be disturbed."

YASME was seaworthy by summer and successfully completed her trial runs. Despite later saying that a trip around the world, alone, was "too fantastic for words," the boat was stocked with canned goods and 200 maps, along with the collection of books on sailing. Departure from England took place on 1 August 1954, with "No turning back but forward to a new life, new countries, new people and ever widening oceans." Danny found it especially difficult to leave his mother behind, knowing the voyage would "part us for an unknown period, maybe forever."

Danny sailed away, south to the west coast of Africa and across the Atlantic. He left Gambia, West Africa, on 20 November 1954, and sighted the British West Indies just 22 days later, having covered 3220 miles, perhaps a record for the time for a boat the size of YASME.

It was a remarkably trouble-free voyage considering this sailor's inexperience and heavy seas and near gale-force winds for much of the crossing, but Danny Weil had literally rebuilt YASME from stem to stern and there was no problem he couldn't handle

Not to mention the unlikelihood of running aground in mid-ocean.

2· Early Years: Lloyd Colvin and Iris Atterbury

"Amateur radio captured Lloyd's imagination and he then and there determined to reach out and communicate with the world." - Iris Colvin, on Lloyd

Danny Weil was a contemporary of the other two major subjects of this book, born, in 1918, just a couple of years after Lloyd Dayton Colvin (24 April 1915) and Iris Atterbury (15 April 1914). The years 1914 to 1918 coincide with the period of World War I. Danny, Lloyd, and Iris all came of age during the world-wide economic depression of the 1930s, and spent the formative years of their twenties serving the Allied cause in World War 2.

Lloyd as a Boy Scout

Lloyd was born in Spokane, Washington, and lived there until 1921. His father, George Rubin Colvin, was from Bennington, Vermont; graduated Cornell, and was voted "most likely to succeed." George met Edna Maria Teeter in Ithaca, home of Cornell, and they were married in 1910. On Lloyd's birth certificate their occupations were listed as "real estate" and "housework." Lloyd's maternal grandparents had been early settlers in New York and traced their ancestry to a veteran of the Revolutionary War.

In a letter enclosed with a QSL in 1991 from a ham in England, the Colvins learned that the name Colvin has both Gaelic and Norse derivations. It appears in land tenure records as early as 1511, which are the earliest still existing, except for a few fragments from the late 1400s. The Norse spelling was 'Kolbein.' The Norse, or Vikings, first invaded, and later settled in the Isle of Man in the year 878 and their military power came to an end in about 1260.

Lloyd's maternal grandparents were separated when his grandfather, James A. Teeter, went west in search of gold. His wife, Emma Kate Teeter, stayed in Ithaca and never saw James again. Lloyd's mother, Edna, stayed, too. Years later, in the 1940s, James visited Edna and Lloyd, in Idaho. James is reported to have been killed in a bar in Alaska and buried there.

Not long after their marriage in 1910, Lloyd's parents, Edna and George Colvin, headed west, where George bought land in Boise. George was involved in many land dealings as they continued west, most of which were being mined for gold. At one time George held deeds to more than 200 mines.

Less is known about Iris Atterbury's ancestry. Her father was Clarence Victor Atterbury, who had been born in Oregon in 1880 and had been a railroad brakeman for the Southern Pacific line. Iris's mother, Cozensa Clark Atterbury, was from Wisconsin.

The Spokane of Lloyd's first six years was a railroad center of about 40,000 people. The Colvin family moved to Boise, Idaho in 1922 and lived there until 1932; Lloyd attended Longfellow Grade School and Boise High School, and for a time was trustee of W7YA, the school radio club. In the first 20 years of the century Boise had doubled in population, to about 20,000 when the Colvins moved there. By then the gold had run out and the area had turned to agriculture.

After Lloyd's death Iris talked about his boyhood interest in radio, saying it began in 1929 when he read about amateur radio in an issue of Boy's Life. "It captured his imagination and he then and there determined to reach out and communicate with the world." she said. Lloyd sent away for a mail order radio kit, the kind that teen-agers were still assembling 30, 40 years later. It worked right away and Lloyd was hooked. He didn't know any local hams so he learned and practiced the Morse code by himself and passed his first amateur radio exam in 1929: in the days of the Federal Radio Commission, pre-Federal Communications Commission. His first license was an operator's permit only, his call sign, W7KG, came a few days later. Details of his first examination are not known; he may have traveled to another city for the exam, or passed it before a traveling FRC examiner.

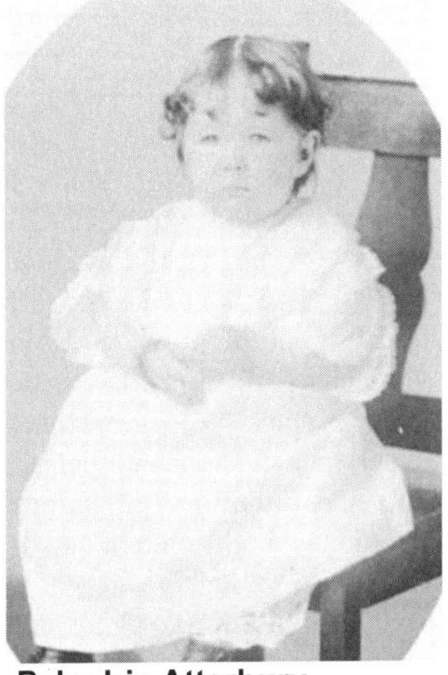
Baby Iris Atterbury

In "The Murder of Roger Ackroyd," one of Agatha Christie's characters, a Dr. Sheppard, says "I am rather proud of the homemade wireless set I turned out." That was 1926, and throughout the 1920s amateur wireless was the most popular of all scientific hobbies.

Grandfather and project

Lloyd worked for Call Radio and Crystal Laboratories in Boise, Idaho, as a "radio service man" from January 1930 to January 1931 - a handy job for a new, teenaged amateur radio operator - and describes his work there in an early resume: "Radio servicing. Wiring and installing of antennas. Connecting and installing radio receivers. Repairing of radio sets, power supplies and loud speakers and building of radio transmitters. Wiring of radio apparatus and equipment. Assisting in the grinding of quartz crystals to specified frequencies. Built and used accurate frequency measuring equipment. Shop management - often left in charge of shop in absence of manager."

The first log book for W7KG begins on 2 September 1930, when Lloyd was living at 1620 North Sixth Street in Boise. Until then he'd operated from the stations of other local hams. His operator's license, No. 24942, was dated 19 November 1930.

From January 1931 to June 1932 Lloyd worked for the Mountain States Public Address Company and Radio Station KIDO, in Boise, as a sound technician and radio engineer, and described his duties as a 16-year-old: "Installed Public Address apparatus. Operated sound truck. Assisted in the installa-tion of sound movie installations. Installed public address-radio broadcast hookups. Obtained experience in sound transmission for broadcast purposes over telephone lines. Designed, built, wired, and tested amplifier circuits. Used radio test instruments, voltmeter, milliammeter, galvanometer, ohmmeter, frequency meter, oscillograph and test oscillators. Obtained experience in announcing, making sound recordings, and operating amplifier mixing panel. Assisted in design, building and operation of new 1000 watt radio transmitter KIDO, Boise, Idaho."

Lloyd's W7KG log book on 10 February 1931 records his first contact from 810 Main St. in Boise; the family later moved across town and his first entry from 1203 North 13th Street came on 17 May 1931. Lloyd recorded his first 'phone contact on 25 November 1931. He made about a dozen 'phone contacts and after that there are no more in this early log book. Lloyd was a CW man (a Morse code operator) from the start.

Lloyd stayed in Boise through his junior year of high school, when he and his mother and grandmother moved to Berkeley, California. George was not with them. Apparently his business affairs went badly and he was drinking heavily, so Edna and her son left him and went to Berkeley. Lloyd was bitter about his father's failure to provide for his family, and in the years Lloyd's daughter, Joy, was growing up George's name was never spoken, she recalled.

Lloyd graduated from Berkeley High School in 1933. He was busy on the radio, too, and played "call sign roulette" with the FRC, for a while being W6TG (first QSO on 21 October 1933), then W6AHI beginning in March 1934. Many years later Lloyd would recount operating from W6ANS, W6IPF and W6KFD.

Lloyd's mother went to work to enable him to stay in college as the depression took hold. As the "man of the house" he worked too, at radio station KRE, in Berkeley. When the call sign W6KG came available, Lloyd asked for and was granted it. 50 years later an article by columnist Dick Barnett in the San Jose Mercury (August 1977) told a story of Terry Hansen, the amateur who previously

Iris Atterbury (L) and her sister Clara.

held W6KG. Barnett said that Hansen got his start in radio in 1909 "when he put up a 100-foot antenna pole at Evergreen without assistance, and became a radio ham." Hansen's father had been a ship captain. Hansen got a commercial radio operator license and served on banana boats to Central America and also on a ship to Australia.

Hansen in 1977 received the Society of Wireless Pioneers "SOS-CQD" award for saving the lives of a ship's crew 63 years before. Hansen was radio operator on the S.S. Isthmian (radio call WKI) when it went aground on 19 December 1914 at San Benito Island off the Mexican coast near Guaymas. Hansen's "SOS" resulted in the saving of the ship and its crew.

Hansen was in the U.S. Navy during World War I, building and operating a radio station at Pearl Harbor. After the war he became W6KG, and was an electrician. The similarities to Lloyd Colvin, who inherited the call sign W6KG, are eerie.

Lloyd's mother worked as the governess to the children of a family being sent to the Far East in charge of the Sprekels sugar plantations with headquarters in the Philippine Islands. In these days before the Pan American Airways Clipper flights to the Far East, a letter took a month by ship from San Francisco to Manila and commercial radiograms were quite costly.

Lloyd's W6TG station was a three-tube TRF (tuned radio frequency) receiver and a 210 transmitting tube in a Hartley transmitting circuit, for CW only, of course. Lloyd had worked very little DX but he wanted to send a message to his mother in the Philippines, and he tried day after day on 40 meters to find a "KA" station. "My antenna blew down, there was heavy QRM and QRN, I would hear a KA but be was busy with a sked, my rig blew up, etc."

After two weeks Lloyd finally raised KA1NE and called it "a thrill never to be forgotten." Lloyd told him about his mother, they made a schedule for the next morning, and KA1NE sent Lloyd's mother a commercial radiogram. The next day, Lloyd got a full report and, later, KA1NE was in Manila and took Lloyd's mom to dinner.

Lloyd, packed for a trip

In 1933, China had been united for five years under Chaing Kai-Shek, although Japan had seized Manchuria in 1931. Lloyd wrote to his mother while she was on this trip:

at 1620 N. 6th ST.
USING 1-210 in Armstrong ckt
300 VOLTS ON PLATE
HERTZ 133 FT ANT.

SEPTEMBER

T	CALL	W C	HIS FREQ	MY FREQ
4:00 PM.	W7AAB	W	7000	7000
5:00 P.M.	W6BQB	W	"	"
9:00 P.M	W6EKE	C	"	
9:05 P.M	W6AQJ	C	"	
6:15 A.M	W9FZ	C	"	
6:20 AM	W5BMC	C	"	

W7KG 1930 log book; Lloyd called more stations than he worked!

> Oh how I envy you on your happiest and most wonderful trip. Think of the pleasures you will experience. Imagine how you will delight in making acquaintances among the Itonates, the Tagalogs, Visayans and the Bicols, not to mention the Malayans, Pampangans and the Mindanaos. While in the Philippines you must explore the southern islands of Sulu. Hunt tigers in Samales, see the fierce Serval in Basilan, eat the raw meat of the Timarau, walk through the muddy marshes of Jolo, meet the Pythan in Siarse and watch the cockatoo fly in Tawi-Twai.
>
> At some propitious time during your delightful stay there, you must visit Sumatra. Meet the Malayans that live on the river Rokan, the Chinese who inhabit the valley of the Muse, and the Caucasian native tribes that live in the lower deltas of the river Jambi. And pleasure of all pleasures you must visit the wild island of Palawan. Become acquainted with the cannibalistic and uncivilized ways of the Sulfas, Yaks and the Buggies.
>
> Oh what I would give to be able to go barefooted and half naked on the shores of the island of Negroes and see the various specimens of Cetacean and Crustacean floating and flopping in the sea. Think of the lazy hours you can spend laying in a tropical stupor, possibly suffering from

some exclusive tropical disease such as Elephantiasis, a disease in which the arms, legs and feet swell up to an enormous size until finally the very life is crushed out of the afflicted victim. Think of the various forms of the poisonous lizard whose bite is supposed by some half cast[e]Malayans to cure the dreadful disease called Barbados leg. Stop and admire the wonderful vines and the strange underbrush, but do not touch it as it might be a variety of the poisonous, creeping Sumach.

I could go on for hours writing of the wonderful pleasures and adventures you will experience, but far be it from me to spoil the fun you are going to have. I will let you find each exquisite little place, each delightful experience all by yourself in the manner of a true adventuress, dreading nothing, enjoying everything and tasting life to the fullest. - Your ever loving son, Lloyd.

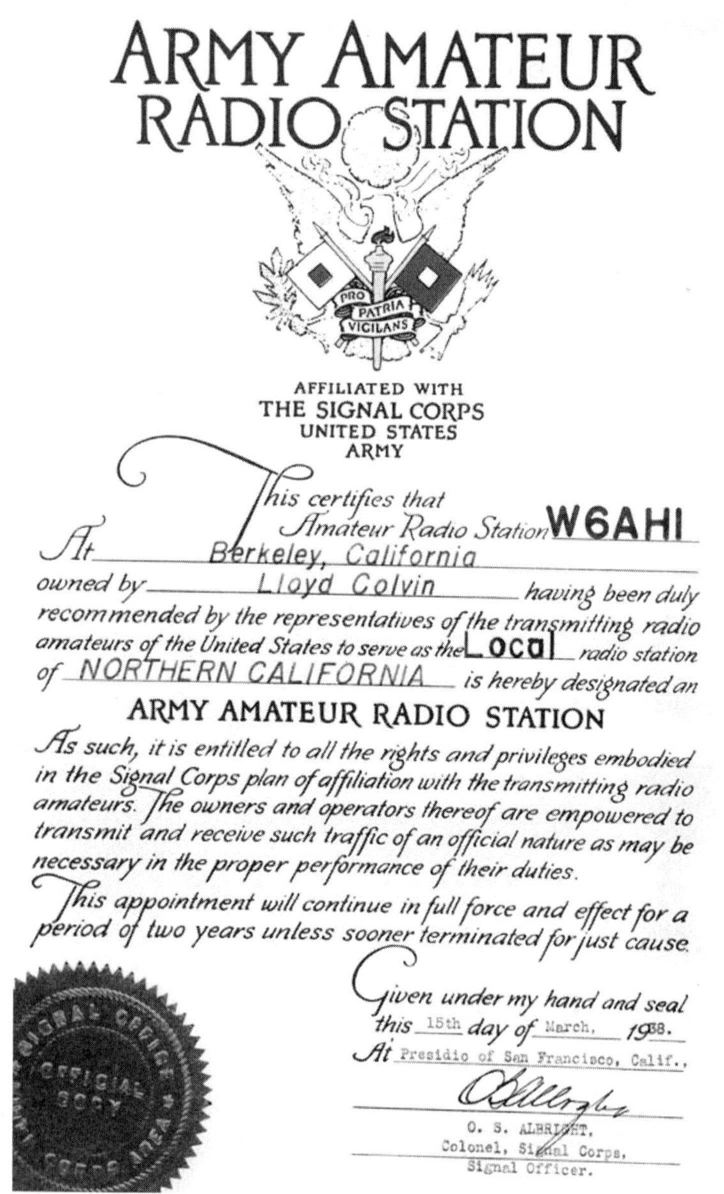
ARMY AMATEUR RADIO STATION

PRO PATRIA VIGILANS

AFFILIATED WITH THE SIGNAL CORPS UNITED STATES ARMY

This certifies that Amateur Radio Station W6AHI At Berkeley, California owned by Lloyd Colvin having been duly recommended by the representatives of the transmitting radio amateurs of the United States to serve as the Local radio station of NORTHERN CALIFORNIA is hereby designated an

ARMY AMATEUR RADIO STATION

As such, it is entitled to all the rights and privileges embodied in the Signal Corps plan of affiliation with the transmitting radio amateurs. The owners and operators thereof are empowered to transmit and receive such traffic of an official nature as may be necessary in the proper performance of their duties.

This appointment will continue in full force and effect for a period of two years unless sooner terminated for just cause.

Given under my hand and seal this 15th day of March, 1938. At Presidio of San Francisco, Calif.,

O. S. ALBRIGHT, Colonel, Signal Corps, Signal Officer.

Lloyd had entered the University of California at Berkeley in the fall of 1933, aiming for a degree in electrical engineering. It took him five years to fin-ish but in those days at the height of the Depression just continuing one's education at all was a struggle. Through college Lloyd worked at Drake Cleaners, in Oakland, a chain of stores. Lloyd did everything from bookkeeping to tending store to delivery. "It was the depression era," his daughter Joy said years later, "and he worked very hard to help support his mother and grandmother and put himself through college at the same time. He considered 'Drake' to be a lucky name, and that was why he took the name Drake Builders for his construction activities."

A 1933 Boise newspaper article "Boise Boys and What They're Doing" reported on the former resident, saying that Lloyd had finished his junior year at Boise High School, then graduated from Berkeley High School, and that studying under three different state school systems did not seem to have set him back.

Boise school mates remembered him as the first trombone player in the high school band. The newspaper also said Lloyd had been "given a federal permit to operate a private broadcasting station, WG3257, in Boise."

Chuck Patterson, K6RK, was W6ATR in 1933, and graduated from U.C. Berkeley the same year as Lloyd, 1938. Chuck and Lloyd were electrical engineering majors and in the Signal Corps ROTC. He remembers

"I first met Lloyd in about 1933, when I went to visit him at his Shattuck Avenue home. He was on the air, so I went to find his roommate Dick Jenkins, W6ANS. I found Dick in the bathroom with a 5-gallon tin of alcohol. He was making bathtub gin."

"Lloyd used to joke that he knew me before he knew Iris."

Despite tight finances, Lloyd managed to maintain a very modest ham station through his college years. His log book in August 1938 notes he was using a "New Skyrider receiver SX-17, (very fine business)." This may have been his first superheterodyne receiver and perhaps his first commercially-made equipment. His log for 10 November 1940 records a "New 814 final, HT-9 xmtr" (a Hallicrafters transmitter).

Forty years after university graduation Lloyd applied to be appointed a commissioner of the U.S. Federal Communications Commission and said, in his application, that he held a BS degree in both Power and Communication Electrical Engineering. He noted that he had worked part time for radio station KRE, Berkeley, California, as combination announcer, engineer and operator, and "assisted in the installation of new transmitter and studio equipment."

Lloyd began his military career while an under-graduate, spending a year and a half of Reserve Officer's Training Corps (ROTC), in the Coast Artillery as a member of the first Signal Corps ROTC class at the University of California. In addition to his basic Coast Artillery Training (military fundamentals, primary coast artillery functions, and materials, fire control and position finding for seacoast artillery); he reported helping to install a 200 watt Western Electric transmitter at the University and acting as a part time instructor in the Signal Corps ROTC. He was elected to membership in Scabbard and Blade; Pi Tau Pi Sigma, both military fraternities.

In 1981, Lloyd told an interviewer that " I got into the Signal Corps through ham radio. In fact, almost everything [Iris and I have] done in our lives has been directly connected with ham radio. I guess I found my military life interesting because much of it was similar to running ham stations."

In later years Lloyd would recall that his ROTC training had included radio code practice, radio materiel, tests and procedure, leadership, aerial photography, tactical employment of radio systems; also wire and radio communications, leadership, training methods, orders and administration, wire and radio traffic, message centers, communication tactics, field problems, mechanization, general signal com-munications, military history, combat orders, wire and radio nets, tactical signal communications, signal staff duties, training, supply and mess management, and O.R.C. regulations. He spent two months in "actual army signal communication work" at Fort Lewis, Washington, and Morro Bay, California.

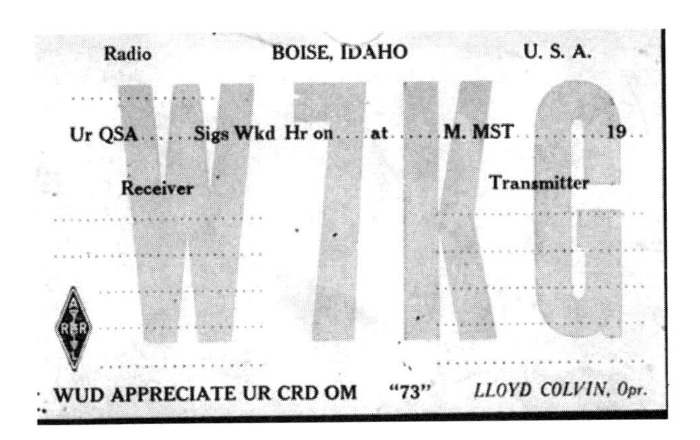

DEPARTMENT OF COMMERCE
RADIO DIVISION

OFFICE OF RADIO INSPECTOR
227 NEW POST OFFICE BLDG.
PORTLAND, OREG.

Nov. 19th, 1930.

Mr. Lloyd D. Colvin,
1620 North 6th St.,
Boise, Idaho.

Dear Sir:

There is enclosed herewith amateur class radio operator's license No. 24942, issued to you as a result of your examination at Boise, Idaho, Nov. 12th, 1930. You obtained a mark of 92% in this examination.

Your attention is invited to the Oath of Secrecy on the reverse of this license, which must be executed in the presence of a Notary Public. The license may then be returned to this office for the signature of the Radio Inspector.

Also, you are advised that it will be necessary for you to return your temporary amateur license to this office for cancellation.

Respectfully,

Stacy W. Norman
STACY W. NORMAN
ASSISTANT RADIO INSPECTOR

50 years later Lloyd received a letter [8 August 1988] from Frank Craig, WJ6U (formerly W6EQV and W3MVU), of Escondido, Calif.:

> Dear Lloyd: I keep seeing references to Lloyd and Iris Colvin in Worldradio and elsewhere. As a traveler and a ham I marvel at your accomplishments in both spheres. What I am wondering is: are you the same Lloyd Colvin who used to work with Jack Prichard and me in the Radio Lab at UC Berkeley in 1938? There cannot be two of you! I remember you worked for a dry cleaner to help pay for your tuition!
>
> I missed the war because of physical - not psychological - limitations, but went to work for the FCC for 9 years, then 32 years in aerospace, involving 16 years in Europe. After 20 years off the air I came back in 1984, and now working on my first DXCC. Let me hear from you when you can take a breath between DXpeditions.

Lloyd graduated in May 1938; working continuously

while in college he managed to finish in five years. For the rest of his life, when he filled out U.S. Army forms, in the square for number of years of college he would write "5." Lloyd's degree plus ROTC experience qualified him for a position as radio operator on ships at sea and he got a couple of offers. He also recalled that he was "one of a limited number of military students" offered a commission in the Regular United States Navy, as an ensign. But he took a job with the Jacuzzi Pump Company, in Berkeley.

At Jacuzzi, the young engineer was the proverbial jack of all trades, inspecting castings, checking and tabulating specifications of new materials and parts, and designing machine components.

A charcoal done by Iris while at Berkeley

Lloyd "Took complete tests of new pumping units. Prepared charts and graphs for general distribution giving capacity, power input and output, efficiency, and other pertinent data. Supervised rebuilding of test laboratory. Assisted in compiling blue prints, plans, specifications, and reports sent to an eastern company to manufacture pumps under patents held by the Jacuzzi Pump Company, Obtained field tests and working specifications of deep well, shallow well, centrifugal, turbine, and injector pumps."

Lloyd then worked for the Pacific Gas and Electric Company in the Plans and Estimating Department.

He designed substation equipment, including inspection and rating of circuit breakers, oil switches, station panels, bus structures, battery banks, current relays, under and over voltage relays, thermal relays, reverse power relays, differential relays and residual relays. He "made reports as to tests and rating of substation apparatus and made recommendations as to new equipment, Inspected new apparatus. Became familiar with operation and mainte-nance of motor generators, rotary converters, synchronous converters, oil switches, circuit breakers, etc."

As a test engineer, Lloyd got his first tastes of supervisory roles, overseeing test technicians and installers in the Bureau of Tests and Division Machine Shops, as well as directing crews linemen. Not only was he doing typical engineer work in designing systems and networks and preparing specifications, he calculated costs, man hours, and materials used. There was radio work, too, namely installing receivers and transmitters for carrier current control of distribution circuits.

It wouldn't be long, however, before PG&E would grant Lloyd an extended leave of absence to join the Army, as war loomed.

3. The Army and The War Years

"I worked all night long grinding new quartz crystals ... and making other necessary changes in order that the fulfillment of the mission would not be delayed." - Lloyd Colvin

Iris graduated from Berkeley in 1937, from the College of Arts and Sciences as a fine arts major. She met Lloyd on a blind date. It isn't known if she worked immediately following graduation. She lived in San Francisco, Lloyd in Berkeley. They were married on 11 August 1939, and moved into 1809 Shattuck Ave, Berkeley, where Lloyd had lived since at least 1933.

Sometime during their courtship the young couple made news when a pleasure boat they were partying on off the coast capsized. An overheated headline writer titled the story "Raging Seas Swamp Tug Near Point Bonita; Coast Guard Makes Four Rescue Trips."

"Twelve members of a gay picnic party, their boat crushed and capsized by heavy breakers in the "Potato Patch" - three miles off Point Bonita - were snatched from drowning yesterday by the United States Coast Guard.

"So treacherous were the heavy seas three miles out from the Golden Gate, that a life boat from the Point Bonita Coast Guard station made four trips to rescue the six men and six young women who clung perilously to the wreckage for more than an hour."

Lloyd was (mis)identified as "Lloyd Colin, 27" and Iris as "Mrs. Iris Colvin, 23." Both were reported to have been treated for "excessive shock.

"The boat's owner, Jack Deal, said 'We were heading into the breakers and I was afraid to take one of the combers broadside. They were so big I thought they would swamp us. A big breaker towered over the boat, crushed the pilot house and flooded the motor. A second later, another caught us broadside and the boat turned turtle under the heavy impact. The gasoline tank was torn loose and floated free. We were all in the water and then grabbed whatever we could lay hands on.

"'Five of us held to the gasoline tank and some others grabbed the roof of the pilot house as it floated by. I thought the Coast Guard would never get there and then when the lifeboat came along the seas were so heavy it could pick up only three of us on the first trip.'"

A follow-up article saying that four Coast Guardsmen received lifesaving medals from "Treasury" identified the Colvins as "Iris and Lloyd Calvin."

In the spring of 1940, with the Depression still in full swing and U.S. entry into the war not yet an inevitability, Lloyd took U.S. Civil examinations and qualified for 15 positions, including engineer, ordnance material inspector, radio technical school instructor, radio operator, and Signal Corps equipment inspector.

The Colvins' first (and only) child, a daughter, Joy, was born on 20 March 1940. Lloyd visited his father in Boise this summer of 1940, for the last time.

A romantic canoe trip

On 7 July 1940 Lloyd, an Army Reserve officer, was ordered to active duty for two weeks, and then again in August was assigned to Radio Station WVY in San Francisco. Lloyd described the station as a "major communications center" on the West Coast for the United States Army. Wire, radio, and facsimile circuits extended to ships at sea, Australia, New Zealand, Alaska, Washington DC and approximately 30 other cities and countries. As, at times, officer in charge, Lloyd supervised technicians, engineers, and operators in building new radio stations and communications facilities, both telegraph and teletype. He also organized Army Amateur Radio System operations, as his active duty kept being extended. He was promoted to 1st Lieutenant in October 1940, "later amended" to 26 August 1940.

The Army shipped Lloyd, Iris and baby Joy to Seattle in March 1941, where Lloyd was Assistant to the Officer in Charge of the Alaska Communications System (ACS)." Lloyd spoke of this assignment in his 1974 application to be a Federal Communications Commission commissioner, down-playing the military aspect while embellishing the civilian implications: "Although this system was run by the US Army Signal Corps, it acted as a semi-civilian organization in charge of all communications in the Territory of Alaska, including the type of control exer-

Sweat equity in Anchorage

cised by the Federal Communications Commission now." Lloyd said the ACS operated telegraph offices in Alaska, gave tests and issued radio operator licenses, and regulated rail and air communication as well as broadcast stations.

Lloyd and family then were sent to Anchorage, the headquarters of the ACS, where he was liaison officer to the Alaska Defense Command, at Fort Richardson. His duties included overseeing radio station WXE Anchorage, "the largest point to point radio station in Alaska." He also represented the Alaska Defense Command and the Alaska Communications System at conferences with the Navy, Coast Guard, CAA (predecessor of the Federal Aviation Administration), and the Canadian government.

Lloyd helped develop radio silence programs for use during periods of enemy activities and participated in radio intercept work, fake tactical radio transmissions to the enemy, and enemy radio propaganda broadcasts. As communication representative, participated in the initial survey of the sites on which many of the army forts in the Aleutian Islands were located, and helped with engineering a multiplex, deep-sea, communications cable system between Anchorage and Seattle.

During the same period, Lloyd helped develop what he characterized as "the first large Single Sideband Multichannel Radio System used for commercial and military communications." That installation was made by engineers for the Western Electric Company.

Alaska was the scene of the Colvins' first business venture, although it was accidental. Iris would recall years later that Lloyd had always dreamed of having a big antenna farm with rhombics [antennas] covering acres of ground. "When we went to Alaska there was [land] near our QTH that belonged to someone in Seattle who never came to Alaska. We sent a telegram offering to buy [it]. He accepted and it took all the money we could scrape together to buy it."

Actually, Lloyd later said they had to borrow most of the money to buy the land. But before they could get the antenna farm built World War 2 began and stopped amateur radio operation, so they sold the land to the Signal Corps for "an immense profit, and no income tax."

That experience launched Lloyd on a career of buying and selling property. To do that, it was obviously necessary to know exactly where particular plots of land were situated - to have properties surveyed. "But in those days there were very few surveyors in Alaska," Lloyd said. "Earlier, while studying for my degree in electrical engineering at the University of California at Berkeley, I had taken one course in surveying. That was enough for me to get a surveyor's license in Alaska. "

Soon he branched out, got a contractor's license, and was building homes, offices, and industrial structures in Alaska and, later, as the Army moved them around, in New Jersey, North Carolina, and California. Along the way, Iris also obtained a contractor's license and worked closely with Lloyd as the projects increased in number, size, and complexity.

Together they continued to build their resources, until 1965: "That year we sold our house, closed out the construction contracts we had going, and went on our first DXpedition."

During the years in Alaska Iris worked for the Bechtel Corporation. On 13 December 1941 Lloyd's mother remarried, to Hal O. White, in Reno. Hal died in 1947 at age 68.

On 14 January 1942 the word from Washington came to Alaska, by teletype:

> By the enactment of public law no three three eight seventy seventh congress approved dec thirteenth forty one the active duty of all officers warrant officers and enl men of all components of the army of the united states including retired officers has been extended for the duration of the war plus six months stop the law includes all personnel who were on extended active duty on dec thirteenth forty one stop individual orders extending the tours of active duty of res officers are no longer required.

In early 1942, Lloyd moved to the Alaska Communications System at Whitehorse, Yukon Territory, Canada, as liaison officer to HQ Northwest Service Command, including point to point army radio station WULO and army broadcast station CKWH. He was promoted to captain on 1 February 1942, and became involved in the planning for the Alaska Highway and CANOL projects.

Alaska and northern Canada were considered strategic in World War 2. An official description of the CANOL project said "in the spring of 1942, after the Japs had overrun the islands of the South Pacific they became a serious threat to America in the North. From Alaska the Aleutian Islands stretch out like stepping stones toward Japan. If the Japs could scramble over these stepping stones and gain a foothold on the mainland of Alaska they might cripple our shipping in the North Pacific and launch aerial and amphibious attacks against the west coast of Canada and the United States.

"Alaska was not wholly defenseless. But its bases were few and inadequate. They could be supplied only by sea or air. By land there was no through road, not even a trail, between Alaska and the most northerly extensions of existing Canadian rail lines or highways connecting with those of the United States."

Whitehorse was the base for building the Alaska Highway from Dawson Creek, British Columbia, to Fairbanks, Alaska, as well as a center for the CANOL project, which built a crude-oil pipeline from Norman Wells in the Northwest Territories to Whitehorse and a Whitehorse oil refinery. The railroad was leased, airports were upgraded, and the war years were a prosperous time in Whitehorse, with jobs galore for military and civilians alike.

The U.S. and Canadians aimed to tie in a chain of airports and flight strips for the use of fighter planes, bombers, and transports, as well as using the Alaska Highway for fuel. The Corps of Engineers, the U.S. Army, along with civilian contractors, would develop the Yukon's Norman Wells oil field to produce at least 3000 barrels of oil a day. At the same time they were to run a pipeline to a point on the Alaska Highway and there build a refinery to turn the Norman crude into gasoline for planes and trucks and other uses.

Iris and Joy in Whitehorse

The requisite France photo

This point was at Whitehorse. CANOL was called the greatest construction job since the Panama Canal and in respect to area covered, time of accomplishment, and sheer pioneering, the pipeline and refinery project, combined with the Alaska Highway, was destined to become the biggest construction program in the history of the world.

Lloyd received an Army commendation for some of his CANOL-related work "in connection" (he later said) "with the planning and providing of radio transmitters for Army Intelligence scouts who were landed by submarine on certain of the Aleutian Islands during the period in which many of the islands were occupied by the Japanese enemy. In connection with this work I made several hazardous airplane trips in the Aleutian Islands, through fog and blind flying conditions, to obtain the equipment and coordinate radio transmission by these army scouts to Navy communications located both on ships at sea and on land. Immediately prior to the departure of the Intelligence scouts on their assigned mission it developed that the Army frequencies which the transmitters were designed to work on were not satisfactory for communication with the desired Navy stations.

“I worked all night long grinding new quartz crystals to the required frequencies and making other necessary changes in order that the fulfillment of the mission would not be delayed."

On Pearl Harbor Day 1941, Danny Weil, 23 years old and serving in the Royal Air Force, had perhaps not even heard of amateur radio. But as an engineer involved in the war effort he certainly came in contact with "radio men." The importance of amateur radio in the war effort was great, since it supplied a trained pool of technicians and operators. Upon enlistment, a person's amateur radio license was a ticket, nearly always, to some communications role. Danny undoubtedly brushed shoulders with many of these people.

But in 1941, Lloyd Colvin had been licensed nearly a dozen years (Iris got her first ham ticket in 1945), but his operating time had been limited and sporadic due to the depression, college, and ROTC responsibilities, and then active Army duty. Lloyd had, however, been a member of the American Radio Relay League, the national organization for amateurs founded in 1914, and received its monthly magazine, QST. In the 1920s and 30s QST was a major source of documentation of the advancement of radio and Lloyd devoured every issue. Perhaps the greatest technical advancement, attributable to amateurs, was the use of the "short waves" for worldwide communication.

Lloyd, while technically knowledgeable, was really more interested in communicating. In the mid-1930s QST announced an award, the DX Century Club, for making contact with amateurs in 100 countries. Some accomplished this before the war; Lloyd would have to wait until operating from Japan in the late 1940s.

While stationed in Alaska and the Yukon during the war, Lloyd held the (Alaska) call sign K7KG. But there was no amateur radio due to the war. Beginning in 1939, as more and more belligerents silenced amateur radio in their countries, the hobby had contracted, with nearly the entire world putting amateur radio on hold shortly after Pearl Harbor.

During the war Lloyd, like tens of thousands of others, kept in touch through QST, which continued to publish technical articles as well as war stories. Lloyd wrote a story for QST that was published in August 1945, the very month the war ended. It was the first of three QST articles he would publish, one each in the 1940s, 1950s, and 1960s.

Hams on the Alaska Highway

Establishing Signal Corps Radio Stations at 72 Degrees Below Zero

BY MAJOR LLOYD D. COLVIN,* SC, K7KG

One question always asked of troops returning from the Alaska Highway is this: "How cold *is* it up there?" Prior to my assignment in October, 1942, as a radio engineer on the Alaska Highway I had spent nearly two years in various parts of Alaska. I thought I knew what cold was — but I had a surprise coming. During the winter of 1942 the weather along the route followed by the Highway has been officially described as perhaps the coldest ever experienced in that area. The temperature dropped to 72° below zero.

At the time of my arrival at Army Highway Headquarters, located in Whitehorse, Yukon Territory, the temperature was 25° below zero. That first night I slept with two other officers in a tent near the airfield. In the morning, after considerable hunting, we found enough wood to build a fire in a small stove near the center of the tent and heated some water. When the water was boiling I took some in a pan to the entrance of the tent, where I could see better, and started to shave. Before I had finished shaving, ice had formed all around the edge of the pan!

My first job was supervision of the building and operation of a large high-speed radio station at Whitehorse to serve the Army Headquarters. The installation consisted of a multiposition operating room near the headquarters establishment, with remote receiving and transmitting stations.

The Army troops and contractors had moved into Whitehorse faster than supplies could follow them. There seemed to be a scarcity of everything except cold weather. Transportation — or the lack of it — was the biggest problem, however. I had been instructed to get the station on the air at the earliest possible date, but work on signal facilities was at a standstill because of the lack of transportation. The distance from the operating room to the receiving station was three miles while the distance from the operating room to th[e] transmitting station was six mile[s]. Neither the buildings nor the an[]tennas at either site had been com[]pleted. With the extreme cold tha[t] prevailed it was imperative th[at] some kind of a vehicle be obtain[ed] to get the Signal Corps person[nel] to and from work.

In my search for transportati[on] I saw everyone from second li[eu]tenants to the commanding g[en]eral. All were very sorry, but available Army transportat[ion] that would run was needed to [] food and clothing to keep the men alive.

In desperation I started a canvass of the st[] in the village of Whitehorse, asking, "Does [any]one know of a civilian who has a car or truck [that] could be rented or bought?" The village fire []

*APO 722 A, c/o Postmaster, Seattle, Wash.

46 QST

The article focused mostly on how cold Alaska was and how hard it was to work on radios, engines, or just about anything in temperatures that had dropped to as low as 72 degrees below zero F. Lloyd said that when he was assigned to the Alaska Highway project in October 1942 he had already been in Alaska for two years, but that the winter of '42 was the coldest on record.

He described his first job of supervising the building of "a large high-speed radio station" at Whitehorse to serve the Army Headquarters, with a multi-position operating room and remote receiving and transmitting stations, three and six miles away respectively.

There was plenty of radio equipment available for the job and the biggest problem was transportation, namely, keeping Jeeps and other vehicles running in the low temperatures.

Lloyd wrote that "Yes, the weather was cold during the building of the Alaska Highway. But, as in so many other parts of the world, the U. S. Army Signal Corps, with its high percentage of amateur radio operators and technicians, is providing communications there of which we can all be proud."

"Lloyd was "temporarily" promoted to major in 1944, later permanently. He described his duties in Whitehorse from

On site in Canada's Northwest Territories

September 1943 to February 1945 as "various staff positions at Battalion HQ from time to time including Radio Engineering Officer; Radio Operations Officer."

He was a member of the War Department Traffic Security Board (a joint Canadian-US Board); Battalion Adjutant; a Battalion Soldier Voting Officer; the Aviation Cadet Board; Battalion Claims Officer, Battalion Investigating Inspection and Personnel Officer; Battalion Radio Censorship Intelligence and Cryptographic Officer; Battalion Traffic and Signal Center Officer. His jobs including installation, operation, and maintenance of some 100 radio stations covering a land area of approximately 3,000,000 square miles (the CANOL and Alaska Highway projects).

"This was a huge project," Lloyd would write, "costing approximately $30,000,000 and covering a large portion of the earth. In addition to the radio stations, a wire-carrier system was installed along approximately 29,500 miles of the Alaska Highway."

He was "cleared for duty with confidential and secret cryptography" on 22 Nov 1943. His later Army records showed he was entitled to wear three "Service Ribbons" (American Defense Ribbon, Asiatic-Pacific Theater Ribbon, and American Theater Ribbon), and to wear two bronze stars (one on the American Defense Ribbon for service outside the U.S. in wartime and one on the Asiatic-Pacific Ribbon, for the Aleutian Islands campaign area.

In September 1944 Brigadier General F.S. Strong Jr. wrote to the officers and men of the Northwest Service Command:

> Your faithful and arduous service in this Command is over. You are going back to the United States - some to your homes, others to new assignments in the service of your country. I take this opportunity to extend to every one of you my admiration and sincere appreciation for all that you have accomplished.
>
> These roads you have built up here, these communications you have set up, and supplies you brought in, undoubtedly had a lot to do with the change in Japanese strategy regarding this hemisphere. From all that we have been able to gather about the Japanese, when their plans go wrong they commit a kind of mental hari-keri, if not a physical one. They don't seem to have alternative plans.
>
> One of the major victories of this war, and I think time and history will bear this out, was the complete and quick clearing out of the Japanese in the Aleutians. We might have had to fight them for

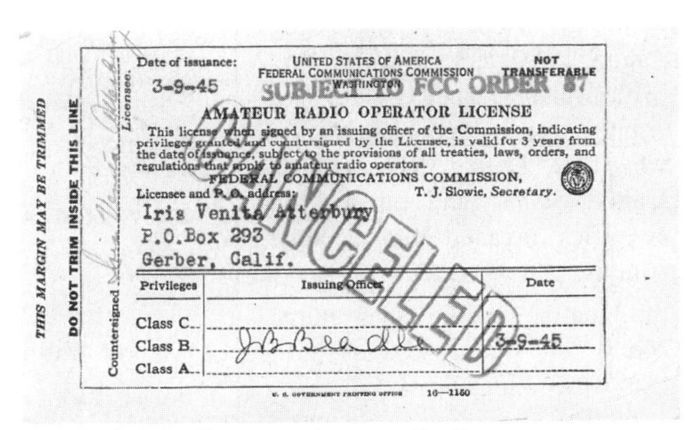

THIS MARGIN MAY BE TRIMMED
DO NOT TRIM INSIDE THIS LINE
Countersigned ... Licensee.

Date of issuance: 3-9-45

UNITED STATES OF AMERICA
FEDERAL COMMUNICATIONS COMMISSION
WASHINGTON

NOT TRANSFERABLE

SUBJECT TO FCC ORDER 87

AMATEUR RADIO OPERATOR LICENSE

This license when signed by an issuing officer of the Commission, indicating privileges granted and countersigned by the Licensee, is valid for 3 years from the date of issuance, subject to the provisions of all treaties, laws, orders, and regulations that apply to amateur radio operators.

FEDERAL COMMUNICATIONS COMMISSION,
T. J. Slowie, *Secretary*.

Licensee and P. O. address:
Iris Venita Atterbury
P.O.Box 293
Gerber, Calif.

Privileges	Issuing Officer	Date
Class C		
Class B	J.B. Beadle	3-9-45
Class A		

CANCELED

U. S. GOVERNMENT PRINTING OFFICE 16—1160

Iris Colvin's first amateur radio license

> years on Alaskan soil, and Canadian soil, and finally on United States soil, but what you men did up here made this unnecessary.
>
> The Alaska Military Highway, our airports and flight strips, made it impossible to invade the American continent on a shoestring. I think it has made that invasion impossible under all circumstances.
>
> All this is largely due to the officers and men of the Northwest Service Command. To them I extend my felicitations on their accomplishments, and my sincere wishes for their future welfare.

Following a three-week leave in Berkeley, in October 1944 Lloyd (now a "permanent" major) attended the 21st Service Staff Course, Command and General Staff School, at Fort Leavenworth, Kansas. After graduating in January 1945 he was assigned to the Signal Corps Officers School at Fort Monmouth, New Jersey.

Iris and Joy continued to follow along, from Berkeley to Fort Monmouth to Asheville, North Carolina and, finally, to Indiantown Gap, Pennsylvania, for processing to go overseas. Iris had "been around" the Morse code for several years, now, and her first amateur radio license is dated 9 March 1945 (the FCC continued to conduct amateur license exams and to issue licenses during the war, although transmitting was of course not permitted). One of Iris's favorite stories was that the first time Lloyd visited her after the birth of their daughter, his first question was "What is dit-dah?" Then, they discussed the arrival of Joy, their new daughter.

Lloyd was assigned to North Africa, arriving in Cairo, Egypt on 30 May 1945. Iris and Joy stayed behind in California. Iris learned French and tried to find employment that would allow her to join Lloyd in Algiers, but she was unsuccessful. Lloyd was senior radio communications engineer for U.S. signal facilities in Egypt and liaison officer to counterpart officers on the British Communications Staff in Egypt. Within just a few weeks he was given a permanent change of station, to Algiers, a World War 2 battleground. France's Vichy government ruled Algeria until 1942, when the United Kingdom, the United States, and other Allied countries invaded and occupied Algeria.

After the war, the Allies returned control of Algeria to France and that's when Lloyd arrived.

From June 1945 to February 1946 Lloyd was commanding officer of the 315th Signal Service Company in Algiers, in charge of what he called "a key radio relay station in the world wide Army Communications System." Radio circuits were operated from Algiers to Washington DC, Paris, Caserta, Italy, Moscow, Cairo, and Tangiers, "to all major headquarters in North Africa," with daily traffic of some 200,000 "groups."

JDOA V JDJD NR 36
JC 24
JDOA V JCYN NR 3040
FROM JCYN 171056Z
TO JDOA
GR NC BT

CLR 7197 RELAYED FOR YOUR ACTION AGWAR CABLE WAR EIGHT THREE TWO ZERO EIGHT DATED FIFTEEN NOVEMBER FOUR FIVE QUOTE IF AVAILABLE ISSUE ORDERS ASSIGNING MAJOR LLOYD DOG COLVIN ZERO THREE FIVE ZERO TWO SIX SEVEN SUGAR CHARLIE TO MIKE ITEM SUGAR WASHINGTON PROVIDING HE HAS SIGNED THREE ZERO JUNE FOUR SIX AGREEMENT PD NEEDED BY TWO SIX NOVEMBER IF POSSIBLE UNQUOTE ADVISE THIS HEADQUARTERS ACTION TAKEN END AMET BT

171056Z

This was high-speed long-distance radio communication, Lloyd Colvin's passion. Earlier, that station had received a "Citation for Meritorious Service" from Major W.H. Jacobs, describing it:

> [To The] 315th Signal Service Company, for superior performance of duty in the accomplishment of exceptionally difficult tasks in the Mediterranean theater of operations during the period from 15 December 1944 to 15 February 1945. Operating and maintaining the Algiers radio relay station with such superior efficiency as to make it one of the most reliable radio centers in the war department world wide communications system, the 3156th Signal Service Company, while operating all existing radio channels to the United States, England, France and Russia at peak loads, was assigned the extremely important task of constructing, establishing and operating new high speed relay circuits for the Yalta Conference.
>
> Outstanding coordination and an all-out effort by the entire organization within the time available. The installation and maintenance section successfully completed the necessary installations ahead of the target date set by the War Department and in time for the traffic operations section to pretest and adjust before traffic developed. Operating continuously from 24 January to 13 February 1945, the traffic section passed large volumes of vital, top secret, world important information without loss, misroute or failure.
>
> The circuit reliability sustained by the company was exceptional; transmission time losses due to readjustments were negligible and reception continuity was superior. Throughout the period the entire personnel displayed a devotion to duty that was exemplary; and the highly successful completion of the mission brought messages of commendation and appreciation to the company from communication centers throughout the world.

A few days after VJ Day in August 1945 Robert P. Patterson, Under Secretary of War, referred to the Algiers operation as a "magnificent contribution to the victory of our nation and our Allies."

Lloyd's Algiers duties took him to Italy, Casablanca, Egypt, Tunisia, and Sicily. As Signal Officer on the staff of the North African Service Command, he prepared signal annexes and operational guidelines, compiled systems data, and coordinated with British and French commands.

Later trips, in 1945 and 1946, took him to French Morocco, French West Africa, Liberia, Gold Coast, Tangier, and Nigeria, mostly for just a few days. Although he had been a licensed ham since his teenage years, Lloyd, now 30 years old and an amateur for half of them, was visiting countries he'd never worked on the air.

By the end of 1945 Lloyd was champing at the bit to get back on the ham bands. U.S. amateurs had begun getting permission to once again transmit, with some limitations. On 8 December, from Algiers, Lloyd wrote to his commanding general:

> Personnel of this company are located in camps many miles from the nearest town. Any form of amusement or entertainment that can be arranged in the camps greatly increases the morale of the entire company. Operation of an amateur radio station at this company will not only provide entertainment but, in addition, will increase the technical and operational ability of radio personnel of

this company.

Lloyd pointed out that many of his company's officers and enlisted men were amateur radio operators, that since war's end ham operations had resumed around the world "at a rapid rate," and that he was hearing many amateur stations in Europe operated by U.S. Army personnel "by authority of U. S. Army Headquarters in Europe." Lloyd said he held Amateur Class A, Radiotelegraph-Second Class, and Radiotelephone Second Class licenses and asked that "a suitable amateur call sign be assigned and that the undersigned be appointed as licensee of the station."

Lloyd got a reply from the Adjutant General's office in Washington on 14 January 1946, addressed to him as "Commanding Officer, 3156th Signal Service Company, NASC" (North Africa Service Command), saying that "The operation of amateur radio stations is favorably considered where approved by the local government concerned."

The Adjutant General's office pointed out that the stations must be operated within amateur bands cleared for use by the military communications boards concerned and that bands below 28 MC were not authorized. It was anticipated that the War Depatment would authorize use of 40 and 20 meters in perhaps six months, and a target date of 1 March 1946 had been set for clearance to 3700 to 3900 KC for amateur use in the United States. For overseas amateurs like Lloyd, amateur radio call signs would have to be obtained from the local governments.

Thus ended Lloyd's first, unsuccessful, brush with governmental obfuscation and buck-passing in amateur radio licensing in a foreign country! Perhaps mysteriously, much later, Lloyd would list FA8JD among his call signs and he would say he had collected the QSLs for DXCC from Algeria. No log book exists for this call sign and DXCC was never issued.

Lloyd returned to the U.S. on 13 April 1946 and was "temporarily" promoted to lieutenant colonel (reduced to major 22 July 1946). He accepted appointment as an officer in the Reserve Corps, and then on 5 July 1946 was integrated into [the] US Army as a 'regular career officer.

There never seemed any doubt in Lloyd's mind that he would stay in the Army. Major Colvin's record was extolled but his letter of commission contained a caveat:

> I hope you will regard your commission in the Regular Army somewhat as the commencement on graduation from college. It is only the beginning of your career in the Army. None of us can afford to rest upon our past successes. The demand and necessity for extensive communications and the consequent enormous increase in the quantity and complexity of Signal equipment has grown to such magnitude during the war that you will find it necessary to constantly study and apply yourself if you are to keep up with our requirements.
>
> The Signal Corps from its birth early in the Civil War has had a brilliant record in both peace and war. I feel certain that we can rely on you to preserve and augment that record. You will never acquire any wealth in the Army, but you will acquire an inner satisfaction from duty well performed that supplies an intrinsic compensation. We are most happy to have you.

Major Colvin was approaching 10 years of military service. His Army papers, from Camp Beale, Calif. (11 August 1946) list his activities as

DOMAIN of the GOLDEN DRAGON
INTERNATIONAL DATE LINE
Ruler of the 180th Meridian
To all Sailors, Soldiers, Marines, wherever ye may be and to all mermaids, flying dragons, spirits of the deep, devil chasers, and all other living creatures of the yellow seas, Greetings: Know ye that on this 24th day of Oct 1946, in latitude 35° longitude 180° there appeared within the limits of my august dwelling the S.S. Central Falls Victory
Hearken Ye: The said vessel, officers and crew have been inspected and passed on by my august body and staff. And know ye: Ye that are chit signers, squaw men, opium smokers, ice men, and all-round landlubbers that Major Lloyd D. Colvin having been found sane and worthy to be numbered a dweller of the Far East has been gathered in my fold and duly initiated into the
Silent Mysteries of the Far East
Be it further understood: That by virtue of the power vested in me I do hereby command all moneylenders, wine sellers, cabaret owners, ______ managers and all my other subjects to show honor and respect to all his wishes whenever he may enter my realm. Disobey this command under penalty of my august displeasure.
Golden Dragon
Ruler of the 180th Meridian

Lloyd first crossed the International Dateline on 24 October 1946

softball, baseball, basketball, skiing, mountain climbing, and his active hobbies as radio, photography, and stamp collecting. He was qualified as Expert, .45 caliber pistol; Expert, .45 caliber TSNG; Sharpshooter, .30 caliber rifle; and Sharpshooter, .30 caliber carbine.

Lloyd was careful with his records, as well as with his money. On 7 September 1946, he paid a $200 deposit to Barrett Brothers, Inc., of Bronx City, NY, for a new 1947 Kaiser automobile, to be delivered in April 1947, at the "prevailing price at time of delivery. Deposit refunded on demand." Lloyd's address was Governor's Island.

Barrett Brothers went out of business on 7 January 1948 and in April orders were being filled for them by Hudson Valley Motors of Newburgh, New York. Delivery time was "about four weeks." Radio and heater optional. By November 1949 Hudson Valley Motors was out of business and, furthermore, told Lloyd that they never took over the Kaiser-Frazer business from Barrett Brothers, and they knew nothing about his deposit of $200.

Lloyd appealed then to the sales department of Kaiser-Frazer in Willow Run, Michigan, in February 1950, and then, a year later, to the Madison Credit Bureau of New York, New York. Madison said that Mr. D.J. Barrett was bankrupt and "that while he would like to pay this debt to you, even though it is not one of his due personally, he is unable to do so."

Flash forward to 6 Feb 1957 - Lloyd, Iris and Joy are in Alameda, Calif. Lloyd writes to D.J. Barrett, now of Mt. Vernon, NY, asking if "you are now able to pay me all, or part, of the $200. I naturally can use the $200 and will greatly appreciate it if you can pay me back my deposit." He apparently never heard from them again.

All this in their archives!

4. Assignment Post-War Japan

"In any contest to find the busiest woman in Tokyo, the winner would probably be Mrs. Iris Colvin." - The Far East Stars and Stripes Review

In August 1946, Lloyd was assigned to the Headquarters, Supreme Allied Command, Tokyo, Japan, where he, Iris, and Joy would spend nearly four years. The Allied occupation of Japan lasted from 1945 to 1951, under General Joseph MacArthur. Lloyd was assigned as Commanding Officer of the 72nd Signal Battalion, communications officer in charge of "all major communications, broadcasting, and radio facilities serving General MacArthur's Headquarters." There he was to give technical assistance to the civilian communications industry of Japan, including broadcasting and radio stations in Japan "capable of or participating in international communications."

Lloyd spent two months on Guam to install long range high powered communications facilities there to communicate from Guam to the Mainland and to Japan and other countries. Japan had captured Guam on 10 December 1941 and the U.S. military, after liberating the island in 1945, took over about one-third of Guam's land.

Iris and Joy left for Japan on 18 February 1947, by ship, to join Lloyd. Iris remembered that when Lloyd went to Japan it was the first time they had been separated "for any length of time" (she apparently did not count, or overlooked, Algiers).

Note at the bottom of this Bikini Island photo reads "To J2AHI QSO 19 at KX6USN Jun 9 1947. Ed Shuler"

We arranged [Iris said] to meet each other by amateur radio. Neither of us knew what equipment or station would be available for use. I went over to [an amateur station in San Francisco] who agreed to assist at the schedule time. We called Japan. A station ... came back with a note from Lloyd saying that he had to go to Guam to install antennas. So we took a chance and called "CQ Guam." A station answered and agreed to call the Army Hq. We heard him asking for Lloyd.

This was not quite so simple on Lloyd's end. He had contacted a ham on Guam who gave him a key to his shack and use of his equipment. Unfortunately, when Lloyd arrived the rig was on a different frequency and the tubes had to be changed in order to get on the [scheduled] frequency. He did turn on the receiver and heard me calling. He rushed out to the Headquarters building, asking for a phone to call the sergeant - who was on the phone - and Lloyd heard him saying "We have no Lt. Colvin here." Lloyd broke in "I am Lt. Colvin." Then he went to the other station and we made the sked, by luck and the helpfulness of hams.

With the Colvin family reunited in Tokyo and Joy in school, and Iris with a never-used U.S. amateur radio license, they dived into the hobby, forming a group they called the Far East Amateur Radio League, with Lloyd as president and Iris as publicity director. Jointly, they were issued the call sign J2AHI. They also used the call sign J2USA as sort of a mini-W1AW (the famous headquarters station of the ARRL), sending amateur radio news bulletins by CW. Japanese who had been amateurs before the war were thrilled to be allowed, once again, to just listen on the amateur radio frequencies. A letter to J2USA from one of those listeners said:

It is our great pleasure to send the following report of your QST (the ham notation for a general news transmission) broadcasting on Oct 27th 1947. On 7 Mc, Heard RST 589X

Receiver, 8 tube super heterodyne with regenerative amplifier at Tsutukawa-mura, Minamitamagun, Tokyo (18 miles south west of your station)

Will you kindly make a verification of above reception to

ex J2NG, Naruo Yoneda, and send a QSL card via JARL (the Japan Amateur Radio League), PO Box 377, Tokyo Central.

We are making the JARL RECEIVING CONTEST in which the all members of JARL and others are making the reception of your regular QST broadcast from J2USA. The publication of this announcement, however, delayed owing to the social condition of Japan right now, and the report of this contest have not arrived more than one, listed above.

By next time, we hope to make a good deal of report all over the Japan Honshu." - Hiroyasu M. Ishikawa, ex J2NF, Temporary QSL Manager of JARL, Director of JARL, Member of editorial staff of JARL official organ, CQ ham radio."

Hams in Japan were eager to get back into the game, just as Americans had been just two years before.

One Lt. General R.L. Eicherbarger commended the FEARL in May 1947, saying

It is my desire at this time to welcome the organization of radio amateurs into the Far Eastern Amateur Radio League. The radio 'ham,' by rendering service in cases of emergency, by enabling overseas personnel to get messages through to their families, and by promoting interest in radio experimentation, can do much to maintain a high level of morale in this command.

FEARL, the Far Eastern Amateur Radio League, can assist by self-monitoring and policing of the 'ham' bands and by familiarizing all personnel with the Eighth Army directives governing Amateur Radio Operation. To all of the amateur fraternity, 73's [sic] es DX..

The Far East Stars and Stripes Weekly Review for June 22, 1947, profiled Iris in an article headlined "XYL, Mother, Wife, Businesswoman, Housekeeper And Ham Radio Operator":

In any contest to find the busiest woman in Tokyo, the winner would probably be Mrs. Iris Colvin.

Besides making a home for her husband, Maj. Lloyd D. Colvin, and seven-year-old daughter Joy at the Sanno Apartments, pretty Mrs. Colvin holds down a full-time auditing job with the Office of General Accounting and shares enthusiastically in Maj. Colvin's hobby of 'ham' radio operating.

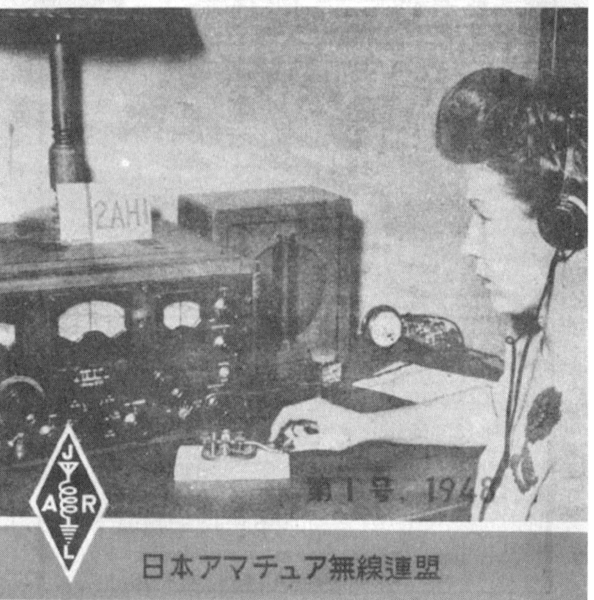

Arriving in Japan on Valentine's Day this year, Mrs. Colvin found herself the only licensed 'XYL' radio operator in Tokyo. In the esoteric language of radio hams 'YL' stands for "Young Lady" and 'XYL' means a wife. Both are much in demand as radio pals by male amateurs the world over.

Both Maj. Colvin, who is Director of the Communications and Operations Division of the 71st Signal Service Battalion, and his wife are enthusiastic and active members of the recently-organized Far East Amateur Radio League, which now has about 200 members. 'What shall we do this evening?' poses no problem for the Colvins when no social engagements are scheduled. Warming up their 750-watt model AT-20 transmitter and Hallicrafters SX-28 receiving set, they flash CQ- CQ- CQ to call any other amateur who may be operating, and settle down to an international confab.

Since J2AHI went on the air, the Colvins have received QSLs from 71 countries and 42 states, proving that they have 'W.A.C.'d' or Worked All Continents. QSL cards are carefully saved by hams, because they are proof of the radio operator's ability. In the States, the FCC issues Class B and Class A licenses. To obtain the former, the applicant just sends and receives a minimum of thirteen words per minute and pass a test. For a Class A license, the station must operate for at least a year, and have cards to prove "W.A.C." as well as passing a more detailed test.

J2AHI, the Colvins' call letters, indicate their 'QTH,' or location. 'J2' stands for the second district of Japan. J2USA are the code letters for the League's station, near Radio Tokyo. 'W' stands for the United States, VE for Canada, and so on.

As is usually the case with XYLs, Iris Colvin became interested in ham radio through her husband, who has been an amateur operator since his boyhood days in Boise, Idaho. At first she used to assist her husband by saying one, two, three, four over the microphone and helping to put up antennas. As the radio bug bit her more, she learned the code and obtained her Class B license.

During the war, all amateur operations were ceased, and her first post-war attempts at operating were efforts to talk by ham radio to her husband in Tokyo. The first time she was successful in contacting a Tokyo ham from San Francisco. She found out that her husband had left that day by plane for a month's trip to Guam. She made several attempts to contact Guam, and finally succeeded in having an "FB QSO," or good conversation with him in International Morse, giving him the news that she was leaving for Japan in a few days.

"The whole family goes to work at eight in the morning," chuckles Maj. Colvin. "In Canada, we used to bundle Joy in furs and take her to a neighbor's on a sleigh while we went to work. Now she is in the second grade at the American School." At this point, Joy, a delightful, pigtailed Margaret O'Brien type, was reading Hansel and Gretel from a storybook to a visitor, not faltering over a single word. Asked if she was going to be a radio ham too, she said proudly that she knew almost the whole Morse code already.

In the jargon of hams, vy fb, hi!

J2AHI was indeed heating up the airwaves and a DXCC, issued on 4 MAY 1948, #193, was Lloyd's (and Iris's) first. The ARRL's DXCC award had "started over" with all operators beginning with a clean slate following the war, and Lloyd's was only the 193rd to be issued. In time, he received other awards and certificates for operations from Japan, including Worked All States and a certificate for achievement in the new CQ magazine's "DX Marathon" in 1948, on both CW and 'phone. Lloyd also held what apparently was a personal call sign, JA2KG, and one day would turn in QSL cards for some awards, although in general he combined his JA2KG and J2AHI "credits."

Early in 1947 Lloyd, busy with business in Guam, found time to increase his "countries visited total" with Okinawa, Iwo Jima, the Philippines, and China.

Before the end of the war the U.S.S.R. had occupied Manchuria. After they left in 1946, the Chinese Communists took over most of northern Manchuria. The U.S.S.R. also set up a North Korean "people's republic." In China, Mao Zedong's Communist troops fought the Nationalist armies of Chiang Kai-shek. The United States gave military aid to Chiang. Beginning in 1947, military forces and their dependents in China were evacuated. The dependents were rushed to Japan where they waited for a month or more before being sent home by ship. There was no phone or other communication between Japan and

Operating J2AHI

China, so to furnish such communication during this critical time, Iris and Lloyd were released from their normal duties in order to operate phone patches between China and Japan for all allied personnel.

Just as these phone patch operations were ending the Berlin Airlift began, and planes were rushed to Germany with their intended passengers stranded in Japan. No regular phone communication existed between Japan and Germany, and Lloyd and Iris were again relieved of regular duties in order to operate their amateur radio station, handling phone patches and relaying messages to family members of military personnel separated half-way around the world.

Forty years later Lloyd received this letter, dated 29 November 1987, from Yukio "York" Komiya, JA1AH, in Tokyo:

> Dear my Radio friend Mr. Lloyd D. Colvin,
>
> Many thanks for nice QSO [contact]. Please pardon for writing to you suddenly. I am 64 years old man. I have taken part in the WORLD WAR 2 since I joined the Japanese Navy as a radio operator in 1942. The first sea battle in which I took part was off Midway Is. in June 4th 1942. and we lost the war.
>
> Since then, the Navy compelled us to take part in the extremely dangerous naval battles, for the purpose of cloaking the real fact that we lost the war. We took an active part in the several battles in the North Pacific severely cold or the South Pacific where many horrible sharks are swimming. During that time, We might have been you unconsciously, I think. Then, I was obliged to leave ship by repetition of sinking of our ships. In April 1945, 1 was obliged to a Marine Corps with a view to protect Japan proper.
>
> Everyday, we were trained to throw ourselves with a bomb in hand under the Tanks landing in the face of our lines. It is not too much to say that the Military Authorities think nothing of man's life. Soon I entered a Marine Corps and greeted the historical day which tells us the end of the WORLD WAR 2.
>
> Just after the war I met you and your XYL at your shack at Sanno-Hotel - do you remember me? "When that time, I was young SWL. Because we Japanese were forbidden on the air by occupation forces. If I remember rightly you were Maj. and belong to 71st Sig. Bn. Also you live in Sanno-Hotel in Tokyo. Your call letter was J2AHI. I think. I was much surprised for your rotary beam and big indicator. We had very good time at there.
>
> Also I have been to SANSHIN-BUILDING" [where the Colvins had lived in Tokyo).
>
> After nice ground QSO, you gave me used Call book (Red Pegasus of cover). It was a only one Call book in JA's at that time. It was like a treasure. Hi! We were using habitually for much value. TNX AGN [thanks again].
>
> Also you open the chest of drawers and show me many JA-QSLs you contact before the War. And you asked to me "How are J2KG doing?" I remember it. I met Mr. Minozuma (ex-J2KG) 4 month ago. His call letter is JF1WKJ now.

HEADQUARTERS EIGHTH ARMY
UNITED STATES ARMY
OFFICE OF THE COMMANDING GENERAL
APO 343

AMATEUR RADIO STATION PERMIT

CALL LETTERS: JA2KG
STATION LOCATION: SANNO APARTMENT BLDG, TOKYO
TYPE OF TRANSMITTER: SPECIAL DESIGN
SERIAL NUMBER: #174
DATE OF ISSUANCE: 25 MARCH 1949
RESPONSIBLE OFFICER: L.D. COLVIN Major

MRS. IRIS A. COLVIN, (C holder of a valid Amateur Radio Operators Licence, is authorized to install and operate the above listed amateur radio station at the above listed location subject to all current Eighth Army regulations and directives. This permit is valid until OCT 1 1949

By command of Lieutenant General Walker

J.W. Donnell
Colonel, AGD

A week ago, I visited to Japanese CQ PUBLISHING CO. President of CQ Co. is my long time friend. and he was remember about your account.

Then we try to searching for old CQ magazines. we find yours at last. That news was better, but paper and bind was very poor. In those days Japan is a penury country by War. I am sending you its copy. Since then, many years passed as a dream. I got license in 1952 as JA1AH.

The Sanno Hotel in Tokyo, where the Colvins lived

Now that Christmas is just around the corner, I extend to you and your family my heartiest greeting for it. Any way, apart from the my desire, I hope we can continue this correspondence, and I shall be very happy to receive letter from you.

My son is also HAM. His call letter is JP1GPP. He is working in the HONDA MOTOR Co. LTD. And his age is 23 years old. If you catch his signals please call to him. If you have a snapshot of HONDA-CAR (include motor cycle). Please send him for HONDA GAZETTE (with background of person and your country sight.)"

Good DX and good luck for you hope that we may continue our radio friendship. PSE QSP TO Iris-san. Sincerely yours, York

Aerial view of a long distance radiotelephone receiving site in Japan

In 1948 a major earthquake occurred in Japan, and more than 1,000 died. For the first 12 hours after the earthquake, the primary communication out of the stricken area was by amateur radio; the Tokyo end of the circuit was operated by Lloyd and Iris Colvin. The following day, their names and a description of the good work performed by amateur radio and the Colvins was featured on the front page of local newspapers.

In January 1948 Iris and Lloyd added further to their "countries total," vacationing in Hong Kong, Manila, and Shanghai [DXCC countries Hong Kong, the Philippines, and China].

Visiting military officers often found themselves in the role of celebrity following the war. While in the Philippines Lloyd was interviewed by The Manila Times, on 24 Jan 1948:

> Ex-Premier Hideki Tojo, now facing trial before the War Crime Commission in Tokyo, Japan, will in all probably get the extreme penalty, in the opinion of Major Lloyd Colvin, of General MacArthur's headquarters, who is a Manila visitor. Traveling with Mrs. Colvin, Major Colvin arrived yesterday morning on the S.S. Washington Mail from Japan on a month's furlough. He [is] leaving on the same boat for Hongkong and Shanghai, thence to Japan.
>
> Major Colvin has attended virtually every session of the trial of the War Crime Tribunal and he believes that Tojo will get “Death.” Major Colvin said that Japanese people are making great strides towards democracy and that they are working hard for the future of their country. Emperor Hirohito mixes with common people now, he said, but the people still hold him in high reverence.

The article referred to the International Military Tribunal for the Far East that ran, in Tokyo, from 3 May 1946 to 12 November 1948. Of 28 defendants, seven were condemned to death and all but two of the others were sentenced to life imprisonment. Tojo was executed.

Around this time Lloyd suffered a major career disappointment. Late in 1947 he had applied to the Far East commander-in-chief's office, to attend graduate school in a civilian university. He was scheduled to return to the U.S. in January 1949. The application was denied. A letter (30 January 1948) from the War Department in Washington said:

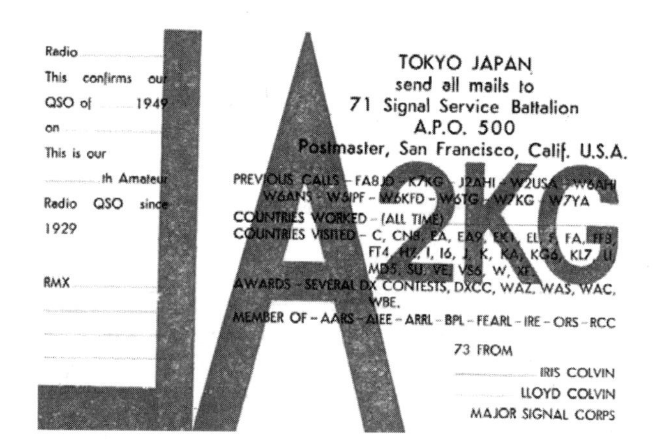

> The military record and college record of Major Colvin have been reviewed together and an attempt has been made to establish the probability of his success in graduate work of this type. While it is realized that there is no well defined formula by which this probability can be determined, it is the considered judgment of this office that his chances of success in this field., when compared to other applicants and to the standards established by the several universities, do not warrant his entrance upon this type of training.
>
> It should be understood that failure to be selected to attend a civilian university for graduate study does not adversely affect an officer's record. Major Colvin's interest in this training is appreciated and it is regretted that his application could not be approved.

In September 1948 Lloyd received assignment to his Army "career field": Communications. This would affect his assignment within the Signal Corps on his return to the States. Lloyd applied for assignment to any "military mission [as an attaché] in any French (or) Spanish speaking countries" but there were no vacancies. His application was "filed."

By July 1948 an "Authorized Amateur Radio Stations" list for Japan shows 70 "J" call signs of Americans in Japan, from J2 to J7. J2AHI is listed as Mrs. Iris A. Colvin

Lloyd's J2USA - the "Headquarters station of the Far East Amateur Radio League" - came under the authority of the Army. A typical report, filed by Lloyd, Acting Chief Operator, this one for the period 26 December 1948 to 1 January 1949, outlined operations for the benefit of the Army's deputy communications officer:

> · Station Location: Roof Finance Bldg, Tokyo

One of Lloyd’s “offices” in Japan

· Equipment: BC 610, Super Pro Receivers, VFO, Beam Antennas

· Operating Personnel: Full time - Major Lloyd D. Colvin, PFC Al Holland, part-time, Major Robert Lantz, Mrs. Iris Colvin.

· Hours of operation: 72 hours.

· China Evacuation Radiotelephone calls successfully completed: 184

· New Year greeting messages handled to the United States, Hawaii and Canada for occupation personnel in Japan: 96.

· Other traffic: 21

· Grand total of messages and radiophone calls handled: 301.

The busiest day of the week was New Year's Day, when 57 radiophone calls were completed. Lloyd reported that a Mrs. Zant had a baby 29 December in Yokohama. "Word of the birth was given to Capt. Zant in Shanghai a few minutes later over the J2USA-C1AF radio circuit. Special arrangements were made with the hospital and Mrs. Zant talked to her husband in Shanghai over the radiophone circuit the next day."

Responding to this report, Signal Corps Colonel J.D. O'Connell praised the Colvins and their operators for maintaining communications with China "at a time when it meant so much to those who were being evacuated. The efforts of Major R.M. Lantz and Mrs. L.D. Colvin in particular were most noteworthy.

"To provide this service, all of you who participated in keeping J2USA and J2HYS on the air worked long and faithfully, and without regard for hours or personal inconvenience. This devotion to the ideals of service in time of emergency is thoroughly in keeping with the highest traditions of our Radio Amateurs and the Signal Corps United States Army, and I wish to transmit to you my heartiest congratulations."

In March 1949 Lloyd traveled from Tokyo to Honshu to attend an Air Transportability Command and Staff Course. Officers attending were advised to take the following: appropriate winter clothing; suitable inclement weather clothing; fatigue clothing and cap; pencil and notebook; one pair of boots, combat; belt, web (pistol or rifle), first aid pouch and packet; and canteen w/cover. The course included weights and balances, loading and lashing, and tactical rides in troop carrier aircraft.

5. Stateside in the Early 1950s

"Any country South or Central America; any country Africa or Middle East; any other country." - From Lloyd Colvin's attache application

In May 1949 Lloyd was assigned to the Army Personnel Center at Ft. Monmouth, New Jersey. He was granted 60 days leave; Iris would accompany him. Arrangements were made for Joy, age 9, to travel from Yokohama to Ft. Monmouth by "rail and government surface transportation." Meandering toward home in the U.S. in the summer of 1949, Iris and Lloyd visited Korea, Formosa, China, Macau, Malaya, Siam, Ceylon, Aden, Egypt, Italy, England, Scotland, Northern Ireland, Ireland, Wales, Jersey, France, Switzerland, Belgium, Germany, Netherlands, Sweden, Denmark, and Luxembourg.

Decades later, when the Colvins would speak at a gathering of hams in a country they were visiting, they would unfurl a wall-sized list of "Countries Visited" and these countries appeared in the first column of that list.

Lloyd attended the Signal Officer Advanced Course at Fort Monmouth from 24 October 1949 to 15 August 1950. As a graduate, he had completed the following "Signal School" courses:

Army Signal Communications; Division Signal Communications; General Staff Organization; Corps Signal Communications; Operations and Training; Organization and Tactics; Organization of Air Force and Tactical Air Power; Communications Center Operations; Military Cryptography; Staff Procedure and Planning;

A triband beam at Lloyd's quarters in Arizona

Theater Signal Communications; Utilization of Air Power; Reading Improvement; Civil Disturbances; English Composition; Intelligence Section; Logistics; Strategic Planning; Naval Operations: Wire Communications; Personnel Section; Military Justice; Tactics (Map Exercise); Psychology and Leadership; Public Speaking; Radio Communications; Signal Supply; Radiological Defense; Special Staff; Communications Intelligence; Map and Aerial Photo; Systems Engineering; U.S. Armed Forces; Utilization of Manpower; Thesis on Signal Comm. Signal Corps Activities at Ft Monmouth; U.S. Amphibious Joint and Combined (Tactics); and Troop Education and Public Information.

During this time in New Jersey in 1949 and 1950, the Colvins built a dozen or more single-family houses in Shrewsbury, including one the family lived in. Some were a prototype of houses that later populated the Drake Park housing development outside Fayetteville, N.J.

In February 1950 Lloyd again applied for "Army Attaché Duty." He was 34 years old. His application noted that both he and Iris had studied French and Spanish. His military background was summarized as approximately five years staff duty and five years command duty, with special qualifications of extensive travel, liaison duties, and international communication. His preferred assignment was, first, language school in Monterey, California, then "any country South or Central America; Any country Africa or

A North Carolina family photo, around 1953

Middle East; any other country."

In March and again in April this application was denied "due to the present and projected shortage of qualified Signal Corps officers of Major Colvin's grade and qualifications."

In the summer of 1950 fighting broke out between North and South Korea, with the North crossing the border into the South on 25 June. In October China joined the war and the U.S. followed. From then until a cease-fire on 27 July 1953 Lloyd was, once again, an Army officer in time of war.

Following Signal School graduation in the fall of 1950 Lloyd was assigned to Fort Bragg, in North Carolina, the premier Army Signal Corps facility in the U.S. Lloyd would be a "Test Officer and Communications Engineer, Army Field Forces Board, Fort Bragg, North Carolina. He held a Top Secret clearance (as did tens of thousands of others at the time). He was "temporarily" promoted to lieutenant colonel on 28 December 1950.

In North Carolina, Iris and Lloyd formed their business partnership, Drake Builders, to build subdivisions. He became W4AHH (later upgraded to W4KE) and in the 1950 CQ Worldwide CW DX Contest he placed third in the 4th call area. Lloyd also operated K4WAB, the base's amateur station, and, during their time in North Carolina, Joy was licensed as W4ZEW.

Lloyd's log book from the period includes operation as W4KE/7 at Yuma Test Station, 17 June to 24 July 1952, an Army desert site. This log book then switches to July/Aug 1955, DL4ZB, then Jun/Jul 1956, DL4ZB (these are Iris contacts). The last page of this log book has one QSO, Feb 22, 1960, W4ZEW/6 (Joy's call sign).

Lloyd's Amateur Extra Class certificate AE-5-2116 is dated January 1953. After joining Lloyd in amateur radio operations from Japan, there is no indication that Iris was active during the years in North Carolina.

6· Danny Weil and YASME in The New World

"Being a bit startled I said 'Where did you come from?' The answer was laconic and very simple, 'a boat.'" - Robert Wilson, on meeting Danny on Barbuda.

Danny Weil completed his solo crossing of the Atlantic in late 1954, at Antigua, where he would spend several months, feeling quite at home in this British Colony. His stay coincided with a visit to Antigua by England's Princess Margaret and Danny earned some much-needed cash taking photos.

Danny was not a radio man. He had taken a tiny, 30-watt transmitter and a American-made war surplus short-wave receiver (a BC-348), courtesy of a friend in England. He did not know a dit or a dah of the Morse code and held no license, amateur radio or otherwise. The receiver worked "but not too well, maybe because it had fallen off its shelf a few times in the Atlantic crossing and cracked the chassis." The transmitter didn't work at all. Danny sailed across the Atlantic cut off from the world.

But Danny knew the potential value of amateur radio to his endeavor. Ellen White, W1YL, remembers a letter from Danny, around 1954. It was addressed to Ed Handy, W1BDI, communications manager at ARRL Headquarters in Connecticut. Ellen, who then was W1YYL, along with husband Bob White, then W1WPO, was newly-arrived on the ARRL staff. "Ed Handy showed me the letter," Ellen said. "Some Britisher was getting ready to sail alone across the Atlantic and wanted information on how to get a ham license."

American Robert Wilson, who now is AL7KK, remembered meeting Danny. In early 1955 he was involved in what he called "secret military projects" as a "soldier of fortune"

> I was operating a radio station on Barbuda, [Wilson said] a remote island which had been an entirely slave island until the British emancipated slaves about 1837. The residents of the island between 1837 and 1955 did not allow Europeans unless special permission was granted by 'The

YASME I docked at Antigua, 1954

> Warden,' the chief government official. But as I was on a special project, and had the rank equivalent to an army Colonel, I was allowed on the island to do my job.
>
> Barbuda is nearly flat 'karst' topography with a slight rise on one side of the island and interesting hidden caves under the rise. The population at that time lived in the center of the island in a number of thatched houses more appropriate to Africa than the New World
>
> I had set up a radio station in a tent between two grave yards in the middle of the island. I had an old U.S. naval transmitter with about 15 watts of power. My antenna was strung between two tents. I had asked "The Warden" if I could operate an amateur radio station there, among the other things I was doing. I also asked what call letters should I use? He said "No problem with operating amateur radio, and use any call you want, we never signed treaties with anyone!"
>
> So I was all set. I operated on the 40 meter band mostly, talking to Europe. However, since I could not tell anyone what I was doing conversations were pretty dull. Mostly, I sat and monitored the military radio, or rode about on a local donkey, my feet dragging the ground.
>
> One day I was working the radio in my tent and felt a presence behind me. A white man, very unusual to be sure, was standing behind me. He must have been there for some time when he said

"I have a radio and I would like to see if it works."

Being a bit startled I said at first "Where did you come from?"

The answer was laconic and very simple, "a boat."

"What boat?" I asked, since the island was only visited by smugglers and surrounded by a nearly impenetrable reef. Only those knowing the secret entrance could get in. The YASME was anchored just inside of the reef and off shore from a very old abandoned fort, as I remember.

"My name is Danny Weil and I just sailed across the Atlantic in the YASME," was the quiet answer. He went on to say "I have a B-19, Mark Two tank transmitter on board and I haven't been able to talk to anyone!" Danny operated his B19 Mk II transmitter on deck under a tarp. I was surprised that he was able to get it all the way across the Atlantic in operating condition under such circumstances

I don't know how old he was at the time, but in my mind he will always be about 40 [he was 37]. Danny and I made an agreement that he would go back to the YASME and I would call him on the 75 meter band. We both did a bit of tuning up and were able to establish good contact, though I had a lot of reservations about talking to a "bootlegger."

So after proving that the old British tank transmitter from World War 2 actually worked, in spite of a trip across the Atlantic under a tarpaulin, Danny was happy. Perhaps happiest because he had talked to a person after such a long solo trip across the Atlantic Ocean. In any case he at last sailed off to destinations apparently unknown to him at the time.

Later on I heard that Danny Weil had obtained a regular amateur call and became friends with many of the Caribbean and U.S. amateurs, including the Colvins. I followed, as best I could while traveling around the world, his progress across the Pacific and the final destruction of the YASME on to the remote parts of the Great Barrier Reef of Australia.

Danny sailed to Charlotte Amalie, the main harbor of St. Thomas, in the U.S. Virgin Islands, in late April 1955, with an aim of getting his radios looked at. "This was my first sight of American territory," he said, "and I was awed with the thousands of large American cars running around the place."

Danny found the Americans helpful and courteous, with paperwork quickly dispensed with and the YASME cleared out of the quarantine area and into the Yacht Basin.

Someone had told Danny to see Dick Spenceley, KV4AA, in St. Thomas, for help in fixing his busted radio gear. Danny found the Spenceley residence and "waited a while to pluck up sufficient courage to knock on the door, hoping Dick wouldn't get too much of a shock at meeting a very scruffy Englishman." Straightaway Dick and Danny met. About a decade older than Danny, "He became a friend," Danny said, "the sort of friend that many people would love to have, the type that money

Dick Spenceley, KV4AA

Danny Weil, left, and Dick Spenceley, KV4AA, right, ran into actor William Holden in the U.S. Virgin Islands.

Dick Spenceley was born in the Boston suburb of West Newton, Mass., in 1905, and lived there until 1924, when he joined the U.S. Navy.

In 1925 he transferred to Navy Radio Station NBB in St. Thomas, U. S. Virgin Islands. He lived there until his death in 1982.

He was licensed in the U.S.V.I. in 1927 as K4AAN. He caught the DX contest bug in 1932, when he took part in the week-long ARRL DX CW Contest, placing third. He achieved the world top score in the ARRL DX CW Competitions of 1951, 1954 and 1956.

His call sign changed to KV4AA in 1947, and by 1962 he held the top spot on the DXCC Honor Roll. After reaching that pinnacle, he continued to work DX, but did not submit any further QSLs to maintain a ranking.

He was DX editor of CQ from 1952 to1958 and fashioned the popular WPX awards during the later part of his stint with the magazine. He was inducted into the CQ DX Hall of Fame in March, 1969.

Dick Spenceley died in 1982.

couldn't buy. I once said to him and his wife that should I have the chance to choose another father and mother they would be the ones; that's the sort of people they were."

Danny Weil chose a special radio amateur for his introduction to the hobby. Spenceley already was world-famous as an operator and as the DX manager for the American amateur radio magazine CQ. Danny was ushered into a large and well-equipped ham "shack," with row upon row of high-powered equipment. Danny "found great pleasure in listening to people all over the world talking to [Spenceley] and often wondered what it would be like to be in his position."

When Spenceley suggested that Danny incorporate amateur radio into his voyage, and described the inner workings of "DXing," Danny was at the foot of the Master.

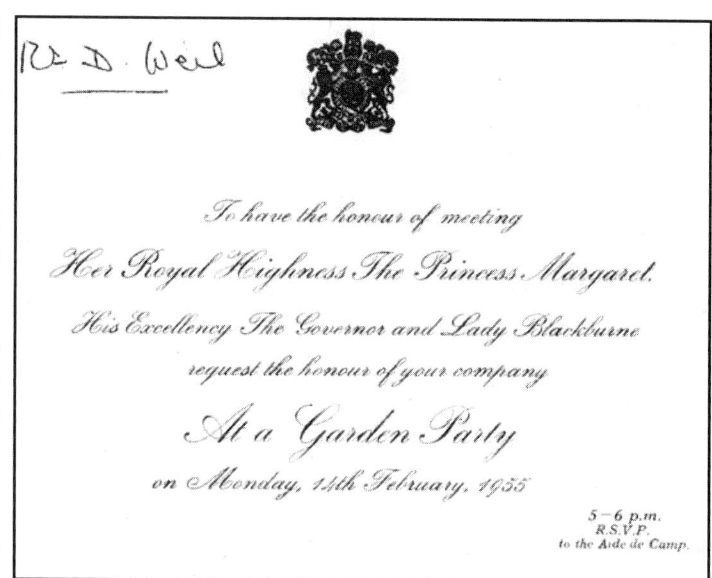

R. D. Weil

To have the honour of meeting

Her Royal Highness The Princess Margaret

His Excellency The Governor and Lady Blackburne

request the honour of your company

At a Garden Party

on Monday, 14th February, 1955

5 – 6 p.m.
R.S.V.P.
to the Aide de Camp.

Spenceley's plan was for Danny to get on the air from as many "rare" islands and other countries as he could. The only snags, Danny thought, were that he didn't have any suitable radio gear, he didn't have an amateur radio license, and he didn't know Morse code.

And, "What I knew about radio you could have put inside a thimble and still had plenty of room to stick your finger." But Spenceley didn't seem bothered by these trivial handicaps. "All he wanted to know was whether I would do the job and leave the rest to him." Danny mused that if hams are a little crazy, "so are single-handed sailors, so that puts me at the top of the tree for being completely nuts."

Danny Weil spent the next several months in St. Thomas, learning radio theory and the Morse code, then took a trip to Tortola, in the British Virgin Islands, where he passed the amateur radio examination and was issued the B.V.I call sign VP2VB.

When Danny arrived in Tortola the Communications Officer wasn't available and so he spent the spare time working on the Administration Building's public clock, which hadn't run for 20 years. "It was fun getting that ancient piece of machinery down from its position high on the wall. I got it working, but two days later it quit again."

By mid-July 1955 most of the radio gear that Spenceley had promoted from Stateside hams had been received and was being installed on the YASME - two Multi-Elmac AF-67 portable transmitters, an Elmac PMR-6A receiver, and a Hammarlund HQ-129X were obtained through the help of the Elmac company and amateurs W9FKC, W5AVF, W8ZY, and K2EDL. These very small radios were designed for mobile use and operated on 12 volts dc. Spenceley also helped fit a forty-foot, semi-vertical antenna to the boat.

Danny was practicing the Morse code, in those days the most effective long-distance mode. "Danny is not the best CW operator in the world," Spenceley wrote in *CQ*, "but he has the makings, and practice is one thing I am sure he will have a lot of!"

Spenceley's *CQ* column in the August 1955 issue tantalized readers with "tentative plans, always governed by prevailing winds, tides and other factors," for Danny to visit British Phoenix Islands, Nauru Island, Solomon Islands, Portuguese Timor, Andaman and Nicobar Islands, Laccadive Islands, Maldive Islands, Seychelle Islands, Aldabra Island, Comoro Islands, St. Helena and Ascension Islands. All "rare" DX islands, many of them with little or no amateur radio activity (or even inhabitants).

Spenceley also contacted radio "mates" in England and obtained a letter from the British Colonial Telecomms, in hopes of facilitating permission for Danny to operate from the various British Colonial Possessions.

"The YASME is in no hurry," Spenceley said, "and Danny plans to stop at any particular spot for the length of time necessary to exhaust its QSO possibilities. He will also pursue his trade as expert watchmaker at these stops and get them back on time!" Spenceley would act as QSL manager and information liaison for Danny.

Finally, Spenceley said in *CQ* that "Funds are necessary to keep this trip on [an on-going] basis and it is requested that small contributions be forwarded to KV4AA or CQ by as many hams as possible.

Cooperation in this matter will avoid the necessity of the "buck a QSL" basis, which, we think, is generally frowned upon."

7· Danny Weil's First DXpedition Voyage

"I think you will understand there were many times when I signed off seated on the floor of the cabin, with the phone cord around my neck." - Danny Weil

After three months in St. Thomas, Danny left on 1 August 1955, headed for the Panama Canal and points west. He and Spenceley had arranged to meet daily on the radio, and for the next nine years these contacts would rivet the amateur radio DX world. Even when Danny wasn't ashore on some desolate island, DXers could find him at sea and eavesdrop on the latest news. Over time, Danny transmitted long descriptions of his adventures, which Spenceley would copy, transcribe, and publish as stories in CQ.

Danny immediately began polishing his Morse code skills aboard ship. "Working the rig on this trip was quite an experience, and for those that were unfortunate in having a QSO with me, I must apologize for my bad CW. To fire up the rig was an easy matter, but to operate the key was another proposition altogether. Try to imagine yourself in a very small cabin which lurches from side to side at odd moments and at angles of 35 degrees. Seat yourself on a very unsteady stool, then try to work CW with an occasional dive into the cockpit to put the boat on course. These conditions are a slight underestimation, but I think you will understand when I tell you that there were many times when I signed off seated on the floor of the cabin, with the phone cord around my neck."

No log book exists for the trip from St. Thomas to the Canal Zone and it isn't clear if one was kept, but Danny said he made about 150 contacts. He was catching on quickly to the pace of being in demand on the radio. "Please remember as much as I like to hold long QSOs, I have to consider my gas supplies. Also the longer I hold a QSO, the less chance I have to natter to others, so make 'em short and sweet and don't think me rude if I suddenly go off the air. One never knows when the old YASME is going to give a lurch."

Docking in the harbor at Cristobal, Panama in September, Danny was greeted by three hams, KZ5EM, KZ5MN, and KZ5LB, "who then whipped me off in a car to Balboa, filled me solid with the finest food I'd tasted for some time, then gave me a bunk, pardon me, a bed to sleep in." Many of the hams in the area offered their assistance in preparing the YASME for her next part of the voyage.

CQ JANUARY 1955
YASME 1

ABOARD THE "YASME" PANAMA TO THE GALAPAGOS

By Danny Weil, VP2VB/P, KZ5WD.

The time spent in the Canal Zone was, as expected, entirely devoted to
out the YASME for the next stage of the trip and any rest I thought I'd get
purely imaginary as every spare moment was taken up. Even whilst asleep I wo
dream of what had to be done the next day. Finally the time arrived when I d
that I was as near ready for sea as I ever would be.

Thanks to KZ5MN, Captain Dick Mann, my stay in the Canal Zone was made
happy one inasmuch as his car and home were made available to me twentyfour

In 1955, the Canal Zone population was some 60,000, including U.S. civilian employees of the Zone and U.S. military personnel and their families, stationed at the 14 U.S. military bases built in the Zone. The rest of the population was composed mainly of people of West Indian descent, whose families had come to work on canal construction in the early 1900s, and Panamanians of both Hispanic and West Indian background. The U.S. residents lived in relative luxury, receiving high pay and generous benefits, in prosperous, well-kept communities. In contrast, the Panamanians and West Indians held the most menial jobs, were paid only a fraction of what U.S. workers received, and lived in separate, inferior-quality communities

This famous early photo of Danny Weil appeared in QST in 1990

Danny's time in the Canal Zone was spent further fitting out YASME for the Pacific. Danny was the guest of Captain Dick Moran, KZ5MN. He was granted the call sign KZ5WD and, while logs for this call sign do not exist, a number of DXers worked KZ5WD and received QSLs from Spenceley.

Danny left the Canal Zone 8 October 1955. Enroute to the Galapagos, the value of ham gear aboard, not to mention knowledge of Morse code, became immediately apparent when Danny found himself unsure of his position. Danny contacted KZ5JW back at the Canal Zone and asked him to telephone the local Air Force base for a position. "All went well and, during the ceremonials, a KP4 (Puerto Rico) station called in and gave me a bearing from Puerto Rico. Thus I got two lines which, although not dead accurate, did at least give me some idea of where I was and that was too darned to close to Malpelo Island."

Danny took this photo of Spenceley in his "second shack"

Passing the Galapagos Islands, the YASME set her sights for Tahiti. Danny reported having little spare time, between acting as everything from cabin boy to ship's captain, and "working the rig continuously, having a good time key-thumping, with an occasional fone QSO." He made more than 1,500 contacts from Panama to Tahiti, using up 175 gallons of gas (an auxiliary Onan generator had been donated and installed on YASME to power the radio equipment and other electrical devices aboard).

Danny's regular radio schedules included Spenceley; Bill, KV4BB; Fred, HC1FS in Ecuador; and Roland, FO8AD, the most active amateur in Tahiti. Thanks to all the radio activity, "I think I conveyed to many of you where I was," Danny wrote in CQ, "and what was happening. It was a great comfort having the rig aboard, and certainly knocked hell out of the solitude, but one has to experience that type of solitude that comes in a small boat from anywhere to really appreciate the value of a rig."

The little Multi-Elmac transmitter and receiver worked fine, as did the Onan gas generator. An unforeseen problem was that swaying of the boat naturally changed the position of the vertical antenna relative to the sea, causing signal fading.

In true ham spirit, Danny said "I intend to experiment with other types of antennas to get better results from the receiving angle."

8· The DX Century Club - 1955-Style

"It's rare if you don't have it." - An eternal DX Truth

The ARRL DXCC program, begun in the 1930s, "started over" in November 1945 and for some years was identified as the "Postwar DXCC." That is, everyone came back on the air after the war with a clean DXCC slate. This was one of those rare ideas that virtually everyone agreed was a good one. Those who had chased DX before the war could rest on their laurels, for a while at least. 1945 was a year for "starting over" in many ways.

So, when Danny Weil shoved off from the Virgin Islands in 1955, British Virgin Islands license in hand, the Postwar DXCC was in its tenth year. The official Countries List in December 1955 QST showed 291 "entities." Most of these at least loosely fit the criteria for listing hammered out at the beginning of DXCC, 20 years before.

These criteria would undergo much scrutiny in the 1950s and 1960s, with the requirements scrutinized, tightened, and defined. For example, in 1955, Germany was one country on the DXCC list, despite the obvious political division of East and West Germany that was reality. All of the "British" islands in the Caribbean using VP2 call signs were lumped into two DXCC entities, the Leeward Islands and the Windward Islands. And so on.

As criteria were refined and the political situation changed, some 1955 entities (places that counted as "countries" for DXCC), would be deleted from the list. Eventually, enough entities were deleted that QST began carrying DXers totals as Current/Deleted, a practice continued to this day.

Lloyd D. Colvin
LT COLONEL SIGNAL CORPS
US ARMY

W6TG, FA8JD, W6ANS, KL7KG, W6IPF, J2AHI, W6KFD, K2CC, JA2KG, W2USA, K4WAB, J2USA, W7YA, W6AHI, W7KG, JA2US, W4KE, DL4ZC
*
WAZ, WASM, WAC, WAA, WBE, WAVE, WAE, WAP, WAS, WAID, KZ-25, BERTA, CAA, DUF, MARS, AIEE, AFCA, ARRL, BPL, FEARL, IRE, ORS, RCC, OTC, DARC, QCWA, DXCC

As I mentioned in a letter I sent you a few days ago I have QSLs for DXCC from 100 DXCC members. I am inclosing a photograph of the 100 cards.

In case you can not make out all the calls the cards are for:

1.	GC2FZC	35.	4X4RE	68.	HC2KJ
2.	VP7NM	36.	FQ8AP	69.	EA9AP
3.	SM5WI	37.	HZ1HZ	70.	CT3AV
4.	KV4AA	38.	ZS2X	71.	SV0WT
5.	ON4CT	39.	KC4AF	72.	CX6AD
6.	XE1AC	40.	TF3SF	73.	CT1JS
7.	VS6CQ	41.	VK3JE	74.	VS1FK
8.	VP5DC	42.	VQ3HJP	75.	KP4KD
9.	OY7ML(OY3ML)	43.	VP6CJ	76.	FR7ZA
10.	OH2RY	44.	ZL1AH	77.	KO6DI
11.	CN8MM	45.	G6ZO	78.	OE1FF
12.	VR4KS	46.	VQ8CB	79.	F9RM
13.	PZ1AH	47.	PJ2AA	80.	OM6MD
14.	EI5F	48	SU1AS	81.	I1BNU/Trieste
15.	OQ5RA	49.	KL7PI	82.	KZ5DG
16.	ZS3AP	50.	HA5KBA	83.	PA0HG
17.	CR3AO	51.	CR6BX	84.	GI3AXI
18.	YS1O	52.	VP9CC	85.	TI2HP
19.	OK1HI	53.	9S4AX	86.	VQ5EK
20.	VU2MD	54.	EA9DF	87.	VQ2GW
21.	LA6U	55.	ET2US	88.	VQ4FI
22.	YU3AP	56.	KP4KAC	89.	ZD6BX
23.	FA4CR	57.	DL7AH	90.	PY2MX
24.	EA8AX	58.	LU9CK	91.	OX3MD
25.	CR7AF	59.	FA8DA	92.	VQ8AD
26.	ZE3JP	60.	W6ITZ	93.	CO2WD
27.	CP5EK	61.	FB8AB	94.	OZ7BG
28.	IS1AHK	62.	DU7SV	95.	ZC4IP
29.	GW5FN	63.	KH6IJ	96.	KT1UX
30.	YI2AM	64.	YV5FL	97.	4S7GE
31.	I1XK	65.	HB9CB	98.	HK3PC
32.	OD5RA	66.	FF8AG	99.	VE7ZM
33.	HP1BR	67.	JA6AO	100.	TA3AA
34.	ZD2DCP				

The above cards are for QSOs I made from DL4ZC. You will recall I hold DXCC certificates for JA2KG (J2AHI), DL4ZC, and W4KE. I have also worked DXCC from FA8JD (not confirmed) and W6KG (confirmations will go in shortly) . At the present time my QSL collection numbers nearly 30,000 QSLs , all arranged alphabetically and arranged in file cabinets in such a manner that I can pick out any card without leaving my operating chair.

73

Lloyd D. Colvin, Lt Colonel Signal Corps
now W6KG

QST's "Hows DX" column once offered to publish photos of QSLs from those DXers who had worked DXCC members in 100 DXCC countries. Lloyd discovered he'd done it from DL4ZC.

In 1955, QST published the call signs of every DXCC member, in the December issue, taking fewer than three pages to do so, as well as running monthly updates as DXers' QSLs flowed into West Hartford for credit to their "accounts." At the end of 1955, out of a possible 291 credits, W1FH led the list with 261, after nine years of work. To put this in perspective - eventually the DXCC Honor Roll would be established to recognize all DXers with totals in the "top ten." If the Honor Roll had existed in 1955, everyone with a total of 252 or greater would have been on it. In 1955, there would have been five DXers on the "mixed" (combined CW and phone) Honor Roll - W1FH (261), W6VFR (257), W6AM (255), PY2CK (253) and W3BES (252).

In 2003, the Honor Roll listings

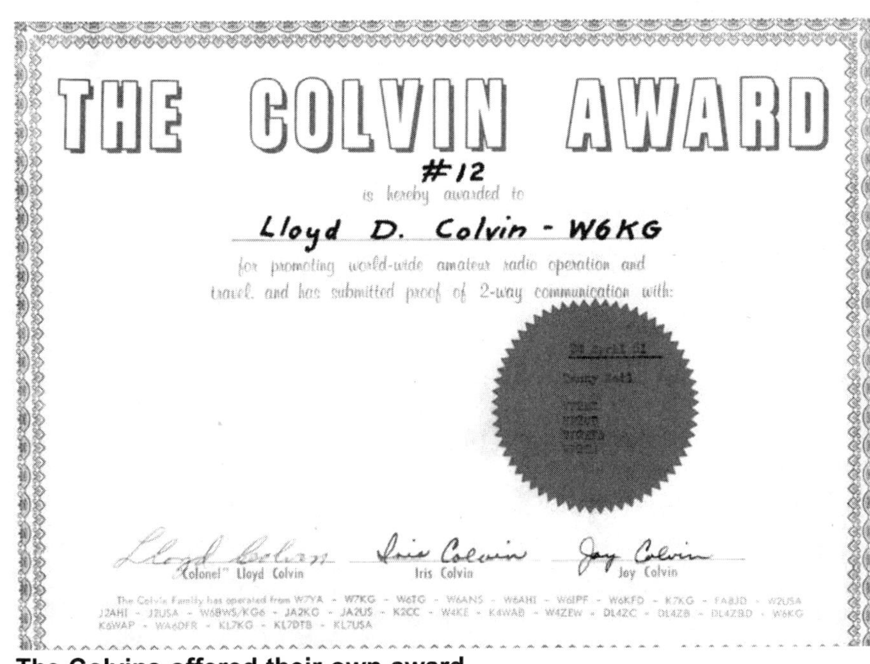

The Colvins offered their own award.

fill pages, with thousands of call signs! But, in 1955, the world's top DXer, W1FH, was still hunting for 30 countries and only 35 DXers were at the 241-or-above level, fully 50 countries below "Having them all."

Like baseball, DXCC is a game of statistics, and one can scrutinize the listings over the years, turning them upside down, sideways, and every which way.

While there was no polling in those days to determine what countries were the rarest, it wasn't really necessary: if W1FH (or W6VFR or W6AM) didn't have it, it was rare!

This was the overheated DX world into which Danny Weil plunged in the 1950s (as well as, a few years later, Gus Browning, W4BPD).

If there was no active amateur in a DXCC country, and an amateur radio license could be obtained, it was fair game. The era of DXpeditions had begun, and a saying in 1955 is just as true today - "It's rare if *you* don't have it."

9. Danny Weil's 1955 Pacific Voyage

"Ham radio was a life saver to me in more ways than one." - Danny Weil

On 29 November 1955, eight weeks out from Panama, Danny sighted Nuku Hiva, in the Marquesas Islands, a spot made famous by Herman Melville in his 19-century novel Typee. Typee, Melville's first book, described the four months he spent in the Marquesas, and was published in 1846 after initially being rejected as "too fantastic to be true." (The Marquesas were the location of one of the American television program "Survivor" series on CBS in 2002.)

The YASME landed at Papeete, Tahiti, 8 December 1955. No log exists from this operation but many QSLs do. QST magazine's "How's DX" column had now caught the Danny fever and, after the unavoidable delays in monthly publishing schedules, weighed in with a note from the rumor mill: "VP2VB/P aboard yacht YASME leaves French Oceania and FO8AN behind in favor of VR [Phoenix Islands], ZM [Tokelau], and CR10 [Timor] areas" in March 1956.

A letter from Spenceley to The Consul on Tahiti, Freddy Devendish, greased the skids ahead of Danny, who soon had a French amateur radio license to operate his ham gear there - FO8AN. This would be Danny's first experience of removing his ham gear and antenna from YASME, taking it ashore, finding a place to set up, and beginning his "Field Day" radio operation. Danny needed to find a spot close to YASME, so he could operate the radio ashore while being able to get to YASME quickly to carry on the inevitable maintenance. A solution came in the form of one Oscar Nordman, who offered the use of his home and storage shed.

"Oscar became more than a friend to me in my subsequent stay," Danny said. "His assistance in filling in all the necessary papers and forms saved me lots of time, and money, too, and he also gave me a dry storeroom to put all my sails in. What more could I ask? He was always available to advise me what to do, where to go, and to tell me all those little things that go to make a person's stay in a foreign port a happy one. I know I shall never forget Oscar. "

Ashore on Tahiti Danny got his first taste of finding a spot for a radio antenna and then actually putting one up. Danny found a 90-foot tree and, with help of a ladder, reached the top, only to realize that he had left the antenna insulators on the ground, 90 feet below.

> Well, I lit a cigarette, sat, and looked around me thinking. what the heck, what am I doing climbing

Arrival in Tahiti

trees like a great big school kid? Below I had my usual audience, but to endeavor to ask one of them to tie the insulators onto the wire if I dropped it down to them was beyond my linguistic abilities, and I realised it would be far easier in the long run to go down again. Oh boy, that trip down the tree was a darn sight more difficult than sliding down the mast of YASME. I have no doubt that many of you have climbed trees and can appreciate what I say when I tell you that the descent is far more difficult.

Dockside in Tahiti

The thought that I would have to do this again didn't make me feel any happier, but the thought also came into my head that all the Hams would appreciate my very fine climbing efforts. I wonder. I bet all you mugs would have laughed your blinkin' heads off to see me in that predicament. By the end of the day I had finally rigged one end of the antenna. Now the next part was to get on top of an adjacent building and fix the other end there. This meant climbing out onto a very narrow parapet, and as usual the inhabitants of the hotel had their wash hanging out to dry, so when I came along with my nice dirty hands to fix this darned chunk of wire their remarks to me, mainly in Tahitian, were definitely far from complimentary. I just smiled in a stupid sort of way, and hoped they understood that there were many more thousands of people like me called 'Hams' in the world and we were all slightly touched. We had to be to do what I was doing.

With both receiving and transmitting antennas in place, it was time to do a little repair work to both transmitters and the Hammarlund HQ-129X receiver from the YASME. Roland, FO8AD, stepped in by hauling all three pieces of equipment to his home to fix them, and 48 hours later Danny was on the air. He spent several hours a day operating thereafter, on both CW and AM (voice). This operation, FO8AN, was really what brought Danny to the attention of the DX world.

Danny Weil, and Marie, a friend it was "very hard to leave"

Danny loved Tahiti, and leaving was very difficult for him. His equipment was back aboard ship. "How I wished at the time the [YASME] mast would snap or something would happen to necessitate my return, but no, apart from the lack of wind, everything went fine, and I could find no logical excuse to return," he wrote in Cruise of the Yasme, a book manuscript Danny wrote that was never published.

At long last it was time to switch on the rig and talk to Dick in the Virgin

Islands. Oh Boy! Someone to talk to, even though it was only in Morse code. That radio certainly takes the solitude out of this deal. My generator started without protest, and within a few minutes I was listening to Dick coming blasting through with his high powered transmitter. Conditions were so good that Dick decided to come on 'phone instead. I think Dick must have known how I felt when he decided that he would talk to me in his own voice rather than the impersonal touch of code. Perhaps you don't know what it is like to hear a friend's voice in my circumstances, but I can assure you, it was worth a million bucks. All my troubles seemed to disappear.

Whilst in this mood I got a whole stack of grub organized, then sat back and waited for the time to come when I should be able to talk to Joe, another Ham in Tahiti I had met. As usual, Joe was late for his sked; he is a real Tahitian so I expected it, but he was there, good old Joe, one of my best friends. When it came my turn to talk to him, my throat became paralyzed, a lump seemed to be blocking my voice altogether, and when he told me he had my girl friend there in the shack, it was quite a few minutes before I could get out what I wanted to say.

Joe's final "bon soir and bon voyage Dannee" and the promise of another contact later put me back again on top of the world, and then I just sat back there and waited for the next schedule with Jock, a good Ham friend of mine in New Zealand. Right on time old Jock [ZL2GX]came through, his cheery voice dispelling all the gloom as he talked to me about the painting job he was doing to his house. To hear his infectious laugh, his jovial way of speaking, how could I possibly remain miserable, and very soon, I was laughing with him. He certainly knows how to bring a body out of the depths of despair.

Danny loved Tahiti.

Docked on Tahiti

That ham radio was a life saver to me in more ways than one, and one can have good friends all over the world, as I did, and yet never actually meet them. One can travel all over the world, too, as I did, and yet still talk to the same friends all the time: distance has no meaning to Ham radio, but it has a comforting way about it that could never be equaled by any other hobby.

Several years later Dave Palmer, W6PHF, said about Tahiti, after he had visited there, that "Although evening propagation conditions were excellent [I] didn't operate for long periods of time because of [my] desire to participate in the famous night life of Papeete, of which you may have heard. (Who hasn't heard of Danny Weil's contact with the same problem ... problem?)

"Yes, it's true! The YLs of Tahiti really do have cooperative and broad-minded attitudes regarding the propagation of the species!"

10. The Colvins in Germany

"There was our battalion commander, Lt. Colonel Lloyd Colvin, operating a big station on 20 meters." - Jim Muiter, N6TP

Lloyd Colvin, after several years in the U.S., mostly at Fort Bragg in North Carolina, found himself headed for his next Army post, in Germany, the Cold War's pivotal spot and flash point. The Soviet Union had crushed a revolt in East Germany in early 1953 and East and West Germany would be declared "independent" in 1955, each joining a side in the Cold War. In 1955, West Germany would join the North Atlantic Treaty Organization, which had been formed in April 1949.

Lloyd arrived at Bremerhaven in early October 1953, for his duty station at Karlsruhe, working in radio communications linking the NATO countries of Europe. His primary job would be in the construction and operation of a common VHF-UHF radio system serving the NATO countries of Europe. Joy and Iris Colvin left from Palo Alto, California, on March 10, 1954, and joined Lloyd in Karlsruhe on 29 March 1954.

Lloyd's assignment from October 1953 to October 1957 was as Assistant Signal Officer, Headquarters European Command, Heidelberg, Germany, and Officer in Charge of USAREUR Multi-Channel Radio Telephone Network, and he was commanding officer of the 315th Signal Battalion. He described his duties as being "directly involved in the engineering, installation, and operation of communications networks throughout Europe to serve the NATO countries in an integrated circuit." In a talk to military brass Lloyd said "in a communications system of this magnitude, one of the major problems was the successful integration of many different communications facilities, manufactured by various countries. The system relied on VHF and UHF radio systems as the backbone of the system with multi-channel carrier systems operating on the radio circuits."

Lloyd's work took him to many of the countries in western Europe, in a project that cost some $30 million and involved the full time work of approximately 32,000 military personnel and another 39,000 civilian personnel, he later said. Several major US communication companies furnished equipment and engineers, including Western Electric, Philco, and General Electric.

Iris at the Colvin's station in Germany

DEUTSCHE BUNDESPOST

AMATEUR RADIO STATION AND OPERATOR LICENCE

for Allied Personnel in the US Zone of Germany and US Sector of Berlin

Mrs. Iris A. Colvin
(Title and name of licensee) (Call sign of FCC-license)

Heidelberg, HQ Fourth Signal Group, Campbell Barracks, APO 403
(Address and location of station)

DL 4 ZB
(Call sign)

is authorized [illegible] operate an amateur radio station under the licensing provisions [illegible] Government (Bipcom) **for a period of one year** from the date [illegible] terminated or suspended by the appropriate authority.

Deutsche Bundespost

5.1.56 [illegible] Karlsruhe, Jan. 5, 1956
(Date of issuance) For the President

Druck Maser, Darmstadt FTZ - 1000 - 7. 55 - (FD 2222) - (IV K - Nr. 1)

In June 1954 Iris accompanied Lloyd to Denmark, Sweden, Norway, The Netherlands, the UK, Belgium, France, Luxembourg, and Finland.

Jim Muiter, now N6TP, crossed paths with Lloyd around this time:

> My first encounter with Lloyd was as a corporal in the US Army Signal Corp. In January 1954 I was assigned to the 315th Signal Battalion in Karlsruhe, Germany. I had gotten my amateur radio license W6KXG in 1951 and had been active on the air with a small station in Alameda California, until I was drafted in February 1953.
>
> The 315th Signal Battalion was headquartered in Smiley Barracks in a suburban area of Karlsruhe. One day I noticed what appeared to be a 20-meter beam up on top of one of the concrete and brick barracks buildings. The Germans built great barracks for their army! Ours were cheap and nasty! So, having some free time and some curiosity, I wandered up the four flights of stairs to the attic and found the door to the radio room.
>
> I opened the door and there was our battalion com-

mander, Lt. Colonel Lloyd Colvin, operating a big station, possibly a BC610 [a famous U.S. World War 2 military transmitter built by the Hallicrafters Company], etc., on 20 meters. Since I was a corporal (actually a civilian draftee in uniform) I was not sure of the military courtesy required in such a nonmilitary situation. I introduced myself and Lloyd greeted me in a friendly manner. I stayed a few minutes and then left.

Shortly thereafter I was assigned to another location within the battalion and our paths never crossed again while in the service. After my discharge in 1955, I moved back to Alameda and got back on the air again. My first ever European contact later that year was with DL4ZC, Lloyd, on 15 meters!

I guess it was a couple of years later, perhaps in 1957 or 1958, that Lloyd and Iris moved to the east end of Alameda and set up a very nice station with a big beam on a telphone pole. I lived in the west end of Alameda and became interested in working DX. I was always somewhat amused by the fact that when something rare was on the air, Lloyd (W6KG) and I (W6KXG) would call the same station. Naturally when the DX station came back it was to W6KG instead of W6KXG. Oh, well.

I did run across a book Lloyd [and Iris] wrote on how to make a milion dollars in real estate, so he and Iris did very well for themselves in Lloyd's post-military carrer.

In August 1954 Lloyd was reassigned to the 4th Signal Group, from Karlsruhe to Heidelberg. Lloyd's notes on his medical papers in October 1954 "Good physical condition. Desire test to determine if mumps left me sterile." In 1955 he qualified on the .45 caliber pistol as Expert (the highest rating).

In a talk in 1956 Lloyd told a gathering of NATO officers that the network was making communication history. He said that using radio circuits instead of cable or open wire circuits to furnish trunk lines between military switchboards was developed and used during World War 2. Those early radiophone trunks consisted of not more than four voice channels over one radio circuit. "Since the war, great improvement has been made technically and the number of channels that can be used over one radio circuit has been greatly increased," he told them. "The Army is in possession of a valuable fixed plant radio facility rather than just paying rent on commercial land lines."

The system, begun in 1953, ultimately was made up of 38 completely self supporting, isolated radio stations. Construction projects included route surveys, the building of access roads, self-supporting steel radio towers, regulated power generator systems, living quarters for personnel, and properly engineered permanent buildings to house the radio and carrier (landline) equipment. In Germany, 16 stations used both VHF and UHF, and served military centers from Berlin to Munich to the French border.

The telephone network contained 80,000 radio "circuit miles." Installations in France, Belgium and England were designed to tie into the system in Germany with control of the system based in Heidelberg. Sounding like a pitchman, Lloyd said [the network] "comprises the largest land operated VHF-UHF radio telephone system in the history of communications. When you dial a long distance number from your telephone or place a long distance call with the operator, your call may be routed either via land line or radio facility and you will normally be unable to tell the difference between the two types of trunks used."

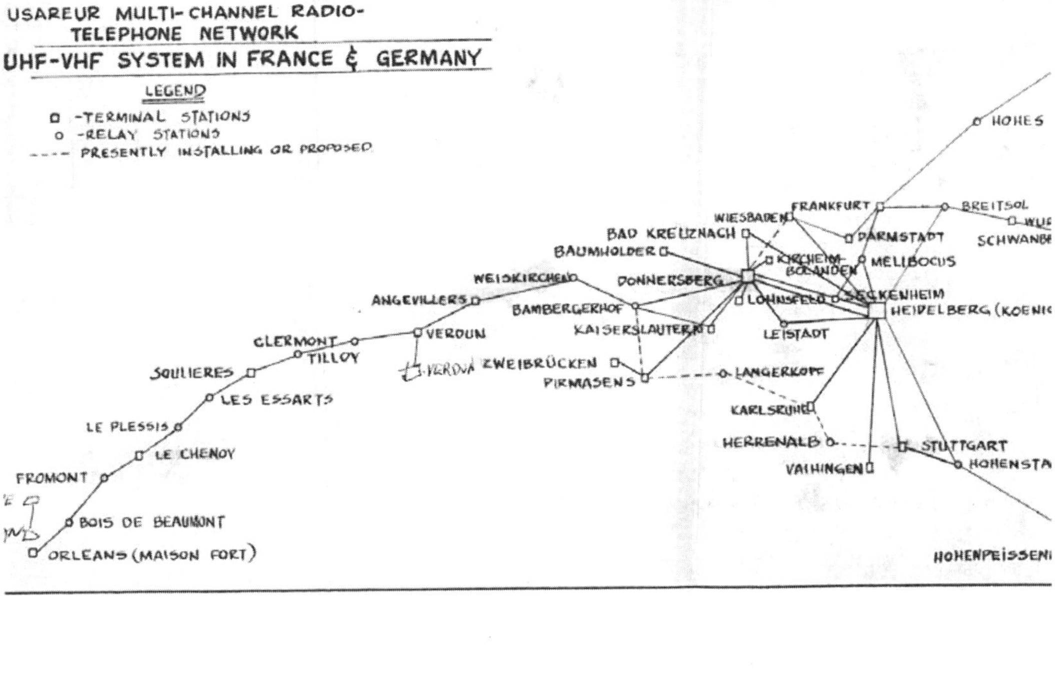

USAREUR MULTI-CHANNEL RADIO-TELEPHONE NETWORK
UHF-VHF SYSTEM IN FRANCE & GERMANY
LEGEND
□ -TERMINAL STATIONS
○ -RELAY STATIONS
---- PRESENTLY INSTALLING OR PROPOSED

The U. S. Army paid for station construction in Germany and manned all the sites. In France, Belgium, and England, the Army and the new U.S. Air Force shared the expense and oversight.

Lloyd was anxious to get on the ham bands, especially after his experience operating J2AHI from Japan. The Allied License Manual for radio amateurs was the guideline; any citizen of an Allied nation was eligible to apply for an amateur radio license to operate in occupied areas of Germany. Lloyd, as a signal officer, was exempt from both technical and Morse examinations for his license. He was issued DL4ZC; Iris, DL4ZB.

As DL4ZC, Lloyd went on a spree in 1955 and 1956, applying for various awards. He was the first DL4 to get the Worked All Europe award (1955); he also received WAC; WASM; WAV(Vasteras); DUF; OH-Award; AJD; WAYUR; All Africa Award; AAA; CDM; Diplomat Espana; WDT; OZCA; WAS; S6S (Czech); DVQ (Quebec). Back in California in 1958 he applied for more awards for DL4ZC, including Award Hunters Club; WADM; AC 15 Z (Poland); OHA-100; In January 1955 his Quarter Century Wireless Club certificate arrived, signifying that Lloyd has been a licensed amateur for 25 years.

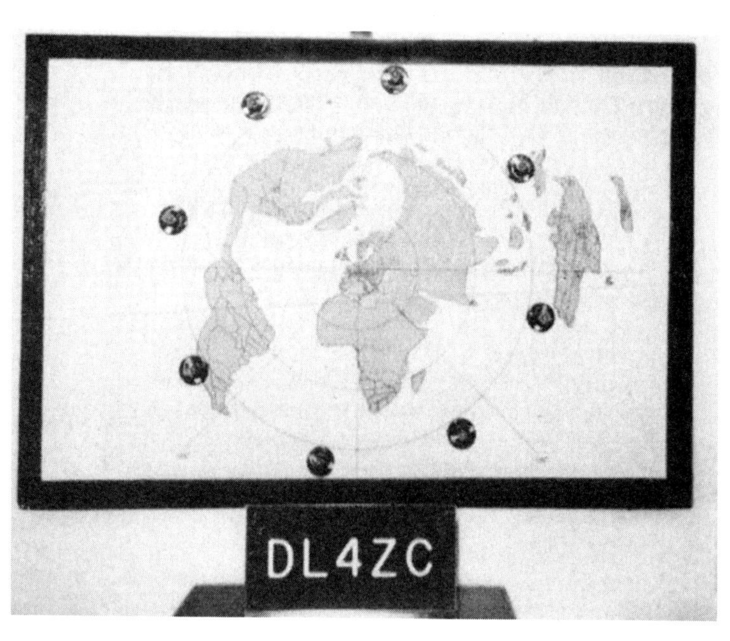

Lloyd's V-beam indicator screen in Germany

Lloyd's DL4ZC DXCC certificates include Phone, 1 Jun 1959, #1,537; and Mixed (170 total), 20 May 1955, #2,125.

A note in QST for October 1955 records that "Three DXCC memberships from three different continents have been earned by Lt. Col. Lloyd D. Colvin, DL4ZC. His previous two DX triumphs were ticked off as W4KE and JA2KG. The colonel has operated under 18 different calls from all over the globe and has had over 65,000 QSOs with amateurs in 242 countries. DL4ZC's wife [Iris] and daughter [Joy, 14] also hold tickets."

In Alaska in the early 1940s Lloyd had dreamed of buying land on which to erect rhombic antennas, but the war interfered with those plans. In Germany, he used several V-beams on the amateur bands, with a fairly ingenious arrangement for switching them that was described in QST for August 1956, his second QST feature article, entitled "Multiple V Beams - High Gain in All Directions with Four Wires.

Lloyd wrote that he was having good luck with a compromise of four V-beams using just four radiating wires and two relays to cover 360 degrees. Unfortunately, the article gave no details of where the installation was (other than Germany) nor of the physical details of the antennas themselves.

Lloyd described using standard 115-volt relays in a "refrigerator box" to switch between wires. To keep the box and its contents dry he sent a low voltage to the relays when they were not otherwise in use, "just enough to generate a little heat in the box."

The article said the length of the antenna wires was not critical but that all of them should be exactly the same, and that feed lines and relay control lines should be as far apart as possible. This was just confirmation of what the textbooks had to say about V-beams; the "Colvin touch" was Lloyd's control head in the ham shack.

While many, perhaps most hams would simply mount a few toggle switches on a panel and label them, not Lloyd. He used a great circle map of the earth with eight lamps spaced around it at 45 degree intervals, in the primary directions of the beams.

"The lights are operated," he wrote, "in pairs and are switched by the station operator at the same time the antenna relays are switched. A multiple-contact switch is used to control the operation of the antenna relays simultaneously with the lighting of lights on opposite sides of the great circle map. A quick glance at the map shows which beam is in use and what two directions it will operate in with maximum gain."

Pretty slick!

Lloyd reported operation on all bands had produced 171 countries in a year. He said that on transmit on 20, 15, and 10 meters the V-beams outperformed a 3-element yagi, while on receive the (bi-directional) V-beams were inferior to the yagi.

QST identified Lloyd thusly: "OM Colvin is well-known in DX circles. Besides eleven different W calls, he has been FA8JD, KL7KG, J2AHI, JA2KG, J2USA, and now DL4ZC. As you might suspect, the list of DX awards he holds is a yard long!"

One of Lloyd's reasons for erecting those rhombics was to work, in 1955, a certain Englishman who had begun appearing on the ham bands from the South Pacific, a long haul from central Europe!

Bob Voss, K1EY, was another U.S. serviceman who crossed paths with the Colvins in Germany, both in Karlsruhe and Heidelberg. Voss also served with the 315th Signal Battalion and then with the U.S. Army 4th

Signal Group. "During that period," he said, "I learned the art of military communication from pole line construction, telephone systems, worldwide HF systems, to my eventual specialty: VHF, UHF and microwave systems throughout Western Europe."

Voss remembers Lloyd's and Iris's amateur radio pursuits as "well-known throughout Signal Corps circles. Many 'volunteer' hours were spent in antenna projects (an accepted practice).

"Professionally and technically," Voss said, W6KG represented the 'best and brightest' of the Signal officers of that period.

"One day in 1956, this signal sergeant was stopped in a hallway of USAREUR Headquarters and thoroughly interrogated, by Lt. Colonel Colvin, regarding his understanding of 'path margin' impact on circuit quality of our Siemens and Halske microwave long lines systems. I must have passed with flying colors, since my promotion to sergeant first class came through shortly thereafter."

Lloyd earned one of his many DXCCs from DL4ZC

11 · A Cold War Tale

"I think that as a goal we all should try to answer QSLs received. At least try to answer one QSL from each station sending you one." - Lloyd Colvin

On his travels around Europe Lloyd (often with Iris) made many amateur radio friends. One he met in Hungary had been caught up in the ugly political crisis there in 1956. He and his family escaped to a neighboring "Western" country and he wrote to Lloyd, via a friend in the States, hoping for help to emigrate to the United States. Lloyd wrote back in December 1956, congratulating him on his escape from Hungary and offering a helping hand:

> I must tell you however that anywhere in the world, including America., a person must work hard to make a living. It will be harder than normal for you as you do not speak perfect English and you are not yet an American.
>
> California is a very expensive state to live in. You can readily find a job, perhaps not in your chosen field but will find that almost all you earn must be spent to live. I understand it will take many years for you to become an American citizen and you cannot obtain an amateur [radio] license until then.
>
> If you are healthy and willing to work you can find employment here. If you use the letter I am enclosing to help you come to America I will of course expect you to eagerly find employment on your arrival in the United States. Any money I [lend] you will have to be small in amount and will have to be paid back starting with your first employment.

Iris and Joy out on the town in Nicaragua, 1956

> My wife and daughter join me in wishing you, your good wife, and your son, the best of luck, and sincerely hope you find a new and better life. It will take much hard work, and you will meet many difficulties, however if you and your family try to meet these problems with the correct mental attitude, I am sure you will some day be very happy that you left Hungary.

Lloyd sent him a letter of reference, saying he was willing to sponsor him and his wife and children, and that he would "furnish financial assistance to the extent that they do not become public charges, or destitute. I have talked to ______ via amateur radio. I met ______ and his family during a ten day period I was in Budapest in 1955. That is the extent of my association with them. Everything I saw during that short period of time indicated they are the type of people who would make good, honest, sincere American citizens, if given the opportunity to make a new life in the United States."

In the summer of 1956 Lloyd, Iris and Joy left Heidelberg, bound eventually for Lloyd's next assignment, in San Francisco. They planned an ambitious trip, with military permission to visit: Portugal, Madeira Islands, Antigua, Guadeloupe, Martinique, Barbados, Trinidad, Brazil, French Guiana, Netherlands Guiana, British Guiana, Venezuela, Colombia, Ecuador, Peru, Netherlands West Indies, Panama, Costa Rica, Nicaragua,

On a cruise, 1956

Honduras, Salvador, Guatemala and Mexico (they did stop at many, but not all, of these).

The Colvins had not been back in California long before they were called to the amateur radio lecture circuit. Lloyd spoke at the January, 1957 "Fresno Joint DX Conference" (now held annually in Visalia, California), about his amateur radio experiences:

> I think that perhaps you would like to hear [about] radio and in particular DX ham radio in Europe. The average European ham is envious of the American hams, and particularly of the W6 DXer with their California Kilowatts. A lot of hams in Europe seem to think that American amateurs are made of money and that next to ham radio the American amateurs think only of money. Of course you and I know this is not true. In this connection I would like to tell you a true story.
>
> On my arrival in this country [the U.S.] a few months ago I rushed out and bought myself one of the new multi-band beam antennas. One Saturday morning I spread it out all over my back yard, and half of a neighbor's yard, and started the rather complicated process of putting it together. I no more than got started when I noticed a lot of the neighbors, and their children, staring at me and obviously wondering what I was doing. Finally two small girls came over near me and one of them looked at the beam and then started to ask me a question. Before she could get the question out she became frightened and the two of them ran away. A short time later they came cautiously back again.

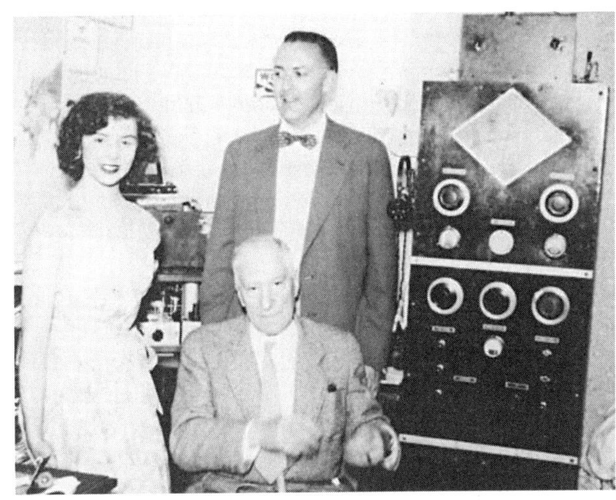

Joy and Lloyd at CT1JS

> I was sure of course that they wanted to ask about the beam, so I asked the girl who tried to speak before what it was she wanted to ask me. The little girl said "M - M ister, can you give us a nickel? Perhaps the Europeans' ideas about money-mad Americans is not as wrong as we might think.
>
> Almost all European DX men use American receivers. There are not many receivers in Europe that will compare with ours, and the few there are are fantastically expensive. This means that the European hams must try to obtain their receivers through American surplus markets in Europe, and through American hams and friends in Europe. Only in the rarest of cases are European hams permitted to import American receivers or other communication equipment, without paying prohibitive import taxes.
>
> Relatively few of the European hams use rotary beams made of aluminum tubing. One reason for this is that aluminum tubing in Europe is as expensive, or more expensive, than here. Figuring the average European's income, this means that a beam there may easily cost three times what we pay here, on a relative basis.
>
> I would like to describe some of the top DXers I have visited in Europe and some personal particulars about them. The top DX man in Germany is Rudi Hammer, DL7AA. The number three DXer in Germany is Dr. Baiz, DL7AB. Both of these hams live in Berlin, and unfortunately they actively dislike each other. It might be better to say they hate each other. When DL7AA became the DX manager for the

Lloyd with KH6IJ and family, 1956

DARC [the German amateur radio association], DL7AB resigned from DARC in protest, and refuses to support any DARC activities. This is an unfortunate and most unusual situation as both hams are fine men when you meet them individually. DL7AA is an engineer in our state department radio station, which beams radio programs behind the Iron Curtain, while DL7AB is a medical doctor.

At HP1BR

I am sorry to say my visit to PA0JQ, one of the top DX men in Holland, proved most embarrassing to him. I went to his house one evening, and on arriving I found that he was conducting classes in radio [for] a group of would-be hams. Arnold holds a position with the government there somewhat similar to an FCC [Federal Communications Commission] inspector here. I had brought with me one of the PA0JQ QSL cards which I had received for a QSO with him when I was in Japan. I thought the group of would-be hams would be interested in seeing one of PA0JQ's cards that had traveled around the world, so I passed it over for them to look at.

All of a sudden there was a terrific uproar from everyone in the room. Everyone started talking in Dutch at once and not knowing Dutch I did not know what was going on. I noticed that PA0JQ's face turned white, then his ears got beet red. Finally he turned to me , and with a very ashamed look on his face, explained to me in English that when he sent me the QSL in Japan he had no idea it would ever come back to Holland., and as he had sent it in an envelope he thought he was quite safe in listing his true transmitting power of 500 watts on the QSL. Actually, he was permitted by law to transmit a maximum power of 50 watts. Thus I had unwittingly put him in a position similar to (what) an FCC inspector would be in this country if he admitted he used 10,000 watts on the ham bands.

The next DXer I want to talk about is of interest to any of you who are single. I refer to Maria, IS1EHM, in Sardinia. She is the leading DXer in Sardinia with a worked total of nearly 200 [countries]. She is young, pretty, holds two college degrees, speaks good English, is quite well off, and she is single.

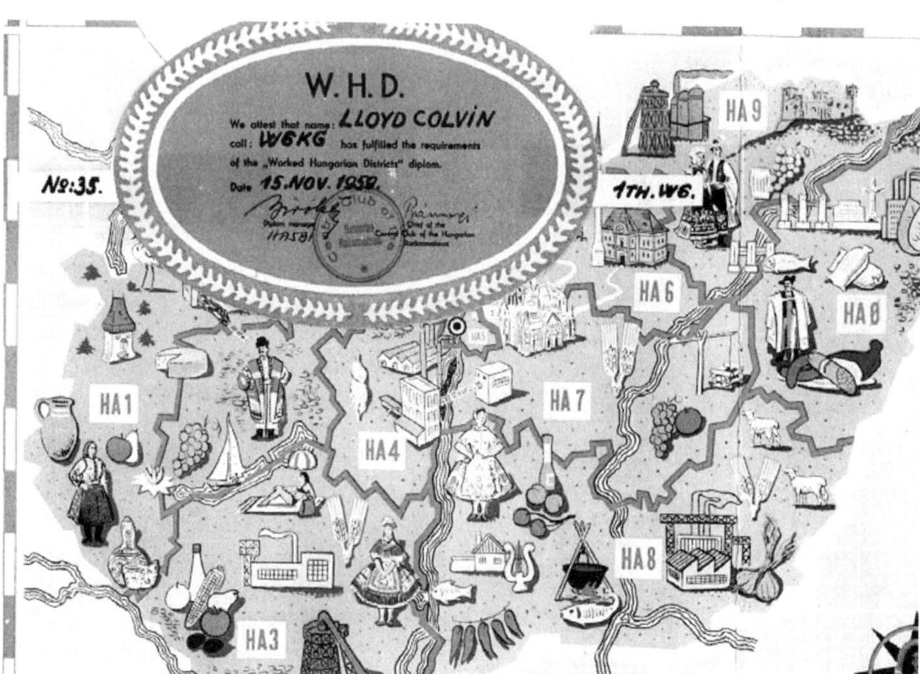

This amateur radio achievement award from Hungary was one of hundreds that Lloyd received.

An interesting sidelight on Maria is that she became a ham through the efforts of IS1EH, who is her brother.

In 1954 the Italian government required all their amateurs, including those in Sardinia, to retake their amateur examinations, including the code examination. This came about because some Italian hams were suspected of obtaining their licenses through [favoritism] or bribery without being fully eligible. Well, Maria, IS1EHM, passed her reexamination with flying colors, but her brother John, IS1EH, was unable to qualify [on code].

Speaking of women operators, I think the most famous woman operator outside of the United States today is Eva, CN8MM [in Morocco]. Her husband Alex has been a ham a long time. After World War 2, however, Eva decided to operate in earnest and today the leading ham in their family is definitely Eva, with Alex taking a back seat. She is a vivacious, talkative woman with many accomplishments. For example: she speaks seven languages, she helps her husband Alex in their import-export business, she runs their household, she looks after their children, which number four with the last two being newly arrived twins, and she finds time to keep CN8MM on the air about eight hours daily.

We are getting close to the ARRL DX Competition and I think some of you would like to hear about how W6s [California stations] come through in contests. I have made a study of my log for last year's ARRL DX Contest [when Lloyd was operating from Germany] and maybe the answer as to who has the best signal from California can be found in that study [it was W6AM on every band, 7 through 28 MHz].

First I would like to mention something that Elvin, W6TT, told me. Elvin says that during a recent trip he made to Hawaii he was told by famous DXer KH6IJ [Katashi Nose] that during busy contest periods KH6IJ would give everyone on a given page of his contest log the same signal report. In that way valuable operating time would be saved, and only one signal report had to be entered for a whole page of contacts.

I do not intend to say whether or not this is a good idea. I will say, however, that so far I have never used such a procedure. I have always tried to give each station worked, even during the busy periods of contests, an honest true report, as I hear it. Thus, to the best of my ability the signal strength reports in my log are correct.

When he asked me to prepare this little talk Horace, W6TI, suggested I include anything derogatory about specific W6s, such as a poor note, poor operating procedure, etc. I am happy to say I do not remember, and my log does not [record] any poor notes, poor procedure, off frequency operation, etc. There is one matter, however, along these lines that I would like to mention and I will name calls. I am referring to QSLs. If a DXer does not want to QSL to all stations worked, that is understandable. I think, however, that as a goal at least we should all try to answer QSLs received. At least try to answer one QSL from each station sending you one.

On the favorable side of QSLing I can say that time after time on my trips to various countries I visited hams and found the most prominent QSL displayed is a W6. Throughout the world the W6s are noted as keen operators, with good rigs, and lots of operating enthusiasm.

12. YASME I in the Pacific: 1956

"Many of you lads can thank Art for this as without it I should never have managed to pick you out of the QRM." - Danny Weil, on the gift of a 75A4 receiver from Art Collins, W0CXX, of Collins Radio

After his first "DXpedition" operation, from Tahiti as FO8AN, Danny Weil's next stop was Canton Island, a long haul from Tahiti with many possible stops and side-trips between. But his goal was not to put rare islands on the amateur radio bands, but rather to circumnavigate the globe, alone, his driving force since leaving England. On arrival at Canton Island, Danny removed his radio gear and was on the air within 18 hours, with the British Colonies call sign VR1B. At the time, Canton, part of the Phoenix group, was under joint control of the U.S. and Great Britain (the Phoenix group was formally ceded to the new island nation of Kiribati in 1983). In 1955 Canton was an amateur radio anomaly: it was possible to sit at one ham station and sign either a U.S. call sign (KB6) or a British territory call sign (VR1). Possible, that is, if one held licenses for both!

"Sometime in the distant past the Americans and the British, after discussing it in the White House and in Parliament for about 50 years, decided to rule the joint together and called it a Condominium. What a name for such a tiny spot. Well, now it is used as an air strip for kites flying back and forth from Australia and the United States," was how Danny described the political situation there.

Danny wrote in Cruise that "My radio friend Howard (KB6BA, an American on Canton whom Danny had communicated with enroute) and British Commissioner Paul Laxton discussed which side of the pass (island) I would like to go to, and whilst I should have preferred to stay on the British side of the island, there was the snag that arose of lack of power to operate my radio gear, so, boarding Howard's car, I went to stay at his place and installed all my gear in his shack, ready to work on the expedition when I had accustomed myself to the place, which wasn't too easy."

Danny found himself "overwhelmed" by the ham radio demand for contacts with him, and he became a Marathon Man at the radio. "My whole life was devoted to key thumping and occasionally spouting on fone, but I must admit I thoroughly enjoyed myself and tried to answer every call that came through." Danny stayed longer than his planned-for one month, operating from the ham shack of KB6BA. Danny said that on at least one occasion he and Howie, "one of the finest fellers I have ever had the good fortune to be associated with," were on the air at the same time to pass out DXCC credits. "We did not continue with this owing to the time factor, but it was fun while it lasted," Danny said.

By now Danny Weil was hot news on the DX grapevine, and he was greeted on Canton Island with a brand-new Collins 75A4 receiver, arguably the best ham-band receiver at the time, at any price. "This had been presented to the YASME expedition by Art Collins, W0CXX," Danny said. "Many of you lads can thank Art for this as without it I should never have managed to pick you out of the QRM."

YASME I docked at Canton Island, 1956

Incidentally, this is apparently the first time in the writings of Danny or Spenceley that the phrase "YASME expedition" appears. It wouldn't be the last!

Danny used one of his little Elmac transmitters to work nearly 100 countries as VR1B. And "the Collins 75A4 did a really super job in picking up all those very weak stations, and by use of the rejection tuning coupled with that 800 cycle filter I was able to split up the twittering mass into separate readable signals. I feel that without this fine [receiver], my number of QSOs would have been considerably less." An analogy to switching from a tiny Elmac receiver, really designed for mobile use, and

Fixing the generator on Papua

fact, there just ain't nothing here, not even fresh water, and as for women, that was wishful thinking. All this joint consists of is a horseshoe shaped chunk of coral about 9 miles by 4 and around 10 feet high if you're wearing elevator shoes. The sun burns down onto the glistening white coral, and if you have any respect for your eyes, you either close 'em or wear sun specs. On this excuse for an island live a certain number of Americans, British, and Gilbertese natives, all who, in my opinion, deserve a great big gold medal and a pension at 30 for their bravery in staying here to man the air strip. Personally, I'd rather cross the Atlantic in a canoe than live here for any time. I must have aged around 50 years in my three month stay.

No log survives for Danny's VR1B Canton Island operation, but enough QSLs reside in files around the world to suggest that this was his first "major" operation in terms of contacts made, both on CW and AM phone. The Collins 75A4 receiver was Danny's introduction to "serious" gear for serious operators. VR1B was definitely "in demand," and Danny, a brand-new operator, faced pileups of stations calling that would have been daunting to the most seasoned of operators. Finances were also a concern, now, as amateur radio dictated at least some of the terms of Danny's voyage. On 30 March 1956, from Canton Island, Danny wrote to Dick Spenceley:

a 75A4, would be driving a family sedan to Le Mans and stepping into a Formula 1 race car.

Later, as Danny sailed westward, Spenceley and a few other insiders beat the bushes for better equipment for him. One company, Eldico, shipped out one of their SSB 100A transmitters and a 500-watt linear amplifier. They arrived on Canton as Danny was preparing to depart but would see heavy use on later stops. W9YFV shipped a step-down transformer and a hash filter for the Onan generator came from W6MUB. "Whilst I cannot possibly write all the names of the very kind types who helped, I must say I do appreciate what they did," Danny said. "Also, I would very much like to give all you lads and lassies a great big hand for all your assistance, both monetary and for your good manners on the job; you all helped to make my work easier."

Danny had no trouble bringing himself to leave Canton Island:

... there just ain't no trees, brooks, waterfalls, in

Dear Dick: I feel I must write you this letter as [conditions] for our usual [contact] are hopeless, and, quite frankly, I feel very annoyed to think I have to bow to all the hams to give up my usual [schedule] with you; however, there it is.

Now, with Howie's assistance, I was able to get on the air less than 24 hours after arriving here, and cleared well over 100 QSO's, wkg about one a minute. I should with a little practice get a little faster, I realise this is slow, but am still feeling a little tired after that hectic trip, and once I can get rested......what a hope.....maybe we shall be able to speed up a bit.

Now you realise, although I want to be preferential to those hams that have subscribed to the expedition, you know I cannot possibly sort them out in the terrific QRM, so, I suggest that you arrange with any of your particular friends to be available on a certain day, and I will make a point of QSOing them directly after our sked. I think if this is properly arranged, I should manage to clear about 4 or 5 of them before the crowd shoots up onto 14080 to QRM the QSO, so, I leave the details to you, and you know I will do my part.

Next item is, what do you advise about fone QSOs for those that need them....such people as Walt (W3RIS) and maybe quite a few others who do not work cw and who have subscribed. I want to do all in my power to satisfy everyone, and although I take no instructions from anyone at sea, whilst I'm ashore, YOU are my boss, and what you say...GOES, so let me know the dope on this point, and I will follow instructions.

Next point. I am sending you the complete log book by air mail. I just haven't the time to copy out all the contacts I am making...it will be well over [a thousand] a week I think, and I am up every morning here around 6 and after a tough day working partly on the boat, and then on this rig, I don't feel much like rewriting all the call signs again. Another point I thought about was, the book will be better for you, as I have entered the approximate times of all QSOs and should there be some small error in my copy...the usual S for H and vice versa then the log will help you clear that point coupled with the QSL from the Ham in question, as the approx time will be there on his card and the log too.

Perhaps its not 'cricket' as the English say, but I am practically ignoring all other calls except the W's and K's [U.S. amateurs]. I am still prejudiced in their favor, and until such time arrives that I can see some other types helping out, I cannot see the point in using the American Hams' money to QSO other countries, when I shall need every moment I can get to make contact with the W's. Let me know your views on that....or perhaps you would rather not commit yourself..Hi.

I have not received my March copy of CQ. Howie is giving all my gear a thorough overhaul. Both TXs have gone up the spout, also the new Elmac power pack had a choke blow out on the way...I fixed that by the same system as the other one..shorted it out...it kept me on the air. He has also given the Hammarlund a good going over, and at the moment it is on test, and seems to be 100% better...he's a good lad and certainly knows his stuff...he should do {too}, he's in charge of the TX [transmitting] station here. He's also having a go at my ship-to-shore rig and getting the loop working for me. It will be a great help among all these damned tiny atolls at a later date.

Next, what do you think of the idea of selling one of the Elmac TX [transmitter] and the RX [receiver], and using the money for a bigger powered TX? I think I can sell the TX, RX and Powerpack here. Should like your views on that pronto.

I hate to admit it Dick, but it's either bad workmanship on the part of FO8AD, or someone in Panama, but the Elmac TX seems to always give up the ghost around the end of every voyage. I don't know yet what the troubles are with either of the TXs to date. Howie is working on them both, and will let you know his findings later. I am not experienced enough to talk on the matter fluently, but it seems to me that they are just not man enough to stand up to the continual usage they are getting...your viewpoint please?

Now to tell you about the Collins [75A4 receiver]. I fully intend to write to Art about it, Oh brother, what [a receiver] it is, and I'm just waiting for the time you come on SSB so that we can really give it a good try out on that angle. I am sitting here writing this whilst waiting for the 20 meter band to open up, then I shall be there till chow time bashing the key. It's tough going Dick, this DX work, but I am getting a wonderful feeling of accomplishment out of it, and I know that I am not letting you down after all the faith you put in me.

I know there are many hams who ridiculed the whole idea in the first place...I can still remember the remark that was passed about putting all this first class equipment into the hands of a person such as myself....something about giving a baby a delicate piece of equipment to play with. Well, I sincerely hope I have partly carried out your ambition Dick, and, from here on, I think it will be easier for me.

The majority of the boys seem to be fairly well behaved. I have on two occasions had to give them all a short lecture about prolonged QSOs, but generally, things aren't too bad, as you have already realised by listening in at your end. There are still the odd few idiots who send "QSO...?" ... they want their brains tested, and of course, we are still getting the odd few who say "QTH?" but we are making out fine, and hope to use up a few log books in my stay here.

Now, I shall be here longer than a month. I just cannot carry out my responsibilities on ham work and also get the boat ready again in one month, so, it looks like a 2 month job. The engine will HAVE to be overhauled, there is no question about it. They have the facilities here, and I have got the permission for use of their equipment. I will have to get some spare parts shipped out by air, but I think that the work can be successfully carried out here. The engine at present has two cylinders not working, and the water jacket is absolutely solid with sludge causing excessive overheating of the engine, and I just cannot chance the next part of the journey with an engine that may fall to pieces any moment.

Would like to know where the sails are being made, and also where they will be shipped to for air transport. We have not succeeded in getting anything organised for their passage here, but will see what can be done in the near future… as you know, I haven't wasted any time since I arrived. There are many people to see here who may be able to assist in different ways, but that all takes time, and, there's only 24 hours in the day in this part of the world. What a pity.

Now there's very little else to say at present except a complaint. Maybe you can guess again who the person is…..W6AM. On the 30th March around 0646, he coupled with W6GAL/7 decided that between the two of them, they would hold me as long as they possibly could in QSO. This meant that between the hours of 0640 and 0720, almost one hour, I made around 16 DX QSOs. The mere fact that I got the rest of the lads to call on another frequency did not deter our two friends, they soon zero beat, and bashed away quite happily with their KWs.

W6AM pestered the life out of me to know when I was going on fone, and, quite frankly, what the hell W6GAL/7 had to say other than giving me an RST of 579, and then telling me he couldn't copy me, I don't know, and really I don't care. As far as I am concerned, to give me a 579 report was sufficient for me to know he had received his report from me, and anything else wasn't important. Now you know Dick I have been working the lads around one a minute, [but because of] these two idiots, it has meant that around 50 of the lads didn't get a QSO, and I feel very sure that many of them had stayed up for some time to make that contact.

Naturally, once I finally got clear of our two sublime idiots, the band had folded up for the W's, and I was so peeved. I didn't feel in the mood to work anyone else in other countries. Now Dick, you can tell these two birds that if either of them ever hold me in QSO again, I will never answer their call again the rest of the trip, and if they have subscribed anything, then I feel that the rest of the Hams will be only too happy to back me in giving them their money back.

Should any verification be required, Howie, KB6BA was present at the time and heard the whole thing. Will write again soon, so until then, cheerio for now and hope you are all happy and healthy up there in St. T.

P.S. Will do my best to get another article writ for CQ directly things settle down a bit. Have also that story for [Stuart Love?] to do too. Oh, well, I can stay on another 3 months Hi.

And another letter to Spenceley from Canton Island, dated 6 April 1956:

Dear Dick: Things going FB up to date, but I must admit that I haven't worked so hard in years. The unfortunate part of this expedition is, the band is best for me only in the God forsaken hours, and necessitates my staying up until 4 am. This is hardly conducive to my health, but I have no option as I realise that it is pointless to fire up the rig earlier in the day, however, the fact that we are getting out here ok is the main thing, and the lads are certainly taking advantage of it.

I have my usual grumble with some of them, but generally speaking, I find that most of the W's are fairly reasonable, and not all bad mannered.

I am still having the usual trouble with the S. American Hams with too much talk, and it seems to me that whatever I say or do, they still blithely come back again and do all the things that you have just asked them not to do ... the JA's are the same. I endeavour to avoid their calls unless they are running about 10 kws, and then I have no option, the same with all the S. American lads too.

I am afraid that I shall have to be on the air later, as I haven't had one opportunity to touch the YASME in any way, and there is hardly time to do any real work in the few hours available in the morning, and the boat is some considerable way from Howie's house, and there is no transport here at all except government and one or two privately owned cars, so I feel that when the time comes, I shall just have to stay off the air for a few days to clear up all the odds and ends that have accrued in the last trip.

Another thing, there is the article to write for CQ and also the one for Short Wave [a British] magazine. Quite frankly, I don't know where I am going to get the time to do all this. I am up around 7am, after breakfast, am writing up the log until around 9, then I have to go batting off to see someone about getting something done to the boat when the occasion arises, by the time these few odds and ends are done, it's chow time,11.30. After chow, then I start to fire up the rig and look around, this goes on until around 5 pm and then chow again .... on again at 6pm, and then I don't stop until around 2 to 3 the following morning...then I attempt to have a short nap, and then we start all over again. I know I'm pretty tough Dick, but am just wondering how long I can carry on at this pace, anyway, we'll see.

Next item is this damned beam [antenna]. As you will know long before you get this letter, the beam has arrived and had to be shipped by air from Honolulu,

Now I am pretty peeved about this whole thing as it was arranged first to be all fitted up by KH6OR. who assured me that he would get the boom all organised before it was shipped here by boat. Well, I find first that the damned thing wasn't on the boat, then, I have to chase over the band trying to contact the right party, and, when I do finally get there, I am informed that the beam arrived too late to be shipped on the boat. On top of this, nothing is done about the boom, except a letter comes through the post from the donor with a very pretty picture showing how to make the boom up out of timber, but in the interim, KH6OR has just sat on his fanny and done nothing,

How long the boom elements would have stayed in Honolulu if I hadn't contacted them by radio I don't know, but now, to crown it all, this morning PAA [Pan American Airways] arrive with a truck and unship the elements all nicely wrapped up, also a box full of insulators, which, incidentally,

Last page of Danny's VK9TW log book

have all the bolts about 3" too short to be of any use, finally, I had to pay out 25.60 cost of air freight here for a load of aluminum tubing which, as far as I can see, is about as much use as an old boot.

Now please Dick, don't think for one moment I am not appreciative for the beam, but really, what on earth use is it to me like this.

I have, since the arrival of the elements, been on the scrounge to obtain some scrap timber .... the local powers that be have none available, so naturally I have had to go tramping all over the place to find suitable material. Well, as you probably know, I have managed to find something to do the job ... my next problem is to build it, and finally, to obtain some coax to use with it. Naturally, I shan't be able to carry the thing with me owing to its bulk, so it means that all the labor incurred will be wasted for the future, and I shall have to start all over again from scratch.

I fully realise Dick that you can't do everything, but you are the only person I can talk to about my troubles, and really the only one who actually cares what goes on. As you have probably gathered by now, the boys all know I am here ... my log shows a little bit of effort; my only regret is they aren't all as keen to really do something solid to enable the expedition to continue successfully. Now, to return to this beam business. What do you honestly think can be done that can clear this matter up. What I require is a prefabricated boom, also a prefab tower or pole that I can put up and brace with wires, This should not be any problem in the States and I feel that the actual cost of the material would be small. The labor is another problem but surely out of all the thousands of hams, there must be another who is prepared to help out such as Howie is doing for us at present for free.

Danny, below, helps with the beam on Nauru

I honestly think an article on this subject by you in CQ should be published, as I am damned sure that 90% of the boys don't realise the work worry and loss of sleep that goes into giving them a rare country.

Now I have been in conference with the rep of PAA [Pan American Airways] here, and told him of my troubles re the sails...I also mentioned the beam, but of course, that's here. I have suggested that he write to head office and obtain something from them that would give me free air transport for all bits and pieces needed on the expedition, and I in turn, would permit them to have use of my name to advertise PAA in any way they thought fit. I also suggested I would write in my articles in CQ some mention of their assistance, and finally, would fly their flag on my voyage. I think they will possibly have some high pressure advertising types that can put this offer to some use.

One thing more, I showed him the QSL card (PAA one) and I think that, with the illustration in CQ, made him a little more enthusiastic. I mention this as I know that other cards have been printed, I also gave him your name and address, to I expect you will possibly get a letter from them ... maybe. The rep required the weight of the sails to put that point before the company.

Now, the next item is the TXs [transmitters]. They continue to give trouble on modulation [for 'phone operation], Howie has spent some considerable time on them, and will have to spend a lot more. All those new batteries shipped out to me in Tahiti are finished...even the new ones only have about 15 volts static...under load, about 10...maybe. Now on Howie's advice, I have ordered from Honolulu two selenium rectifiers which we shall couple into the power supply to get the necessary negative voltage.

I feel that Elmac should know also that the output coupling condenser pi-network C58-A-B has insufficient plate spacing, and also the plate spacing is offset too. This is causing arcing. This was cured by removal, cleaning in carbon tet and replacing, but this will occur again owing to the salt air and moisture always present.

Now I think if a wider spaced plate could be fitted, and also something to dispose of this B battery, it would go a long way toward efficiency in the rig, and also save unnecessary maintenance which, to me, is not so good. I have asked Howie to write a short letter giving you his findings and also opinions, so that will save my trying to write about it here in my very unskilled way.

Here I must say that Howie has been 100% for me on the expedition.

Well Dick, I think I have covered everything, so lets .... You can do something about it. I might refer you to an advertisement .... The CQ Mag of January of 1956 pertaining to a firm namely, Rex Basset Inc, Florida, who are apparently manufacturers of the type of equipment we seem to require. Perhaps a word in their ear relating to a spot of advertising may assist without impairing Onan's reputation in any way...it's a point worth considering if Onan cannot assist.

Well Dick, I think I have blabbed enough now, and I sincerely hope you haven't gone to sleep in the processing of reading it, but I had to get all this off in one go, and I have been working on the rig pretty solid, and whilst this is a form of work, it's a change from pounding a key.

Am also enclosing list of QSOs to date. I don't think I have made any errors in H or Ss', except in one doubtful case, but I think you will be able to check that easily enough, so will put all the dope on another piece of paper and get this off tomorrow.

Raising a mast on Nauru

YASME I at Nauru

By the way, you owe me some money for fotos and also cash for cable to Art Collins, roughly 15 bucks, but I suppose all this cash is in the same bank, so it doesn't matter which way it comes in, as long as we have enough to keep going OK.

Well Dick, I would like to know exactly what was said by the G's at home in respect of your msge re they were worried about me. I am very intrigued about that, so don't forget to mention it in your reply. When my own countrymen say that [they] are worried about me, it usually means they are in preparation to ask a favor, or maybe want to borrow some American bucks, and I'm not being bitter either, but it's the first time on record that anyone other than you or my mother has had any concern about my well being, so naturally, I am a little suspicious. Can you blame me? -- Cheerio for now, All the Best, Danny"

Nauru Island - VK9TW

"When I tell you blokes I spent 12 hours a day on that rig it's no lie." - Danny Weil

Danny wasted no time upon arrival at Nauru Island, visiting a Mr. Cameron, the Island Manager. "Seeing important people always gives me a feeling of apprehension," said Danny, "but I need not have felt that way, because directly I got into his office, he made me feel at home, even inviting me to take morning tea with him." Danny was promptly fixed up with a place to both live and operate his radio gear, and Cameron even arranged to have an antenna put up.

Nauru had been occupied by the Japanese during World War 2, from 1942 to 1945. In 1956, Nauru was under United Nations trusteeship, administered jointly by Australia, New Zealand, and the United Kingdom. (It gained its independence and was proclaimed a republic on 31 January 1968.) Danny said of his antenna there that

> ... this antenna wasn't anything cheap or shoddy, but the real thing done in the right way. Some chaps came along and dug a couple of deep holes around 200 feet apart, and into them were cemented two 40-foot steel poles. The usual wire was strung between them, and finally a long metal earthing rod was sunk into the ground, leaving me with nothing to do except bring all my gear ashore and couple it all up to the two wires coming into the house. Little time was lost after that and I immediately started work contacting the Hams all over the world. Being a completely new country from the Hams' viewpoint, I was overwhelmed with calls coming through the receiver, and the first day I stuck there for a solid 14 hours just sending out reports etc.

Contemplating, at Port Moresby

Most of the radio gear had suffered in the last voyage, mainly through the terrific beating it had had in the Pacific storms, and while there was nothing very serious wrong it all took time to fix. "One of the chaps on the island. Les Wright, VK9LW, the Commercial Radio Operator of the island, came forward and offered to overhaul every piece of equipment on the boat in his spare time," Danny said. He had been a ham when in New Zealand and his interest in the project was fantastic. Hour after hour he spent with me in the evenings testing and helping me with my work. Had I been a King, I couldn't have been treated better."

Wright didn't have a transmitter, so Danny sought the help of the famous Bill Bennett, W7PHO, in finding the necessary parts for him to build one, "so that we shall have a permanent ham on Nauru. The likelihood of another ham ever going there is pretty remote, so it would certainly help the DX boys if Les was on the air regularly."

Danny tackled the radio pileups after both YASME and his radio gear had been overhauled. "For good transmission one must have good antennas," Danny observed.

Danny's wire antenna was doing "a fine job," but he wanted a beam antenna, and Ken, W7FA, had shipped the necessary materials to build one. Using drill presses and other needed tools on Nauru, a 20-meter yagi was put together. Although he was due to leave Nauru in two days, Danny naturally had to put the beam up and try it out.

> First a 40-foot 2-in diameter pole was put up and stayed, then came the problem of taking the beam to the top and fixing it. The beam is 33 feet wide and 12 feet deep and although very light it is quite a handful in any wind. Now the pole being only two inches in diameter gave me plenty of misgivings as to its strength to support the beam, let alone me at the top to fit it, so being one of the brave types, I looked

around for volunteers to climb the pole. The lads here had no fears whatsoever, and in a few seconds one had climbed to the top with a rope, and there, swaying in the breeze he looked like a blob in the sky.

He didn't seem to worry; the end of the rope was attached to the beam and the lad at the top started to pull. As it rose in the air, it narrowly missed a high tension wire which had it touch would have frizzled the lad to a cinder. We managed to avoid having roast meat that day. Part way up, the beam jammed in the bracing wires, and before I could say a word, another lad had shinnied up the pole to help the other. That really worried me. The two of them got the beam to the top, struggled for quite a while, but couldn't get it to sit in the correct position. A call went to the crowd below, and in a few seconds a third member was at the top of that 40-foot pole. Certainly it swayed, and I shuddered and closed my eyes a few times, but they still remained at the top, complete with beam, rope and the electrical leads hanging down. Finally, with the three of them fighting the wind, the beam dropped into position, and from below came a hearty cheer for a job well done.

"I sure breathed a great big sigh of relief." Danny wrote, "when those lads were on solid ground again, but the beam was up, and when I connected it to the radio, it worked 100% fine."

Between DXing periods, Danny, with help from the locals, fixed a quarter-inch, foot-long leak in YASME, and he was off.

Danny's next stop was the Solomon Islands, a 7-day, 700-mile sail from Nauru. He was not impressed...

Customs lost very little time in coming aboard, then I was left to myself. I rather expected to have a few visitors, but not a soul came out to greet me, and really, I felt far lonelier then than I had been in all the past thousands of miles at sea. Knew from the moment I stepped ashore that I wasn't staying long enough to even get a haircut.

Most of the Solomon Islands had been occupied by the Japanese during the war and they are remembered as the scene of some of the bloodiest fighting, especially the battle for Guadalcanal. After the war portions of the Solomons (including Honiara) were administered by the British, and gained independence in 1978. The Australian-administered Solomons became independent in 1975 as part of Papua New Guinea.

"Mooching around," Danny said, "trying to find a suitable location to rig up all my radio gear was a tough job. No one seemed to care whether I had arrived or not, and for the first time in many months I felt really despondent. After my wonderful experiences in Nauru and then to arrive in a dump like this was a real shaker, but I kept up the search for a couple of days until I actually found one good natured bloke in the entire community who offered his help.

"Hewton Amos, the secretary of the local Copra Board, was my angel there, and with his help, I was able to get all my radio gear installed and operating. He did all he could under the circumstances to help out, but the unsociability of the local inhabitants decided me to make my stay as short as possible. It may seem silly but I hardly took my camera out in the entire stay, but devoted practically every moment to my radio work."

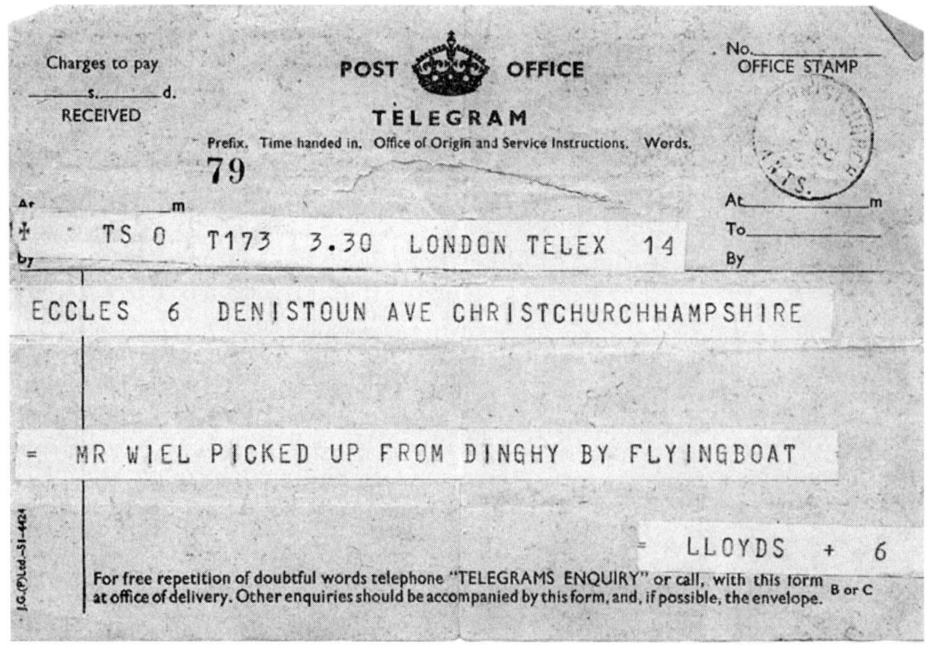
Charges to pay s. d. RECEIVED

POST OFFICE TELEGRAM

No. OFFICE STAMP

Prefix. Time handed in. Office of Origin and Service Instructions. Words.

79

TS 0 T173 3.30 LONDON TELEX 14

ECCLES 6 DENISTOUN AVE CHRISTCHURCHHAMPSHIRE

= MR WIEL PICKED UP FROM DINGHY BY FLYINGBOAT

= LLOYDS + 6

For free repetition of doubtful words telephone "TELEGRAMS ENQUIRY" or call, with this form at office of delivery. Other enquiries should be accompanied by this form, and, if possible, the envelope. B or C

Amos offered Danny the use of his office for a ham shack. It was fairly close to YASME, had electrical power, and space for a mast and beam antenna.

Within a few hours, up went the beam and I was on the air as VR4AA. What a wonderful reception I had from the boys when I bashed the old key. It far surpassed my expectations there for DX, and I know for a fact I made more QSOs here than Nauru, yet it was not such a rare spot. Apparently I was putting out a fair signal and the lads simply flocked in for a report," Danny wrote in CQ magazine.

When I tell you blokes I spent 12 hours a day on that rig it's no lie. I was happy working there and it took the bad taste out of my mouth of the place itself. Old Hewt certainly pulled his weight there on fixing me up with the shack, and later I found he was an old pal of ZL2GX (whom I sked daily), so by the time we had finished, we were good friends. Quite frankly I don't know how many countries I contacted in the short time I was there, but it was the ham's delight from a good QTH. Everyone came through, and I felt if I had stayed another week I should have made my DXCC. Fone QSOs were made to Europe as easy as the Ws and I really had a good time there on DX work. My plans to stay for two weeks were strictly adhered to, not because I hated the place, but because of the bad weather ahead of me. After 12 days' solid DX I pulled the big switch.

After 13 days in Honiara and 1650 QSOs Danny "just had to get out and feel the freedom of the sea again." He left Sunday morning, headed for Papua Territory, New Guinea, and disaster.

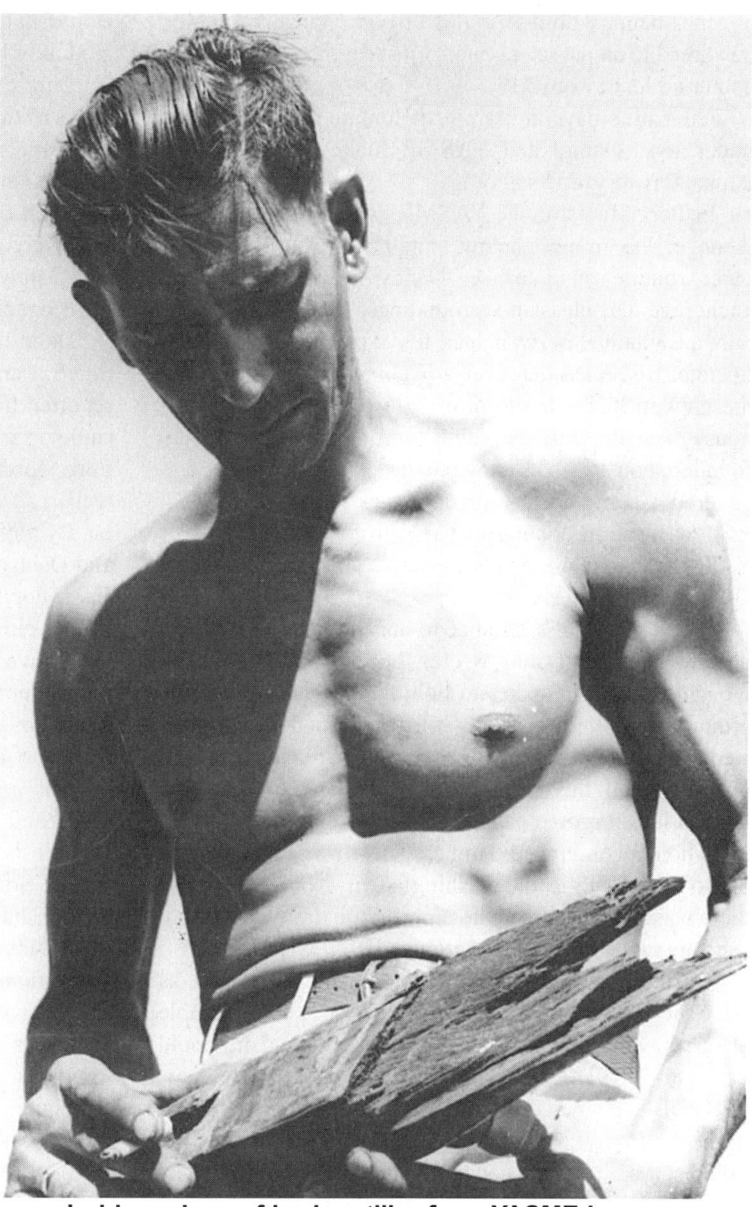
Danny holds a piece of broken tiller from YASME I

Papua Territory - VK9TW

The trip from Honiara, Guadalcanal, to Port Moresby on Papua wasn't far and might not have taken long, but a typhoon enroute turned the trip into the longest nine days so far in Danny Weil's attempt to be the first Englishman to sail around the earth alone.

The typhoon struck when YASME was literally within sight of Port Moresby, perhaps a dozen miles away. Danny was stretched out for a rest when "One second I was lying in a horizontal position, the next I was vertical, actually standing on the inner sides of the cock pit coaming. For some unaccountable reason the wind had swung a complete 180 degrees, and I was getting the full blast of a hurricane, typhoon or what the heck you care to call it with all canvas up and sheeted in hard. YASME was thrown absolutely flat, and the cross-trees on the mast were actually in the water; the engine, which had been running steadily suddenly roared as the prop was brought clear of the water. Water poured into the cockpit and flooded into the cabin, and I could hear the water pouring in through the forward hatch which was wide open.

"Panic stricken," Danny wrote, "I clambered out of the cockpit and ran along the side of the cabin to free the jib halyard and sheet. I had already let the mainsail sheet go. The wind screamed through the rigging and the towering seas broke right over YASME, trying hard to sink her as fast as possible. Above all this noise I could hear the gush of water as YASME gradually filled up."

At this very moment it was time for the daily radio schedule with Spenceley. The generator was still running and the radio equipment came to life, but Danny's telegraph key was too wet to work - shorted out. Danny used gasoline and a rag to dry off the key's contacts and finally hooked up with KV4AA. Danny wanted to switch to 'phone, and they did (Spenceley always preferred CW), and Danny told him he had "almost lost YASME.

Despite being in radio contact with VK9FN, in Port Moresby, as well as with Australian amateur stations, Danny was unable to get his bearings in the storm, despite at times being within 10 miles of Port Moresby. Finally, he agreed to an air-sea rescue. "I just sat there in the cabin and cried like a baby."

After nine days at sea since Honiara and nine hours under tow, Danny and YASME make Port Moresby, Papua Territory, and safety.

In Port Moresby, the YASME, while free of serious damage, was "a mess below," and need revarnishing and other routine maintenance. Danny described "a good anchorage and pleasant surroundings," and the port captain, a Captain Hawley, helped find a place from which to operate. It seemed that every dwelling was occupied but the captain had a brainstorm. "Stowed in an old warehouse was the salvage cabin from a ship which had foundered on a reef," Danny recalled. "Directly it was discovered, a crane was organized to pick it up and drop it at its new location about 100 yards from the moorings of YASME. Power was made available and the rig was set up" - instant ham shack!

"I had plenty of assistance to hoist the 40 ft. steel pole up into vertical," Danny wrote, "but the fun came when we had to bang in spikes to hold the bracing wires. The ground was solid rock, and whilst we bashed away with a sledge hammer, all that happened was that the spikes bent up. While all this was going on, several of the New Guinea lads forgot to hang onto their guy lines and decided to hold a conference. Timber! There was a mad scramble to dodge as it came crashing down. No one hurt, but there was a nasty bend in the pipe; a little unsafe for future use. Everyone had a good laugh and up it went again."

During his stay Danny made some 600 contacts as VK9TW from Sept 24 to Oct 16 1956, despite incredible electrical interference. At first the only signal he could hear was VK9DB, who turned out to be very close by. "VK9DB explained that there was a strong copper deposit in the air from the local mines and owing to dry weather it had settled on every power insulator in the place, causing this infernal noise. Wait for rain? They've never heard of rain here. That put the lid on serious DXing."

Danny spent time with another local, Frank, VK9FN, "playing around" with the Eldico SSB rig, and Frank himself made some SSB QSOs, despite the electrical interference.

It was time to leave, with the "North Westerlies" due any time. Danny knew that "meant that I should have a lousy trip through to Darwin" [on the northern coast of Australia - Danny was heading due west]. As soon as a new mainsail arrived by air from New York, Danny left Port Moresby behind, but not for long.

After clearing the infamous Gulf of Papua, Danny entered the open sea and what he thought was calmer, safer waters, but YASME was off course and struck a reef. "I thought we might be lucky and get off later when the tide rose, so I scrambled over the sloping deck and cast out the anchor," Danny wrote in CQ. I prayed something would happen, but all my prayers went to naught and gradually [YASME] settled firmly on the reef."

Danny's story of the final days of YASME I, in the pages of the April 1957 CQ, sealed both his amateur radio fame and future. Danny told of his daily schedules with ZL2GX in New Zealand and VK9FN at Port Moresby, and then of how his ham gear saved his life. The heavy seas were breaking over the bow of the lifeless YASME and Danny's dinghy/lifeboat damaged beyond use. Even his life jacket was useless in the shark-infested waters.

Danny used the last of his generator's life to call an S.O.S., answered by VK2AUR in Australia. Danny reported his position as best he knew it. Frank, VK9FN, came on frequency to say that a rescue plane would leave Port Moresby at daybreak and perhaps a rescue boat as well.

By morning YASME had gone further below the water and Danny knew it would be difficult to spot from the air. Just after 8 a.m. the rescue plane found the wreck and began circling, and signaling to Danny that a rescue launch was on its way and should arrive about 4 p.m. The plane departed, as Danny considered his plight ... he could not survive another eight hours; YASME was sinking/breaking up too quickly.

But, a couple of hours later, another plane began circling, and Danny recognized it as an R.A.A.F. 4-engined Lincoln bomber. "What use could that be?" he thought, when "Something fell from his bomb bays and hit the water. Thank God ... it's a dinghy! It gradually took form as its tiny cylinder of CO_2 inflated it to full size, and there, floating toward me, was my life saver."

The date was 24 October 1956, and, somehow, Danny Weil was saved. Exhausted, he managed to swim to the dinghy, and passed out. When he came to, a Catalina Seaplane from Qantas Airways was landing nearby in the heavy seas, and soon Danny was aboard. Danny arrived in Port Moresby with nothing. He was helped by the Red Cross, particularly one Don Ethridge. "Both he and his wife were kindness personified," Danny said.

The YASME was not insured and not worth salvaging. Danny soon repaired to Australia where, after talks with KV4AA and others, he decided to fly to the United States and mount a campaign for support for a new boat, and another DXpedition. "If I am successful in finding a boat there it will mean crossing the Pacific again," Danny wrote, "but also it will give me a chance to call at some of the rare spots I missed on the last trip, such as the Kermadecs and Wallis Island. Maybe you boys will think out a few new ones for me, anyway, that is all in the future."

As Rod Newkirk, W9BRD, QST's DX editor put it in January 1957, "VP2VB/P after losing YASME on Papuan reefs, was reported torn between continuing westward under new sail, visiting the U.S.A., or returning forthwith to England."

13. Danny Weil on the Lecture Circuit

"I believe if Danny were to visit the States he might be able to aid his cause by personal appearances" - Dick Spenceley

At the end of 1956, Lloyd Colvin, then aged 41, Iris (42) and Joy (16) were just arriving in California, to begin their new lives in the San Francisco Bay area. Simultaneously, Danny Weil, 37, was arriving in southern California. Danny, still a bachelor, was broke and jobless. The good ship YASME (hereafter referred to as YASME I) was only a memory, his sailing dream crushed on a reef off Papua New Guinea. Danny, 6,000 miles from home, was literally forced to beg, and American amateur radio friends came to his rescue.

Dick Spenceley began the campaign even before Danny left Australia, with a letter to radio clubs on 3 November 1956:

> As you may now know the DXpedition of Danny came to grief on October 24th when the "YASME" ran onto an uncharted reef, 130 miles Southwest of Port Moresby, while enroute to Darwin, Australia. Danny was able to send out a distress call on 14130 kc which resulted in his rescue, by Catalina aircraft, some seven hours later.
>
> Danny is now a guest of the VK2s in Sydney, Australia, and is determined to acquire a new boat, with or without outside help, and continue his voyage.
>
> At the present moment we are exploring ways and means whereby he could obtain another boat similar to YASME in size and type. It would seem that our best and fastest bet would be to interest some large firm who would provide backing in return for whatever advertising or publicity value which would attend such a venture. Several "feelers" have been put forth and a couple of them seem promising altho, at this early date, no replies have been received.
>
> The YASME II could probably be financed by contributions from the DX fraternity alone, but it is felt it would take a considerable time to raise the necessary amount. Also, over the past year and a half, considerable efforts have been put forth, scrounging ten dollars here and five dollars there which, along with contributions received with QSLs, have been keeping the expedition on the move. To say I am a bit tired of this would be the understatement of the year!
>
> Acquisition of a new boat seems to be the biggest problem. Much of the radio gear, I am sure, would be furnished gratis. Much depends on whether a suitable craft may be located, if it is for sale at a reasonable price etc. etc. The specs call for a boat of the "Bermudian Sloop" type. About 40 feet long, enclosed cabin, no bowsprit, sails inboard, etc. If such a boat could be located on the West Coast Danny is quite agreeable to sailing again across the Pacific and would touch at some rare spots, like the Kermadecs and Tonga, which were missed on the first go-around. In fact Danny says he wouldn't mind sailing the Pacific fifty times if he could get going again!
>
> I believe if Danny were to visit the States he might be able to aid his cause by personal appearances. He has a gift of gab and I am sure his talks would be most interesting and give impetus to contributions, etc.
>
> My main reasons for writing this is as follows: Would your club be interested in extending an invitation to Danny for a short visit. During this

Man!

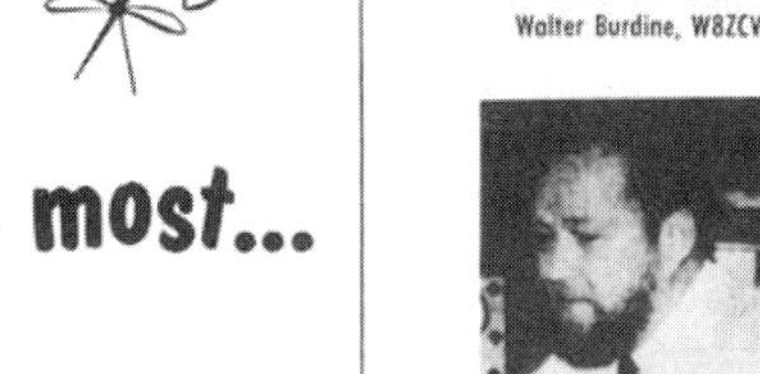

Dayton Hamvention

Walter Burdine, W8ZCV

Sam Harris, W1FZJ

John J. Willig, W8ACE

Danny Weil, VP2VB

time it is expected that Danny would be a guest of the Club. Funds now on hand will just cover his traveling expenses and no more. In this way Danny could travel across the country and pay interesting visits to a great number of hams with whom he has QSOed.

Should all efforts fail to provide the necessary amounts to start him off again, Danny would then continue home to England. While Danny has his own ends to serve on any trip I think it will be generally agreed that he has done his utmost to give all of us new countries and, literally, has lost "his all" when the YASME went down.

We, of course, would be interested in any suggestions from any members which would further his (and our) cause. A series of lectures for the general public or radio and TV appearances should not be impossible to arrange. A reply, at your earliest convenience, would be greatly appreciated so that his plans and itinerary may be formed.

CQ and the Wayne Green Connection

If any one person aside from Spenceley had helped make Danny's name in the DX world, it was Wayne Green, W2NSD, then editor of CQ magazine. CQ, then less than a decade old, was the only major publication competing with the ARRL's QST, and it did so by focusing on DX. Spenceley, of course, was CQ's DX column editor. Green took a chance on Danny after seeing some of his writing, promising to publish regular reports of his adventures, which he did in 10 lengthy installments, from October 1955 to April 1957. The final installment, "With YASME to the End," was accompanied by the following editorial, written by Green:

There is hardly one of us who hasn't imagined ourselves on a small boat cruising the beautiful South Pacific without a worry about when we had to be where. Somehow it has always sounded like the last word in freedom and serenity. If you have been following the YASME series in CQ you will have put that dream away with your Santa Claus and Easter Bunny. The principal characters in this epic have been the 40' sloop YASME and Danny Weil, the skipper. Danny's dream was a lot bigger than just sailing the South Pacific. He wanted to make his mark in the world, to go down in history as having accomplished something difficult and unusual. His plan was to be the first Englishman to sail around the world alone. And for this dream he was willing to sacrifice everything. Perhaps you'd be interested to hear how this got started.

Danny Weil lived with his mother in the outskirts of London and worked every day, six days a week, in his small watch repair shop, getting deeper and deeper into the rut of life. Then one day he separated himself from all the rest of us by making a decision. Sounds simple doesn't it? That's all it really takes to change something: a decision. He spent two years finding the proper boat, rebuilding and outfitting it for his goal, and getting things in order for his departure.

Early in 1955 he set sail from Poole, England on the first step of his voyage. To North Africa, the Azores, and then across the stormy Atlantic to St. Thomas in the Virgin Islands. His little homemade ham transmitter, more for emergency than regular ham communications, had fizzled out so Danny sought out Dick Spenceley, KV4AA, to see if he could get it back in working order.

Dick was glad to help, but as the story of Danny's trip unfolded he began to see some possibilities that Danny hadn't considered. As long as Danny was going around the world why not take along a good set of ham gear and make a super DXpedition out of it? He then could stop off at those many small islands which were so badly needed by the DXers and furnish contacts to everyone who needed them.

This sounded like a good idea, but

Danny plays the game in Cincinnati

brought up the problem of how to finance the trip. The original plan was to stop at the larger ports so that he could set up shop for a month or two as a watchmaker and earn enough for gas and food to the next port. But any port big enough to pay a watchmaker already had plenty of hams and ... if he spent his time working he wouldn't be able to get on the air much.

Danny does his best in Cincinnati

After reviewing the expenses involved Dick called me and explained the whole plan. He thought that if CQ were willing to buy a series of articles by Danny, that between that and donations from DXers it could be kept going. I agreed to go along with him on the game. I could just see the short boring travelogues coming in from here and there for us to publish. Ugh. You know ... "sailed for so-and-so on May 20th, arrived so-and-so May 30th, found shack, set up equipment, made 2500 contacts. 73."

There followed the most exciting series of stories to ever appear in a radio journal and one of the most popular features in CQ. We paid top dollar for them ... but gas was darned expensive, the sails were getting worn to a frazzle, and the donations from DXers were dwindling. It was decided at this time to go to the dollar-per-QSO system of financing. Thus once you contacted Danny you had a choice of sending a dollar and getting a QSL immediately by air mail or else waiting for him to get through and take care of you via Bureau.

This succeeded and enough money was finally saved to send a completely new set of Dacron sails plus an extra mainsail and jib when he almost got sunk coming into Port Moresby and lost the others overboard.

Danny's DXpedition was something that had never been seen before in ham radio. He was usually on the air 12-14 hours a day and worked five to six thousand stations from almost every spot he landed: Tahiti, Nauru, Canton, Guadalcanal, Port Moresby. Certainly no DX minded amateur lacks any of those countries today. Even I managed to work him at almost every stop with either my DX-100 on CW or the Phasemaster II on SSB and the trusty folded dipole antenna.

Much of the radio equipment on the YASME was donated to the expedition as a service to ham radio by the manufacturers. The participating manufacturers are to be commended for they get dozens of requests for free equipment every day and it is often pretty hard to tell when a really legitimate need is at hand. All of us thank Art Collins for sending the 75A4, which made it so a lot of us weaker signals could get through and particularly helped the SSB gang. Elmac sent an AF-67 which was responsible for thousands of the contacts. Later in the trip we began to get the benefit of the Eldico SSB-100 as Danny became familiar with it.

Then came catastrophe. Suddenly Danny was left without anything but a torn pair of shorts and his life. Gone was the YASME and all that radio equipment ... torn to shreds on an uncharted reef off Papua. Gone was a $50,000 investment. Danny had nothing left in the world.

The Red Cross flew him to Australia and the VK gang put together enough to get him back to the States. Danny is only interested in one thing today: getting another ship and continuing his trip. All he has to sell is his story, which is a hair raiser. He has been visiting ham clubs throughout California and getting on as many television shows as possible. He needs about $12,000 to buy a used boat of the type that could do the job, and that is what he is trying for.

By the time this is in print he will have arrived on the east coast. If any radio clubs in the area are interested in an exciting evening they can drop a note to him via CQ. If any of you have any contacts for helping him get on TV or know any manufacturers who might be interested in some sort of mutual aid proposition (he will tow watches around the world underwater, give paint, fiberglass, sails, etc. the test of tests) please get in touch with me or Danny.

If you or your club would like to help Danny get YASME II you can send cash, check or money order made out to "Danny" c/o CQ. Let's get him back to the sea again so he can visit more of those islands we need to contact. Let's help him fulfill his dream of circling the globe alone.

The Rubber Chicken (Boiled Ham?) Circuit

Danny arrived in the U.S. that winter and wrote to Spenceley from Los Angeles on 10 March 1957:

As per our [schedule] of Sunday night, I mentioned all the facts that needed to be said at the time, but I am very much like you at the moment inasmuch I have very little time in which to write to anyone.

Major items that I would like you to attend to for me as I find that telephone messages etc, have no effect on the parties concerned. Let's start at the beginning.

I spoke before the Mountain View Radio Club on 7th Feb. The sum of $25 was collected but the president told the me they would have another raffle at the next meeting and all the proceeds would be sent to you ... if the collection from the members didn't come up to expectations, the club would make the total sum up to $50 and send direct to you. By the time you get this letter they should have had the meeting and sent the $50 to you.

Secondly, I spoke before the Downey Radio Club on 12th.Feb and collected $100, they promised to send same to you. I rang the president and he said it was on its way .... that was about two weeks ago...Hw?

Finally, I spoke to the "050" Club of Los Angeles, and the President decided at the time that he would not ask for a collection as, Quote "The most we should get at the moment would be a lousy 50 bucks, but I will talk to them at the next meeting and we should be able to push up the amount considerably" unquote. This club is quite exclusive, the members are all wealthy and have much influence, as the number of members are limited to 50. My speech was made on 26th Feb., so you may not hear back from them until say about the 30th of this month.

Forgot to mention that all members of "50" club are hams. Can only suggest that if a phone call from Steve, K6AWL fails then you may have to write. I feel that my style of writing will upset someone. Re Albuquerque, Mid collected $76, I think, at the meeting and also another $5 at the dinner, but I know he won't let us down as he said he would send his personal check to you immediately. One member of the club who I feel would have contributed handsomely was "DOC" Howard Merideth (W5PQA) but for some reason unknown to me was unable to attend the meeting. It might be worth while mentioning this fact to Mid when you write him.

"Doc" and his wife were more than kind to me in my stay there, and he arranged to have my eyes tested, also footed the bill for a pair of Polarized sun specs for me; they weren't cheap either. Another point about Doc is, at the time YASME was sunk and I was in Australia, he was planning to fly over to help me out but, time was insufficient ... anyway, I give you this picture to aid you in your letter to Mid .... I will be writing to Doc maybe tomorrow, but of course won't mention cash at the time.

The talk here will be on 13th March and I will leave for Houston the following morning (14th) and will arrive same day. Will check and let you know by radio what time plane arrives if I go straight through. Naturally, if I have to stop at Dallas for the Collins bloke, that will mess up my time of arrival, but let's wait and see what happens, can always give you dope later. Would like to know date I am supposed to talk at Austin.

Danny noted donations through 10 March 1956 of $1062, and his expenses at $332.

What a life. Been here in the United States for over three months and all I have to show is just over $1,000 … how heck can I do better? We still have the air fare to pay to Texas and St. Louis etc. Lets hope we get some good donations down Texas way ... Hi.

Well, Dick, this is my statement of accounts in case anyone wants to know...the new typewriter took a large chunk ($120)...the gas for car...well, the boys were good enough to loan me the car...I had to pay for the gas....must pack up smoking...$32 for 3 months is too much and that $10 for renewing my visa every 3 months is a racket, also, will have to stop eating too when I'm out...it's a bad habit. Aw to heck with it...I'm going to bed. - Danny

Danny wrote again to Spenceley on 22 April 1957, from Washington, D.C.:

Got your letter OK re Bill Halligan [the president of radio manufacturer Hallicrafters] etc and have noted all the other things in your letter, Wanted to answer them in a little more detail, but damned if I can find the letter now.

Haven't received the letter dealing with the Australian bloke, expect it will arrive in a day or so, but regardless of what he has said and what he has offered, this has to be done by me and me alone Dick, of which you are fully aware. I can appreciate the attitude of the average Ham in this deal, but what he doesn't know is what can happen when two people are alone on a boat for long periods, and in any case, this bloke, whoever he might be will not stand for me going to all these odd spots where he can't have a good time, and that's for sure. I have in my mind should this happen, the two of us ready to cut each other's throats when we are about 200 or 300 miles off shore, and should that happen, it will mean the entire expedition will be washed out. The rest of the Hams won't like that one little bit, particularly after they have all tried to get the thing going, and I hate to think what Bill Halligan etc. would say to you if the whole deal flopped after they had bestowed all the gear.

Well Dick, I won't discuss this matter any further. You know how I feel about the whole thing, and I think you know me better than any other bloke in the USA. 'Nuff said.

Suppose I will write to the bloke concerned when I get his letter, but that's about all. If I find I am in a ticklish position, will send the letter I have written to him down to you for your opinion, if its OK, then you can post it from your end.

Pressing the flesh in Birmingham, Alabama (YASME NEWS)

Re the other bloke in Kansas with the lecture deal. This sounds FB, but don't forget Dick I am here in the States with a visitor's permit and I am not permitted to work. Whilst I am getting contributions that appears to be fine, but once I start to get paid lectures then it will have to be figured out in a different light altogether, anyway, I will wait until I get the letter and reply to it as I think fit.

Had a talk with one of the boys here at FCC ... the big chief, but although I have searched through my note book, just can't seem to find his name or call sign. Anyway, he said he will write you with some suggestions of getting me the OK to operate in American Samoa when the time comes. Seems to me the only way is for me to get a VE licence and then the FCC will play, so am sending you some cards they gave me already signed, and if we are lucky enough to be able to wangle a VE licence then all you have to do is to send the licence up with enclosed card signed. I don't have any bright ideas on acquiring this VE deal, but maybe you might know someone up there in the right position who could swing it, anyway, that's another chore that's been thrown in your lap.

Had a chat with Bill Halligan over the phone and he tells me he has written you, but he told me that his son is fitting us up with all the gear we need and it will all be tropicalised and tested to army specs. That's all from Bill except I did ask him if (a) he could get a lecture fixed up in his area and (b) if he would care to contribute a few handfuls of $$$ besides the equipment. He says he might do that, but feels the amount he could personally donate would be pitifully small!!! I suggested that however small we should be very glad to accept it

... that's about all for Bill.

Enclosed find resume I am sending to all concerned. If you can give me a list of addresses pronto I will see they all get sent out. I have done this 'cos I am getting really fed up with all these idiots telling me they had no time in which to prepare a talk for me, and they have no dope to give the papers, Well, in this resume, they have the entire works and they will have no excuse at all… I think .... still think they will get fouled up somewhere along the line also. it will give the papers more opportunity to mix up their story...they are damned good at that.

Didn't want to write as much as this but seems that so many things are happening that I have to keep on.

Now Dick, don't ever stick me at a place for a whole week again, at least, not unless they can give a damned good reason for it. I can find much better use for my time than wandering around ham shacks and sleeping in a different bed each night. This stay here would have been FB if I could have stayed at one place, but the boys started bickering about Marvin keeping me all the time and now I have to keep packing my bag and moving to a different QTH each day...its enough to drive me nuts, and if I have to keep on the move, let's make it profitable. I am presently trying to arrange a talk-up at Silver Springs or Silver something, can't remember offhand for Friday or Saturday of this week, but seems time is a little short for them, and they already have a speaker organised, they are trying to push him off to another date, but more of this on the rig.

Published in the YASME NEWS with the caption "Who? Us? Danny, VP2VB/ Jake, W8LNI, Miami Convention"

Don't know exactly what we picked up at the Old Timers deal in Trenton. I just haven't had the time to check it thoroughly. It's all small bills and loose cash, but made a rough check and its about $80 ... lousy deal I think for an audience of 250 or thereabouts, anyway, transport there was NIL, or partly so. Sorry if I am a bit vague about a few things Dick, but the way I am pushed around is enough to make even you go a bit haywire and forget the whole works, I do try to write all data down, but sometimes I forget to do that.

Gave Jake in Cincinnati $800 to send you....hope you got it OK. Am also enclosing check for $572 which closes the account I have in LA. Have a few more bucks with me and will send down when I get all things cleared up. For the record, Sam Rowly W6VUP ... his wife borrowed $30 from me and I told her to let me have it when I next hit LA again. Loan made by check through my a/c on Feb 18th, check No. l.

May as well enclose bank statement and cancelled checks then you will have the lot. The pink form is for a check that bounced but that is all squared away.

There is a check to come from [W8]YIN Mickey for the expenses incurred on the Detroit deal...the amount is $64.00, also Dick Hatton W8LKH borrowed $15.00 off me, We were coming back from W3RIS and he had a couple of burst tyres and no money on him. I had to get back to Cleveland so loaned him the cash ... he promised to send it to K6EWL's place. I have written him and now await results ... will let you know what happens,

Also Press Richardson W0OBJ bought two tubes from me that I had [been] given to raise some funds ... he offered $40 and took the tubes but didn't give me the cash... promised to send to me also to K6EWL's place...I wrote him for the money. Let's see what happens.

Well Dick.. think that covers about all the dope to date, but would like you to send me a copy of that statement of accounts I sent you ... have lost mine and I want to get everything all squared away before it gets beyond me. I can't tell where the last lot finished and where the new ones begin, so I have to have that copy. Well, Dick, once again, this is it, You have all the cash to date except the bit I have with me, and that is almost nothing, and you know about the money that is owing. This clears me up from the cash angle, now let's see if we can double it a few times, Hi!

Re the book, Charlie W6GN, that's Steve's Pop, has given it to the author bloke and we should be hearing something about that pretty soon .... I hope [Danny's on-going manuscript, Cruise of the YASME]. Have just found your letter, but can't find anything to comment on except the fact that we seem to have ALL the gear we need radio-wise, now where is this RUDDY boat .... I would give my right arm to be back again at sea .... OH for a nice quiet hurricane.

Hope to hear what happens about that 700 pound deal in Australia, the bloody rats...excuse my French....but it's not the Aussy Government, it's that bloke in Morse! -- Danny

On 24 April 1957 Danny responded to someone named "Hedley," who offered to accompany Danny on his next voyage (if there was one) as well as an infusion of cash, and, apparently, a 37-foot boat.

Dear Hedley: Thanks for your letter which was passed on to me by Dick Spenceley the other day. I daresay you have received his letter in reply, and can only reiterate what he has said in this matter. Inasmuch as I should love to accept your offer there are many circumstances which make this impossible at the moment.

First I have an intensive lecture tour to complete here and on completion will endeavor as I did on the last trip to run an Amateur Radio Expedition at all the odd spots of the world. This would necessitate going far from the normal shipping routes, and also, the spots chosen for this work can hardly be called exotic. My work at these places would hardly amuse you, and I feel very sure that were you to invest your money in any sort of a partnership deal with me, you would live to regret it afterwards.

I hate to throw a damp rag on your enthusiasm, and were it not for this expedition which I have promised to undertake I should be only too eager to accept your offer. Let me try to explain a few factors here. First the boat would be filled with radio gear which, although it didn't bother me, would probably make your life hell aboard, as it takes up practically all the available space, and leaves very little for us as individuals.

YASME was 40' overall, and still I found it difficult to stow all the gear needed for the job. Can you imagine what it would be like on a boat 37' overall with two of us aboard?

Another point I feel I should mention is I am dedicated to making this trip alone, and am more than ever determined to carry this through, more so since the incident off New Guinea.

I do want to thank you for your very kind offer. It is the first of its kind since I left England 3 years ago, and I want you to know that should you require any advice, please feel free to write me any time either at the above address (Los Angeles, which Danny used as a mail drop) or c/o Dick Spenceley.

One last word of advice. Try to make it alone if you can, I have seen many yachts of all sizes littered around the world where crews have fallen out through being alone together for long periods. It needs two special sorts of people to maintain good relationship on a small boat, and they are hard to find. Don't be one of them stuck in some obscure place; you may find yourself the owner of half a boat, and that is pretty useless to get any place, I have found with sailing alone for 2-1/2 years that to be ALONE is usually the best way...think hard before you take a partner.

I feel I have said all that is necessary, so will close down. Thank you again for your most generous offer. I wish you all the luck in the world in your venture, maybe we shall meet at some spot in the future, this world of ours is a small place, -- Sincerely, Danny Weil.

And another letter from Danny to Spenceley, from Washington, 24 April 1957. It appears Danny does not savor Rubber Chicken:

Enclosed copies of letters sent out today…you know about them, so hope these replies are in keeping with your thoughts on the matter…Hi!

Don't think we shall be taking advantage of this lecture bureau in the future as I sincerely hope we have got something else organised by that time, and I should hate to think that I had to spend the rest of my life talking….what a horrible thought, even though its for the benefit of the hams. You weren't wrong when you gave me your opinion about this particular area and when I came to count up the cash on the deal we had the grand sum of

$74. The stingy rats, and there were about 90 of them there....good job. I had free transport from the last place else we should have been out of pocket on the deal. Checked on all that loose cash I picked up at Trenton and we have collected $96. Suppose you have received the registered letter I sent earlier today with the check in it...that covers the account in Los Angeles...you have it all now except a few bucks which I am carrying for fares etc. Will get a few more off to you later when I get sorted out.

Realise that you cannot tell what may happen at any place when I arrive, but try not to let me stay too long in any one place if you can avoid it...I think this place stinks with a capital S and for all the help we are likely to got here, I may as well go to sleep for a week and forget the whole joint .... Have been invited to a dinner this Saturday with a few "Big Nobs" and they may be able to offer a few suggestions that may prove fruitful...I doubt it, anyway, its a free meal..Hi!

Re. Bill Halligan. Don't go much on what he says about giving us $2000 of gear and the stipulation that we don't use anyone else's. The way I see it,

Danny and his mother, 1954

We have a Globe King offered, NC 300, Elmac's chaps I met at Dayton said they would help out etc etc, and they made no stipulations. As far as I can see, if Hallicrafters wants the monopo1y, then they should be prepared to offer a little hard CASH. What say you, anyway, if you want to write to Bill on this score. I leave it to you. Quite honestly, I don't think much of their offer ... do you? Not that I expect to ever get an answer from you on this, that is far more than I could ever expect, but it eases my mind to tell you what I think, not that it makes any difference to you either.

What I want to know, and so does everyone else I meet, What has been the results of this very weak and very mild appeal put in the April CQ brought? Quite frankly I did think Wayne could have done a little better than that, and is another appeal likely to materialise, or what? Seems to me that the whole deal is being shared between you and I. How about a few others sharing the work? What a hope we've got.

I am slowly learning to hold my temper at these meetings when I get these damned idiots stand[ing] up with their inane suggestions; quote "You need a manager...You need a publicity man....you need a good firm to back you..." They make me sick with their remarks, but when I throw it straight back at them and ask for volunteers to do this work, they all clam up. Seems to me people are the same the world over. PHOOEY!

I've done enough writing now .... and talking ... to hell with it ... see you on the air.

P.S. Got Leo to give me his personal check for $150. I had a lot of loose cash in my pocket. - Danny.

Danny and Spenceley were looking into his marketability as a lecturer to general audiences. Danny wrote to a lecture bureau in Kansas:

Your letter dated April 20th has been forwarded to me from the Virgin Islands, and I do appreciate your interest in my future welfare as a lecturer. I would also like to thank Mr. Edwards of Greenville, N.C. for suggesting to you that I am sufficiently competent to carry out this type of work.

Unfortunately I do not possess any newspaper writeups ... quite frankly I pay little heed to these things, and certainly do not wish to have my name in the Hall of Fame as a lecturer.

My life for the past few years has been devoted to sailing a small boat, and I feel that when the time comes, and another boat becomes available, I shall continue to do this sort of thing. Seems to me I have a love for the sea that is very difficult to describe to the layman but whilst I can sail and write of my adventures to those less fortunate, I shall be content.

As you are possibly aware, my talk is entirely on my travels and does not become technical at any time, regardless of the fact that I have been undertaking an Amateur Radio expedition.

Hello There!

The show on which you appeared will be heard on radio, KFI, Sat. afternoon, April 20.

The television show will be the following Thurs., April 25, KRCA, 8:00.

Regards,

Groucho

"YOU BET YOUR LIFE"
On NBC Radio and Television
For DeSoto-Plymouth Dealers

I have in the last few months here in the United States made many TV and radio broadcasts through the various States I have visited, and these short interviews seem to be welcomed by the studios concerned. Naturally there have been many newspaper writeups too. I am enclosing a copy of the resume of my talk which I normally give to clubs etc, before I arrive hoping it will assist you in some little way.

Mr. Spenceley is instrumental in my movements, and should you care to contact me, I feel it might be advantageous to write him first. Thank you again for your interest. I sincerely hope I can meet up with your high standards when the time arrives.
-- Danny Weil

Danny wrote his own "Tour Press Release," which he referred to as his "resume":

"THE CRUISE OF THE YASME - THE FIRST ENGLISHMAN TO SAIL AROUND THE WORLD SINGLEHANDED"

This trip to date proves that ANY man can do the same providing sufficient interest is given to it. Guts unnecessary - only abilities required, knowledge of mathematics for navigation, and an ability to sail a boat whilst being seasick-this is the hardest job of all. Linguistic abilities are unnecessary. Drawings, gestures, plenty of Scotch Whisky and American cigarettes smooth any rough paths and awkward customs officials.

I started to build YASME around 1950 using an old rotten hulk as a pattern. "YASME," a name derived from the Japanese word "Yasume," meaning FREEDOM, took 4-1/2 years to build, was 40' overall, 11' beam, 5' 6" draft Bermudian rigged with 10 H.P. gas engine for auxiliary, and built of pine and oak frames. It was completed and ready for sea August 1, 1954. I gave up my job as watchmaker and set out from Christchurch, a small place near Southampton. My ambition was to be the first Englishman to sail around the world single handed. (The first man to do this was Joshua Slocum of Boston and America about 50/60 years ago in the Yacht "Spray"). I set out with no knowledge of deep sea sailing or navigation, learning on the way - the hard way.

The English Channel was tough sailing, bad weather necessitating many calls into Southern British Ports to repair sails and recurring seasickness, The first foreign port was Vigo, North West Spain. The crossing from Falmouth, England to Vigo took 6 days - gale in Bay of Biscay lasting three days enroute.

Route from Vigo: Lisbon, Portugal; Gibraltar; Balearic Islands in the Mediterranean; Tangiers Casablanca.; Canary Islands; Dakar; Gambia; Gold Coast; then came the crossing of the Atlantic. Strong winds on voyage making this the fastest crossing of Atlantic yet made single handed and with a boat of this size. The distance of 3200 miles was covered in 22 days.

The first island was Antigua in the West Indies where I stayed 4 months, sailing next to St. Thomas Virgin Islands, U.S.A. Here I met Dick Spenceley, world famous "Ham." He suggested I undertake an Amateur Radio Expedition whilst sailing around the world. This necessitated calling at all the odd spots of the world that either had little or no amateur radio activity. The purpose - I was to set up a radio station ashore at these places and transmit to all the Hams world wide. Signal reports and other data would be exchanged, and confirmation of these contacts made through a bureau in writing.

All radio equipment for the expedition was donated by American manufacturers and Hams and consisted of 6 transmitters, three receivers, two 110 volt generators and a large quantity of spares. I encountered "Connie" the hurricane in the voyage from St. Thomas to Panama. I transited the canal at a cost $5.40. The longest voyage of the whole trip was from Panama to Tahiti the distance covered being 6,300 miles in 60 days.

Route after that: Canton Island, Ocean Island, Nauru Island, and Guadalcanal. Two days out from Guadalcanal a cyclone struck YASME, capsizing her, All sails were cut adrift and I managed to right the boat. She took on about 7/8 tons of water but I pumped it all out in an hour with engine pump. After the cyclone, a gale took over and my position became uncertain for the next five days. On the fifth day I obtained a position and communicated it via ham radio to Port Moresby in New Guinea. They sent out an Air Sea Rescue Craft and towed me 30 miles through 40' winds, in pitch darkness. The journey took 12 hours.

Departing from Moresby 23rd October 1956 noon, the YASME had covered 135 miles by the 24th. At 6 p.m. on the 24th of October, the boat struck an uncharted reef in the Papuan Gulf. I tried to sail the boat over the reef into deep water but the tide fell away leaving the boat stranded with big seas breaking over the entire length of the boat. I sent out a distress call via Ham Radio which was picked up a few seconds later by an Australian Ham 4000 miles away. No aid was available until the following day. Twenty minutes after sending the call the boat filled with water and sank on the reef. All radio communication ceased. All lights went out.

I stood inside the boat with water around my waist. Salvaging the gear was impossible, the fast rising tide made me climb up the mast. The tide rose through the night and I gradually climbed higher. Around 10 A.M. on 25 October a Royal Australian Air Force Bomber dropped a rubber raft and flew very low over the scene of the wreck to scare numerous sharks away. I swam to the raft reaching it just in time. The raft had part of the side torn away by sharks but maintained its buoyancy O.K., drifting all day until almost dusk when a Catalina Aircraft landed on the water and picked me up.

I returned to Port Moresby, spent one day in hospital to have numerous coral cuts treated and spent the following week trying to get the boat salvaged with no success. I went to Sydney, Australia and from there, flew back to the United States at the invitation of the "Hams". The Hams desired that the Radio Expedition be continued, if possible and suggested that I travel the entire United States lecturing to all the clubs with audiences contributing toward the fund for YASME II.

A YASME II Committee has been formed by top radio men of the U.S.A. led by W8YIN, "Mickey" Unger of Michigan. Dick Spenceley of the U.S. Virgin Islands plans the routing and speaking engagements via ham radio. Most radio equipment has already been offered by numerous American radio manufacturers for the next expedition, also diving equipment by "Healthways" of Los Angeles. In the last expedition, underwater photography and exploration was undertaken extensively. A book on the voyage has been written with the possible publisher "Little Brown," New York, New York [this is the manuscript Cruise of the YASME; it was never published]. The manuscript is now being edited. I have lectured to approximately 100 clubs. The public is invited as the lecture IS NOT a technical talk but pure adventure and travel. The public is under no obligation to contribute.

Facts about myself: Born London, England, Age 34 [that's what it said] , single, height 6 feet, qualified mechanical engineer and watchmaker, family trade - watchmakers. Served 9 years Royal Air Force, leaving the end of 1946 and opening store as watchmaker. Found that work interfered with pleasure....decided to take world voyage and work at various ports enroute to build up funds to continue. This system being highly successful, but hard to make a fortune. Expected time to depart on next voyage according to statistics is approximately August/September 1957 from the U.S.A.

As Danny Weil traveled the country in the spring of 1957, appearance dates were few and very far between, and finances were pretty much imaginary. More than 40 years later, an "old timer," speaking to the author, would refer to Danny Weil as a "ham bum." Apparently, this ham had already forgotten how many "new countries" Danny had given him. On 13 May 1957 Danny wrote to Spenceley:

> Mickey Unger should send to either you or I the sum of $64 which he told me De Soto would pay to cover my expenses to and from Detroit. I have not got that money yet although I sent the bill a few days after returning from Detroit.
>
> Expenses to and from Birmingham were paid by Ack, but in the first instance Wayne had already bought the tickets with money he had taken out of what CQ owes me for my articles, so I still have the checque given me to Ack W4ECI for these expenses.
>
> Press Richardson W0OBJ owes me $40.00 for payment of two tubes that had been given to me to auction. He gave the highest in words and took the tubes, but he hasn't paid up yet, so I think I shall have to chase him up...Have already written him a letter with no response, but maybe as I have him K6EWL's address, the letter is floating around the USA somewhere unless he posted it to you. I just don't know, but I am getting a little fed up with people who don't pay up on the nose.
>
> Another bloke is Dick Hatton W8LKH who borrowed $15.00 from me when we were driving to Pittsburgh. He had a couple of blow outs and hadn't the cash for a tyre so I had to fork out under the circumstances. I have written to him for the dough, but am still waiting for it.
>
> To try to boil all of this down. I sent you two checks around 25th April when I was with Leo W3WV...one of them was closing the account in my Los Angeles bank and the amount was for 572 bucks, and also enclosed with it was Leo's personal check for $150 which he gave me for a whole load of loose cash I had with me....that letter with these in hasn't arrived according to you, so how do we go from here?
>
> Am enclosing the following checks which I would VERY MUCH LIKE ACKNOWLEDGED if you ain't too tired as I hate to think of a few hundred bucks chasing around this country without anyone but me caring about them;
>
> Checks issued by: Ack W4ECI $140; Ditto $100; Art Brothers $20; W3TM Ed Laker $25; Frankford RC [Radio Club] Jesse Bieberman $132.19; QCWA [Quarter Century Wireless Association] David Falley $50; Birmingham RC $50; Wayne Green, Check for loose cash $200; Brooklyn RC $25. Total $742.19.
>
> When you get the two checks I have sent you for 572 and 150, these with the $208 sent by the 3 clubs mentioned in the accounts, you should have received the sum of $1672. I will wait a little while in case you get some mail from K6EWL; the letters from W0OBJ and W8LKH might be in them. I gather you will send any mail direct here or wherever I may be, but by then we should have all the cash straight, I hope.
>
> To heck with all this accounting, it gives me a big pain. Will be glad when I can be away from it all. - Danny.

14. The Colvins Settle in California

"I will put a lot of pressure on my last 'hold out' stations to QSL." - Lloyd Colvin

Iris and Lloyd used this 1957 photo for many years in their applications for overseas licenses.

Finally "settled" back in California in 1957, Lloyd (and now, increasingly, Iris) organized his amateur radio files, log books, and QSL cards. In 1958 he sent in a photo of his 100 QSLs for "DXCC Squared" from DL4ZC, noting that "I hold DXCC certificates for JA2KG (J2AHI), DL4ZC, and W4KE. I have also worked DXCC from FA8JD (not confirmed) and W6KG (confirmations will go in shortly.)

"At the present time my QSL collection numbers nearly 30,000 QSLs, all arranged alphabetically and arranged in file cabinets in such a manner that I can pick out any card without leaving my operating chair."

From the fall of 1957 to August 1960 Lloyd was Regular Army Advisor to the 49th Infantry Division, Alameda, California. He "advised on all matters of communications, including technical, tactical, and special communications needs, of a highly rated US Infantry Division, and directed the design, installation, and operation of a radio communications network linking together all of the National Guard Armories in Northern California."

Lloyd joined the Army MARS (Military Affiliate Radio System) in early 1957, and participated in its activities. He got a certificate from the Department of Defense Armed Forces Day 1957 for "copying the International Morse Code by receiving and transcribing without error the Armed Forces Day message of the Secretary of Defense transmitted via military radio at 25 w.p.m. on May 18, 1957."

On 21 June 1957 came a commendation "For the reception of the Military Affiliate Radio System (MARS) Broadcast in Continental Morse Code at 20 words per minute from MARS station A6USA on 21 June 1957. This Broadcast was part of the celebration of the 97th Birthday anniversary of the United States Army Signal Corps. By this participation, he has demonstrated his interest in the furtherance of the art of radio communications, especially reception of Continental Morse Code." This was from Steven S. Cerwin, Colonel, Signal Corps, Signal Officer, Sixth US Army.

Despite these awards, Lloyd was in the hunt for enduring amateur radio fame, and he intended to earn it by being the first to qualify for a brand-new award. CQ magazine had concocted a new award, for working amateur radio call sign prefixes; the idea was from none other than Dick Spenceley, KV4AA, CQ's DX manager. It was announced in 1956, and the starting gun for making contacts and collecting QSL cards was 1 January 1957. Lloyd was in a house, with a beam antenna and the finest of equipment: Johnson Ranger and desk kilowatt amplifier, and a Collins 75A4 receiver. W6KG came on the air in January, 1957 - just in time!

By spring Lloyd had the needed 300 prefixes worked and by summer he was biting his nails as the QSL cards dribbled in. He wrote to Spenceley in July that "I am getting pretty close to qualification" and asked if the prefixes SM8, LU0, and XE0 could be counted for a regular WPX (not limited to a "mobile" WPX; these were calls used in Sweden, Argentina, and Mexico for mobile stations). "You inferred (but did not positively say) in a previous letter that they counted for a mobile WPX, but not a regular WPX." Lloyd wrote.

Lloyd also asked about prefixes he'd worked, and had QSLs from, during 1957 Armed Forces Day and Signal Corps Anniversary activities - call signs such as NAA, WAR, AIR, A6USA.

Lloyd was pioneering new rules for amateur radio operating, something he and Iris would do again, 10 years later, when DXCC was in question. Spenceley admitted that "In the rush here [to structure a new award] we have not pinned down all contingencies (which we should have done). However I am sure all these may be ironed out. (Spenceley allowed the mobile call signs but ruled out the Armed Forces Day and Signal Corps Anniversary stations, since they transmitted outside the amateur radio frequency bands.)

In August Lloyd again wrote to Spenceley asking "Has anyone made WPX yet?

"I am enclosing a stamped self addressed envelope. Please answer immediately. If your answer is "no" I will put a lot of pressure on my last "hold out" stations to QSL. If the answer is yes I will relax and be mad at myself for not trying harder-hi.

"I worked WPX in the first seven weeks of 1957. Getting the QSLs has been another problem. I am very close now however."

Spenceley replied "Not that I know of! W1BFT has 358/188 confirmed. CQ in N.Y. handles this. - Dick"

Lloyd wrote himself an amateur radio resume recounting his 28 years in the hobby, beginning in 1929 when a few stragglers were transmitting with spark transmitters. As of 1957, the resumc said, "In addition to the many moves by the Army, the Colvins have traveled a lot on their own. All [United] states, all [Canadian] districts, and countries totaling 86 have been visited. Over 75 certificates have been received including WAC (19 minutes), WAS (46 hours), Diploma Espana, WAZ (#89), DXCC (#193), WAP (#1 for Asia), DUF 1234, WAE 2&3, WASM 1, 2, WAA, CAA, WBE, WAVE, BERTA, WAJD, KZ-25, CDM, WAVE, AJD, H-22, DPF, AAA, OHA., WAYUR, WDT, WGSA, and numerous other DX and operating awards."

Lloyd poses in Germany in 1956 with his QSL collection

W2 WESTERN SECTION OCTOBER 30, 1957

This House Fits Narrow Or Wide Lot

THIS striking home was designed to fit on either a narrow lot, or a wide one. If your lot is no more than 37 feet, you still have clearance of 12', usually more than enough for any area.

The exterior puts the emphasis on fieldstone, V-jointed siding and sparkling glass areas at the front.

As you enter, there is a pleasant foyer, separated by a wrought-iron rail and the back of the fireplace from the living room proper. You can reach the living and dining rooms, the kitchen, the bedrooms, the bathroom and the stairway to the cellar — without turning the living, service or sleeping rooms into a corridor.

This is especially desirable when there are young children in the family, especially if it's hard to convince them that the back door should be used to come into the house.

There is a "back" door, too, of course, opening to the kitchen, with a counter-top handy to set packages on.

The dinette corner will take a built-in unit. A sliding door is a great convenience leading into the dining room, since, when it is closed, there is complete privacy from the kitchen, and an unbroken wall effect on both sides.

The L-shaped entertaining area measures almost 20 feet in both directions, with the flagstone fireplace a focal point for both living room and dining room. A curved bow window is at the front and a picture window at the side.

Three bedrooms, bathroom and master lavatory and six super closets complete the plan.

Overall dimensions: 25'4"x48'4. Square feet: 1100. Architect: Lester Cohen.

Decorators Give Advice

This 1957 *Army Times* house plan is very close to what the Colvins would build for themselves in Richmond, California.

Lloyd listed memberships in MARS, AIEE, AFCEA, ARRL, BPL, FEARL, DARC, IRE, ORS, RCC, OTC, QCWA, Al-OPR, ORC, CAA, GAARC plus various military organizations. He had worked DXCC and WAS [Worked All States] from four continents, with certificates for JA2KG, DL4ZC and W4KE. "The QSL collection for the various calls held numbers nearly 30,000 cards all of which are arranged alphabetically in metal QSL cabinets.

"W6KG was put on the air in January of 1957. The first seven weeks of operation resulted in 110 countries and 304 prefixes worked!"

Lloyd applied for DXCC for W6KG on 23 Aug 1957. All the QSLs (104) were from 1 January on. In late December, Lloyd sent ARRL the QSLs for his fifth WAS certificate (W6AHI, JA2KG, W4KE, DL4ZC, and now W6KG). "You will note that all of the enclosed cards are for QSOs made this year!!" he said.

Late in August Lloyd had collected QSLs for 301 prefixes, and he immediately hopped a non-stop flight (a big deal in those days) to CQ magazine's office in New York. This is the stuff of dreams for public relations types! Lt. Col. Colvin would get the first important exposure for his accomplishment not in the amateur radio press (not even in CQ), but rather in the October issue of an area National Guard publication, The Argonaut, with even a photo of Lloyd with a beauty queen, Miss Mary Ann Sky, a "contestant in the Miss Forty-Niner Contest."

15. "Ham" Colonel First to Win World-Wide Award

"This will give you some idea of the enthusiasm with which W6KG tackled the job of winning WPX Number One." - CQ magazine

The Argonaut, a publication of the Army National Guard, heralded the achievement in the fall of 1957:

Lt. Colonel Lloyd D. Colvin, Division Signal Advisor, has communicated from his amateur radio station, W6KG, since the first of the year, with other amateurs in 330 call areas located in 130 countries throughout the world. For this accomplishment he has been given the first WPX Certificate in the world. WPX stands for Worked All Prefixes.

Throughout the world there are more than 200,000 radio amateurs or "Hams" as they are called. They communicate with each other via both phone and code (CW).

Practically all countries in the world, including the USSR, permit their radio experimenters to obtain government licenses to operate on the ham bands. Lt. Col. Colvin has been a radio ham for 28 years. His wife and daughter are also licensed amateurs and they can be called an 'all-ham family.'

The Colvins have traveled extensively, having been in 86 different countries. They find their hobby a great help when visiting a new country as there is always a local ham they can visit whom they have previously talked to by amateur radio, perhaps from the other side of the world. Col. Colvin states that the armed forces recognize the potential Military value of hams who spend their own time and money on their hobby of radio.

During emergencies, they have frequently provided valuable communications. During World War 2 they were leaders in many phases of military signal activities. The Armed Forces urge all active amateurs to join stations on military frequencies and obtain valuable training on the proper military procedures and practices.

In the 49th Division there are approximately 40 units which hold amateur calls and are members of MARS.

"It is hoped that there will eventually be a MARS station in every armory," says Col. Colvin.

Another important event took place in October. Iris and Lloyd clipped and saved an article from Army Times magazine; a house plan entitled "This House Fits Narrow or Wide Lot." It was an ingenious, three-story house plan with a very small "footprint." Iris and Lloyd, as Drake Builders, would duplicate it in just a couple of years, at 5200 Panama Avenue in the San Francisco suburb of Richmond. It would be their home for the rest of their lives and the site of the Colvins' famous July Fourth amateur radio parties.

Lloyd appeared on the cover of CQ for December 1957, with an article inside, CQ having rushed the story to press as quickly as a monthly could in those days. The editor said "When Lt Colonel Lloyd D. Colvin, W6KG, obtained his 301 confirmations for WPX he immediately took an airplane ride (non-stop) from San Francisco, California to New York City and hand-carried his proof of WPX to the editor of CQ. This will give you some idea of the enthusiasm with which W6KG tackled the job of winning WPX number one. After hearing his story we asked him to write this account of 'how he did it.'"

Most contest winners, [Lloyd wrote] have gone through many sessions of seeing someone else beat them out before they finally perfect their equipment and operating techniques to the point that they finally come through a winner. The winning of WPX No.1 is not an exception in this respect.

When the rules for WAZ [CQ's premier award, for working all 40 "zones" of the world as defined by CQ], were first announced, I was J2AHI in Japan with a KW, all-band station. I started out to win the first WAZ and had worked all zones some time before the first award was issued. I had 39 zones confirmed when WAZ No. 1 was issued. My missing zone was No 23 [Mongolia and part of China, close to Japan but with only a handful of hams - ed.]. My three letters to AC4YN [in Tibet] had not been able to convince him of the necessity of a prompt QSL (took one year to receive it). On the basis of my WAZ experience I realized that winning a similar first award requires:

A good station with well designed equipment and antennas, permitting operation on all DX bands; Good operating ability; and Some special way of getting confirmations (QSLs) in a hurry from the stations worked.

When the WPX rules were announced in the January 1957 CQ, I decided to try to win one of the first awards. I had a slight advantage over some stations in that my call

Lloyd D. Colvin
LT COLONEL SIGNAL CORPS
US ARMY

PO Box 30
Alameda, California
2 August 1957

Dear Dick,

Here I am again asking questions on WPX. With good luck this should be my last question before I obtain WPX.

The question is, " Has anyone made WPX yet ??"

I am inclosing a stamped self addressed envelope. Please answer immediately . If your answer is " no" I will put a lot of pressure on my last " hold out " stations to QSL . If the answer is yes I will relax and be mad at myself for not trying harder-hi.

I worked WPX in the first seven weeks of 1957 . Getting the QSLs has been another problem . I am very close now however.

Thanks and 73,

Lloyd

Lloyd D. Colvin
Lt Colonel Signal Corps
W6KG

was brand new and I was not faced with the problem of working any station a second time or obtaining his card a second time for WPX. All stations I worked would be original contacts. I had a good KW transmitter, a receiver with good sensitivity and selectivity and one of the new trap type, tri-band, rotary beam antennas. I realized that if I were to compete successfully with other stations in a similar position I must concentrate on a method of obtaining confirmations faster than they did.

Lloyd's method was to send a QSL promptly, continue to work and send QSLs to the same prefix, whenever possible to explain during the contact why his card was needed, send a second QSL after two months of waiting, then send a special International Reply-Paid Post Card, and, finally, send a commercial cablegram asking for a QSL when the goal was in sight.

In the years to come Lloyd and Iris would be consumed with QSLs; they were the world champions, as anyone who ever worked them can attest. An amateur radio contact is worthless for obtaining awards until the required "confirmation" comes in the mail. The following recounting of Lloyd's negotiations with the Postal Service is therefore presented, just as it appeared in CQ - Lloyd must have been an expert, since he was the first to deliver 300-plus QSLs to CQ!

> My initial experience with the use of International Reply-Paid Post Cards was discouraging. These cards cost 80 cents. Any US post office can obtain them for you. It is a double card. You print your QSL on one section and on the other section you print (a stamp can be used if desired-or even made by hand) a QSL from the foreign amateur to you. I found very few US post offices have the cards in stock. A great number of US postal employees are unfamiliar with the rules and regulations pertaining to the use of the cards. The percentage of postal employees in foreign countries who are unfamiliar with these cards is even higher than in this country. I found that many foreign amateurs were unable to convince their postal personnel that the reply card could be sent back to me without any local postage stamps affixed to the card. I reported this, and some other problems, to the office of the Postmaster General in Washington.

The Postmaster General replied that the reply cards were at each country's option, that they were designed to be sent in open mail (not in envelopes), the prepaid return feature was valid only if both halves of the card arrived together, the cards could be sent air mail if additional postage was affixed, and that there would be "no objection" if the sender rubber-stamped the cards with additional instructions. Lloyd stamped his cards with "No additional postage required," and said "I consider this [scheme] excellent in view of the fact that these cards have only been sent to hold-out stations who did not answer my regular cards. I found it a wonderful way to obtain QSLs from stations like XF1A and others who do not QSL often."

Lloyd also said if you have any military rank, do not put it on your cards.

Lloyd told CQ readers that "Between 1 January 1957 and 20 August 1957 I worked stations in 331 different prefixes and received confirmations from 301 different prefixes to win WPX Award No. 1 on CW. There are still many more WPX awards to be won, including several 'firsts.' I wish all of you much luck and hope this article may be of some assistance to you in your WPX endeavors."

The importance to hams of QSL cards simply cannot be underestimated. In late 1957 and early 1958, an active amateur, J. Bruce Siff, W2GBX, who was a member of the Niagara Frontier DX Association (Buffalo) circulated two letters that were widely discussed among amateurs. One of them discussed the matter of foreign licensing; the other QSL cards. Both letters were of particular interest to Lloyd. In the year 2002, the American Radio Relay League was developing a way in which contacts (QSOs) could be confirmed "electronically," much in the way that many people now do their banking, paying of bills, etc. In 1957 paper, and the mails, were of course the only way. In August 1957 Siff wrote:

> It appears to some of us that certain phases of DXing are not keeping in step with the advances being made by amateur radio in general. The increasing number of DXers has created a situation that certainly is worthy of some thought on our part. This situation is the much discussed, but seemingly never acted upon - the problem of QSLing.
>
> First, can and will the foreign amateur QSL when his card is needed for our country credit? If the QSL is sent will it be properly filled out or is there a good chance that the card will be rejected as DX credit? Finally, if the QSL bureau handles the card when, if ever, will it reach me? Let's look at each of these questions separately.
>
> It ought to be about time for us to do something to alleviate the burden that QSLing may put on some foreign amateurs. Credit for DX contacts has been handled almost the same way for many, many years and it may now be obsolete. We all like the actual QSL card but this should certainly not be the sole requisite for DX proof.
>
> A few remedies to this problem come to mind and certainly one of you has a solution that is worthy of extensive consideration. Take time to do something about it, now. The answer to this problem may also solve the difficulty encountered when a DX station accidentally fills out his QSL improperly.
>
> No doubt in 1935 it was reasonable to expect one man, with little or no help, to receive, sort and forward the foreign QSL cards sent to him for distribution to U.S. amateurs in his area. Even today every QSL Bureau Manager is happy to spend most of his hamming hours in helping others. Certainly the QSL Bureau Manager deserves a lot more credit than most of us give him.
>
> However, it's not 1935 and the ranks of DXers have grown tremendously in the past years. It is time to consider the QSL Manager and the fact that his burden, even though he complains not, may be unreasonable and that the method of handling DX QSL cards in this country may be entirely unsatisfactory to the majority. Again let me emphasize that I do not wish to detract from the efforts of the QSL Manager but only impress on the DXer the magnitude of the QSL bureau in certain call areas...and what better example than the 2nd and 6th districts.
>
> The QSL managers in these areas are certainly as prompt as possible in forwarding the QSL cards. Yet, as the number of DXers grows, it is only reasonable to assume that QSL cards must take longer to be handled. Now is the time for the new and old DXer to take steps to solve this ever growing problem. How about splitting the K and W calls into separate QSL bureaus? Why not have the QSL Manager supervise a group of assistants at a local radio club or at a veteran's hospital? What is your solution?
>
> While working new countries and exchanging QSL cards are only a part of DXing, we are presented with a problem that is worthy of discussion at your DX meetings and with your SCM (ARRL section communications manager). Request your division director to present your solution at his meetings. If we want our phase of amateur radio, DXing, to advance with the hobby as a whole, we must take action ourselves. No one else will do it for us.

Naturally, advertisers were hungry to cash in on Lloyd's success, through endorsements. Hy-Gain, the largest maker of amateur radio antennas in 1957, carried a special ad featuring W6KG - Lloyd was using a Hy-Gain beam antenna. "I have built and used many beam antennas," Lloyd said. "Considering weight and size limitations, I have never before had a three band, rotary antenna system that compares with my present Hy-Gain three band, single transmission line, antenna," Lloyd gushed. "During DX contests and other operating competitions, fast band switching is a great help. The use of the Hy-Gain 3-band, one transmission line, antenna goes a long way toward accomplishing fast band switching and the working of more stations than the fellow struggling with 3 transmission lines."

Sadly for DXers, the burden for Dick Spenceley, KV4AA, of handling DX matters for CQ, writing a monthly column for them, plus acting as liaison for Danny Weil, VP2VB, and, finally, shepherding the new WPX award through its first year, must have been too much for him. He bowed out of his CQ duties at this time, and here's what the magazine's editor, W2NSD, had to say:

"Dick Spenceley, KV4AA, has done a magnificent job of running the DX Department of CQ, making it one of the most interesting sections of the magazine. We were thus very sorry to learn of Dick's decision to finally give up the reins. All of us will miss his perceptive wit and kind hearted tolerance.

"We are indeed fortunate in finding an excellent fellow to take over the DX Department, one that all of you DX men know quite well: Don Chesser, W4KVX. Don devotes altogether too much of his time to DXing and is in close touch with most of the top DX men around the world. He is the editor of the Ohio Valley Amateur Radio Association DX Bulletin and a past president of the OVARA. Don is primarily a CW man, that being where most of the DX lies. I see by the latest Bulletin that he has 4 countries on 160M CW, 75 on 80M, 131 on 40M, 239 on 20M, 112 on 15M, 19 on 11M, and 84 on 10M. That's a pretty good list," Green wrote.

Don Chesser greeted his new CQ readers:

> My sudden elevation to DX Editor of CQ is a big surprise to everyone-including myself. It seems impossible to successfully succeed Mr. DX himself, Dick Spenceley, KV4AA, for Dick has done such a tremendous job. But we must carry on. With your usual fine assistance in supplying me with DX news and tips, I'll do my level best. I'll be on 14010 kc most evenings at 0000 GMT looking for hot news items. - Don Chesser, W4KVX.

And Chesser also wrote of the change in his DX Magazine:

> Trying to fill Dick Spenceley's shoes is somewhat like following Judy Garland's act at the Paramount. Anything less than a colossal performance will seem anticlimactic. Dick's tremendous DX columns, his topflight DXing on his own, and the precision flow of DX information in and out of KV4AA give him proper rights to the title Mr. DX himself. Dick rates an overwhelming ovation for his dedication to DX and for his many efforts on our behalf these many years. There are few in the ham fraternity who don't know him and his achievements, and none who won't grant him DXer of this decade, if not of our time.

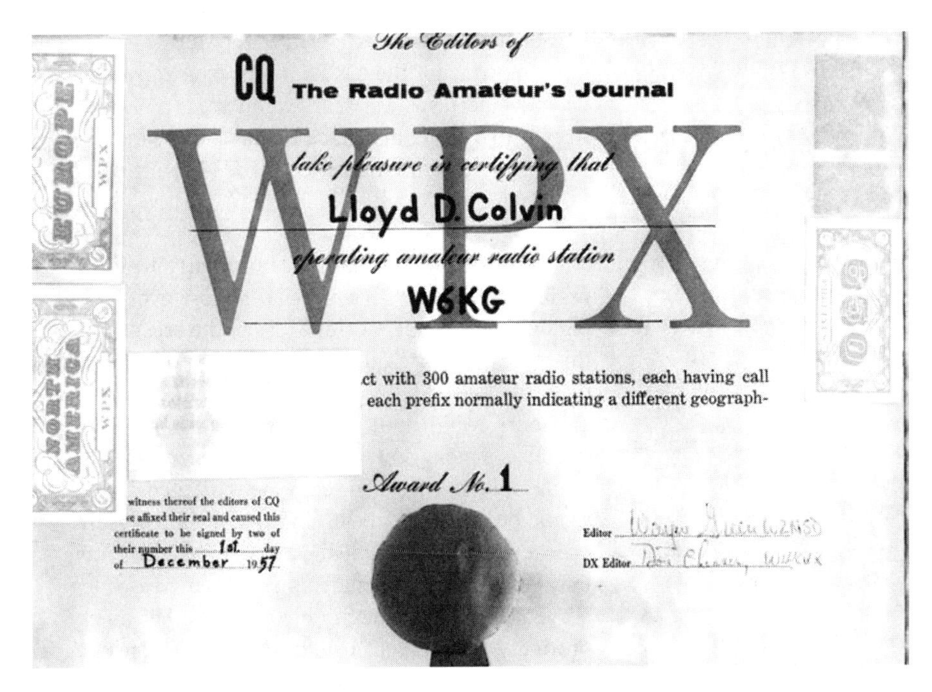

The Editors of
CQ The Radio Amateur's Journal
WPX
take pleasure in certifying that
Lloyd D. Colvin
operating amateur radio station
W6KG
ct with 300 amateur radio stations, each having call
each prefix normally indicating a different geograph-
Award No. 1
witness thereof the editors of CQ
e affixed their seal and caused this
certificate to be signed by two of
their number this 1st day
of December 1957
Editor
DX Editor

Lloyd made sure his bosses were aware of his WPX accomplishment, too, as the following letter (16 December 1957) from the commanding general at Fort Monmouth, New Jersey, attests:

> Dear Colonel Colvin: I want to congratulate you on the prominent place that you have in the December 1957 "CQ" and the fine work that made this possible. The article on page 28 shows that obtaining WPX Nr. 1 was not an easy matter and I cer-

tainly commend you for your efforts. This distinction for an officer in the Signal Corps reflects credit on him and his organization as well.

We have a number of hams at Fort Monmouth, but I don't believe any one of them has seriously endeavored to enter a WPX contest. They are on the air a great deal, however, and you may hear from them from time to time. - W. Preston Corderman, Major General, USA, Commanding.

The Colvin's daughter, Joy, recalls that amid the flurry of amateur radio activity in 1957, their first full year back in the U.S. after years in Germany, Lloyd, with Iris as his business partner, began their Drake Builders activities in California. For a time, they operated a Builders Control Service, in which funds of other builders on specific contracts were dispersed through Drake Builders. Building activities were all of larger types, including apartment houses, commercial buildings, and hospitals. Many of the building projects were bonded with several national insurance companies.

The matter of being able to obtain an amateur radio license, or at least a temporary operating permit, in a foreign country, was always of paramount importance to DXpeditioners. In January 1958 Bruce Siff weighed in again, with a letter to DX clubs around the country on the subject:

> Each of us has our own separate part of amateur radio that we prefer. However, in the overall picture we are all the same … amateur radio operators. Therefore, whether you are a traffic handler, a VHF man, a novice or an old timer, it behooves you all to take an active interest in the problem confronting us.
>
> Since many of us like to work new countries regularly, it's about time that we took some steps to help guarantee that our new country list will continue to grow. One of the best ways to "work a new one" is through a DXpedition that sets up in some practically uninhabited foreign land. Many cases come to mind immediately such as Clipperton, Brunei, Easter Island and the Cocos.
>
> Almost every DXpedition is composed of a group of American amateurs trying to provide you with a new country for your DXCC list. In the past, fortunately, the governing bodies of certain foreign lands were willing to grant a license for amateur radio operation to a non-citizen. But think for a moment how much smaller the DXCC list would be if these governments took the attitude that is prevalent in the United States regarding the licensing of a foreign amateur. With the recent exception of Canada, the chances of a DXpedition from another country operating in the United States or one of its possessions are nil. How lucky we are that this was not the case with other governments when American amateurs requested a French license to operate FO8AJ; a British license to operate VP5BH; a Dutch license for PJ2MC; and many others.
>
> As every traffic man knows, third party traffic is prohibited between the US and almost every foreign country in the world. Most of the reasons for this restriction are quite feeble, including the one that states that a loss of revenue would occur if amateurs could send messages in and out of any country. Think how many written messages or phone patches are of extreme emergency and would be sent at any cost. Five or even ten percent, you may decide. Obviously, the amount of revenue that is lost because amateurs send that five or ten percent is certainly negligible and the other ninety or ninety five percent would never have been sent at all if any cost was involved.

Maybe this and the other reasons given are not the real barrier that prevents us from handling traffic to foreign countries including American servicemen operating from France, Germany, Morocco and elsewhere.

Admittedly, a phone patch between an American serviceman and his loved ones back home is one of the greatest morale builders that the armed forces have ever known. Yet, every day, many a serviceman is turned away from a DL4 station because of this ridiculous third party restriction. If he is lucky he can stand in line at the already overburdened MARS station. He knows that there are many W or K stations near his home in the states but most probably there is no MARS station nearby and his call never gets through. Yet, I don't think that any of you blame a foreign country for not permitting us to handle this type of traffic that we desire when we treat them as inconsiderately as we do.

When, at the ninth ARRL [national] convention, the question of reciprocity was asked, the answer was to the effect that this country has never issued amateur radio licenses to non-citizens. Further investigation showed that the reason for this non-licensing policy was supposedly one of "national security." While this reason can cover a broad scope, it would seem that its specific application to amateur radio would be to prevent foreign agents from using amateur radio to send data to the homeland. Are we to believe that if an espionage agent wants to transmit messages he will apply to the FCC for a license to do so? Hardly. But the sincere foreign amateur who attempts to get a license in this country is refused on no uncertain terms. How then can we expect other countries to broaden their views toward further amateur radio operation between their country - and ours?

If a situation like this continues, there will no doubt be forthcoming a program of tighter licensing regulations in all countries. The DXpedition will then become a thing of the past. The majority of you must agree with many other enlightened amateurs that if DXpeditions are to continue and third party traffic between all countries to become a reality, we must extend to others the same courtesy that we have been receiving and would like to continue to receive. In a world sorely lacking hospitality to neighbors, a reciprocal agreement between this country and all others interested is certainly worthy of extensive consideration.

A recent editorial pointed out the tremendous hardship that this lack of reciprocity puts on foreign aircraft entering this country. The pilot, who is no doubt not a US citizen, is refused a radio license. Yet, the aeronautical authorities prohibit his flying without a radio set that he can put on the air. He can't fly without a radio. We won't issue him a radio license. He is now ready to take his business to a more hospitable country next time. Our lack of courtesy toward strangers entering this country should put us all to shame.

There are few amateurs amongst us whom, if given the opportunity, would refuse to license a fellow amateur who held a license somewhere else and now resides here with the intent of becoming a citizen. No doubt you all agree that now, at a time when the United States' prestige is reaching an all time low, we must go out of our way to tip the scales in the other direction. If a group of British amateurs desire to set up a DXpedition in the American Virgin Islands, let us welcome them ... not turn them away. They didn't refuse us at Brunei and the Caymans but who would blame them for turning us away next time if our selfish policy persists.

Now is the time for us to return equal courtesy and consideration to our fellow amateurs throughout the world and, at the same time, protect our own DX interests. We may even push the United States up a notch in the esteem of our foreign neighbors. Billions of dollars are spent yearly by the US government to try (to) accomplish this. Why not use amateur radio to help?

All around you are people who can make this possible. Individual and club letters to the SCM, division director and ARRL will let them know how you feel. Since this is a program that will ultimately be resolved by federal action, you should contact your congressmen and representatives. Tell them of the many benefits to be gained through the simple expedient of extending the courtesy of an amateur radio license to a foreign amateur in this country. It's as easy as that. If we want to keep amateur radio in the high esteem that it already holds, we must do it ourselves. No one else will do it for us.

Lloyd was often involved in the "DX Quiz" at the joint Southern-Northern California DX Clubs Conventions, both as a participant and a few times, as a past winner, in devising the quiz.

Here's a sample, from 1958 (with answers in parentheses); it should entertain recent DXers as well as non-amateurs; it's tough!

Dickson Island is part of what country? (Asiatic Russia)

What prefix does Dickson Island use? (UA0KAA)

The prefix for the Molucca Islands is? (PK6)

Aldabra Islands are nearest to what other country? (Comoros)

What prefix do these islands use? (none assigned)

Fridtjof Nansen Land is better known as (Franz Josef Land)

MB9 is the prefix of what country? (Austria)

The prefix for Jarvis Island is? (KP6)

Siam is in zone -7 time. If it is 1100 on Jan. 19 in Fresno, what time and date is it in Siam? (0200 Jan 20)

We are permitted to work stations with the prefix XV; True or False? (False)

List the prefix of the country of which Sana is the capital. (4W1)

About 12 countries count toward DXCC which lie wholly or partially within the Arctic circle. How many lie likewise within the Antarctic Circle? (one)

You have just had a QSO with a station whose call letters are within this block of prefixes: YMA-YMZ. What country would it be? (Turkey)

16· The FCC Comes to Visit W6KG

I think fast and point out to [FCC] inspectors that my plate current meter also reads the screen current and will they consider the screen power at 75 watts? - Lloyd Colvin

As a military man, Lloyd had plenty of experience dealing with government bureaucracies. Lloyd got a visit to his station from the Feds, and after receiving an FCC citation for his amateur radio operations he sought relief. Here's the reply from the FCC (29 March 1958). Lloyd wanted the infraction to be downgraded by being recorded on a different form. The FCC said:

The "Gil" cartoon that appeared with Lloyd's story in QST

> Your request that our Form 793 be changed to Form 790 for operating your amateur station W6KG on 14200.48 and 14200.56 Kc. with A-3 emission, can not be granted. Our Washington Office has indicated that citations of this type shall be Form 793.
>
> Such things do happen to amateurs in spite of considerable care to avoid incorrect operations. Our main interest is to call the matter to your attention so that it can be corrected and extra care be taken to avoid a similar situation. The Form 793 will do no particular harm to your amateur record unless the same violation is made again within a period of twelve consecutive months. Quiet hours will quite likely be imposed from 6 to 10:30 PM for the repeat violation. - Very truly yours, Tom B. Wagner, Engineer in Charge.

This FCC visit took place during the ARRL International DX Competition, CW, in February. It was such a rough weekend for Lloyd that he sent a story about it to QST, and it was published in May 1958 under the title "A Hot Contest."

Lloyd wrote his story in stream-of-consciousness form, quite a divergence from the bureaucratise in which he was steeped. It's pouring rain at the beginning of the contest - and of the story -- and his rotator isn't working. His usual lineman has just had surgery and is still in hospital. Lloyd calls the local power company and "After talking to several people, convince them this is a real emergency.

It's Friday afternoon and Lloyd offers to pay double time plus bonus. The lineman climbs the pole and reports Lloyd's brand new $350 rotator has jammed gears. Lloyd throws in the towel but at 11 p.m. changes his mind, and gets another line man to climb the pole, cut the rotator loose, and attach a rope to the boom. "I can pull antenna around with rope from ground but must walk through a pool of water one foot deep."

Lloyd points the beam west and sits down to operate when he smells smoke. He's left the rotator motor on with the gears stuck and fire has spread to the rotator housing and cover. Lloyd says "Rain has conveniently stopped for a few minutes."

Lloyd gets a garden hose but it has a leak and the stream of water won't reach the rotator anyway, so he calls the fire department. They leave at 3 a.m. Lloyd gets a lineman back for the third visit and they cut loose all the control cables and take down the remains of the rotator. An insurance agent arrives to tell Lloyd that the loss is not covered by insurance.

Lloyd operates Saturday until the FCC arrives at the door in the form of two engineers, to make power measurements. "He enters house and asks me not to operate my antenna rotator (Ha - as if I could!)" or touch any transmitter controls.

The engineers take their measurements and determine that Lloyd's transmitter was running 1075 watts input. "I think fast and point out to inspectors that my plate current meter also reads the screen current and will they consider the screen power at 75 watts?

Inspectors talk it over and reluctantly agree to put on report that my power input is exactly 1000 watts (whew)."

Lloyd replaced the Johnson Ranger with a Central Electronics 200-V

The FCC men agree to overlook that Lloyd had been recording the wrong date in his logbook (this is really un-Lloydlike!).

Then, Lloyd has to turn the shack inside-out to find his license (also very un-Lloydlike). The FCC inspectors agree to look the other way while Lloyd signs the license

The FCC visit concluded, Lloyd went back to work. Contesting was very important to him in those days, as he continued chasing amateur radio awards in earnest, combing logs back to his earliest operations in the 1930s. Among those applied for and received were Worked All States for W6AHI, the Awards Hunters Club, for DL4ZC, and Worked All Zones, for W6KG (all 40 CQ Zones worked since coming on the air on 1 January 1957).

17· England, and a New Boat for Danny Weil

The whole deck blew straight off the yacht with me on top of it. The next thing I knew I was in the water and I started to swim like heck in case she blew up again. - Danny Weil

Just as Lloyd Colvin was flying to New York City to claim WPX Number 1, in late August 1957, Danny Weil flew to England to search for a boat. He wrote to his mother from the Palmer House, Chicago, in August 1957 (Danny may be forgiven for having apparently embellished his fame just a little):

> My Dearest Mamma,
>
> As you can now see I am at the above address and I can now tell you that I have reserved my seat on the plane and will leave on Monday from here. Before I tell you about what is happening here etc I must get the times of the plane down as I want to get this letter away this morning...I will also send you a cable AS WELL so that there will be no mistake.
>
> I shall fly Pan American Air Lines from The O'Hara [sic] Field in Chicago departing 12 noon my time here on the 2nd September, Monday. The plane is due to arrive at 9.20 am Greenwich mean time at the London Airport North, and the flight number is 58. It will arrive on the Tuesday Sept 3rd.
>
> Now you will notice that I have given you the time in Greenwich Mean Time for my arrival as I am not at all sure whether you are on Summer time or what ... maybe you are back to normal winter time by now, anyway you can figure that out for yourself. I will send you a cable that will arrive on Saturday of this week to verify the above, so look out for it.
>
> I don't really know how long I shall have to myself to carry on with this letter as I am hardly my own boss in this part of the world. It is difficult to tell you of the fantastic programme they have lined up for me here. Not a day ... even an hour goes by without I have to see a newspaper man, do a TV or radio show, visit a factory or sit and have my photo taken. Last night had several taken with a blonde professional model; what a smashing piece of homework to take along with me. You will see the shots when I get home.
>
> I have been publicised in every top paper in Chicago, been on two top TV programmes, and there are many more lined up to date...in fact right up to the time I leave here I shall be surrounded with cameras. Hallicrafters have allocated their publicity man who stays with me all the time, all except [to] sleep with me. He tells me where I have to be at certain times, arranges things so that I'm not late, fixes taxis to be available and pays all

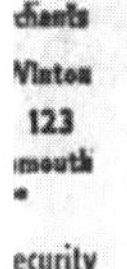

BOURNEMOUTH TIM

No. 2367 3d FOUNDED 1858 FRI

Another yacht from Christchurch?

DANNY WEIL SAGA TO START AGAIN

DANNY Weil, the Christchurch yachtsman who sailed part way round the world in a 50ft. sloop and narrowly escaped death when it was wrecked on a coral

he went on, "but this time it must be quite modern."

Danny, who is now 37, set out from Bemister's boatyard three years ago in the 50-year-old Yasme to attempt to sail round the world single-handed.

Until he bought Yasme for £200, she was an old hulk lying on a Southampton mudflat.

It cost him £2,000 to convert her

ordinary. I shall see a bit of the world and meet interesting people.

"I have been called a fool, but there have been scores of other 'fools' who have asked to go with me."

Danny, who lived with his parents at 6 Dennistoun-avenue, Somerford, moved his watch and clock business from Pridge-street, Christchurch, to Boscombe several years ago.

He was running the Precision Watch Company in Christchurch-road before he left for his unknown destination.

Danny's voyage ended suddenly last October, when Yasme was wrecked on a coral reef near New Guinea. He was rescued by flying boat, and went to America.

the bills as I go along. I have been given a very large and VERY expensive three room suite in Chicago's finest Hotel, but there are so many people there all the time it is usually about three in the morning before I have the entire place to myself.

Today I shall be having lunch with the President of Bell & Howell. They are America's top movie camera manufacturers, and lots of negotiations have been made for them to supply all the movie gear. I am also getting a big Diesel driven generator that will give me all the power on the boat....no more petrol any more. Also a big deep freeze is being installed when I get back to the States on my return trip.

Danny Weil's ill-fated second boat

The Promotions manager has really been working hard to get as many companies in on this new trip and of course I have to meet the presidents of all of them.

On Friday I suppose I shall have the greatest man in America next to the President attending a cocktail party in my suite. General [Butch] Griswold. He is the chief of the Air Force, you will remember. I spoke with him over the air from Dick's place some little while ago. [Griswold was a ham - ed]. Also on Friday I shall be flying with him and many other prominent people, all of which are Hams, in fact the entire crew of the plane including all the passengers are Hams, and they are being given a two hour-trip over Chicago. Naturally, for publicity.

This whole thing is beginning to overwhelm me, and it is all I can do to keep pace with the everyday happenings. On the programme, they have managed to have me attend TWO dinners at exactly the same time. You will see this if I can manage to get my schedule sheet to you with this letter ... this is on Saturday. I know the Americans can do anything, but how I am able to be in two places at once is something I cannot figure out. However, that is their affair. Guess they will be able to fix it some way...they can do anything here.

I am going to take a chance and send you my schedule sheet on the off chance I can get another. As you will see, it is well and truly messed around as there [sic] extra things popping up daily that I have to be squeezed into. I hope to have the entire programme all laid out neatly for you to see by the time I arrive home....IF I get the time to write it out...Hi!

Now, the time is 9.25 and in 5 minutes I have to be interviewed by the Daily News Feature Editor so I will HAVE to close whether I want to or not. Will try to get another letter off to you before I leave, but don't rely on it. Must run now, they are waiting for me. All my love to you both [Christine and her second husband]. - Dan.

On 16 September 1957 Chesser reported Danny in England "on the verge of plunkin' down hard cash for his boat. Hallicrafters has sent him an HT-32 and SX-101 for immediate use. Danny plans to return to the Virgin Islands and KV4AA, with the new boat, in about a month."

Danny's "hometown paper" the Bournemouth Times, near the port of Christchurch, reported his homecoming in the following story, which included a photo of Danny the day he left Christchurch, on YASME I, in August 1954:

"DANNY WEIL SAGA TO START AGAIN - Another yacht from Christchurch?"

Danny Weil, the Christchurch yachtsman who sailed part way round the world in a 50 ft. sloop and narrowly escaped death when it was wrecked on a coral reef in the Pacific, may soon be embarking on the second chapter of his saga. And his new boat, successor to the ill-starred YASME, will again come from Christchurch.

Mr. H.G. Bemister, whose Bridge Wharf Yard virtually rebuilt YASME from a hulk, has been asked by Danny Weil to find another boat capable of completing the second phase of his mammoth trip. Danny, who is now 37, set out from Bemister's boatyard three years ago in the 50-year-old YASME to attempt to sail round the world single-handed.

Until he bought YASME for 200 pounds she was an old hulk lying on a Southampton mud flat. It cost him 2,000 pounds to convert her into a sea-worthy craft, and at least another 1,000 (pounds) to refit for the world voyage.

At that time he had only been sailing four years, and had spent the first 12 months, as he put it, "capsizing 12-foot dinghys." Before starting out, he said, "...I am doing nothing really out of the ordinary. I shall see a bit of the world and meet interesting people. I have been called a fool, but there have been scores of other 'fools' who have asked to go with me."

Danny, who lived with his parents at 6 Dennistown-avenue, Somerford, moved his watch and clock business from Bridge-street, Christchurch, to Bascombe several years ago. He was running the Precision Watch Company in Christchurch-road before he left for his unknown destination. Danny's voyage ended suddenly last October, when YASME was wrecked on a coral reef near New Guinea. He was rescued by flying boat, and went to America.

After 45 years, one of the eternal questions posed by Old Timers is "Just how many YASME boats did Danny Weil have?" The answer is, simply, that it depends on how you count them.

You see, Danny found a new YASME in England, christened her YASME II, but she was destroyed before she and her skipper broke the surly bonds of England. In Danny's mind, to this day, there were four YASMEs. But ... for this book we do not number the second YASME, the one that never left England.

DX Magazine broke the sad news on 19 Sept. 1957: "DANNY LOSES BOAT AGAIN! W8JIN reported by telephone this evening that Danny Weil, VP2VB/etc., again lost his boat, this time the new one just purchased last week, but this time by fire. It is reported the accident happened while refueling the new boat, in a Scottish port. The boat is a total loss, but was covered by insurance, it was noted. Danny is reported uninjured (physically) in the accident, but this will undoubtedly delay resumption of his world tour."

Two weeks later Chesser published "MORE ON DANNY WEIL. The following newspaper article appeared in a prominent British newspaper [probably, again, the Bournemouth Times] shortly after Danny's unfortunate accident with his new boat. The article was forwarded by G3ESY, and appeared in the current issue of the Southern California DX Club Bulletin":

The Burned hulk of the YASME II that wasn't

An "Around the World" yachtsman was catapulted high in the air when his 40-foot sloop blew up underneath him last night. He landed in the sea 15 feet from his yacht, the YASME II, lying in Holyhead harbour.

The explosion shook the town and sent the fire brigade, police cars, an ambulance, and hundreds of people racing to the promenade. Holyhead lifeboat was also called out.

The yachtsman, Mr. Danny Weil, 34 [newspapers continued to get Danny's age wrong, and to misspell "Weil"], of Christchurch, Hampshire, had just returned from the United States, where he had been giving a series of lectures on how he sailed around the world single-handed.

He was picked from the water by another boat and taken to hospital, but not detained.

Afterwards he said "I had just brought the yacht into Holyhead for a refill of petrol. After I had got the fuel on board I filled one of the two 20-gallon tanks. I had just finished filling the second when there came a terrific explosion, and I was blown into the air. The whole deck blew straight off the yacht with me on top of it. The next thing I knew I was in the water and I started to swim like heck in case she blew up again. I had been in the water for about 15 minutes before a tender from the Trinity House depot saw me. Just before she came up there was a second explosion and the water boiled around me, but luckily I was not hurt."

Firemen were taken out of the burning yacht in a borrowed tender. Flames up to 50 feet transformed the sloop into an inferno, and for the first 10 minutes the heat was so strong that firemen had to work from about 20 yards distance.

Mr. Weil, who earlier this year completed a 43,000 mile world voyage, only recently purchased YASME II to refit her for another world cruise.

Chesser relayed in November that Danny was reportedly "delighted" with the purchase of another boat, this one a yawl built entirely of teakwood - this is what we henceforth refer to as YASME II. Danny was set to sail for the U.S. Virgin Islands in mid-November, 1957. The new YASME was built in 1912 and drew 7 feet of water, a bit more than Danny wanted, but "apparently that's its only deficiency." It had two new masts, a lead keel weighing seven tons, a new mainsail, and a deck winch, "the very latest in hydraulic winches. All he has to do to haul in the anchor is to move a lever. And, best of all, it has a huge diesel engine that allows the 50-foot boat to cruise at 7-1/2 knots."

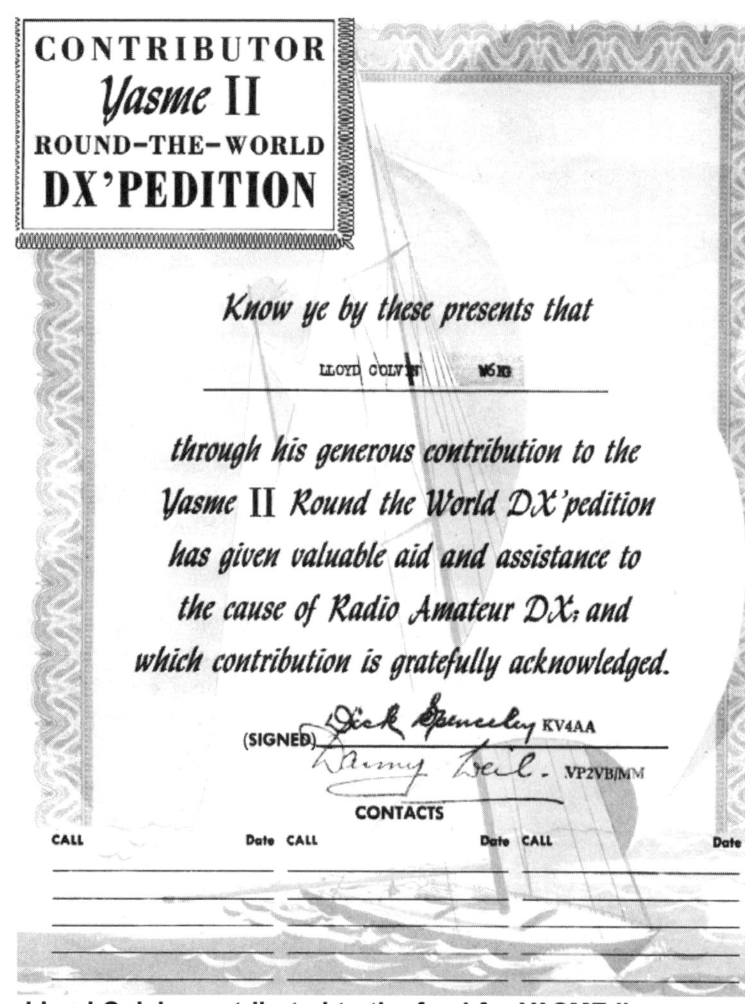
CONTRIBUTOR
Yasme II
ROUND-THE-WORLD
DX'PEDITION

Know ye by these presents that

LLOYD COLVIN W6KG

through his generous contribution to the Yasme II Round the World DX'pedition has given valuable aid and assistance to the cause of Radio Amateur DX; and which contribution is gratefully acknowledged.

(SIGNED) Dick Spenceley KV4AA
Danny Weil VP2VB/MM

CONTACTS

CALL	Date	CALL	Date	CALL	Date

Lloyd Colvin contributed to the fund for YASME II

Danny reportedly had a new Hallicrafters ham station awaiting him, an HT-32 transmitter and SX-101 receiver. In later years 1950s and '60s gear like this, each piece weighing 80 to 100 pounds, would be avidly collected and restored, and affectionately referred to as - "Boat Anchors."

But, before he could get away, Chesser reported on 18 February 1958 that YASME II had broken from its moorings "to the tune of $2000 but it was covered by insurance." Then, a leak was found in the hull. "Danny plans to go to Navassa first. Galapagos and Clipperton are also mentioned in his early sailings," Chesser said.

After this, the amateur radio press is quiet regarding Danny Weil. His second trip alone across the Atlantic, in YASME II, apparently was uneventful. Dick Spenceley was free of the grind of editing a magazine column. And Gus Browning, W4BPD, was beginning to travel the world and put many rare DX countries on the air, including islands. He was a business owner and could afford it (although he did have significant support from manufacturers.) These operations would significantly narrow the list of countries from which Danny Weil could operate (and generate support). When a DXer has worked a new one and made a contribution as a result, he isn't likely to do so again!

Salvation for Danny came in the form of changes to the ARRL's DXCC Countries List, on 1 June 1958 - and it involved islands. Previously, nearly a dozen Caribbean islands had comprised just two DXCC "entities": the Windward Islands and the Leeward Islands. As England loosened its grip on these British colonies, the DXCC rules required a reconsideration and, as a result, the Leeward and Windward entities were removed from the DXCC list. The following new DXCC entities resulted:

Anguilla, Antigua, British Virgin Islands, Dominica, Grenada, Montserrat, St. Kitts, St Lucia, St. Vincent, and Jamaica/Cayman.

While all these islands had resident hams, few were able to satisfy DXers' demands for contacts; some did not work CW, some did not work phone, some did not send QSLs. Here was a built-in itinerary for YASME II: many islands, some within sight of each other, close to St. Thomas (where Danny had just arrived), and DX demand. A Caribbean expedition for YASME II was announced, as a warm-up for the ambitious Pacific trip to come.

Chesser's new magazine would spread the word and provide timely (for 1958!) reports on each operation. By June 1958 The Ohio Valley DX Bulletin [i.e., DX Magazine] was at #34, its first anniversary, and Chesser was DX Manager for CQ magazine.

"Starting with 50 local members, plus 80 more recruited in the VP5BH campaign [an expedition in which Don Chesser participated], in this one year the mailing list has climbed to 700 top DXers in the world. The Bulletins are now going into 44 of the 48 states and to several dozen countries on all continents," Chesser wrote.

Don Chesser, W4KVX, and his DX Magazine

Until 1957, publicity for Danny Weil among DXers, indeed all information available to DXers, was pretty much limited to CQ, QST, a few mimeographed "tip sheets" circulated in small geographical areas with a high concentration of DXers, and, of course, the on-the-air rumor mill. In 1957, enter Don Chesser, W4KVX. Chesser, who lived in Burlington, Kentucky, near Cincinnati, was a well-known top DXer and associated with the Ohio Valley Amateur Radio Association. Chesser was a talented writer and what began as a small, general-interest club newsletter morphed, in June 1957, into the DX Magazine.

Chesser always put DX tidbits into the OVARA newsletter, since that was his main area of interest, and the word spread. Soon, Chesser was receiving requests to be put on the mailing list from far-away places, like Dayton, and then farther still. The list grew and grew … Chesser's rag was the only game in town, and, besides, it was good. For the next half-dozen years Chesser would have as much to do with Danny's fame as anyone, and excerpts from his writing appear often in this book. The first known Chesser reference to Danny appeared in August 1957:

Don Chesser, W4KVX, visited 4U1ITU in Geneva in 1962 and published this photo of himself in his *DX Magazine*.

Don Chesser, W4KVX, visits Hallicrafters president Bill Halligan, W9AC, at the Hallicrafters factory in Chicago. After the fire that destroyed Chesser's *DX Magazine* operation, Halligan not only donated an HT-32A transmitter to help Chesser get back on the air, he committed to regular advertising in the magazine, where this photo appeared.

"Danny enroute to Chicago for ARRL Convention, after which he will journey to England to purchase new boat, then back across the Atlantic to KV4AA where he hopes to start his next world tour in October."

Don Chesser was born on 7 January 1917 in Byesville, Ohio. He was first licensed in 1933 as W8KVX. Sometime around 1936 he enlisted in the U.S. Coast Guard, was accepted as a radioman, and was sent to school in New London, Connecticut. After graduation he was assigned to the USCG cutter Tahoma, based at Cleveland. A January 1937 QST note identified him as one of three amateurs aboard.

Chesser completed his enlistment and moved to Cincinnati before the war. During the war he worked as a civilian radio operator for American Airlines and was married at the time.

At some point he landed a job in the Cincinnati Police Radio Section; the chief was Jake Schott, W8FGX. Later he worked as an engineer for WKRC in Cincinnati, then left to start his own television and radio repair shop.

He died 16 June 1985 while on a motorcycle trip in Ohio.

18. YASME II Begins With Controversy

"It is a fact that things go much more smoothly when Danny is unaccompanied at his DX spots ... that the work necessary to launch these trips and the minor hardships encountered are not apparent to the uninitiated." - Dick Spenceley

Danny spent little time in St. Thomas, stowing the Hallicrafters radio gear aboard and sailing off. His first operation, from Aves Island, a Venezuelan possession, produced 2250 contacts. YV0AA/YV0AB was on the air from 7 to 14 July 1958. Aves had become a DXCC country on November 1, 1956, and operations since then had supplied only a handful of contacts. Unfortunately, this first major operation produced serious backlash which was never publicized and appears here for the first time.

It was a condition of the Venezuelans (represented by the Radio Club of Venezuela) that any amateur radio operation from Aves would include Venezuelan operators. Complying with this, Danny picked up a number of them for the trip to, and operation from, Aves. Two log books exist from this operation (Spenceley had fitted Danny with a quantity of ledger books that would become his official log books for all future operations. One log book records 20 or so contacts by YV0AA, a call sign the Venezuelans used from Aves to work other Venezuelans. The other log book, for YV0AB, recorded the "main" operation, by both Danny and the Venezuelans.

Following the operation, on 25 October 1958, the Radio Club of Venezuela (RCV) sent an official letter of protest to the American Radio Relay League (ARRL), publishers of QST. In fact, the Venezuelans, according to their letter, had sent a cable to ARRL on July 9, just two days into the 7-day operation and, again according to the RCV, had received a letter from ARRL dated 22 July. Unfortunately, neither the cable nor the letter survive.

Most countries of the world have an official national amateur radio society. These societies are banded together as the International Amateur Radio Union, going back to the 1920s. The ARRL, as the largest of these societies, has always been the de facto "leader" of the IARU. The headquarters operations of the IARU are administered from the same building as the ARRL (in 1958 it was in West Hartford, Connecticut; since 1965 it has been in Newington, Connecticut). In this case, the overwhelming reason why the RCV protested to the ARRL was because the ARRL administers the DXCC award, that award being the sole reason for an amateur radio operation from Aves Island. In fact, the letter was addressed to the ARRL "DXCC Committee."

The letter apparently was sent to ARRL, and no one else. The Venezuelan society wanted the letter published in QST, and were prepared for same (it was never published). The ARRL sent a copy of the letter to Hallicrafters, a company not only associated in the public mind with YASME, because of its equipment support, but also a major advertiser in QST. No correspondence on this matter remains in the ARRL files. Tom Stuart, W0REP, a Hallicrafters executive, wrote to Spenceley, on 2 December 1958:

> I trust that by the time you receive this you will have the crystal which we sent you to give additional coverage on the 10-meter [band] and for the HT-32.
>
> The enclosed letter was forwarded to us from QST. I am enclosing both the original letter in Spanish and the translation as furnished us by QST. You will note that the radio club of Venezuela has requested QST to run this material in a forthcoming issue of the magazine, and they feel so strongly about it that they are willing to pay for its publication.
>
> We will be very interested in hearing any comments you may have relative to the information supplied in this material. -- Tom Stuart, W0REP

Here is the letter from the RCV to the American Radio Relay League:

> In relation to our cable of July 9 and again that of August 3 in answer to your letter of July 22, we wish to bring to the attention of the Committee of DXCC the reasons that prompted our first cable.
>
> The proceedings to put on the air Isla de Aves, after our first expedition, received the backing of the Board of Directors as well as the approval of the National Government through the Ministry of Communications. Our desire to cooperate with KV4AA and Mr. Danny Weil led us to negotiate with the Government for the granting of a new station YV0AB through the (aid) of our Treasurer Sr. Falkenhagen, YV5GO. Thanks, then, to these actions of good will and cooperation on our part it was made possible for Sr. Weil to operate YV0AB, since our laws and regulations prevent, for persons who are not natives of Venezuela, the installation and operation of stations.
>
> We never thought that all this work and effort carried out by us would not result in the proper under-

standing and gratitude on the part of Sr. Weil. After what we have found out through Julio Pena, YV3BS, and Fernando Falkenhagen, YV5GO, in relation to the conduct and actions of the above-mentioned gentleman we haven't the slightest doubt about his piratical activities with commercial ends within purely amateur operation.

If we had known that this would happen, Sr. Weil would never have succeeded in putting foot on Isla de Aves nor in using station YV0AB to do business. It is notable that in our first expedition to Isla de Aves the RCV sent a total of thirteen operators, all Venezuelans, and it never sought even the endorsement of IRC for the shipment of QSLs from YV0AA, but rather added to these costly QSLs signals alluding to said expedition.

At this time also nearly 2,000 contacts were made. Our operators, Julio and Falke, brought U.S. $50 for the purchase of food. 100 U.S. dollars were invested in all, and yet they were hungry the greater part of the time that the expedition lasted.

YASME II

According to their information, Sr. Weil transferred to the YASME II leaving them on land operating for the space of 50 hours without food or water, in spite of having promised them that he would return in 20 minutes. He clogged or delayed the QSOs with Latin America, saying that he had a schedule with KV4AA or that there was nothing on. In a special way when work was going on with amateur YVs he interfered with said QSOs, starting an electric drill, which caused such interference in the reception that it was impossible.

On one occasion he said to YV3BS that the QSOs with Latin America were not of interest since they did not remit the "voluntary contribution" for the expedition. When they had been operating three days from the Isla, KV4AA informed Weil that there had already been received in St. Thomas hundreds of QSLs and also some $600 in U.S. money. The YASME II went well-provisioned with {???} gallons of water to drink, besides 288 bottles of soda water; nevertheless he rationed the water and the drinks, returning to port with 120 gallons of water and 48 bottles of soft drinks.

He used the same method with the food saying on occasion that when he was traveling, he did not eat often.

Of all this bad treatment to our representatives, there was another victim. Sr. Jules Wenglare, KP4AIO, who accompanied them on the expedition and who is ready to furnish a written account if he is asked to do so. Our operators at first operated in a clandestine manner when Sr. Weil was not on land and he had forbidden them to handle the equipment. When we received these complaints we sent Sr. Weil a message which seems to have made him change his conduct a little and also his demands on our representatives as to the operation of the equipment.

This Sr. Weil refused to handover the cards of YV0AB and a copy of the Logs showing that they were in charge of making the remittance. Thus he ignored the fact that our representatives were the legitimate [licensees] of the stations of Isla de

Aves and that he had been allowed to use one of the stations as SECOND OPERATOR. He disobeyed and was disrespectful of our representatives who were traveling on an OFFICIAL mission of the Venezuelan Government as their official passports showed, besides representing in our territory beyond the sea the RCV. From the point of view of human relations the comportment of Sr. Danny Weil was abominable as he paid no attention to radio treaties and the simplest rules of social behavior and education.

He never used courteous words or decent manners in addressing our representatives, his speech being rough and coarse. He ignored the fact that he was the one who had been allowed to go on the expedition, that it was carried out in Venezuelan territory and with Venezuelan operators representing it and that he was granted the opportunity of operating a new station and that help of all kinds was given him, including money, in order that he might make the trip, bringing only his boat and equipment. This last anybody can. Add the fact that much before this expedition and right after his flop in the waters of the Pacific, the radio fans of Venezuela and the RCV helped him with U.S. 196 dollars, without counting those who did it directly on their own account.

We never imagined that in his activities in the field of radio amateurs all that he was after was profit and business with the QSLs, with the pretext of that which they call "voluntary contribution" which is necessary to pay for the expedition. We understand that there are many firms ready to defray costs of the expeditions as advertising for products without its being necessary or lawful to charge dollars for the cards of confirmation since in accordance with the aims of amateur radio and its laws and international rules, any form whatever of profit or propaganda of a commercial nature deprives it of its amateur status.

For these reasons NEITHER THE RADIO CLUB OF VENEZUELA NOR SR. FALKENHAGEN, grantee of YV0AB, CAN ACCEPT ANY RESPONSIBILITY IN ANY TYPE WHATEVER OF CLAIM OR JUDGMENT THAT CAN BE DERIVED FROM THE SYSTEM OF OPERATION CARRIED OUT IN A COMMERCIAL FORM BY SR. WEIL AND SR. DICK, AS FAR AS YV0AB IS CONCERNED.

The Radio Club of Venezuela represented by its Board of Directors did not consider it suitable to accept a proposal made by one of its members in a General Assembly held to consider the problem which our representatives were facing at that time in Isla de Aves, that the authorities should be asked for the cancellation of station YV0AB as a protest and reprisal against Sr. Weil for his conduct and actions, out of the very desire for international collaboration since it decided that those who had worked at {sic}YV0AB were not responsible for the situation created by Sr. Weil.

In accordance with procedure established from our first expedition in June of 1956, the cards of YV0AA will not have any validity, no matter in what expedition the QSO was carried out, unless they carry the official seal of RCV and the corresponding signature of the secretary of functions. Those who have received one from this second expedition should send it to the RCV for authentication.

We should like to see this report published in textual form in QST and we are ready to pay for it in order to make known to the amateur world who Sr. Weil is and how he operates and likewise in order that this official protest that we are making to you about the absurd behavior of the above-mentioned Sr. Weil may be recognized.

/s/ Pedro Joes Fajardo, YV5EC, President
Julio Peña, YV3BS-YV0AA
Fernando Falkenhagen, YV5GO-YV0AB

In response, Dick Spenceley sent a three-page letter to Tom Stuart. Only pages two and three of this letter survive, but the gist is obvious:

Prior to YASME's Aves Island trip ALL funds had been exhausted. To make this trip possible I was forced to dip into a fund, set aside for insurance, to the extent of about $500 which I personally guaranteed.

An electric generator was purchased for $195. This was considered a necessity for the simplest comforts in a tropic clime and with the Venezuelans in mind. Other items include: Diesel fuel $60, Gasoline $30, Food $150, Electric cable, Galv. Pipes, other fittings $110.

After the trip 1342 QSLs were sent out by airmail. Postage and envelopes were in excess of $115. During the trip, damage to the dinghy and main shrouds was estimated in excess of $100. Added up these figures give us a total of $760 for expenses. This amount might be considerably greater if I had time to back and check all items.

True, we still have the refrigerator and other items but, money-wise, our effective expenses were the above figure. Now let's see what we can produce on the credit side of the ledger.

Contributions accompanying Aves Island QSLs totaled $540.25. The YV food donation of $50 raises this total to $590.25. Acceptance of these figures gives us a deficit on this trip to the tune of $169.75.

It may seem, therefore, that this particular venture was not a howling success if so-called "Commercial Profits" were our goal. Fortunately Danny's subsequent stops at the British Virgin Islands, St. Kitts, Antigua, Montserrat and Anguilla produced contributions which absorbed this deficit and now, I am happy to say, we are somewhat in the black.

Due to the rough weather at Aves Danny spent the greater part of his time on the YASME. This will be proven by a glance in the log book which shows that he made only 513 contacts at YV0AB. The other operators made 1863! This should tend to refute any accusations that any limitations were imposed on the operating times of the YV boys. Interference caused by electric drill operation on the YASME was not intentional, as indicated, but necessary to effect immediate repairs to the end of a boom which had broken.

Due to the excessive rolling of the YASME at anchorage the Venezuelans showed a marked preference for staying ashore. The YASME's dinghy most certainly made at least one trip ashore daily and most days two trips. Another dinghy, belonging to the Schooner "Water Baby," which was at Aves during all of the YASME's stay, except the last two days, also made daily trips ashore from the two boats. The Aves Island log book will easily bear out that there was no period anywhere near 50 hours that operators were not changed. The claim that the YVs were "marooned" for a fifty hour period can only be taken as a "flight of imagination." An abundance of turtle meat was offered to the YVs and was refused.

Yes, 288 bottles of soda were taken along. 48 were returned to St. Thomas. That doesn't sound too severe to me! The question of water, of course, was one which had to be closely watched as none was available on Aves. Failure to ration water in a situation like this would be a dereliction of duty on Danny's part. The YASME's main water tank has a capacity of 50 gallons. 15 additional gallons were taken along in bottles. Statements that 120 gallons of water were returned to St. Thomas is ridiculous. 65 gallons were taken on the trip and approximately 10 gallons remained when the YASME docked in St. Thomas. At all times there were two 5 gallon demijohns of water ashore on Aves. I can state with absolute assurance regarding YASME's fresh water capacity as we have just replaced the old main tank with a new one.

Again, regarding food, the Venezuelans refused to lift a hand toward preparing or cooking food. Nor would they do any cleaning up. This attitude hardly fitted in with the scheme of things and Danny's growing reluctance to cook for them and serve them may be understandable. Their sufferings were hardly as indicated in their letter and they had access to food at all times on a par with other members of the expedition.

Their statement that Danny refused to hand over the cards for YV0AB and a copy of the log is not understood here. QSLing procedures were clearly outlined to YV5GO with agreement on all points. At no time did he ask me for a copy of the log which I would have, and will, be glad to furnish him.

Before the trip, for identification purposes only, the YV boys agreed to use the call YV0AA on Aves while Danny and KP4ATO were to use YV0AB. This was not adhered to and only some 90 contacts were made under the YV0AA call sign. It was felt this was done to avoid QSLing at the YV end.

YV5BS took a roll of Danny's color film with the promise to develop same and return immediately. All our efforts to have this film returned, developed or undeveloped, has met with no response.

All of this sets forth one side of the story which may, or may not, be accepted at its face value. It is a fact that things go much more smoothly when Danny is unaccompanied at his DX spots. It appears that the work necessary to launch these trips and the minor hardships encountered are not apparent to the uninitiated.

Of course the ARRL is free to take whatever action it [deems necessary]; to print, or not print, the

RCV letter. I hardly see that any useful end may be gained, nor ham radio served. By so printing. If printed, I think we should be given space for a reply in defense.

Present authenticity of YV0AB QSLs should be obvious and, undoubtedly, a great many have already been processed toward DXCC credit. Should a ruling be made that these cards are not legal until a RCV stamp appears on them would only lead to further questions and unpleasantness.

Also, as they do not have the YV0 logs I do not see how this stamping can be done with any accuracy. Should any YV0AB contactee wish to submit his card for a Venezuelan Award I imagine they would accept it and stamp it at the same time.

I sincerely regret that this expedition was marred by this clash of personalities. Such happenings cannot be foreseen but we can learn by experience. It is also regrettable that the Venezolano Club sees fit to pursue this matter. I would be most amenable toward writing this club, in conciliatory tone, should such action be thought worthwhile. Sincerely yours, with 73, Dick Spenceley, KV4AA

Well, as they say, it's only hobby. None of this was ever published (including the RCV letter); the YV0AB operation was properly licensed, Spenceley had the log books, and the ARRL credited YV0AB QSL cards for DXCC. The ARRL took no action in the matter.

19. On the Rocks in the Caribbean

"Danny could get very high strung and I didn't want to get him mad at me and not feed me!" - Joe Reisert, W1JR

Five days after the Aves Island operation ended, Danny opened up from the British Virgin Islands using, for once, his own call sign, VP2VB. 11 days of operating , July 19-28, 1958, netted 4800 contacts.

Joe Reisert, W2HQL (now W1JR) joined this operation. In June 1957 Joe, age 20, along with Fred Capposella, W2IWC [now K6SSS], had accompanied the Coast Guard on one of their bi-annual trips to Navassa Island, where the two set up and made some 300 quick contacts.

Spenceley asked Joe to go along on the first operation from the new DXCC country of the British Virgin Islands. Joe accepted, and Rudy, W3CXX (and his unlicensed son), Fred, W3BSF and Doc, W5PQA joined the group. Joe remembers making about 5,000 QSOs, a big total in those days, with one transmitter on 160 through 15 meters.

"I still remember Danny's cooking after 40 years," Joe said. "He cooked all our meals on the YASME and everything seemed to be cooked in oil in the same frying pan, even peas! He introduced me to tongue (ugh!). We had the radio set up on Buck Island , about a mile off Tortola, and had to shuttle to the YASME in a dingy for our meals.

"One night the young boy Danny hired from Tortola to be our bus boy took my shoes up to the house while I was eating dinner on the YASME. Later, I had to climb up the hill in the dark and got many cactus spines in my feet. Fortunately, Doc, W5PQA, a world renowned throat surgeon, was with us and knew what to do. He had to carefully remove all the spines from my feet and then dug into his doctor's black bag and gave me some antibiotics.

"After a stop overnight to pay a visit to the governor

of BVI on Tortola in Road Town, we sailed on the YASME back to St. Thomas. On a subsequent night in Charlotte Amalie, Dick received a phone call to expect visitors and, soon, in walked Iris and Lloyd Colvin and later yet, Reg Tibbetts, W6ITH, joined the party. I had not yet met any of them before, although I knew of them and had worked Lloyd from his DL4ZC operation in Germany."

Joe Reisert recalled that during the VP2VB operation "Danny was not too active and let the rest of us do most of the ham operation. He also had a rig on board which he used to talk to KV4AA on 40 meters daily and that was our only relay line of communications. I remember Dick Spenceley reading me a letter from my girl friend over 40 meters to me while I was on the YASME!

"One funny thing I can remember (that Danny didn't think was funny) was my friend Paul, then K2UME, tried to relay a message to me from home one night on 20 cw while Danny was operating and Danny got real angry at him calling us and ignored him. I just decided to let it rest. Danny could get very high strung and I didn't want to get him mad at me and not feed me!

"Danny was still a wild man when I met him. We never discussed it but he really was a woman chaser. However, once he met his eventual wife, he seemed to be a changed man. I visited with him and his wife when I was in San Diego in 1961. "

VP2KF, St. Kitts, followed Aug. 28 to September 4, with 2300 contacts, then VP2AY, Antigua, Sept 8-Oct 14

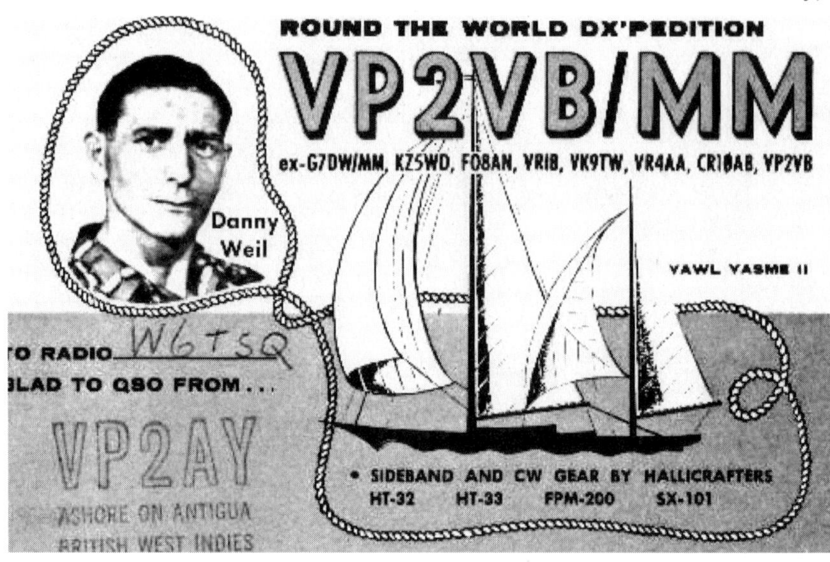

1958, and another 3400 QSOs. Chesser reported that Danny was " having ear troubles (perforated ear drum) and may have to come to the States for treatment." QST said "VP2VB & Co. unlimbered heavy multiband Antigua artillery to batter the ether with fierce VP2AY fusillades in early September. A Dominica demolition detail followed. "

Danny landed at Montserrat and produced 1950 VP2MX contacts, 22 to 27 October 1958, followed by 2520 contacts from Anguilla as VP2KFA 1 to 8 November 1958.

Danny celebrated Christmas 1958 with a VP2DW Dominica operation of 2400 contacts, then produced another quick 2750 from St. Lucia as VP2LW Jan. 4-11, 1959.

Danny Weil was on a roll! The press, in typical fashion, was quiet - there was no bad news to report. But the YASME Expedition was still, as always, operating on a shoestring. On 2 January 1959 Spenceley wrote to Mr. V.G. Gilbertson, "Manager Publicity," D.W. Onan and Sons Inc., Minneapolis, saying that a new Model 205 AJ-1P/1E Onan gas generator was now installed on YASME II. Danny was on St. Lucia by then, with stops scheduled at St. Vincent and Grenada. After that he would return to St. Thomas, around February 20, to be fitted out for "his long Pacific trip."

Spenceley said a second gas generator was still needed, and that if Onan would not give him the best price, he "could have the Hallicrafters Co. of Chicago order this machine for us as it is apparent that they are more favored with trade discounts." The cost to YASME of the generator had been $389, while a previous generator that Hallicrafters had given to the expediton had cost them only $234. Spenceley, and YASME, had already received seven generators from Onan, he reminded them.

Danny continued on a roll, handing out DXCC credits for the new "countries" in the West Indies. An operation from St. Vincent, VP2SW, netted 2800 contacts Jan 15-24 1959. He then headed for the last island on the Caribbean itinerary, Grenada. QST's DX editor Rod Newkirk, always battling the long lead times of a monthly magazine, submitted his copy for March QST just before everything changed....

"Following VP2KFA action at St. Kitts, VP2VB/mm and YASME, abetted by W8VDJ, fetched Dominica, St. Lucia, and St. Vincent in early January for sizzling multiband sessions as VP2DW, VP2LW and VP2SW respectively and in that order. Danny is now QRX for YASME repairs." - QST Mar 1959.

Poor QST, always trying mightily to keep up with news while laboring under the long lead times of a monthly magazine. April 1959's "How's DX" said "In late January VP2VB's round-the-world DXpeditionary marathon struck a snag on Union Island in the Grenadines when YASME II became rockbound en route Grenada.

Joe Reisert and Danny Weil on Tortola

Dan retrieved his radio gear and sustained painful lacerations in the excitement. W8YIN has VP2VB's possible Pacific itinerary as including the Galapagos, Clipperton, Cocos (TI9), the Marquesas, Pitcairn, the Kermadecs, Tonga, Nauru and Samoa when the situation is shipshape once more."

Unfortunately, it was much worse than that. Danny, on his way to St. Vincent, did indeed "strike a snag," but YASME II was destroyed. Danny wrote to Spenceley after the wreck:

> Dear Dick: To write this is the most difficult thing in the world. I feel as though I would rather be dead than face all the hams again. I would love to be able to tell you that the weather was bad, or the engine failed at on awkward time. It has often been said that any excuse is better than none, but in all our associations I have never yet lied to you, regardless of circumstances.
>
> The whole truth of the matter is, I went to sleep instead of being on watch. It is unnecessary for me to tell you I was tired. I have been over-tired for some time, and should have stayed over in St. Vincent for a few days relaxation. [but] I chose to move on. Call it what you wish - over enthusiasm -or just bloody stupidness-the facts still remain that I had no right to be sailing YASME with so much at stake, feeling as I did.
>
> I am feeling so badly now that I often wonder why I didn't finish off the job and drown myself, instead of inflicting myself on you and many others, again. I don't intend to go into the gory details and try to make a hero of myself. I know that I have not only let you down, but all the gang that

have supported me. Your association with me will also be the cause of many nasty remarks from the 'know-alls', and to say I am sorry is more than an underestimation.

The feeling of despondency is growing every hour. YASME is in a position where it is almost impossible to extract her. Her damage at the moment consists of a large hole in the starboard side where she has been lying on the rocks. Heavy seas and winds have caused the hole to get increasingly bigger. The latest report tells me that it is about 10 feet by 3 feet. I am dependent entirely on reports from others, as my foot injury prevents any real action which might produce results.

The generator has suffered beyond any possible hope and I doubt if I were able to remove it from the boat it could be made to operate after being submerged for so long.

If it is humanly possible to look on the brighter side of things, I have managed to salvage the following items, which, if they haven't been damaged in transit from the scene of the wreck to this QTH, should be o.k. One HT-32 and SX-101, two mikes and bug and a damaged cone speaker; both 2-1/2 KW generators (thank heaven I disposed of [the 1-1/2 KW generator] in Dominica), my salvage pump and 12 volt gas generator (I doubt if either of these will work, having been submerged); this also includes my scooter; tower and beam are ashore; my two Leica cameras are o.k. Owing to my inability to move around, the Police have taken over and are getting off as much stuff as possible. I have learned today they have managed to do some fine work, there.

I have to face the ultimate, which is that YASME cannot be salvaged. I shall be stuck here with a stack of odd gear . . . Naturally, I should strip the wreck of everything which is saleable . . . But, what to do with it all is something I cannot figure out in my present frame of mind.

With the radio gear I have available I shall endeavor to get on the air, but, not having the circuits available (I cut all the wires adrift with cutters, to speed up removal of the gear) nor do I have any relay. In fact, I have strong doubts if I have any wire to string up a long wire, if, by some wonderful streak of luck (whatever that is) I can get on the rig, then things should be easier all around. But in my present state I doubt if I have enough strength, or common sense, to even start up a generator without blowing it up. Incidentally, I forgot to mention I also got the lathe off, with the motor, and practically all of my tools (I thank heaven now that you didn't send down the compressor).

Whatever you do, Dick, please do not mention my foot to my mother. I know this is unnecessary, but I have to say it anyway. Personally, I would happily cut off my foot to have YASME back in the water again.

I had in mind chartering a boat to bring everything to St. Thomas and disposing of it there . . . the boat gear I am talking about, in the event of non-salvage here. What it would sell for, I haven't the slightest idea, but somehow I shall try to scrape up enough to buy some sort of boat without resorting to the hams. Somehow I am going to finish this expedition; how, I shall never know, but I still haven't lost the urge to carry on where I left off, even though I know I leave myself open to every form of ridicule it would be justified, too!

I only wish I had a salvage expert available here, as I am confident YASME can be saved. (but) Speed is essential to save the boat, and I am afraid that by the time you get this letter, unless I have managed to get something worthwhile done,

YASME will be lost.

No more now...Many thanks for cheering cable. -
- Danny"

By this time, Wayne Green was gone as editor of CQ; Don Chesser published Danny's story of the sinking in his DX Magazine, over several issues in 1958:

"YASME II ON THE ROCKS"

Met a guy who told me he had a ball pen which wrote under water ...... Dunno why!

Just can't figure why my Remington won't do the same after all, it's only been in the sea for a couple of weeks. It sure needed a new ribbon! Stuck it in a barrel of fresh water to wash the salt out then stuck it in the sun to dry out. After this, a pint of lube oil worked wonders except the carriage wouldn't move. The darn spring had broken.

Guess I couldn't have picked a worse spot for this to happen. Union Island in the Grenadines is about the most primitive of the entire West Indies Islands, and they sure don't carry typewriter spares in their one and only store. Just had to fix the thing, but my complement of tools consisted of a hammer, screwdriver and pliers. The rest had taken the deep six. Removed the old spring, narrowly avoiding decapitation of a couple of fingers and bashed a hole in the end with a nail, also bashing same two fingers in the process.

Ever-present audience of local natives showed great admiration for my skill both with nail and full command of profanity. Needless to say, the typewriter ultimately worked, dunno why! Guess I'm getting a little ahead of myself in this tale, what with busted typewriter and tools in the drink; so, if you're interested, may as well get on with the story.

Some, no doubt, will remember I wrote a short story a while ago entitled "YASME SAILS TO HER DOOM". Little did I realize then that I should be rewriting the same story, but with a slightly different twist. As the actress said to the parson, 'History repeats.'

DX-wise, St. Vincent had been a good spot. Managed to clear around 2500 QSO's in a week and decided it was high time to pull out. Pulling the big switch on Sunday nite, I pulled the rig down. Most of the night and Monday was spent stowing it aboard and, by the afternoon, hands had been shaken and the usual adieux made ashore. I was ready to pull out.

I felt rough. Physically and mentally tired, I should have hit the sack, but the thought of the last and final QTH, a mere 75 miles away, made me cast discretion aside. I moved out. With sublime faith and little knowledge of the future, YASME and I crept slowly out of Kingstown Harbor at 10 p.m. As we neared open sea, the sails went up to fill immediately with a fine north-easterly breeze.

It was pleasant to look astern at the faintly twinkling lights of the town. I doubted if I should see St. Vincent again, where ignorance is bliss. As Kingstown disappeared into the night, Bequia, its silhouette sharply defined against a sky already beginning to lighten with the rising moon, came into view. Bequia Harbor looked calm and peaceful. I yearned to go in, but call it over enthusiasm or just plain stupidness, I pushed on. (Bequia is one of the northern Grenadines Islands.)

It could have been an exhilarating night. YASME heeled under the freshening wind, her bow wave forming an arrowhead of luminescence which gradually closed at the stern, leaving a long silvery wake as we sped on our way. As the moon rose to its full brilliance, the scene ahead changed from one of vague outlines to one of intricate beauty.

Have you ever watched a print develop in its tray - to see a blank piece of paper transform itself into a picture? So the moon developed as it rose to its full brilliance, an apparently blank seascape into a scene of ethereal magnificence. The entire group of islands came into bas-relief. Every valley and bay, every tree, even the ridges in the mountains became sharply defined as the moon's power quelled the darkness. I regret to say it had little effect on my mind. The serenity of the view acted as a sedative to my already dulled senses. I have little recollection of events after sighting Bequia. General fatigue had taken over. Oblivious to reefs, currents, and course, I slept.

The first grinding crash awoke me instantly. Massive seas broke over the stern, sweeping into the cockpit. For seconds nothing could be seen as the spray blinded me. The grinding and crashing of YASME on solid rock left me under no illusions as to what had happened. Jumping out of the cockpit, I threw the engine astern, opening the throttle wide. A mad rush along the decks to release the sails. As they tumbled onto the deck, I dashed back to the cockpit.

The wind was dead astern, forcing YASME further and further onto the rocks. For seconds she would be afloat; then would come the heartrending crash as the mountainous seas picked her up and crashed her down on the unyielding rocks. How long she would accept this treatment was doubtful, I cast an anxious eye below, but saw no signs of water. Suspecting the worst, I stuck the engine bilge pump on, hoping I wouldn't need it.

The moon, where it had so recently shown peace and magnificence, now showed desolation and horror. Vicious black rocks appeared to surround YASME, as she fought for her LIFE, dead ahead rose a sheer cliff towering into the sky, acting as sentinel to its myrmidons of small fry. For those brief seconds I thought I had landed into a nightmare.

YASME trembled as the spinning propeller fought with the sea and wind to drag her clear. The interminable crashing and grating as nature strove to destroy her almost drove me insane. All my hopes were tied up in the engine.

I kept looking at the rocks alongside. With each rise and fall of the seas, YASME moved an infinitesimal amount astern. Could it be possible she would get off under her own power? My body dripped sweat, I trembled like a leaf as I stood at the wheel leaning astern as though that alone would assist the screaming engine.

The first rock slid out of sight into the seething spray. For those few moments my hopes rose. She was coming off slowly but surely. Something was amiss. An undercurrent of fear pushed itself up into my feelings of elation. I glanced quickly astern. A mountainous sea was roaring in. Its high breaking crest appeared as jagged white teeth as it swept in to engulf YASME in its maw. Petrified with fear, I gripped the steering wheel, unable to take my eyes from this monster.

Suddenly it struck. YASME rose into the air as though she were a matchbox. As it receded, she came down with her thirty tons dead weight. The wheel jerked itself from my hand. A demoniacal scream came from below as the gear box tore itself apart; then, the engine stopped.

The rudder had been smashed and jammed the propeller. A deathly silence pervaded, broken only by the breaking seas as YASME was swept back onto the rocks. Without power or steerage, she was helpless and I knew her time was limited.

With a superhuman effort, I threw the dinghy over the side. I wanted to get a rope and anchor out astern to stop her slewing around broadside. As the dinghy struck the water, it was immediately swamped. Attempts to bale it out were futile. Several times I attempted to get into it, but the seas swept it from me. Within a few minutes, it started to fall apart, then it was gone. Only the painter tied to the rail and a small piece of timber hanging from the end proved there was actually a dinghy there at one time. As the oars floated away in the surf, my hopes vanished with them.

I ran below and fired up the rig. I had done my best to save YASME and now it was time to save me. Hanging on to the lurching cabin door, I anxiously awaited the warm-up period. The seconds seemed like hours. It was tuned on 7mcs and the band was wide open. I snatched up the mike and almost passed out. The entire rig was alive with 110 volts and I was standing in a foot of water.

YASME was holed. I had a choice. Drowning or electrocution. YASME was taking water fast. I stood there fascinated. I could hear many of my friends talking. Any one of them could have organized aid in a few seconds. The transmitter was

working and I couldn't even switch it on. What an ironical position to be in! I hated the thought, but had to accept the fact that YASME was finished.

I cursed myself for being all sorts of a fool, but realized that recriminations would not help. I had to act. I had to do something, but WHAT? I clambered out on deck and watched YASME being forced high onto the rocks. Every crash bit deep into my body like a knife being inserted and twisted. I wanted to scream and pray to God. I wanted to jump overboard and pull her off with my bare hands. To stand there and do nothing drove me frantic and made me feel life just wasn't worth living. I thought of all the work, worry and effort that had made the expedition. All of it wasted through my utter stupidity.

I was ready to give up. It seemed pointless to save even a tube. Without YASME, I was finished, and yet, it seemed so utterly crazy to let all that gear be lost. The old brain box started clicking into high gear, and I thought hard and fast. Moving along the lurching deck to the' bow, it appeared I could get ashore with slight difficulty. There were many large rocks partially covered which might be used as stepping stones to the beach. I swung over the bow, my feet fumbling for a foothold, With the jerking and swaying of the boat, coupled with surging seas, it proved an impossible task.

I tried to pull myself up. My body 'hanging full length was too heavy. My strength had gone. I had little alternative but to drop into the water and hope for the best. It was a rough decision to make but there was no alternative. With a prayer on my lips, I let go. I tried to time the drop to coincide with a receding sea, but nature played one of her dirty tricks on me, and a double wave came in when it should have been going out and picked me up like a matchstick. For a few seconds I was completely submerged as I rolled in with the wave. I expected to feel the cruel bite of the rocks any moment and knew my chances were 99 to 1 against survival in that maelstrom of angry water. Strangely enough, I felt nothing. My head broke water and I struck out shoreward, wondering all the time if I was doing the right thing. Guess I was going to have to meet up with those rocks sometime and it may as well be now, as I was pretty weak and couldn't hold out much longer.

Seems we all get that little extra strength in times of need. I found myself a big rock and, swimming to its lea, managed to climb its rough sides with little damage other than minor abrasions. From this vantage point, I was able to make a complete survey. The actual shore was twenty feet from me. Timing my dive right, I did reach the shore O.K. My prayers were surely answered, and I thanked God to be on dry land and safe.

As I lay sprawled on the stones, spitting water and trying to get some breath into my bursting lungs, I decided this was hardly the place to take a nap and decided to drag the old aching body higher up the beach. The wind had increased in intensity and was biting into my half-clad body. I shivered and tried to stand, but my entire body had become cramped it was rough. Using my elbows, I managed to get to the base of the cliff into a niche.

Fellers I would have given my right arm for a cuppa coffee and a cigarette at that time. There was no one around at the time to take me up on my offer, but the way I saw it, my arm was as much use as a hole in the head right then.

Suppose I must have laid there for half an hour. It's sure difficult to judge time under those circumstances. Minutes seem like hours. Gradually, with the attendant pain, life came back to my cramped limbs, and I staggered upright, trying hard to maintain a footing on the stones. The moon, still full, swept the entire scene, and I could see YASME stuck between two large rocks. As the seas lifted her, she gradually bored a hole through her planking. She was sluggish in her movements and I knew she was full of water. It was only the bigger seas which moved her.

As though prearranged, YASME had moved further inshore and a large rock had placed itself beneath her bow. It was fairly easy to get aboard. Going aft, I went into the cabin with water up to my waist. She certainly wouldn't go any lower, and it helped to know that fact. The big rig, HT 32; 33, and SX 101 were completely submerged. In the main saloon I noticed the standby rig, 32 and 101 only partially submerged..... maybe they would work again. I just didn't know.

I cut the leads away and dragged out the HT32 first. That was the heaviest and I had to use the little strength I had with some sense. Wrapping it in an old sail-bag, I staggered along the steeply sloping deck toward the bow where I rested it. Swinging over the bow, I reached up, grabbed it and took it ashore. Number one OK. Next came the SX 101. Everything worked fine till I reached the bow, then it happened! A giant sea, the granddaddy of all of them, swept in and over the decks. My feet slipped on the deck and I went over backwards still holding the 101. As I fell, an intense pain jagged through my body as my spine hit the head of the deck winch. I slid flat onto the deck, the lower part of my body completely paralyzed, still holding the 101 as though it would save my life.

Stupid as it may appear, my direct concern was to hang onto the 101 and hold the canvas around it. Slowly the pain subsided and I struggled to my feet. Things were tougher now as YASME's decks were at a 60 degree angle. That big sea had shifted her and the big rocks supporting her had penetrated the planking. It was tough getting over the side. The SX 101 reached shore safely.

The 'beach' was around 12 square feet, consisting of a variety of rocks hardly a place for relaxation and sunbathing, but it was dry and a safe spot for the gear. Several journeys were made to get loose equipment, etc., ashore, but there remained one thing which I had given up any hope of removing my lathe. It had taken two of us to get it aboard and installed. What chance had I alone under these circumstances of getting it ashore? It was my pet. The thought of seeing it lost was too much. It just had to come out. First the motor was disconnected and taken ashore with all the other odd bits and pieces. Managed to get the counter shaft off, too, without cutting the belt. Felt proud of my self over this deal, as it meant dismantling the entire shaft. I didn't even lose the Allen screws, either.

Finally I tackled the big job. All the nuts off, the holding down bolts first and then all the bolts came out except one. As the last one came out, the lathe swung downward with its weight. I think the hardest job of all was to remove that last bolt and take the weight of the lathe at the same time, but I did it. Without a lot of detail to explain how, I did get it on deck and slung out over the bow, ready to be taken off. Now came the final test of all .....to get it ashore.

I clambered onto my favorite rock and reached up. Gradually I pulled it over until it overbalanced and slid into my arms. Just at that moment, one of those lousy seas came roaring in. With one foot on the rock and the other in shallow water, I braced myself for the shock. With a resounding crash, it hit YASME's stern. As I stood there, the bow lifted and she slid forward. As the seas receded, the bow came down smashing my instep. I screamed in agony as the bones crunched. A thunderbolt slid up to my brain. I was passing out.

As quickly as she held me, so I was released. I bit my lip to hold my senses. The foot was numb as I expected, but would it support my weight and the lathe, which was still clutched firmly in my arms? Gingerly, putting my weight on the foot, I moved shoreward. It felt normal, but I knew the worst would come and I hustled as best I could to get the lathe into a place of safety.

I dreaded to look at my foot until I had rested and lit a cigarette. My shoe had split right across and the instep was badly gashed. As I watched, the veins bunched themselves into a pile and the foot swelled visibly until it resembled a balloon. Then the pain began. It started with a dull throb until I was almost screaming with the agony. I sweated for a while and took deep drags on the old cig [and] it helped.

As with all pains, so this one abated until I could stand it without too much discomfort, but it appeared to me my night wasn't finished. Wet canvas and two wet blankets were used in a vain effort to make myself at home without any success.

Never did I realize it could be so cold in the tropics. After the initial effects of shock had worn off, I tried to rest, but found the seas were rising considerably and were lapping at my feet. This was most disconcerting, as I had little room left in which to move back. It was obvious to me the tide was rising. Knowing the rise and fall is slight in this part of the world, I was uncertain how much the sea WOULD rise.

We had an onshore wind and a full moon, both of which make bigger tides. It was time for me to look around but first I had to consider all the gear strewn about the beach. It would be heartbreaking to have it submerged after all my work to salvage it. I had to save it, yet was scared to move in case the pain from my foot made me pass out what a spot to pass out in with a rising tide!!! Regardless of this, that stuff HAD to be moved.

An island friend washes the bottom of the Collins 75A4 with fresh water.

Slowly I slid backwards until my back touched the cliff. Keeping my left foot from the ground, I used the right to gradually hoist myself up. Just the slightest weight on the left foot sent a twinge of pain up to the old brainbox, nearly causing my collapse. I clutched at the cliff face for support, hoping I could hold onto my senses. Tears came to my eyes with the pain, but the inevitable cigarette helped a lot.

I stayed upright much to my surprise. Looking at the cliff face, I could see a series of crevices which would hold the gear, but the problem was to get it there with a bum foot. As before, I started with the heaviest piece and by maneuvering my back against the cliff, I was able to stow the whole pile until the beach was cleared .... one small snag appeared. I had omitted to leave any space for me.

The moon was sinking fast and it looked as though I was STILL in trouble. I examined the face of the cliff for a way out, but no joy at all...it had to be UP. Have never tackled mountain climbing in my life before, but it looked as though I was going to have to learn fast, as the sea was now lapping around my ankles. How to climb an almost sheer cliff in the dark with a useless leg didn't occur to me at the time maybe it should have, but I was too intent to get away from that rising sea to even think of the stupidity of attempting such a climb under those conditions.

I looked around for the first hand hold and started up. The cliff was rotten and every second hold broke away soon as I put any weight on it. Every inch of the way had to be taken with my arms and an occasional push with my right foot ... the left leg just trailed along for the ride and became a damned nuisance. One of the things I learnt from that climb was: it is far easier to go up than down, particularly when the cliff is sheer.

I'll try not to dwell on the agonies of the climb. My brain was set in one groove and to even think of what was going to happen next was impossible. I just took it as it came and hoped. Suppose I must have been about 100 feet up when I decided to look down. Just that slight twist of my head caused me to lose a handhold, and for those few seconds, I thought my days were over. Several small rocks were dislodged, which dislodged greater ones as they descended, causing quite a landslide. The sound of those rocks hitting below sent a shiver through me, and it wasn't because I was cold either.

To say I was frightened is the understatement of the year. My whole body trembled with apprehension of what would happen if I followed those rocks down. I reached a point where the cliff started to curve over into a steep incline. I could see it was covered with some sort of green foliage, so I grabbed a handful. It was cactus which was hardly rooted and it came away quite easily. I might add, its spines did not leave my hand so easily, and the intense pain almost made me drop. I scrabbled again with my hand, digging into that shallow soil trying hard to make a hand hold. It was rough going as my arms were aching with the strain and I knew I couldn't last much longer at this rate. I rested a few minutes and then made a concerted effort. It had to be ALL OR NOTHING. Using my hands., nails, elbows, chin .... even my useless leg, I scrabbled forward and up like a limpet. With the concerted action of every part of my body, I moved up several feet and found a deeply embedded tree which I draped my body over, head and feet hanging down.

In just a few days the crashing surf had taken its toll of YASME II

The blood rushed to my head and the pressure on my stomach made me retch. I pulled myself around and used the tree as a foot hold. Just above it was a narrow ledge about 18 inches wide. It didn't lead anywhere but it was a place to stretch out and rest. Sticking out from the rock was a deeply embedded vine. I removed my belt, lashed it around my wrist and then to the vine. If I passed out, I knew I couldn't fall far.

Once able to relax, I hauled out a cigarette and drew the smoke into my lungs as though it were the elixir of life. From this position, I could see the YASME, now only a vague outline in the dusk which was enveloping the area. The base of the cliff was invisible so I was unaware of the position of my gear. I hoped the recent landslide hadn't damaged any of it or covered it up, but my main thoughts then were for my own safety as in my short stay, I couldn't see a way out from this spot.

With only a pair of shorts, I soon chilled. The perspiration soon dried and I started to stiffen up. The space was too confining to move around or even wave my arms. There was no argument about it. I was going to have to move as the chances of someone sighting me were remote .... remote! They were impossible the way I saw it! I figured the noise of the sea would drown my shouts if anyone arrived guess I figured lots of things all of which decided me to move on. It didn't matter a damn whether it was up, down, or sideways. I just had to get to heck away from here fast before I froze to death.

Reconnaissance was futile with the light almost gone; it had to be a hit and miss effort. Two attempts were made along the face of the cliff until I saw the outline of a ledge six feet below. It looked OK and I slid down not caring two hoots if I slid all the way to the bottom. I had reached the stage where nothing seemed to matter anymore. I was bitterly cold, hungry, and thirsty. My throat ached with every breath from excessive smoking. I hit the ledge with a shower of stones around my head, also landing on a large cactus. This did not improve my temper in any way. From here on in, things panned out a lot better. I discovered a vertical crack down the face of the cliff which, had I the use of both legs, would have been child's play to descend; but it had to be done the hard way. The rock face was stronger here, and I had little trouble with miniature landslides. Ultimately, I reached the bottom, the word ultimately means it took me the best part of two hours with the attendant frights, etc.

The crack apparently finished very close to YASME. I tried hard to figure out why I hadn't taken this way up. It wasn't until daylight I noticed, that it appeared an impossible climb looking up ... guess that's why I by-passed it in the first place. The gear was all safe and I found my half filled bottle of water. I rationed myself to a mouthful being a little doubtful of the future. I could have managed a gallon right then. My luck changed for a few minutes when I discovered a canned cake on the beach. Need I say that, without a can opener, I soon disposed of half of it which improved my spirits considerably.

My two blankets were soaked and I rigged them to form a screen as the wind was like an Arctic blast. It appeared to strike the cliff and funnel down with increased intensity just at the spot I lay. The tide had risen and started to fall off again since my rock climbing escapade, so I stretched out full on the beach and slept. It must have been around 8 a.m. when I awoke. The rising sun was hot... damned hot. I went to arise but my entire body had become rigid with cramp. I had been lying on one arm and it had gone dead. I felt like nothing on earth and were it not for the fact the sun was burning my eyeballs out and the rocks on the beach were doing their darndest to bore holes in my body, should have thought I was already dead. Gradually I moved around a bit and thoughtlessly went to climb to my feet. The left leg gave out and I fell down with a thump. Needless to say my language at that time would not be found in any dictionary, but it helped my feelings but not my confounded leg.

Looking down at the offending limb, I discovered my foot resembled a balloon. My shoe which had been on the large size now looked twenty sizes too small, as my foot spread out over the edge like a fat man's paunch over his belt - My shin and thigh were cut and I had a lump on my hip which comfortably supported my pants without the aid of a belt. My chest, legs and arms were one mass of minor cuts and abrasions, but my hands seemed to be the worst. They were bunged solid with cactus spines and somewhere along the line, I had lost a finger nail .... tried to get the spines out but they all broke off and the pain was something to write home about.

YASME had now swung broadside onto the seas and lay with her masts toward the cliff. Her main mast was touching the cliff and as though it had been prearranged for this moment, a sea lifted YASME, making her mast head light disintegrate. It also smashed the truck which supported the Parkstone Yacht Club's Burgee, and in the light breeze of the morning, it fluttered down onto the beach. Tears sprung to my eyes for those few moments I realized it would never fly again on YASME II. I limped over and picked it up. It is one of those things I shall always treasure although the memories will hardly be pleasant.

I had not maintained my usual skeds with Dick and wondered how long it would take him to decide something was amiss. It was far too early for him to start anything in the form of a search) but I knew that eventually he must realize something was screwy. I found quite a bit of driftwood on the beach and formed it into a small pile. I also discovered a can of gas which had floated ashore. I had the bright idea to make a fire to attract attention. This is fine when you read it in a book, but take it from me, it is strictly BUNK.

Around 10 or 11 a.m. what difference does it make anyway) I sighted a boat around three miles off. She was making good time and broadside on. I poured a liberal supply of gas over my sticks and lit it. For a few seconds there was a magnificent blaze which almost burnt my eyebrows off, then it fizzled out, leaving a faint smoke which wouldn't have made an ant sneeze .... quite frankly, I could make more smoke with my language. I thought once again about what I had read in books that damp wood would create lots of smoke. About all it did was to create a darned stink.

The boat came nearer and I staggered to my feet, grabbing the "white?" blanket. It was no easy matter to wave that blanket around, balancing on one foot but I may as well have waved my handkerchief for all the good it did. Those guys must have been blind and I cussed them good and hard for twenty minutes solid. Morning drifted into the afternoon. I had sighted six boats in all and even managed to get my fire going. The little smoke it made was blown up the face of the cliff with the increasing trade wind. Later I realized and saw for myself that YASME was almost invisible from the sea, owing to the high breaking surf and the cliff which did a fine camouflage job.

My eyes continuously swept the skies for a search plane. I knew I wouldn't hear it with the roar of the surf. I wondered if one would come at all.

By this time I had given up blaming myself for this predicament and blamed the entire world for not coming to this precise spot and rescuing me. I hadn't the faintest idea where I was other than I had covered 35 miles since leaving St. Vincent-this conveyed precisely nothing.

The old brain kept buzzing with ideas all of which were useless. Seemed to me I was going to have to stay put until some bright guy found me, but when that would happen was the $64,000 question.

I looked longingly at the rigs ashore, then I looked at the two gas generators still lashed to the deck of YASME. If only I could move around and get one of them ashore. It would be several days before I could move around the way I saw it, and those generators may as well be at the North Pole for all the use they were. The receiver seemed to mock me as it sat there in its little alcove in mute silence. The transmitter, the key, the microphone) all this and ample wire too for an antenna, but no power, and I thought of the future and how I would be able to get a generator ashore. How I would rig it up and send my message. I even had the wording all figured out as to what I would say.

The joke was - by the time I would be able to get that generator ashore, YASME would be broken up anyway .... what a laugh!!

I had another swig of water. Had to take it easy as it was almost gone. Tried to figure how long a man would live without water, not that it was too important anyway. Questions, questions, and more questions bombarded my brain with no answers. I lay back at the foot of the cliff and pulled the blankets over me and slept the sleep of utter fatigue.

I have the faint recollection of glancing at the luminous glow from my wrist watch which seemed to give me that faint ray of hope which is always prevalent when things look black .... tomorrow was another day!!!

Something knocked my shoulder and I unconsciously moved closer to the cliff, thinking a rock had fallen. Once again came the knock and something grasped my shoulder like a vice. Still half dopey with sleep, I pulled the blankets from my head and was immediately blinded with a white light shining directly into my eyes. Slowly coming to my senses, I heard the mumble of voices. In my semi-stupid state, I couldn't understand the language although it faintly resembled English. Some damned idiot kicked my bum foot. That really woke me. I screamed with agony and a flow of adjectives streamed from my cracked lips. I told whoever it was to turn out their bloody light and not to kick the hell out of me. The light continued to shine in my eyes, so came to the conclusion that the natives did not understand English .... or profanity either. I put up my hand and pushed the light away.

With the reflection of the light from the cliff, I could see there were several natives standing there, all speaking a brand of English which even after five weeks on the Island I have never comprehended. Seeing I was obviously alive, one of them thrust a piece of paper into my hands and then shone the light back into my eyes, just to be sure I couldn't read it. After a few more ejaculations, I did finally manage to convey to this guy that it would be a bright idea if he shone his flashlight in a place other than in my eyes. The message was short. It told me to keep a stiff upper lip and all that sort of thing and that help had arrived, etc.

I asked where I was .... Union Island. This conveyed exactly nothing and I was just as wise. The guy pointed to the piles of covered gear and asked how many bodies there. He looked real disappointed when I told him there were no bodies. Well there it is. There could be a lot more to this little story which I may tell you later if I live that long, but it's enough for you to digest at one sitting.

As a final to this little escapade, I am now in Grenada operating with an RX and TX which have been well and truly dunked in the sea. How they are working is something I don't figure but to date, I've cleared about 1,500 QSOs. Hope by the time you see this, I shall have worked most of you from Trinidad. Getting around in a native sloop from one island to the other is hardly my idea of pleasure, but it is making the expedition carry on and I hope the future will produce a YASME III and I can work you all from all those other rare ones throughout the world. -- Thanks for all your support, Danny VP2VB/MM, etc.

20· Genesis of the YASME Foundation

We all know of the misfortunes you have had with your various boats, but I think you will agree that from a businessman's point of view further investment looks particularly risky. - Bill Halligan, W9AC

YASME II was history without ever making it out of the Caribbean. Danny carried on, completing his Caribbean schedule, with 1300 contacts from Grenada between 25 March and 13 April 1959 (VP2GDW), then made 400 contacts as VP4TW, Trinidad. QSLs for these operations are stamped "LESS YASME."

The DX rumor mill was of course in full swing, with talk of a YASME III. QST's DX editor W9BRD, in Chicago, may have had an inside connection when he reported in May that "ZL1AV, visiting K9ECO [in Chicago] would like to wend his way homeward in the company of VP2VB/mm aboard YASME III. Danny now takes a breather as VP2GDW after salvaging some twenty crates of paraphernalia from the battered hull of YASME II, abandoned after becoming rockbound in the Grenadines early this year." QST May 1959.

In part of a letter that remains, Hallicrafters Company president Bill Halligan, W9AC, wrote to Danny in Port of Spain, Trinidad:

> I received your letter last week and I've been waiting for my son Bill's return from the West Coast before I undertook to send you an answer. I was reminded by Trav Marshall that we already have spent something in excess of $15,000 in helping you, and all of our fellows were of the opinion that the big effort and the greatest return in the form of publicity to our company for expenditures would be after you got out into the Pacific.
>
> Well, of course, we all know of the misfortunes you have had with your various boats, but I think you will agree, Dan, that from a business man's point of view further investment looks particularly risky.
>
> We are associated with a MATS (Military Air Transport Service - ed.) flight now going around the world which will last 45 days wherein we merely loan the equipment and have one of our own field men aboard, and I'm inclined to believe - and so is everyone here - that this will be tremendously productive for us from a point of view of public relations, and is rather the kind of thing on which we as a company should be spending our money.

President of the YASME FOUNDATION, Richard C. Spenceley - KV4AA

here is your first edition of the **YASME NEWS** . . .

> Now I know you say that I should view this as an individual, but as the founder and owner of this company....

Spenceley, as Danny's agent, received a copy of this letter. Spenceley's only publicity outlet was Don Chesser's DX Magazine; the two continued a cordial relationship (Danny was always good copy!) and Spenceley said for the benefit of Chesser's readers:

> After unloading all salvaged gear in St. Thomas, Danny Weil flew to Dayton to attend the Hamvention on May 9th. Some progress has been made towards YASME III, and about $7,000 would put him on the road again. ZL1AV, now visiting in the States, can accompany Danny on his trip should a boat be obtained within a reasonable time. His presence would assure ham tickets for such spots as the Tokelaus, Kermadecs, and any other spots under New Zealand jurisdiction.
>
> There is also room for another ham on the

YASME Foundation attorney James J. Lindsay, left, accepts the foundation's charter from Hillsborough County, Florida judge John F. Germany

> YASME, preferably an American. Should any one wish to make a substantial contribution toward YASME III it is possible that this contribution could be repaid at least in part. At the present moment Danny stands ready to speak before any radio club or other gatherings upon invitation. Such dates may be coordinated via QSO with KV4AA, who is on nightly, on 14082 kc, between 0100 and 0145 GMT. Please act now!

Chesser said "Danny, did, indeed, make the 1959 Dayton Hamvention with hours to spare, and he made an impromptu and unscheduled appearance at the DX Session Saturday morning, during which he described the harrowing job of salvaging the equipment from YASME II and the equally harrowing trip aboard a chartered vessel from the scene of the accident to KV4AA. It makes interesting listening for any club or DX meeting.

"Dave, ZL1AV, a quiet and shy lad, was also present at the convention. He indicated his willingness to go as far as New Zealand with Danny. In private conversation later, Danny said his next DX operation would probably start in the Pacific."

Charles Biddle, W6GN, makes his first recorded appearance on the YASME scene in the following letter, dated 12 May 1959, apparently sent to the amateur radio press and probably to major DX clubs.

Biddle was a prominent DXer, and a lawyer. Informal talks among prominent amateurs had been taking place through the spring, leading to the formation of a non-profit foundation to support yet another YASME voyage.

> At the request of Dick Spenceley, KV4AA, I am sending the attached photographs and the story of the last voyage and wreck of YASME II as written by Danny Weil [the story, above, that appeared over several issues of DX Magazine]. As one of Danny' s close friends, I feel privileged to write this letter, and believe that I can speak with a good knowledge of the problems Danny is facing.
>
> Amateur radio operators and other friends the world over share the shock and sorrow at the dismal end suffered by Danny and YASME II. This sorrow can only be buoyed by the "never say die" spirit of Danny, who, in spite of personal injury and the discouragement brought about by the unfortunate turn of affairs, is still willing and anxious to carry on.
>
> Danny will continue his trip and will make all of those interesting DX spots provided he can receive sufficient cash donations to obtain a new boat and necessary equipment. Past donations to Danny have never been more than adequate to meet the expenses of the expedition. Equipment salvaged from YASME II will bring some cash recovery, but the large part of the funds for another start must come from your new donations.
>
> Danny has investigated available boats and believes that he can secure one for about $12,000.00. This is an absolute minimum that will be needed to start again. Some equipment will be donated by manufacturers, but the balance will have to be purchased. Danny will take at least one other crew member with him the next time, so that he will be able to insure the new boat.
>
> I am certain that there are some of Danny's friends that are willing and able to make a substantial donation. Some of this larger type of donation will be absolutely necessary. There are many who would like to help that can do so only in a smaller way. Any donation, no matter how small, will be appreciated by Danny and all of us who are trying to be of help to him.
>
> Danny, Dick, and I will appreciate all the publicity this plea can be given by you. Donations should be sent to R. C. Spenceley, "Dick", KV4AA, Box 403, St. Thomas, Virgin Islands, U. S. A.
>
> Danny will attend the Dayton Ham Convention, and after its close, will be available for speaking engagements. Please contact KV4AA on 14082 between 0100 and 0145 GMT for Danny's time schedule. Let's get Danny' s show on the road again! 73 - Charles Biddle, W6GN.

A Bold Venture -- The YASME NEWS

"I shall try to inject just a little of the adventure and excitement of the voyage into your lives with three of us

along to bring rare DX into your homes." - Danny Weil

For a writer - both Danny Weil and Dick Spenceley were writers - there is no thrill like starting one's own publication, then waiting to see if it will get off the ground, if it will produce some income. The inaugural issue of The YASME NEWS was mailed to some 4000 addressees (Vol. 1 No. 1 September 1959). The issue, dated September 1959, was mailed in mid-summer (for longer shelf life).

Smart businessman Spenceley and his fellows allowed themselves enough time to generate some advertising and a couple of friendly letters for what was, basically, a fund-raising vehicle. Always a good sport, Chesser refers to portions of this first issue of YASME NEWS in his DX Magazine in June. The masthead announced "The YASME Foundation, A Corporation not for Profit, Box 13165, Tampa 11, Florida; "Fla. Stat. 1941, Chapter 617."

Officers were President: Richard C. Spenceley, KV4AA; Vice President and Treasurer: Edward A. Stanley, W4QDZ; Secretary: Golden W. Fuller, W8EWS; and Directors Charles J. Biddle, W6GN, Golden W. Fuller, W8EWS, William J. Halligan, W9AC, J.E. Joyner Jr. W4TO, Edward A. Stanley, W4QDZ, Richard C. Spenceley, KV4AA, James H. Symington, K4KCV, and Danny Weil, VP2VB.

Curiously, this first issue of the foundation's official organ announced the foundation's genesis in a "Foundation Report " buried on page 14. The time and place of the beginning of the YASME Foundation was put at The Club Boyar in Chicago on the 20th of May, 1959. At dinner with the guest of honor, Danny Weil, were Mike Hexter (W9FKC), Fritz Franke (Hallicrafters), Pete Morrow (W1VG, ARRL), Bill Halligan, Sr., (W9AC), and Spenceley.

"The honored guest was not in too good a frame of mind," Spenceley said. "The YASME II was lost, expedition funds were low, and the sheer magnitude of the job to be done pressed heavily on the minds of all of us. Things just didn't look good. The discussion rolled back and forth all evening. The past operations were taken apart and viewed from all angles. The future possibilities of continuing the DX-peditions were pro'ed and con'ed. The evening ended on a note of "look and see."

Danny and Spenceley made the rounds in Chicago, meeting hams at lunch, at the Radio Parts Show, in the ARRL suite. "We decided one thing," Spenceley said. "The Hams Wanted The DX-pedition to Continue. And that was the one thing we had to know."

Spenceley and Danny returned to Tampa, "having successfully avoided any and all reefs," and went to work on a plan for a foundation. The "conferences up north had helped bring into focus just what was needed to put the YASME III in the Pacific by Christmas 1959."

They decided there would be no more lecture tours or hat passings, and that a foundation operating as a non-profit organization, governed by a board of directors, with its business affairs being conducted by a Business Manager, would be the best way to get the new expedition underway. A Charter was drafted with the advice of the Foundation Attorney, and it was then sent to its subscribers, who were: Golden W. Fuller W8EWS, Flint, Michigan; Charlie Biddle W6GN, Los Angeles, Cal.; Dick Spenceley KV4AA, Saint Thomas, V I., U. S. A.; Bill Halligan, Sr., W9AC, Chicago, Illinois; and Ed Stanley W4QDZ, Tampa, Florida. The charter made the rounds and on 31 July 1959, it was approved in the Circuit Court in Hillsborough County (Tampa) Florida, by Judge John Germany.

Meanwhile, in the six months since Danny had left the air, YASME was rapidly fading from the consciousness of DXers. Some data-crunching was called for, and the log books from the West Indies operations were used, along with a Callbook, to generate a list of names and addresses, along with a record of contributions that had been made. Of some 34,000 contacts made, around 2600 hams had contributed money, at an average of $5.50 each. "On this known information, Spenceley said, "we had something to go by. There were 2,600 active supporters of the DXpedition who had put their best into making the project go. Odds on, they would do it again." These 2,600 contributors would likely constitute the new foundation's initial membership.

In order to finance Issue One of the YASME NEWS, "loyal supporters in the jobbing and manufacturing field" were approached, and advertising was sold. Danny Weil, an expert and fast typist, typed the mailing labels.

Issue One said the purposes of the YASME Foundation were "very clear": to "contribute to the advancement of amateur radio as a scientific and educational medium, to assist those handicapped hams in need of the therapy and pleasure which ham radio can give as nothing else can, to create international good will, to con-

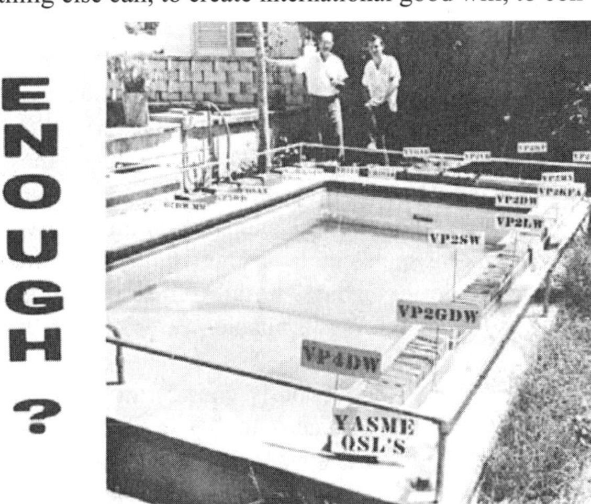

This photo accompanied Spenceley's description of how the foundation would handle QSLs

duct scientific explorations in the fields of oceanography and hydrography, and to share its findings with every one regardless of race, creed or color."

The main project would be the "care and feeding of the YASME DXpedition." This time a crew would be aboard (memories of two trashed YASMEs still in DXers' minds), namely 22-year-old Joe Reisert, W2HQL, of Wantagh, New York. The YASME III was announced, berthed in Clearwater Marina, near Tampa, and had "accepted Danny's recommendations on her." The asking price was $17,500 and some modifications would be needed. "With fast response from our request for contributions we can have her South of Panama by Christmas."

One of the new foundation's greatest assets was Danny Weil's writing talent, immediately put to use in the first installment of "Danny's Diary":

> In my opinion, diaries are for the past and strictly for the mermaids. Right now we sure need a little of the future to whet our appetites. Facts about new developments and fabulous spots scheduled for our next voyage which will even invoke interest to the XYL.
>
> We can't all go along but, through this column I shall try to inject just a little of the adventure and excitement of the voyage into your lives with three of us along to bring rare DX into your homes. Working 24 hours a day on all bands - - yes, even 6 [meters] too, I know that many will think it easy to work all spots but, I can assure you from bitter experience that many still fail to make the grade even though a KW is available.
>
> Incidentally I am seriously considering applying for a license to operate from Tampa. It was never intended to be a rare one but, with the great rains, we are slowly becoming an island. Volunteer operators please bring bathing trunks. Suggest hip boots too. Heck! Better bring aqua lungs and be absolutely sure.
>
> It's rough typing with water up to my waist - can't even keep a cigarette alight with the rain pouring through the roof. Am wondering if I can finish this before the shack drifts away. Just in case of emergency my present position is Lat. 270 52 N. Long. 820 29 W.
>
> To get back to what I started to say before I was rudely interrupted by Ed swimming for his pants. Assuming by some stroke of genius we reach St. Thomas [and] KV4AA, we shall head for Panama complete with the hurricane which invariably follows me. With luck we may even manage to traverse the canal which will undoubtedly do its worst to break every line and sink the boat as it did once before. As an Hor doeurves we'll begin with the Galapagos (HC8) where we shall enjoy ourselves with the dragons on Albemarles. They are quite tame apart from the annoying habit of biting off a leg or an arm. I shall ensure that all crew members can operate with either hand before starting the trip and a goodly supply of wooden legs taken aboard.
>
> Naturally we shall fool with a little DX as a warm up for really serious work on Pitcairn or Easter Island. I know that few of you really need the Tokelaus as there have been natives operating there for generations but, as there is no anchorage and the island is surrounded by reefs, I thought it would be a good spot to test out YASME III and see if she could stand up to a spot of reef work.
>
> Tahiti is next on the list but we shall only be there for a short sojourn of five years or so to refuel. Really I hate to waste time at such an uninteresting spot and I feel sure the crew will be terribly bored. The Kermadecs might be useful to some and perhaps Wallis Island. They tell me it is owned by the French and I sure don't wish to lead the crew astray.
>
> I think the real fun will begin when we get through the Torres Straits and reach Timor. I know that somewhere along there is Bali and they do have some wonderful carvings ... They prefer wood but sometimes when the mood gets them, an odd ham or two makes them feel good. To Borneo via the Malacca Straits could create a diversion for the crew. I do sincerely hope they are well trained in the use of firearms and have been inoculated against every known disease. Am just wondering if there are any volunteers who will stand by in

Ceylon to replace the crew who didn't quite make it. By keeping the boat afloat after getting this far ... we shall install 6 motor driven bilge pumps as I know the hull will be almost eaten through by teredo worm by now, we may reach Chagos Island. Here we may need a helicopter to get ashore so will put a call through to "Whirlybirds, Inc." express delivery.

Seychelles, Aldabra, Zanzibar will all come in their turn and finally Ascension and St. Helena ... which will be enroute back home again. More about the future later. - 73, Danny.

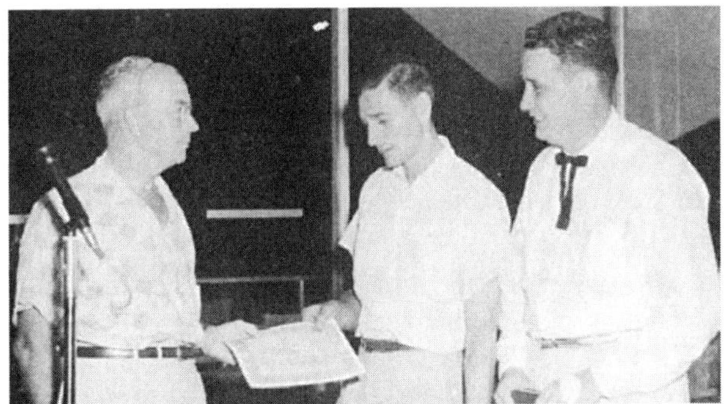

Danny gives Bill Halligan a YASME souvenir. Foundation treasurer Ed Stanley, W4QDZ, is at right.

Friendly letters to the editor are always good for drumming up support (especially in the first issue), and here are a couple:

Thank you for the autographed scroll on the YASME II. I have typed in all the contacts made with on this expedition, including the one from the boat at sea near Madeira. This makes a complete log of your trip, and I worked you from every stop you made. Just how many other fellows accomplished this I don't know, but I don't suppose too many at that, Danny! This scroll is being framed, and will take its place on my wall with numerous other SSB Awards.

I am enclosing . . . for a membership in the YASME FOUNDATION, and think it's a heck of a good idea. Too many amateurs sit on their rear ends and gripe all the time, but don't do anything constructive. I always find that those who do something are always being criticized by somebody who does nothing.

Sure, Danny, this is an unusual deal for you to go around the world in a boat and give all the hams new countries, etc., and you have ALWAYS QSL'd the boys, and handled these contacts in a fast, efficient and gentlemanly manner. You can't say that about a lot of the other expeditions, as preparations were never like your trips, and BESIDES - you always had a powerful signal that we could hear and work

I am not in the running with the top DXers, as I don't have the time. I know I have the signal and the equipment to do it, but it is still a hobby with me, and I want some FUN out of this ham radio. I have 132 countries confirmed on Two Way SSB, with DXCC, WPX, CQ100 Country, WAC, TPA, 101, CCC, and sending for BERTA, also. I have numerous other awards, too, so I am not too bad regarding SSB, and I have only been on SSB for about three years. I have been on the air since 1913, and I have had lots of fun during these years and active in many, many things including our new Amateur Division at FCC, when four of we amateurs appeared before the full FCC Commission in 1948.

I should be happy to receive your YASME News, and if I can help out with some news-it would be a pleasure. Do not be discouraged with adverse reports on the YASME Foundation, which many amateurs will call sour grapes commercial interests-and nasty names. Just go ahead and do a job-get your boat. I am sure with prompt response to the QSL cards as you have always done in the past, and Dick at KV4AA helping out, that whatever venture you make will be a success.

I just ran over a bunch of wire yesterday with my new mower-I am sure my wife doesn't need a new skipper at the house, just because I had an accident, too.

Best of luck to you and your crew, and it does take $$$ to do a job, Danny, so don't be bashful about the YASME FOUNDATION-let the world know they can HELP, too. There are too many amateurs who are willing to take all but do not care to help It's SO EASY to sit on the sidelines and call the shots, but it's a heck of a lot different when you are in the game. Chas. W. Boegel, Jr., W0CVU, Cedar Rapids, Iowa.

. . . and another friendly letter:

The YASME Foundation is a good idea and think it is the best way to get going again. I go along with Dick on the idea of not trying another [speaking] tour ... considering all phases I think you will do better with the Foundation idea. ... You need some publicity in the DX columns of QST, CQ and all the rest of DX club bulletins throughout the country. There are still plenty of places to be visited and with fall and winter coming along it would be timely to spread as much news about the new YASME as possible. 73 - Jim, W8JIN.

On page 18 of this slick, 20-page magazine, Spenceley pictured the new YASME III. 3,500 copies of Issue One were mailed to hams, distributors, and manufacturers. A minimum contribution of five dollars was asked for, to cover the projected $20,000 cost of fitting the new boat having it underway to the Pacific by Christmas 1959.

Danny, meanwhile, made his way to the Bahamas (keep in mind he was "without boat"). He put VP7VB on the air Sept 28 to Oct 5 1959, and the log book shows 850 contacts.

The YASME NEWS Vol. 1 No. 2 October-November 1959, opened with a controversy over Danny's license.

The controversy erupted in the fall of 1959 as the result of bogus information that found its way into print. An item in the September 1959 issue of Short Wave Magazine (published in England, G2DC, editor) was, as is unfortunately often the case, quoted in other publications, including in the Sept. 30 issue of Don Chesser's DX Magazine (under the unfortunate headline "Danny Reported in Trouble Again").

A brief in Short Wave claimed that "just as we were going to press" the British Colonial Office had withdrawn all amateur "facilities" from Danny and that he was no longer allowed amateur operations under his British Colonial call sign, VP2VB. The brief note said the reason was "Flagrant abuse of amateur privileges," and Short Wave admitted to not having "official confirmation."

G2DC was asked to "run this rumor to earth" and he attributed it to a California ham. "This W6 has been quite peeved at Danny ever since he signed VP2KFA during his DXpedition to the Island of Anguilla in November 1959," Spenceley said in the YASME NEWS. Spenceley said that "the W6" considered Anguilla his private stomping ground and that he had "pirated" the call sign VP0RT from there in 1958. "At that time we were quite content to let him "have" this island," Danny reported, after visiting Saint Kitts, which governed Anguilla, that he was told the authorities had no knowledge of the VP0RT call sign, and that they had authority to issue calls only in the VP2K, and were furthermore not authorized to grant licenses to U. S. Citizens.

Spenceley advised Danny at that point to "proceed to Anguilla with all speed. Our connections with the YASME trips have been close and we are in a position to state that his licenses have been granted by the proper anthorities and his operations have been in keeping with the best amateur traditions," Spenceley said.

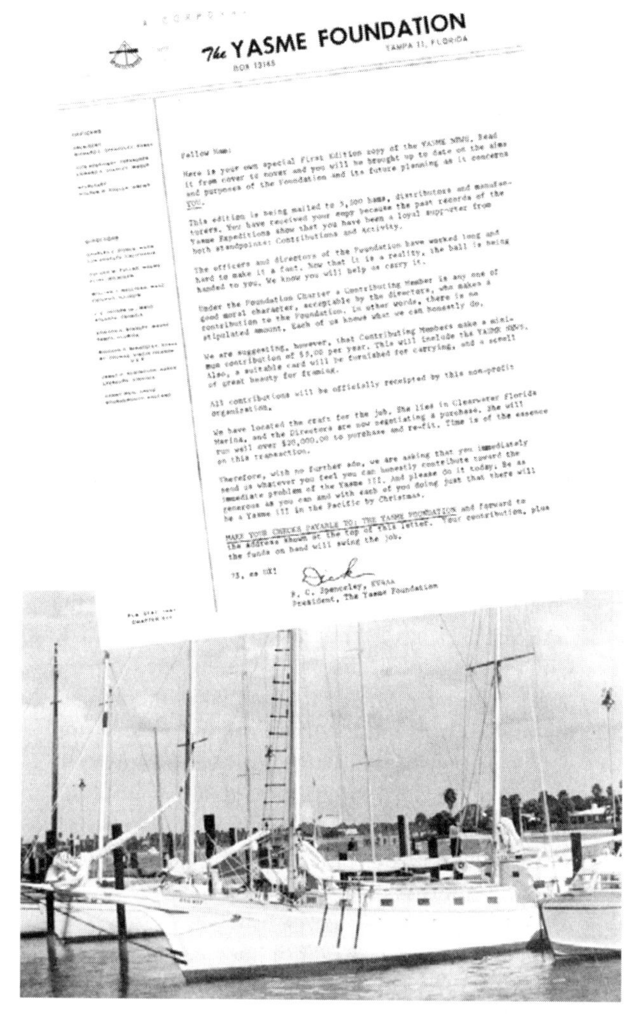

The YASME FOUNDATION
BOX 13145
TAMPA 11, FLORIDA

Fellow Hams:

Here is your own special First Edition copy of the YASME NEWS. Read it from cover to cover and you will be brought up to date on the aims and purposes of the Foundation and its future planning as it concerns YOU.

This edition is being mailed to 3,500 hams, distributors and manufacturers. You have received your copy because the past records of the Yasme Expeditions show that you have been a loyal supporter from both standpoints: Contributions and Activity.

The officers and directors of the Foundation have worked long and hard to make it a fact. Now that it is a reality, the ball is being handed to you. We know you will help us carry it.

Under the Foundation Charter a Contributing Member is any one of good moral character, acceptable by the directors, who makes a contribution to the Foundation. In other words, there is no stipulated amount. Each of us knows what we can honestly do.

We are suggesting, however, that Contributing Members make a minimum contribution of $5.00 per year. This will include the YASME NEWS. Also, a suitable card will be furnished for carrying, and a scroll of great beauty for framing.

All contributions will be officially receipted by this non-profit organization.

We have located the craft for the job. She lies in Clearwater Florida Marina, and the Directors are now negotiating a purchase. She will run well over $20,000.00 to purchase and re-fit. Time is of the essence on this transaction.

Therefore, with no further ado, we are asking that you immediately send us whatever you feel you can honestly contribute toward the immediate problem of the Yasme III. And please do it today. Be as generous as you can and with each of you doing just that there will be a Yasme III in the Pacific by Christmas.

MAKE YOUR CHECKS PAYABLE TO: THE YASME FOUNDATION and forward to the address shown at the top of this letter. Your contribution, plus the funds on hand will swing the job.

73, es GL!

Dick
R. C. Spenceley, KV4AA
President, The Yasme Foundation

The back cover of the first issue of YASME NEWS. The promotional letter from Spenceley is accompanied by a photo of the boat they hoped would become the next YASME.

Spenceley bemoaned the "unilateral vindictiveness" of this W6, saying "It is unfortunate but it is a risk we will have to put up with."

Spenceley further said he had contacted the administrator of the British Virgin Islands and was told that Danny's VP2VB license "is quite secure and may be used whenever Danny chooses to operate from that spot. Danny had no trouble whatsoever in procuring his VP7VB ticket under which he operated from September 29th to October 5th. These facts further refute the validity of this item" (in Short Wave). Spenceley, in the pages of YASME NEWS, took the editors of Short Wave magazine to task for "printing such an, admittedly, unsubstantiated item when they could not but be aware of the short range derogatory effect it would have on the work we are doing toward the YASME III trip.

"We doubt that the influences of this W6 extends into the British Colonial Offices to the extent that they would close down Danny without informing the principals in the case and without an equitable hearing."

Spenceley also reprinted a reply from Austin Forsyth, G6FO, managing editor of Short Wave, who said that the report was "accepted in good faith. It reached us as a last minute item, and was written into "DX Commentary" by me - as I often have to do - just as the issue was closing for press. A check subsequently with our Colonial Office here in London showed that the information was not correct. We had therefore already decided, before your letter was received, to publish the necessary correction in the next available (November) issue of the magazine.

"I regret it if the paragraph in question has in fact caused you any difficulty - but if this is so, I am sure the correction in the November issue will put it right. May I just add that difficulties of this sort would not arise if we were given, at this office, factual information about VP2VB and his plans and movements."

Chesser dutifully reprinted Spenceley's letter, as well as G6FO's reply, saying he "concurred" with Forsyth's last paragraph.

Finally, an apparent end to the saga came and was reported in the December YASME NEWS, Spenceley reporting that he had received a visit from "VP0RT, who "Assured us that he did not write Short Wave magazine, that his call, VP0RT, was legally granted, and that he bears us nothing but goodwill."

Spenceley said "we accept this in the spirit it was given and apologize for any inaccuracies which might have appeared in our letter."

Despite the time taken up with this fiasco, the second issue of the slick, new YASME NEWS came out with a new department for women radio operators, penned by Harriett, K5BJU, who appears in her radio shack on the cover.

Spenceley reports on a "policy meeting" of the foundation there in St. Thomas September 11, 12, 13. The foundation's by-laws were finalized, and the directors put their stamp of approval on the handling of income based on the recommendations of director and founding member Charlie Biddle, W6GN. A maximum of 25% of income would be allocated to administrative functions, with 75% of contributions and other income going to a special account "for the purpose of purchasing the boat (YASME III) and furthering the forthcoming DXpedition through subsistence and maintenance."

The by-laws allowed the foundation to set up "Field Chapters, for the purpose of creating and maintaining strong membership activities which will contribute to the well being of the YASME Foundation." This was said to be in response to "several requests" for such chapters. The chapters would operate under "Special Charters. Each charter would be required to "be in full compliance with the Charter of the YASME Foundation and also must conform with the rules and regulations affecting such Chartered organizations in whatever geographical or political subdivision in which such a special charter may be granted."

Several directors offered loans to the foundation, to be repaid on the basis of 50% of the income until paid. Negotiations were underway to buy the "Sans Regret," a boat similar to YASME II, for $14,600. The "Sans Regret" was chosen because she would accommodate spare parts, sails, and so on that were salvaged from the YASME II off Union Island. Danny found the boat in Ft. Lauderdale, and approved it following a "survey." Ten percent of the price had been paid down, with an additional payment on delivery to bring the balance to $10,000.00. The terms of the purchase permitted the new owners to bring the boat to Tampa for outfitting. It was not expected that it would be possible to get insurance for the boat. "Our own insurance will be its crew," Spenceley said.

The previous issue of the YASME NEWS, the first, went out to 3,500 persons, at a cost for printing and postage of some $1300. It produced advertising income of $630. Obviously, more advertising was needed, and the second issue was, indeed, fatter.

Foundation income ending September 30, 1959, was $6,031.13 (including contributions of $5,922.73). Disbursements were $3,077.48.

In this issue, Danny Weil wrote from the Bahamas that

> Dick has shipped my old rig up from Saint Thomas as it was admirably suited for this type of work of being fully operative whilst submerged, but we decided that the trip to Nassau was the real thing.
>
> My passage through the red tape of officialdom at Nassau was smoothed out considerably by the all out cooperation of customs, and within 30 minutes after arrival in Nassau all the gear was stowed aboard Charlie's (VP7CN) Austin, the smallest of all British Cars. It was night and no street lighting. I sat astride the hood and Charlie complained he couldn't see. Neither could I with my eyes full of flies, etc., but leaning to one side on my part, and Charlie sticking his head out of the window, we ultimately arrived at the 'Shack.'
>
> I have operated from practically every type of QTH in the world, from a cave to a coffin makers workshop, but never, never, before have I worked from an Ice Plant. A friend of Rd's had offered us facilities there, and apart from the slight thumping of a few hundred ton compressors, everything worked out swell. The ice storage room was offered as a shack, but I had failed to bring my igloo and fur lined pyjammas.
>
> Stringing up my special long wire, trapped antenna, sent by Bill Lattin, W4JRW, I started punching

out a few holes in the ionosphere as VP7VB. All bands were wide open, but skip was rough from Europe and few QS0s were made in this direction. QSOs were not made in quantity, but the quality was FB and for the first time I had an opportunity to rag chew with my pals. It was great to be back on a rig again and there are a few we have to thank for this.

VP7CN set the ball rolling, and Allan, the Telecommunications Officer assisted with the license, and last, but not least was Chappie Chapman at the [power] plant, who gave us a fine QTH and power for free. Returning to Tampa I had anticipated a few days rest, but the call to go to our new boat came through and off I went to Fort Lauderdale. A trial run had been arranged, but a part of the gear box had jammed up and broken. The yard mechanics removed it and then came the snag. Steel strike, no material. Much scrounging around at the local scrap yard produced a chunk of steel and at the time of this writing, the work is going ahead. I did want to tell you about the trials, etc. but the tight publishing schedule has prevented these plans, so will give you a word picture of your new DXpedition boat.

She is the sister ship of the YASME II. Ketch rigged, with fifty foot mainmast and a fine diesel auxiliary power plant similar to the YASME II. Her accommodation is slightly different too, a little better than YASME II although I hate to admit it. Let's board her and see how she is laid out:

In the stern there is a hatch leading into the lazarette. This is the place where all those things that might be useful are stowed and is rarely opened other than to stow more junk. Next hatch is very uninteresting as one can only see the steering mechanism. Then comes the cockpit only used on fine days and holidays as all steering is done from the deck house. She has a small deck house with windows all around and fitted very well with ship to shore rig, depth finder, wind speed indicator and other gimmix and gadgets too numerous to mention. Here we steer the boat and do all chart work. And, I think we can fix our sea going rig with room to spare in this combination chart-room, deckhouse and ham shack.

Going aft from the deck house we go below to the after cabin. This is the luxury cabin of the boat. There are two bunks ... one port, the other starboard. Up the ladder into the deckhouse and down again we came to the forward saloon. Here we have two more bunks, a table, and thence forward is the galley where I and the crew will do our worst operating. Through another door we head still further into the forepeak which has two bunks. This part of the ship has the reputation of having the most uncomfortable berths. (Joe and Dave are already allocated bunks here to toughen them up during the first year of the voyage).

The fore peak has a variety of uses, but mainly holds the anchor chain and sails. I feel that we shall turn it into a workshop, as the existing workshop aboard is now in the engine room and could be slightly warm.

Incidentally, the engine room is below the deck house and is entered through a sliding door from the saloon. I have managed to take a few fotos of the exterior but, with the interior in a state of survey, I am reluctant to shock you with it. After survey, if she passes, I will bring her back to Tampa for fitting out. You are all welcome aboard at any time to view and work on your new craft. So, I am looking forward to many visits and the chance to fill up many visitors books.

As time is short and there is lots to do, I will pull the big switch now. Thank you all for what you have done. With our crew, Dave and Joe (ZLIAV and W2HQL), we shall try to justify your faith and helping hands. When I bring the YASME to Tampa my love for all of you will ride the helm with me. - Danny VP7VB/MM [maritime mobile].

21 · More Crew, and Radios, for YASME

One of these fine days I shall be able to do things in a leisurely way and have many people at my beck and call to carry out instructions but, until that day arrives, I shall carry on doing things myself and getting them done. - Danny Weil

Since Danny's agenda was no longer to sail the globe alone, some help aboard YASME III was sought. The first to sign on was Joe Reisert, W2HQL (now W1JR), who had accompanied Danny on the VP2VB British Virgin Islands operation in July 1958, "where his work was observed by the chief op, Danny Weil," as Spenceley put it. Joe was 23, the oldest of 10 brothers and sisters, and lived on Long Island, with already plenty of amateur radio credits to his name. Joe was a communications technician in the Naval Reserves and worked for IBM in Poughkeepsie, New York.

Joe was an Eagle Scout, first licensed at age 14. He joined the Naval Reserves at 17 and entered Long Island Agricultural and Technical Institute, about the same time he got his DXCC, a Second Class Radio-Telegraph and First Class Radiotelephone license, and his Amateur Extra license (a tough one in those days, that carried no extra privileges over a General class ham ticket).

Also signing on to YASME III from Down Under was Dave Tremayne, ZL1AV, age 22, from Auckland, New Zealand. Dave was first licensed in 1954 as ZL1ANM, then transferred to Wellington, where he picked up the call ZL2CV. In the latter part of 1957 he moved back into ZL1 territory and was given the present call ZL1AV.

Dave described his main interests in life as amateur radio, DX, and traveling. He had been in the New Zealand Air Force and was a commercial radio operator for the New Zealand Post and Telegraph Department. Like Reisert, he was a young hot-shot operator, another "born ham."

Dave was visiting in the States when he got wind of Danny's misfortune on Union Island and offered his services on the coming cruise. In May 1958 at the Dayton Hamvention he met Danny, as well as Ed Stanley, W4QDZ, the foundation's treasurer. Dave subsequently went to Saint Thomas, where he helped KV4AA sort out the gear salvaged from YASME II. He returned to the mainland, where he visited prominent DXer K9ECO and waited for YASME III to materialize.

The YASME NEWS Vol. 1 No. 3 December 1959, came out quickly, the editor noting that the foundation was finally getting some mention, in CQ and QST in particular. A paid mailing list was taking shape and an Addressograph machine was found (high tech in 1959).

The Financial Statement through October reflected payment of $10,000 on the total cost for the "Sans Regret" (YASME III) of $14,600. Income from contributions and advertising was about $1,400. This was boosted by a loan of $5,000 from one foundation officer, $1,500 from another, and $1,000 from an equipment manufacturer (none of these was identified, then or later). The foundation pledged the boat to the note holders as collateral, and the final payment due December 1, was made as the issue went to press. Expenses other than purchase of the boat and printing and mailing of the YASME News were inconsequential: "Office Expenses" were listed as one dollar and thirty seven cents!

Addressing the meager response and income from contributors (now referred to as "Contributing Members"), Spenceley said "We are sure that this has been in many cases due to the feeling that we had picked out an impossible task. Well, I am

August 2nd 1960

Mr. Travis Marshall
Sales Manager
The Hallicrafters Co.,
5th and Kostner Aves
Chicago 24, Ill

Dear Trav,

Refer my letters of July 28th and 29th regarding payment for ham equipment for the YASME expedition.

Please find enclosed a Foundation check to the amount of $1420.00 in full payment thereof.

Please return my personal check for $355.00 which was enclosed in my letter of July 28th.

Attempts to reach you on the telephone have revealed that you will not return to the plant before August 8th. I trust that full instructions have been left with your assistant so that our order may be air-shipped with a minimum of delay.

Very truly yours,

R.C.Spenceley, KV4AA
Pres:

Copy-W.Halligan Sr.

SHIPMENT GOES VIA AIRFREIGHT TO:

CAPTAIN DANNY WEIL, YACHT YASME III
RODMAN NAVAL STATION, CANAL ZONE.

sure that the presence of the boat, with ZL1AV and VP2VB aboard, fitting her out like mad, will certainly be incentive to all of us to roll out the barrel, and get the show on the road."

In addition to YASME NEWS, members received a membership card and contributors scroll. The scrolls were delayed because they had to be over-printed: YASME II had become "YASME III."

Writers were not paid. A Novice editor was sought because "it is our feeling, along with a large percentage of our membership, that the time to shape the future operating activities of today's new hams is to get them in the swing of things, and there's no better way than through DX, traffic, and those other activities which can be brought to their attention."

Joe Reisert, W2HQL, was forced to bow out of the upcoming Pacific cruise, citing his new job with IBM just not allowing the time. In his place Jake Sharp, W8LNI, volunteered. Sharp, from Detroit and a medical doctor, was 28 and single, and had been a U.S. Air Force flight surgeon. He would later back out, too.

Joe recalled that "In 1959 Dick asked me if I would join the YASME on the trip to the Pacific. I also found out that Dave, ZL1AV, was also enlisted. Dave had settled in at K9ECO's home for a long visit so I took a flight out to K9ECO's house to meet Dave and see what would transpire. I can't remember what happened next but I was now employed by IBM and doing well. I eventually did not go. Dave did go but got testy with Danny on their second stop, the Canal Zone, and Danny accused him of theft and had him deported back to New Zealand!"

"W4KVX Burnt Out"

Spenceley's headline said it all. Don Chesser, W4KVX, had suffered a devastating fire that destroyed the outbuilding at his home in Burlington, Kentucky, that housed his DX Magazine. The fire, on November 28, 1959, destroyed everything. This was a blow not only to the tireless Chesser, but to the YASME Foundation, since Chesser was a vital conduit for foundation news that went to subscribers who did not contribute to the foundation.

According to a letter to Spenceley from several of Chesser's local ham friends, they heard the town's fire siren, called the telephone operator, and learned the location of the fire. The information was immediately put out on the group's 6-meter calling frequency and the locals hurried to Don's place, but "there was nothing to do and nothing to save. All that man had worked for and saved in more than twenty years was rapidly being consumed in a roaring fire."

Those same locals said "W4KVX, Don Chesser, is a Ham's Ham. He was licensed in 1933 and since that time has given much to amateur radio. He has undoubtedly given as much as any other one amateur and, perhaps more than most. About two years ago, Don conceived the idea of a weekly DX magazine. He felt that most of the gang would be more active as Radio Amateurs if they knew when and where DX could be heard and worked. So, as always, Don gave his all to improve Amateur Radio."

Danny Weil and Bill Halligan, at the Dayton Hamvention.

The YASME News urged everyone to "send what you can" to help Chesser get back on his feet. Meanwhile, the only genuine DX newsletter in the world was not appearing in DXers' mailboxes.

"Danny's Diary" continued to be a regular feature of the early issues of YASME NEWS, and in this issue Danny related sailing the new YASME III from Fort Lauderdale (where they had found it) to Tampa, via Florida's inland waterways, in company with Ed Stanley, W4QDZ, the foundation's treasurer. Danny's tale was mostly about fixing all the niggling problems such as seized and rusted hardware, not to mention "a big hole in the deck." They found three cases of canned goods below deck but it turned out to be dog food.

What's a membership journal without letters to the editor? Not much. While the choice of letters to publish can be self-serving, Spenceley generally tried to show both sides of an issue and, as always, to fuel the DX Fires. Here's a letter from December 1959:

> Relative to your request for contributions to the YASME FOUNDATION, may I suggest action on your part which should assist you greatly in your effort.
>
> At the present time there is considerable, and rather strong feeling that the expedition is just a racket to support Danny and Dick. Running under the guise of a FOUNDATION does not alter the picture. I travel all over the East and this issue has been discussed frequently with groups of top DX men. The general feeling is one of disgust over what appears to be commercialism which dominates his adventures. Specifically, the return of

one's own QSL card, rubber stamped "contact confirmed" unless a contribution had been made. No other expedition has ever refused a QSL even though assistance was "welcome and appreciated." At least a card came even if via bureau.

I have contributed a dollar for each QSL plus $10.00 toward the purchase of his second boat. Perhaps you can imagine my feeling when the answer to one card with another dollar resulted in my getting my own card back marked "no contact" when the QSO was solid as a rock with both reports 599 and the time in GMT to no excuse - he kept the dollar. I know of two hams who sent cash to help purchase his second boat but who had their own cards sent back marked "contact confirmed". He will not get any more from them.

In my humble opinion, if a man were to be given a credit toward future QSLs for a contribution now, it would help you raise the money needed at this time, Think it over. I know the boys think that 1000 contacts a week at one buck a card can make a pretty good living so they will not contribute. If you will clearly state a definite policy regarding QSLs it [would] do a lot to regain confidence. I hope you will take this in the spirit it is given as I do not want to get on Danny's black list so he will not answer a call. I also feel acceptable QSLs are owed the boys who are the customers of the manufacturers who have assisted with equipment. I hope your venture is a success.

Unfortunately, this writer was identified only as "Mac, K2." But publishing it gave Spenceley the opportunity to address issues he knew to be hanging around.

Foundation treasurer Ed Stanley, W4QDZ, replied:

The remarks in the above letter bear out the facts that the various points covered have been discussed among the Fraternity. And, although this editorial [sic] considers it bad taste to even discuss the matter of integrity which was touched upon regarding Dick and Danny, it is logical to assume that such remarks would come up in a project of the magnitude of the YASME Expeditions. Whether they are justified or not does not affect the basic issue, which is that there is a lack of understanding in the matter of QSLing, and the YASME Foundation certainly cannot permit this type of thinking to be carried into the future DXpeditions, which it is sponsoring with its Membership.

I have talked to Dick and Danny both about this matter, and the Directors have been kept thoroughly advised. We have recognized the probability of errors in the matter of confirmations. This is bound to happen when so many are involved, and all that can be done is the best possible. The logs are to be transmitted from the expedition logs to KV4AA, then worked from there for QSLing. So far the percentage of error has been rather low. But errors will occur.

It is the policy of the Foundation that any funds received are strictly donations to help keep the project going. No one has been or will be assessed. In other words, QSLs are not for sale. If a contribution arrives, it is considered as a contribution, and if the contact cannot be confirmed, for whatever reason, the contribution is retained as a contribution, and effort made to eventually run down an error in time or date.

The QSL policy would be to send QSLs with at minimum an SASE, to Spenceley, and "whenever possible, support the continued effort of the Expedition with whatever cash donation you feel you want to make. All QSLs not supported by costs of mailing will be sent via Bureau. An official QSL will be sent in all cases.

On a happier note, K0PCG wrote to thank Danny and Dick "for helping me achieve the first goal in my brief career as a ham - DXCC No. 4419 dated July 9, 1959, which arrived a few days ago (I have been licensed since April 1958). Thanks for the exploration of the Caribbean and excellent QSL service. You have accounted for eleven new countries in my application."

Given a new forum from which to speak, Spenceley editorialized occasionally, confirming that old adage about "He who owns the printing press." No topic in 1959 more divided amateurs, and DXers, than CW vs. 'Phone (unless perhaps it was AM Phone vs. SSB). By international law then, as now, some band segments were exclusively for CW. For some time stations, mostly in South America, had been operating AM between 14.000 and 14.100 Mc. This was particularly aggravating to DXers, and Spenceley weighed in with a plea to move them.

One thing as true in 1959 as it is in 2002 is if you want to work DX and have to choose one band, it's 20 Meters. And, in 1959, if you had to choose a mode, it was CW. Not only were you able to work more and farther with a modest station on CW, compared to 'phone, but CW was where the DX was, with only a few exceptions. 20 Meter CW was the place to be. Four years later Paul Rockwell, W3AFM, wrote a classic series of articles for QST entitled "Station Optimization for DX" that pretty much said, over many pages, "if you want to work DX, be loud, and be able to hear, on 20 Meter CW."

Spenceley, a die-hard CW man, began something of a

crusade to combat the "resurgence of phone stations, mostly of Latin-American origin, which, for their own selfish interests and obvious disregard for their fellow amateurs on CW, make use of the accepted CW portions of the 14 MC band for their phone transmissions. Fortunately these offenders are few in number," he wrote, suggesting that radio clubs around the world mount campaigns to educate their members to the "gentleman's agreement" of no 'phone operation below 14.100.

The Hallicrafters Connection

If there was one thing Danny always needed it was radio gear. Anyone who has lived in the tropics or near salt water can describe what the salt air does to electronic equipment. Add to that the rough-and-tumble of hauling receivers, transmitters, linear amplifiers, and beam antennas from a boat to land and back again, and it's no wonder new equipment or repairs were always needed.

In the early expeditions the Elmac company, as well as one or two amateurs who had Elmac gear available, lent or gave the company's very compact rigs to Danny. But by 1960 the state of the art had advanced on shore and more was expected from the other end. The Elmac rigs were just not up to the tasks of pile-ups (and the 75A4 donated by Collins was apparently out of the picture). Hallicrafters had been forthcoming with "loans" of equipment, but an apparent run-in between Danny and Hallicrafters president Bill Halligan, W9AC, (noted in Halligan's letter of 28 April 1959, to Danny) had caused the company to rethink its commitment to future voyages. The following exchange of letters refers to a meeting at a place not identified but can be inferred to have been a convention sometime in 1959, perhaps the Dayton Hamvention. Someone prodded Danny into apologizing to Halligan, and he wrote:

Dear Bill:

I feel an apology is due to you and am acting accordingly. It has been said there can be no excuse for one losing one's temper, but, even though the excuse will not be accepted, an explanation is necessary which I hope will clear the air and also salve my conscience. I had waited patiently for Ed [probably Ed Stanley, YASME's treasurer] to finish his rag chew with you [in a hotel] so that we might have a few words about things in general. I am not a patient man under the best of conditions.

I thought I had found the time, hardly the place however, to have that very short talk but, as you noticed, Ed decided it was high time to bring a friend of his into the party for an introduction to you. I think you will agree this was hardly good manners and it affected me in a disastrous manner as you noted.

I will not go into the high pressure under which I have been working. You are well aware of this but, I suppose it may explain my sudden flare up. I leave you to judge. I did later discuss this matter with Frits and I was pleased to hear you had realised how I had felt and accepted the circumstances for what they were.

One of these fine days Bill I shall be able to do things in a leisurely way and have many people at my beck and call to carry out instructions but, until that day arrives, I shall carry on doing things myself and getting them done. I am very fortunate in having Dave, ZL1AV, who has been a wonderful help. Jake, W8LNI has only just joined us and I feel he will also be a great help in getting things done. I sure need all the man power I can get at this stage of the game.

With a few strokes of luck, there is no reason why we shouldn't be away this month. I am presently awaiting some news from the Marine Section of the University of Miami. One of the Directors will be flying up to see me this Friday to discuss the equipment they will be installing and, from what I can gather, there are several other fields which are now under discussion in which we shall be able to assist. Perhaps when you are down here I shall be able to tell you more. It is fine thing for us to know that we have been recognized by such a body and entrusted with their work. Things like this not only boost my moral[e] but also aid the Foundation in its aims.

Hope to see both you and your wife when you visit this part of the world. Sincerely, Danny.

On 12 Feb. 1960, Halligan replied:

[I] was puzzled when you turned on your heel and stomped out of the hotel that evening. I thought here, indeed, is a dog in the manger, but anyway life is too short to indulge in any ill will for, as a matter of fact, I'm wishing you all the very best. I know things aren't easy and I really wanted to sit down and talk to you, but unfortunately you slammed the door and I couldn't figure out why.

I'm going to try to get over to Tampa before you leave. It happens that Florida, though it appears like a vacation, is also an arm of our business activity because we're always entertaining customers and things like that. W4CF and K4JG want me to visit them in Bradenton and I'll try awfully hard to get over there and see them and visit you. I'll do my best.

The YASME NEWS Vol. 2 No. 1, Feb-Mar 1960, featured Danny's Diary, the story of the conclusion of his voyage to Tampa from Ft. Lauderdale on YASME III.

Colonel J.M. Drudge-Coates, G2DC, became a YASME Foundation director and would play a pivotal role in the adventures of Iris and Lloyd Colvin beginning in 1965.

With the February-March 1960 issue Spenceley acknowledged to foundation directors that high cost of publishing the slick YASME NEWS were becoming a burden, despite most of the work being by volunteers. After buying YASME III (mostly through loans from officers) the foundation was "effectively broke." While original plans did not call for what Spenceley called a "flamboyant effort" in publishing, Spenceley began a last-minute effort to save the YASME NEWS in its "slick" form. "We visualized just a modest mimeographed sheet to inform our members of the expedition moves. Probably we shall have to return to this idea," Spenceley told the directors and volunteers.

YASME Newsletter Number One June 1960 sported a new format and a new title: YASME Newsletter appears in June 1960, as Number One. It's a simple, mimeographed, legal-size sheet, a bitter, but necessary, pill to swallow after the slick magazine format of the YASME NEWS. Spenceley called the NEWS a "gallant effort" on the part of Ed Stanley, W4QDZ, but says it "proved just too rich for our blood."

Spenceley published this photo of new YASME Foundation director G2DC in the YASME NEWS

Even a move by Stanley and Howard Hogan, in Tampa, to take over publication in return for the advertising revenue did not work. Revenues could not cover the costs of operation and that they could no longer publish the magazine unless the foundation could subsidize it.

More bad news came when Ed Stanley became seriously ill. Spenceley noted "His substantial efforts in bringing the YASME Foundation into being."

22· The First Voyage of YASME III

Brother, if we operate from here I doubt very much if there will ever be another expedition to this place! - Danny Weil

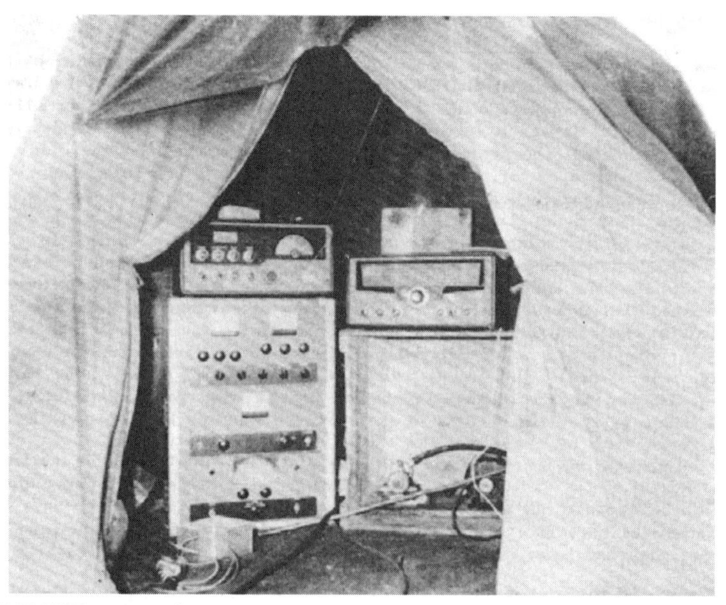

VP5VB operating tent.

YASME III, with Danny and Dave Tremayne, ZL1AV, were finishing their first island operation with the new boat. YASME III left Tampa at midnight on 1 May 1960, for Kingston, Jamaica. It was estimated that the trip would take a week. Lack of wind from Tampa to around the western tip of Cuba necessitated full use of the ship's engine. Danny reported to Spenceley by radio that "Plenty of breeze was encountered as the YASME then headed east, but all from the wrong direction, thus her speed was cut and the engine clanked on. As the Isle of Pines was neared the question arose as to whether their diesel fuel would hold out until Kingston was reached." To play it safe YASME called in at the Isle of Pines and secured an extra drum of diesel fuel. Again setting forth, fighting wind and current, they arrived in Kingston at midnight on 13 May, thus making nearly a two week trip of it.

The call sign VP5VB was quickly obtained and on 17 May at 0749 GMT, VP5VB was all set up in a tent on the dock and the first contact was with W5RDA on 14 Mc. SSB. Spenceley observed that "Things would have been easier if the station could have been supplied from the YASME's diesel generator as she was tied off to the dock just fifteen feet away but to satisfy the DXCC requirements of having everything ashore, a gasoline generator was set up on the dock and powered the station. This "put-put" could be clearly heard during the many phone contacts. "

Danny and Dave operated until June 4, 1960, and made 3,200 contacts. Spenceley noted that while Jamaica was not too rare a spot, for many it was a new one on 'phone, and the operation was seen as a good "shakedown cruise" for the expedition to come. While they operated, Spenceley scanned his maps of the Caribbean looking for spots that might qualify as "new" DXCC countries. Baja Nuevo looked like a good possibility, and inquiries in Jamaica first identified Baja Nuevo as British (it is visited by many Jamaican fishing boats) but in reality it was the property of Colombia. Spenceley said that, according to the way he read QST reports on the DXCC status of the San Andres/Providencia group and the Serrana Bank/Roncador Cay group, Baja Nuevo should count separately for DXCC and an operation from there would be with a Colombian call sign.

Spenceley cabled the minister of communications in Bogotá, stating his case for permission to mount an operation from Baja Nuevo. With cooperation, stemming from QSO's with HK3SO and HK3RR, and an assist from Luis Caicedo, HK3AO, president of the Liga Colombiana, permission was granted, and The YASME Newsletter speculated that "The YASME expedition should be active from Baja Nuevo by June 7th or 8th and remain there for a ten day to two week stay. (FLASH! from W1LLF comes word that ARRL says Baja Nuevo looks good to them providing there is no USA claim or control. This would take about a month to find out - we will, of course, go ahead.)"

Since the YASME had Colombian permission to visit Baja Nuevo there was hope that a landing at Malpelo might also be possible. "Landing on this forbidding island is a matter of guts, luck and fair weather," Spenceley said. Operators from a recent unsuccessful HK0TU Malpelo attempt, including W4CVI, said there "is a spot on the northeast side of the island marked by a seaward outcropping of rocks. From these you can get on the island, walk up a hundred foot slope and that's it. You then find yourself up against the base of the cliff which entirely circles the island. Any ham station would have to be set up on this slope. The island, some 700 feet high, is fairly flat but to get to the top you would need a helicopter.

"This adverse weather at the time of the HK0TU venture deposited fifteen to twenty foot surf on this rock outcropping and, although a couple of the boys actually did get ashore with a generator it was impossible to proceed further. A mighty disheartened bunch of hams finally had to give up and return to Colombia and they deserve a mighty big hand for the efforts that were made. None of the much advertised coral snakes were seen on the small area covered," Spenceley said.

Danny's "definite" next stop was Guayaquil, Ecuador, and from there to the Galapagos, if he obtained Ecuadorian operating permission. After that, it was hoped a Clipperton operation would follow.

The YASME Newsletter Number Two for July 1960 reported that Ed Stanley, W4QDZ, died on June 10. This news, combined with the demise of the YASME NEWS, must certainly have crushed the spirits of Spenceley and everyone who had helped put out the magazine.

But there was good news, in the form of a very successful operation from Baja Nuevo, HK0AA, from June 11 to 17, 1960, that produced 2,500 contacts. It would not be until May 1, 1961, that Baja Nuevo would officially be added to the DXCC Countries List (for contacts back to 1945) but at the time everyone was reasonably certain that this would indeed become a New One.

Spenceley, drawing on radio reports from Danny and Dave, wrote the following run-down of the operation:

> Sailing from Kingston, Jamaica, at 10 PM on June 6th, YASME III headed southwest. Weather was of the usual 'YASME variety,' with high seas running and a southeasterly wind. At 5 PM the next day the Podro Bank (Jamaican) was reached and they were able to enjoy some small respite from the weather by anchoring and fixing up their first meal since departure.
>
> Little data regarding Baja Nuevo had been available and this was of the oral type obtained from fishermen who had visited the vicinity. It seems that Baja Nuevo consisted of a seven mile stretch of small sandbars with the main marker being the wreck of a tramp steamer with its bow and stern broken off. The mid-section extending some 25 feet above water.
>
> On the afternoon of June 8th, with indefinite sights available, they estimated they were close to the island (they could smell it!) but nothing was to be seen. Better sights obtained on the morning of June 9th put them about five miles to the northwest of Baja Nuevo and the search continued. At 5 PM Dave sighted the wreck and then, with liberal use of depth finder and very cautious approach, YASME was able to anchor some 300 yards off the island, at 7 PM, and morale was on the upswing.
>
> On June 10th southeast winds of up to 40 knots persisted but Danny managed to make a landing on the island after a few [dunkings] in the 10-foot surf. On subsequent landings they were able to set up the tent, using quantities of driftwood to act as a windbreak, and, following this, got the beam up. No attempt was made to get any radio gear ashore on this day as it surely would have suffered immersion and damage.
>
> On June 11th the wind abated and the surf dropped considerably. First the Onan generator came safely ashore followed by the HT-32 and SX-101A. At 2100 GMT, June 11th, HK0AA came on the air bestowing his first QSO on KV4AA. An hour and a half QRX was then called for to further organize things. At 2230 GMT HK0AA was back in business and the first ten contacts, on SSB, were W4SKO, W8PUD, W8DPF, W4AZK, W8MG, W1BAN, K2JGG, W3JNN, TI2RC, KZ5WZ, and W3PGB. By 2400 GMT, June 12th, 837 contacts had been made!
>
> Operations continued until June 16th when the weather worsened and the wind shifted to the northwest. Such a wind would pile YASME III up on the island if the anchor chains were to snap. This wind also built up the surf conditions, which were always a landing problem, and a decision was made then and there to get the gear back to the ship at the first lull and before weather conditions made abandonment of the gear on the island an alarming possibility.
>
> HK0AA continued until 1942 GMT, June 17th, conditions at this time looked good for getting the gear back aboard so, after a QSO with W9TKR, the big switch was pulled and a mad rush ensued in dismantling everything and getting it to YASME. We are happy to say that this transfer was accomplished successfully and YASME III immediately set sail for Cristobal and arrived in that port, fairly uneventfully, at 2 PM on Tuesday, June 21st.
>
> Danny's first remark to me upon setting foot on Baja Nuevo was "Brother, if we operate from here

VP5VB operating site at Kingston, Jamaica.

I doubt very much if there will ever be another expedition to this place!" The main island, or rather "sandspit," is only about 160 feet long and 60 feet wide. The average height of the island is from two to three feet. There are a few birds there (probably bad actors expelled from Aves Island!) and a goodly number of king-sized turtles."

On the daily trips from YASME to the shore in the dinghy, reefs had to be navigated which made the trip impossible in darkness. Also, the dinghy's outboard motor could not be used as the surf would have certainly broken it up on the coral formations near shore when the dinghy was flipped over, as it invariable was. As things stood the dinghy took a terrific beating and probably would have not survived a few more days operation.

Most times one man was aboard YASME to look after things. Danny usually took the day watches while Dave took over at night. One can picture Dave, in pitch darkness, save for a bulb over the operating position, sitting in this windswept tent, on this god-forsaken island in the middle of nowhere, knocking off contacts by the hundreds in a beautiful display of operating know-how.

Thus, of necessity, HK0AA activity was cut short a few days but over 2600 contacts were made and we can only say - hope you got him.

Unfortunately, Colombia did not grant permission for YASME III to land and operate from Malpelo Island. "This spot is being saved, " Spenceley said, "and probably rightly so, for another try by the original HK0TU expedition, which may make the attempt later this year or early in 1961."

Following the Baja Nuevo sojourn, Dave and Danny returned to the dock at the Rodman Naval Base, Balboa, Canal Zone, an excellent place to make repairs. The aim was to fit out YASME for the long Pacific voyages ahead. A troublesome leak had to be found and patched, and new radio gear located and installed; at the time, the only transmitter on board was a single Hallicrafters HT-32.

Rodman Naval Authorities were very cooperative, the Newsletter reported, in placing the facilities of the Base at Danny's disposal. And Danny's call sign, KZ5WD, assigned to him on his first trip through the canal in 1955, was reinstated. Danny and Spenceley maintained daily radio schedules on 20 Meter SSB. The itinerary remained Ecuador/Galapagos, Clipperton, Marquesas.

Spenceley and the YASME directors now faced in earnest the task of begging or borrowing more radio equipment. Looking at what they thought could be a three-year stay in the Pacific, with the usual equipment failures, they thought that they would need three 100-watt SSB/CW exciters; three receivers; and one or two linear amps of 500 or 1000 watts. Portability was also a consideration.

"You may recall," Spenceley wrote, "that the expedition was promised Hallicrafters gear to this extent. We are sorry to say, through misunderstandings, or what have you, this deal was called off. It was climaxed (when) a representative of Hallicrafters walked aboard YASME in April (in Tampa) and removed the HT-37 and SX-111. This left YASME, at the time, SSB-less. At this point Hal Sears, K5JLQ, came to the rescue and loaned us an HT-32 and SX-101. This was the gear used at HK0AA and VP5VB (Plus a Globe King rig at the latter spot).

"It seems to me that ham gear used on an extensive DXpedition of this type would reflect publicity value to its manufacturers. The Foundation would also extend to such gear whatever favorable publicity that it could."

The Newsletter called on "all foundation members" for help. "Possibly some of you are in a position to donate or loan such gear to the expedition. Perhaps you could approach manufacturers or dealers of such gear. You might be able to locate some used gear, in good condition, which we could obtain at a low price. We cannot chance a continuance, successfully, of the expedition if 100% reliance is placed upon a single HT-32. This gear is needed for onshore operation. We have enough non-portable gear to assure communications with YASME itself. Please see what YOU can do about this and write KV4AA. I will be glad to coordinate all efforts in this direction."

Also announced was the engagement of Danny and "Miss" Naomi Kay, of Tampa. Tentative plans were for the two to get married in the Canal Zone on 22 July 1960. Danny listed Naomi's assets as "good cook, boating experience, expert secretary, considerable nursing experience." Spenceley observed that "she is obviously a kindred soul to Danny in the adventure department as she is going along with him on the voyage. No doubt she will lick the Morse code in nothing flat! I know that all Foundation Members will wish them all the luck and happiness in the world and most will agree that they will have enough stories to tell which will keep their grandchildren spellbound in the years to come!"

Don Chesser's DX Magazine scooped the YASME Newsletter on this news, in the issue of 29 June 1960, and confirmed the rumor the following week. However, Chesser in that same issue managed to publish the following bit of misinformation:

"During a QSO with HB9J from HK0AA, Dave, ZL1AV, said they have received a license to operate from Malpelo Island, and they plan to tackle that difficult landing, weather and seas permitting."

Chesser also noted that the Dutch publication DX-press had characterized the HK0AA Baja Nuevo operation as "disappointing" for European DXers, "because

Danny and Dave ceased their Baja Nuevo activities too soon after satisfying the top layers of W/K/VEs, leaving the European DXers gasping in mid-air."

In his DX Magazine for 13 July 1960, Don Chesser reprinted a newspaper clipping from the Panama American, which he said was the Panama Canal Zone's daily newspaper. Chesser headlined his coverage "MORE TROUBLES IN YASME EXPEDITION." The newspaper had headlined its article "Young New Zealander Up For Larceny Aboard Ship," and it read:

> Following a complaint from the master of the radio research yacht YASME III, Daniel Weil, Canal Zone police arrested a 22-year-old New Zealander, David Trevalyan Tremayne, for the attempted theft of clothing and equipment from the vessel. Today Tremayne appeared in Balboa Magistrate's Court charged with attempted grand larceny. The case was continued to Monday, with bail fixed at $500.
>
> The YASME III, on a world radio research cruise, reached Balboa last month after its crew of two, Weil and Tremayne, had spent a week on an isolated Caribbean atoll, Baja Nuevo [sic], communicating with radio 'hams' who have sponsored the cruise.
>
> After spending some weeks in Balboa refitting the vessel, with an augmented crew including Weil's new bride, due here from Miami on July 18, was to sail for the South Pacific, with its first stop at the Galapagos Islands.
>
> Tremayne, who claims to be a former R.N.Z.A.F. pilot officer, is charged with attempting to steal a sports shirt and trousers worth $8.50 from Weil and radio tubes and relays worth $71.75, which is the property of the USAF but in the legal custody of Weil. Part of the radio research work, according to Weil, is being undertaken for US and British authorities and is classified.
>
> The New Zealander signed on the YASME some eight months ago, after spending a year in the United States. He was unpaid apart from expenses but eventually would have reached his homeland aboard his yacht.
>
> The YASME is Weil's third radio research vessel. Of two previous yachts, also sponsored by amateur radio enthusiasts, one was wrecked in the South Pacific, the other in the Virgin Islands. It is Weil's second visit to Balboa. His first visit was made some six or seven years ago.
>
> During his stay in Miami earlier this year, Weil became engaged to Naomi Kay, whom he expects to arrive in Balboa on July 18. Her arrival, however, depends on whether she can sell her Miami home. The 40-year-old Britisher and his American bride are to spend their honeymoon aboard the YASME.

In publishing the clipping, Chesser said "Danny Weil's attempt to circumnavigate the globe is still encountering obstacles, according to the following news item published July 8th in the Panama Canal Zone's daily newspaper, The Panama American. The David Trevalyan Tremayne mentioned in the clipping is, of course, Dave, ZL1AV, noted for his very expert job of operating as VP7VB, VP5VB, and HK0AA, most recently.

"He appeared with Danny at several amateur radio conventions in the United States during the past year, before departing on the current YASME III world tour."

Well, daily newspapers don't always get things quite right on the first try. Chesser did not mention any of the clearly questionable parts of the newspaper's story. For one, Danny was 42 years old, not 40 (although he may have just said "I'm 40"). And YASME II sank in the Grenadines, part of St. Vincent, not the Virgin Islands. Furthermore, the 8 July newspaper story referred to Danny's "bride," when a few paragraphs later it said that Naomi was still in Florida, and Chesser already had published information that a 22 July wedding was planned to take place in the Canal Zone.

Spenceley was magnanimous, saying "We regret that Don Chesser saw fit to reproduce the Canal Zone newspaper clipping in his DX Bulletin (sic). We are of the opinion that this clipping gives

Young New Zealander Up For Larceny Aboard Ship

Following a complaint from the master of the radio research yacht Yasme III, Daniel Weil, Canal Zone police arrested a 22-year-old New Zealander, David Trevalyan Tremayne, for the attempted theft of clothing and equipment from the vessel.

Today Tremayne appeared in Balboa Magistrate's Court charged with attempted grand larceny. The case was continued to Monday, with bail fixed at $500.

The Yasme III, on a world radio research cruise, reached Balboa last month after its crew of two, Weil and Tremayne, had spent a week on an isolated Car-

worth $71.75, which is the property of the USAF but in the legal custody of Weil.

Part of the radio research work, according to Weil, is being undertaken for US and British authorities and is classified.

The New Zealander signed on the Yasme some eight months ago, after spending a year in the United States. He was unpaid apart from expenses but eventually would have reached his homeland aboard his yacht.

The Yasme is Weil's third radio research vessel. Of two previous yachts, also sponsored by amateur radio enthusiasts, one an erroneous slant to the whole affair. Suffice it to say that all is now cleared up and Dave sailed for New Zealand, on July 17, by commercial transport. Dave did a tremendous job of operating from VP5VB and HK0AA as witnessed by all who contacted him."

Meanwhile, Chesser continued doing generally good for the YASME Foundation, such as saying (in the same issue as the reprinted newspaper clipping) that the expedition also faced equipment shortages and that, "from other sources it is rumored Bill Halligan, president of the Hallicrafters Company, has resigned from his participation in the YASME Foundation and sponsorship of the tour."

Chesser went on: "It should be pointed out that manufacturer's equipment is usually loaned, not given, to expeditions planning all-out operations from a remote spot of the earth, for the publicity value of the event and for the purposes of field research in the ruggedness and reliability of the equipment under severe usage. Responsibility for the condition of the equipment, or its possible loss due to an unavoidable accident, however, is usually born by the manufacturer. The equipment is often available for sale at a considerable discount after the event and upon completion of the research."

YASME Newsletter 23, August 1960, reported that YASME III, at the Rodman Naval Station in the Canal Zone, was just about seaworthy again, her leak found and filled and the rudder repaired.

Naomi Kay and Danny were married in a civil ceremony on Jul. 21, 1960. They planned a church ceremony later, when time permitted. With the departure for New Zealand of Dave Tremayne, Joe Emmert, K4IGR, joined the expedition. Joe was connected with the Marine Division of the University of Miami and hoped to conduct "oceanographic work" in the South Pacific. "His technical know-how and ham operating assistance will be an invaluable asset to the expedition while the oceanographic work will lend important scientific values to the venture, not contained in former trips," Spenceley said

YASME III in dry dock, Balboa, Canal Zone, enroute to the Pacific.

Spenceley also asked to suggest how "scientific interests might be served from the radio angle, such as propagation reports" by the crew.

The YASME itinerary was further set out, although readers were reminded that "to predict the appearance of YASME in any particular spot has been a tough job. Unexpected delays always seem to crop up which set our timetable back." Best estimates were departure from the Canal Zone August 9, arriving in Guayaquil, Ecuador around the 18th. After that, an eight-day, 800-mile trip to the Galapagos, arriving September. Word was received that the Ecuadorian government had granted the call signs HC2VB and HC8VB (Galapagos).

From the Galapagos it was hoped that Danny, Naomi, and Joe would go directly to Clipperton. "Many hams need Clipperton contacts." Spenceley wrote, "and while YASME will make all attempts to go there, certain formidable problems arise which may make this impossible." Of primary concern was gasoline (for YASME's generators) and diesel fuel (for its engine) might not be available in the Galapagos, and for sure not available on Clipperton and the Marquesas. Danny would have to haul, from Guayaquil, a supply of both fuels, large enough to enable him to get to, and operate from, the Galapagos, Clipperton, the Marquesas, and then travel a further 1000 miles to Tahiti where the next fuel would be available. And Clipperton also presented unique and dangerous challenges, including winds and currents.

Spenceley continued to beat the drums for DXCC status for the Marquesas Group, where YASME would go after Clipperton and where Danny presumably would be able to reactivate his FO8AN call sign from 1956.

Response to the appeal in the previous YASME Newsletter for equipment, Spenceley and crew put on a "crash program" and "via four contributors, we took advantage of a very generous discount offered to us by Hallicrafters" and bought some gear from the company. In addition, Doc Woods, W4GJW, donated an Elmac AF-67 transmitter, which would be used as an excellent emergency rig to operate from the ship's batteries.

On 2 August 1960 Spenceley sent Hallicrafters $1,420 for equipment for the expedition, to be sent by air freight to Danny in the Canal Zone.

For the first time (at least in print) Foundation members were advised that they would be "assisted" in contacting the expedition. They were advised to contact KV4AA, who would set up schedules for the next day. This was to be for CW contacts only, at first, and if the scheme worked it would be expanded to include SSB schedules.

Spenceley, always attuned to the DXCC Countries List and its possibilities, noted that the ARRL had, on July

14, 1960, added five "new ones": KG6I Marcus Island; 6O1/2 Somalia; 9U5 Ruanda Urundi; FF8? Mali Federation; and FF7? Mauritania. (There also were five deletions, and the Cayman Islands, then VP5, reverted to the separate status they enjoyed before July 1, 1958.)

Spenceley commented at length on these changes, saying "it seems that the sorting out of the African situation is not complete as several questionable areas remain to be passed on." His knowledge of the international situation as it related to Amateur Radio DXing at that time may have been second only to the ARRL DXCC Administrator, Bob White, W1WPO (later W1CW).

There was no word yet from the ARRL on accepting Baja Nuevo for DXCC status. Spenceley said he believed "this is due to ARRL's checking as to whether the USA has any claims on this island. We were told that this would take a month or so."

After years of handling Danny's QSLs, Spenceley announced that YASME Foundation director Golden Fuller, W8EWS, had taken over and would handle all cards after HK0AA. "This humanitarian action relieved KV4AA," Spenceley said.

A window into Don Chesser's sources appeared in the DX Magazine for 24 August 1960: "Danny Weil, VP2VB/MM, has left the Canal Zone for Ecuador, and was heard Tuesday about 200 miles south of the Canal. During his SSB transmissions on 14298 kc from aboard ship, piped to 'DX' [Magazine] via telephone by our Burlington sleuth, K4OCN, Danny was heard to say he expects to arrive at Guayaquil Friday, August 26, and to operate from that port city for the next 10 days to two weeks. Presumably, after that period Danny & Co. will continue to the Galapagos Islands where he will set up operations as HC8VB."

YASME Newsletter Number Four, September 1960, focused on Danny's day-by-day radio reports of his voyage from the Canal Zone to Ecuador in the repaired and outfitted YASME III. Here's the report, as transcribed by Spenceley:

> At last the big day had arrived and YASME III was ready to start on the first leg of her long Pacific voyage. All gear, fuel and food were aboard and a million last minute details had been taken care of by her overworked crew. Last goodbyes were exchanged with the very cooperative Rodman gang and, at noon on August 20, Danny cast off and set his course for Cape Mala, with final destination, Guayaquil, Ecuador.
>
> The waters of the Gulf of Panama, and beyond, are the meeting place for various Pacific currents and a rough trip, such as YASME I had through these waters in 1955, was expected. Needless to say, no one was "disappointed."
>
> Capa Mala [Cabo Malo] was passed at noon the next day and a course was set for Ecuador's Cape San Francisco, directly south following the 80th Meridian.
>
> At this point cross-seas tossed YASME III around like a chip and rendered the automatic pilot which, like AVC, will hold only to certain limits, practically useless. At this time a thoroughly sleepless, miserable and seasick crew expressed wonder as to why they had ever let themselves in for this and wished that some of the militant dissenters of the venture could have been along to see and feel for themselves what a "sleigh ride" it was.
>
> Things improved somewhat during the next few days as sea-legs were acquired but winds were adverse, or non-existent, and the engine ran continuously, as was necessary for the entire trip.
>
> Nearing the northern coast of Ecuador the wind blew up considerably and ten to twelve foot seas were encountered. This made the preparation of food impossible and it was decided to put into the port of Esmeraldas for a day's rest, a decent meal and a chance to clean up the ship a bit. It was also hoped that the seas would calm down a bit.
>
> They sailed into Esmeraldas on August 27 and anchored. At last YASME was on an even keel and steady, a strange sensation to its crew. But even this did not pass without incident for the tide went out (Danny said "nobody told me about the tide") leaving YASME sitting on her side at a 45 degree angle. They were now steady as a rock but the angle was not conducive to easy movement around the boat although it did not unduly interfere with rest and all hands got a measure of shut-eye wedged in their bunks. At 7 PM the tide returned, YASME righted herself and was moved to a deeper anchorage.

PO A 196 AEC34 QUITO MR DICK SPENCELEY BOS 403
ST THOMAS VIRGIN ISLANDS USU DQ NR 344 DTC PERMISSION TO
OPERATE HC-2-VB AND HC-8-VB SEND AS TODAY CARE GUAYAQUIL
RADIO CLUB SIGN DIRTECNICO COMUNICACIONES ENDQUOTE RAFSO

At 10 AM August 28, after an underwater caulking of a small leak near the stern, YASME sailed from Esmeraldas heading for Cape San Lorenzo. The Ecuator was crossed at 2 AM the next morning but none of the usual amenities [sic] were observed as King Neptune was not awakened from his slumber some 340 fathoms down.

Dave Tremayne, ZL1AV, eyes Bajo Nuevo from YASME III.

Cape San Lorenzo was passed at 5 PM on the 29th and YASME continued ... southeast into the Gulf of Guayaquil.

Danny had maintained radio contact with hams in Quito and Guayaquil, who planned "a red carpet reception. Cesar, HC2CS, manager of a company that operated a steamer between Guayaquil and the Galapagos, had shipped two drums of gasoline, two drums of diesel fuel, and two special fresh water drums to the Galapagos on 20 August. Danny planned to pick these up for the next leg of the voyage, be it Clipperton or the Marquesas.

Spenceley said that new crew member Joe Emmert, K4IGR, planned to collect plankton at various South Pacific spots, study underwater noise levels, and do meteorological work for weather maps.

Danny set up HC2VB ashore in a section of the Guayaquil Yacht Club building, very close to where YASME III was moored. "This was most convenient," he said "with the only drawbacks being QRM from the numerous motorboat engines and, on a couple of occasions, where festivities within the Yacht Club itself reached a volume over and above the capabilities of the SX-101!"

Dick Spenceley appeared on the cover of DX Magazine, 21 September 1960, and Chesser said "In the conventional words of masters of ceremonies, the subject of our front cover picture this week needs no introduction, for who in the DX world doesn't know Dick Spenceley, KV4AA, former DX editor of CQ and Western Radio Amateur magazines, and President of the YASME Foundation. 'Talk about the Drooling Corner!' writes W3AYD, who snapped this picture during his recent Caribbean cruise. 'That wall in the background has QSLs from 285 countries, and some of them are the juiciest! Dick stopped hanging up the new ones some time ago, and he now has 15 more to mount!"

Spenceley now was publishing YASME Newsletter religiously every month (later, when YASME was between ports, inactive, etc., the newsletter would become more sporadic). Newsletter Issue Five for October 1960 wrapped up a momentous stay in Ecuador, including a church wedding and a three day side trip to Quito (home of HCJB, then and now one of the biggest shortwave broadcast installations in the world), and predicted that Danny would open up as HC8VB from the Galapagos around 14 October 1960. The voyage from Guayaquil to San Cristobal Island in the Galapagos Group was 750 miles, about eight days good sailing time. A "very tentative start date" for Clipperton was set for November 15.

Readers were invited to tune in and listen to daily radio schedules between KV4AA and VP2VB/MM on 20 Meter SSB. Today's internet generation probably would call this a "wireless chat room." Hundreds of hams at a time, around the world, listened to these radio communications.

The bad news was that Joe Emmert, K4IGR, was forced to leave the expedition due to the illness of his father. Joe flew back to Miami, from Guayaquil, on September 16, "expressing hope that he might rejoin YASME but this seems improbable," Spenceley said. "A native deckhand may be picked up in Guayaquil to go along with Danny. This is considered a necessary addition and, among other things, it would allow Danny more operating time at each QTH."

It would turn out that Danny Weil and Naomi Kay were, in the '90s term, soul mates. Naomi dived headfirst not only into the world of ocean sailing but into amateur radio as well.

Chesser's DX Magazine continued to pick up the slack in publishing photos and stories about Danny, Naomi, and YASME III, including the wedding photo on his October 12 cover.

In a note to Don Chesser, Naomi said that she "is no ham operator and she didn't know if she'd prove to be a good sailor, but she did boast a fine cup of coffee, and she promised to keep Danny gurgling with the stuff so he'd remain awake to work more of us than usual from their many DX sites." His appetite whetted, Chesser was "dying of curiosity to see the beauteous creature who could bring such a popular, dynamic, adventurous man as Danny to his knees."

23· "I Married DX"

Naomi Kay Weil wrote this remembrance of her wedding to Danny, which Don Chesser published in his DX Magazine.

Naomi identified this photo "September 11, 1960, Junta Misionera Foranea de la Convencion Bautista Surena, Guayaquil, Ecuador. Joe and Ellen Megin, Rev. Howard L. Shoemaker"

"I MARRIED DX (or, Gone with the dots and dashes)"

I never knew of anyone who tried as many times to get married as Danny and I. In church, that is. They tell me it is done every day in Chicago, but you just try to find the time, while living on a boat, or residing in Panama. We had set the date no less than three times, and always something interfered. We finally gave up trying while in Panama, and said "for sure" in Ecuador. We finally made it, but only by the help of the Guayaquil Radio Club.

Joe Magen, HC2OM, President of the Club, is a born organizer, and what he can't do, his very lovely XYL, Elsie can. However, we did not know of his talents when discussing with him our sad, sad tale of trying to get married. In case anyone is wondering about my status in the interim, I forgot to tell you that we had already been married. You are probably wondering why, if we were already married, we were trying so hard to get married. As you can see, it is all very confusing, and I am still confused.

Danny writing, in Tampa, 1960

Just for the record, we were married on July 21, by a local judge, in the Canal Zone, but somehow it just didn't seem right. Maybe, because his office was next door to the local jail. Anyway, even though everything was very nice and legal, I just didn't like the atmosphere and Danny didn't think that the judge had been loquacious enough. He really hadn't left anything important out but he went through it so fast we were still saying the first "I do" when he had finished with all the rest. This was the story we were relating to Joe and he decided that something must be done. We set the day, Sunday, September 11, and he promised to do the rest. Joe then grabbed a telephone and went into action. In no time he had everything set up. His fluent

Spanish helped. Danny and I would still be trying to get the operator.

Now, since I was finally getting married in a church, I realized it was time to change my appearance from that of an able bodied seaman into something a little more resembling a female, so, on Saturday, off to the beauty parlor I went. Little did I know that every other woman in Guayaquil had the same idea. They do not have a telephone so I couldn't make an appointment and probably wouldn't have found the place if it hadn't been for Elsie Wilmot, Joe's sister-in-law, the wife of another club member. She took me by my sweaty little hand and led me through side streets, alleys, parks, etc., until we found the place. Me, still being a little green from the States, expected to be able to sit down and rest my aching feet when I arrived there, but with only three chairs in the place, it was just a thought. After all, the whole room was about 12 by 24 and that included the storeroom. They did, however have seven operators who worked in assembly line fashion.

All the customers, who stand while being tinted or shampooed, were in various stages of having their hair done. I really set them in a dither when they discovered that I only wanted my hair cut and some of the YASME dust removed. In other words, I didn't want to be turned into a raven-haired beauty. I never saw so much black hair in all my life and the amazing thing about it was that they were all having it made even blacker. I really stood out with my golden, tho slightly tarnished tresses. Finally, at the finish of the relay race, I had been trimmed and shampooed in very cold water (hot water doesn't exist) and was ready for the finishing touches.

Naomi, probably around 1950.

It was quite warm and I was practically asleep when it dawned on me that my hair was beginning to feel like a bird's nest. I opened one of my glamorous blue eyes and peeked into the mirror. I nearly fell off my chair with what I saw. Every hair

the weekly amateur radio news magazine

From:
DX Magazine
Don Chesser
Burlington, Ky.
U.S.A.

FIRST CLASS MAIL

First Class Mail
U.S. Postage
PAID 4 Cents
Cincinnati, Ohio
Permit #9770

JOHN T. LANEY III, K4BAI
3500 14TH AVENUE
COLUMBUS, GA.

on my head had been brushed into reverse and little straggly ends sticking out all over. I came to in a very few minutes and realized I was getting a South American hair-do. I had wondered, ever since arriving in Guayaquil, just how that 18 inch dome was acquired … now I knew.

I frantically looked around for Elsie but she had slipped out somewhere while I was napping. Through much use of the hands and my pidgin-Spanish, I finally made it clear that I still preferred the old fashioned coiffure of the slightly wind-blown variety. Then Elsie walked in and explained to her that I was slightly eccentric and that it was better to humor me. As it was, she did win a little because she managed to slip in a few puffs on top of my head and had made some spit curls in front of my ears. I was scared to go back to the boat but I don't have any other place to sleep, so back I went with my heart in my mouth.

I slipped in the back door, or what passes for the back door on our boat and ran … smack into Danny. I expected to hear a blood-curdling scream but all he did was to grin with delight and told me that I must get one of the Spanish style dresses that are skin tight to the knees and flare out from there on. With the addition of a rose between my teeth and a tambourine …I had it made!

As I mentioned earlier, Joe Magen said he was going to take care of all the details and all Danny and I had to do was to be ready at 11 a.m. We took him at his word and, about an hour before time, we went through the usual ordeal of two people trying to dress in a very cramped area. Danny, who usually outruns me, made it out in record time, all decked out in white. Since I was still dazed from my experience of the day before I really didn't recognize him and thought for a minute that he had been sent by the hospital to pick me up. The maroon tie convinced me however that he was my intended and everything would be alright. No kidding, he really looked handsome and I was gladder than ever that he was going to be the one.

I was rushing like mad and every few minutes Danny chirped out "Aren't you ready yet, darling?" That didn't help my ruffled nerves and besides, I couldn't find my garter. The blue one ... for my "something old, something new, something borrowed, something blue." My very good friend and neighbor back in the States, Ethel Smith, had given it to me and I wasn't going to be married without it. After practically taking the cabin apart, I found it.

Por la lumbrera existente a proa de la cabina de derrota, surge la cabeza de una mujer de cabellos claros y ojos azules. Nos mira y sonríe.

—¿Qué tripulación lleva?

—Bueno; mi esposa —y señala a la dama— un marinero ecuatoriano de apellido Quinde, que ha decidido dar la vuelta al mundo, y yo. Nadie más.

—¿Qué ruta seguirá?

Hace un gesto y nos invitá a subir a bordo. Penetramos en el puente, que es al mismo tiempo Cuarto de Radio, Derrota y timonera. Observamos un radiogoniómetro, un transmisor de gran potencia, dos receptores superheterodinos y un maremagnum de planos, transportadores, escuadras, tablas y relojes. El compás de gobierno aparece cubierto por un cubichete de latón.

Danny Weil extiende sobre el planero una carta de navegación y nos señala con un lápiz la derrota prevista. Comienza a hablar con voz cortada y ligeramente monótona:

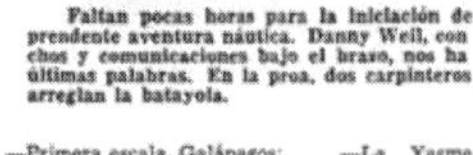

Faltan pocas horas para la iniciación de prendente aventura náutica. Danny Weil, con chos y comunicaciones bajo el brazo, nos ha últimas palabras. En la proa, dos carpinteros arreglan la batayola.

—Primera escala, Galápagos; después Clipperton, islas Marquesas, Tahiti, Cook, Tonga, Fidji, Samoa, Nueva Caledonia...

Cesa de hablar y la punta afilada del lápiz sigue trazando sobre la carta la línea de una fantástica navegación, que nos lleva a la primera juventud, cuando las novelas de Jules Verne o los relatos de Stevenson, Alain Gerbault o Vito Dumas, constituían las fuentes inagotables de las noches blancas; esas noches perdidas entre las tumbas de los recuerdos que todo joven ha presentido a través de las mágicas recreaciones de la imaginación.

Es una navegación de unas 28.000 millas, por la Polinesia, la Micronesia, Indonesia, las costas achicharrantes del Océano Indico, las caletas oscuras de Africa del Sur, las arribadas a los fondeaderos mudos cuajados de leyendas de navegantes, de piratas y de negreros: Nicobar, Seichelles, Madagascar, Zanzibar, Ascensión, Santa Elena, la isla rocosa del Gran Corso... Mar, siempre

—La Yasme
patrocinadora p
Woods Hole Inst
York, entidades
investigación icti
pruebas de radio
tal electrónico
gación. A este fo
ter científico se
hasta la fecha 32
el mundo, incluy
dor.

—¿Qué trabaj
hace sobre la pes

—Recogemos el
tuamos las manc
do y recogemos
especies raras.

—¿Y sobre rad

—Verifico las
de origen solar
nético sobre las o
buscando una int
estos fenómenos
diciones meteorol
tes, estáticas, etc

Hacemos un p
quebrado por el
martillazos y las
obreros que prep
mes III" para es

A Guayaquil newspaper, Vistazo, featured Danny and his adventures.

In Guayaquil, as published by Chesser

This made me a little late but that is the habit in S.A. and in a very short time Joe and Elsie arrived to pick us up. They had indeed thought of everything, even to a bouquet for the bride. The flowers were lovely, pale pink carnations with just yards of delicate ribbon holding them together. As Elsie says, "Whoever heard of a bride without flowers?" We all arrived at the Junta Misionera Foránea de la Convencion Bautista Surena, Baptist Church to you, and there the Dedication of our Marriage to God took place. That is the name given to a ceremony which follows a civil ceremony. We have learned that this is a custom here as marriages performed only by the church are not recognized.

The tiny church was almost filled as there was a large representation from the Radio Club plus a number of neighborhood children to see the Americanos. As we were being showered with rice upon leaving the church the USA didn't seem so far away.

We returned to the Yacht Club where YASME III is anchored, and the bride and groom were toasted in the approved manner and a lovely luncheon was served to all the guests. Joe and Elsie had done it all. When YASME sails Danny and I will sincerely say, 73s and 88s to the Guayaquil Radio Club.

Newly-married Danny and Naomi took a side trip from Guayaquil to Quito, and Danny wrote about it in the YASME Newsletter. Danny noted seeing "The Middle of the World. Here we saw the monument with Latitude 0.00 and we could straddle the northern and southern hemispheres. It was a great moment for Naomi and me." After two days of sightseeing he said they were ready for their next stop, the Galapagos.

YASME Newsletter Number Six, November Nine 1960, reported that with Joe Emmert's departure, Danny and Naomi were sailing alone. They cast off from Guayaquil on 6 October, 1960, headed for the Galapagos. Danny continued his daily radio reports to Spenceley, the first of which reported diesel engine troubles; the intake water filters were clogged with refuse from the Guayaquil River and Danny reported it took an all-night job to clean them.

But off the coast of Ecuador both the main engine and the Onan diesel generator stopped for the same reason. "Danny had no alternative to strip down the entire cooling systems of both engines and get to the seat of the trouble," Spenceley wrote. "It was found that the refuse had permeated the entire cooling system of both engines even to the extent of springing a couple of oil seals and breaking a water pump piston. After 16 straight hours of Spartan effort, parts were machined on the lathe, the engines dismantled and thoroughly cleaned and then the entire mess reassembled. All this was done with the YASME rolling at 40 degree angles and, as may be imagined, this was no ball!"

After that, the generator's water exhaust pipe broke and was replaced with "various lengths of pipe and the liberal use of pipe couplings." Then, with fair winds and favorable seas, YASME III made good time, with San Cristobal Island sighted on 15 October.

Danny poses at the Guayaquil, Ecuador club station

Because of fog and strong currents Danny chose to wait until the following morning to approach the island, dropping anchor on 16 October.

The commandant of the islands provided Danny a shack to operate from, one mile from the anchorage. Gear was lugged to the shack - an HT-32 transmitter weighs nearly a hundred pounds - and HC8VB started operations at 0030 GMT, 19 October. From then until 30 October 3,200 contacts with 63 countries were logged, on CW and SSB plus a few on AM phone, according to the YASME Newsletter (no logs survive for this operation).

24· Off to Clipperton Island - No Picnic

What a place for a land-lubber XYL! - Dick Spenceley

Danny and Naomi left the Galapagos on 2 November 1960, headed for Clipperton. The last amateur radio operation from there had been by Bob Denniston, W0NWX (later W0DX and VP2VI), more than five years before, signing FO8AJ. The 1200-mile trip to Clipperton was expected to take 12 to 14 days, against the currents.

Clipperton would ultimately be the rarest ham radio spot Danny would operate from, and it is a lousy place for a vacation. It's uninhabited, for one thing, and the weather is very unstable, featuring sudden and violent squalls. Spenceley noted that "sailing instructions describe Clipperton as a very hazardous spot to visit even under the best conditions. We are confident that Danny and Naomi, who by this time could, no doubt, qualify as an able bodied seaman, will overcome these dangers as Danny has in other treacherous spots such as Aves Island and Baja Nuevo. We will, however, breathe a considerable sigh of relief when YASME heads Marquesa-ward."

It took much longer to reach Clipperton that two weeks. Chesser listened in on Danny's radio schedules with KV4AA, with relays from W4OMW and W6AM, and said "Like another and memorable stab at Clipperton Island (FO8AJ) some five years ago, it was just one flamin' thing after another recently when Danny, VPZVB/MM et al, made his scheduled attack on that same island. It would appear landing conditions on the island haven't improved a bit since the W0NWX saga, when almost constant heavy seas were reported in that area with 40 knot (that's right - 40 knot!) currents swirling past the island. If caught and held in one of those currents, it would require a motor-torpedo boat at nearly full throttle just to maintain a stationary position!

From the Galapagos's San Cristobal Island YASME had sailed to Santa Isabel Island where Danny and Naomi had rested, cleaned up, and repaired the Onan diesel generator. From there to Clipperton they used the ship's engine sparingly, to save fuel for the trip they would have to make from Clipperton to either Tahiti or the Marquesas. The winds cooperated for a couple of days, then diminished to a dead calm.

By 17 November the winds, and seas, were back and building and "YASME was taking a pounding," Danny later wrote. Things worsened and both the fresh water drum and one diesel fuel drum "broke loose and ran amok on deck before crashing through the guard rail and disappearing into the sea. One especially large wave swept over YASME and carried away the fuel tank for the outboard motor. One of the outboard booms was broken in the middle and the sails had suffered many rips."

The next two days were even worse and "the bruised and groggy crew were debating whether, or not, to jump overboard." The sails were ripped, cooking was impossible, and a leak in the stern of the boat made pumping necessary every three hours. "Finally the fore stay, a half inch steel cable running to the top of the mainmast, snapped and ripped away the forward guard rail. This imperiled the mast itself and prompt action was necessary to save it."

"No 40-knot boat, the YASME III took a terrible beating in those seas. What a place for a land-lubber XYL!" said Spenceley.

Chesser called Danny's transmissions on 27 November "terse and dramatic." They were recorded by Don Wallace, W6AM, verbatim. On 27 November Danny transmitted:

> Sails gone. Can't use engine, so guess we are drifting about. We are about sixty miles south of Clipperton. Doesn't look like we will get there. Gale has been blowing for over two days, now.
>
> Weather very bad and everything

Danny and Naomi have dinner with members of the Quito Radio Club before leaving for Clipperton.

aboard is in rough state. We are all soaking wet. Our safety rail has been ripped off when our fuel drums broke adrift. They broke one inch rope and I couldn't handle them. Two drums went over the side. We are tired because this is the third day without any help. We are rolling about 45 degrees. Wind speed up to 50 miles per hour. It's easing now to about 30. It's been heavy squalls for 4 days and we don't seem to be making headway. Also got bad leak in stern, but we seem to be able to handle it with electric pumps. Let's hope we still got juice because- I'm in no fit state to pump by hand. We will do our best to get to Clipperton, but must think of safety of boat and my wife and me. Will head to Acapulco (Mexico) when we can get boat to go in that direction. Wind now from Northeast and dead against us. Seas are about 18 feet high.

My time is taken up in keeping boat afloat and my wife and myself alive. I'll try to be on this frequency (21093 kc.) every day on the hour. I have skeds with KV4AA three times daily. Now I'm going to try to get some chow because we have not eaten for 36 hours. My wife is ill, and I'm anxious to reach an anchorage. But guess the only place is Clipperton, and that's about sixty miles away to the north, and the wind is ahead.

Chesser listened in again on 3 December to a badly garbled contact between Danny and W6AM:

... because all my anchors are gone and my . . . (QSB) [fading signals] ... squalls made it very dangerous ... (QSB) ... we could lose YASME ... (QSB) ... we are on our way to the United States but (encountering) very bad head seas and winds and don't think we will have enough ... (QSB) ... to get back. All our sails except main are torn to pieces and we hope to get some ... (QSB) ... to help us. Heading southeast because headwinds in direction of ... (QSB) ... Return to Clipperton one day but my YASME ... (QSB) ... repairs and fix our leak, then we will see what transpires.

Chesser later wrote that "Despite these hardships, however, Danny did get to Clipperton, and he got on the air two nights for a total of 775 QSOs during the two sessions, according to KV4AA. There (might) have been more had he not been forced to put to sea between operating sessions and cruise about, probably because of weather conditions. With fuel supply seriously depleted, he headed for Acapulco, Mexico (estimated a six-day sail), and arrived there safely. At this writing he and Naomi are still there, with plans to sail next for Los Angeles. KV4AA reports there's a good chance he will make another attempt at Clipperton in about two months' time."

Dick Spenceley, meanwhile, wrote and mailed Issue Seven of the YASME Newsletter while Danny was on Clipperton. Spenceley did a masterful job of being on the radio for schedules with Danny, and others, taking notes, and then publishing a report before the operation had even ended:

As we write, Danny is at Clipperton Island, signing FO8AN. He dropped the hook, in sixty feet of water, at the island's only anchorage at 2220 GMT on November 28th and YASME III came to a comparative rest, barring considerable motion from the surf, with 30 feet of water under her and just 300 feet from shore.

Every time YASME sails we die a little down here. This trip was one of the roughest ever and, at times, it looked as though they were not going to make it. But, thanks to Danny's navigation and perseverance plus his ability to stay awake for prodigious lengths of time, on a very sketchy diet, YASME came through. One may think, when reading stories of these voyages, that the attending rigors are overemphasized in an underhand effort to arouse sympathy, raise contributions, or just to sound interesting. In my opinion, exactly the opposite occurs as mere words cannot begin to present a factual picture of the hardships encountered to those hams sitting in their cozy shacks unless they have been participants in trips of similar nature.

YASME's voyage from Santa Isabel Island, in the Galapagos, took a total of nineteen days to cover the 1200 miles to Clipperton. At no time during the trip was YASME blessed with favorable winds or currents. In fact, contrary to his charts, there were no currents which were of a speed to be taken into consideration in navigating. The last week of his voyage was attended by almost a full gale with twelve foot waves and winds up to 50 MPH. Spenceley said

On November 29, at 2335 GMT, Danny had set up an HT-32 and SX-101 ashore, strung a long wire antenna between two sticks, and F08AN/Clipperton was in business with a substantial signal. Operation continued, with a few hundred more contacts until 0325 GMT on the 30th.

On the morning of the 30th Danny discovered that

his anchor winch has rusted up and was inoperative. This was a five hour job and the subsequent happenings are a story in themselves which we hope to cover in the next issue. Briefly, the anchor had to be raised by hand and was lost in the process. YASME then had to put to sea and spent the entire night of Nov. 30th cruising off the island, much to the disappointment of the many stations waiting for him. Early the next day another anchor was rigged and YASME returned to her anchorage. All this added up to another sleepless night for Danny who, by now, had just about "had it."

Danny promises to do his best at Clipperton and rack up 3000 or more contacts, but it must be pointed out that his operations may be sporadic as dictated by weather and surf conditions. YASME must be watched closely at all times. Should the anchor chain snap from strain or from being snubbed around coral YASME would quickly be driven ashore unless the engine was immediately started to head her seaward. W0NWX, of the 1954 Clipperton expedition (FO8AJ), states that their ship, the BARCO DE ORO II, lost three anchors at the island. I think that all will agree that this spot is not the best place in the world to be stranded on altho castaways could probably exist there for a considerable length of time.

Some W8 advised Danny that Clipperton is some 5 miles southwest of its charted position. Danny refutes this. Clipperton is right smack where it is supposed to be.

Sometime in the week or two after leaving Clipperton Danny found time to write a lengthy story and mailed it to Spenceley. Not having space in the single-sheet YASME

***** **********
YASME NEWSLETTER
***** **********

****** *****
NUMBER SEVEN
****** *****

******** ****
DECEMBER 1960
******** ****

******** ***** ** *** ***** **********
OFFICIAL ORGAN OF THE YASME FOUNDATION
******** ***** ** *** ***** **********

DIRECTORS: W6GN W8EWS VP2VB G2DC KV4AA
EDITOR: KV4AA
QSL MANAGER: W8EWS

MEMBERSHIP $5.00
PER YEAR

SPONSORING THE YASME III DX EXPEDITION.

As we write, Danny is at Clipperton Island, signing FO8AN. He dropped the hook, in sixty feet of water, at the islands only anchorage at 2220 GMT on November 28th and YASME III came to a comparative rest, barring considerable motion from the surf, with 30 feet of water under her and just 300 feet from shore.

Everytime YASME sails we die a little down here. This trip was one of the roughest ever and, at times, it looked as though they were not going to make it. But, thanks to Danny's navigation and perseverance plus his ability to stay awake for prodigious lengths of time, on a very sketchy diet, YASME came through----.

One may think, when reading stories of these voyages, that the attending rigors are overemphasized in an underhand effort to arouse sympathy, raise contributions, or just to sound interesting. In my opinion, exactly the opposite occurs as mere words cannot begin to present a factual picture of the hardships encountered to those hams sitting in their cozy shacks unless they have been participants in trips of similar nature.

YASME's voyage from Santa Isabel Island, in the Galapagos, took a

Newsletter, Spenceley again passed the story along to Don Chesser, who published it in the 30 December 1960 issue of the DX Magazine. Chesser said "We are most pleased to present this description of the problems and adventures encountered during the recent FO8AN/Clipperton Island operation, written by Danny as he and his bride rested at Acapulco, Mexico, and forwarded by Dick, KV4AA., President of the YASME Foundation. This is by no means the end of Danny's troubles, as you will learn at the end of this article."

The story established Danny Weil as the gold standard for DXpedition writing, something Wayne Green had understood five years before.

"CLIPPERTON!
by Danny Weil, VP2VB/MM, FO8AN, et al"

Arrival at Clipperton Island brought no hearty cheers from the crew of YASME III - in fact, the sight of the breakers which surrounded the entire island and the long run of reefs tended to create great despondency to all aboard.

We had been assured there was one anchorage only available and its position was made very clear to us both in the sailing directions and also by those who had been there before us-and yet, when we saw what was to be our anchorage for the next week to ten days, a slight feeling of the shivers crept up our spines.

It was a lea shore. For the uninitiated, a lea shore is a place where the prevailing winds and seas break continuously, and there is NO shelter; but it was the only place near the island where the water was shallow enough to drop an anchor. Elsewhere, as I discovered by a long and arduous search, the edge of the island dropped clean away to no apparent bottom, making it impossible to anchor in any case. Here on the northeast side it was a steeply shelving beach, which gave us an anchorage within about 200 or 300 feet from the edge of the "beach". As little as 400 feet off shore there was no apparent bottom, according to our depth finder, so we were forced to approach the shore too closely for my comfort to drop the hook.

I chose the biggest of our anchors-56 pounds-and with it went out 60 fathoms of "1/2" tested chain. As the anchor dropped into the depths, I slowly went astern, with the breakers and wind helping. Naomi went forward and called to me when the chain went taut, and I threw the engine into neutral and waited for YASME to assume her own position. She swung inshore for a few minutes and then broadside to the surf, creating a fantastic roll which threw everything not secured onto the cabin floor. I swung the helm, hoping she would head into the wind a little more, but the current and wind placed her in the worst and most uncomfortable position possible.

The sailing directions stated very clearly that northeast squalls were prevalent in this area, that it was advisable to clear out immediately when they came, and to let the anchor and chain go without waiting to haul them in. This is fine if one has an unlimited supply of anchors and also a sufficient quantity of available rope and buoys to drop over the side at each and every occasion when the squalls come .... We weren't equipped this way, And it meant hauling that anchor in every time and hoping we would make it before the squall reached its peak. Our hydraulic winch had been a little troublesome in Galapagos, which 1 thought was caused merely, by air in the system and which should ultimately clear itself.... How wrong I was!

Before I say another word, you will have to excuse me if I get a little fouled up with dates and times. Those things were of little importance to us under the circumstances, and no notes were taken other than to enter the correct dope in the log for OSOs. The ship's log remains very blank even to this day, as our experiences at Clipperton could never have been put into words which could ever be believed by any normal person not accustomed to the sea.

Even now, as I write, I have to speak with Naomi and try to figure what we did on certain days. Even she cannot remember, but I do know we spent one ghastly night at anchor, and the following day I made my first run ashore.

I had been advised that the wrecked vessel ashore would form a lea for landing, and one could get ashore very easily, in water' up to one's neck, while watching for the innumerable sharks that abound in this place. I made three attempts to land at this point, but the breakers were far too big to attempt it. Also, an occasional glimpse of rocks or coral dissuaded me from that particular sector, so a search was made farther along the coast for a better landing.

Farther northward I discovered a shallow ridge, which I can only assume was at one time an entry into the lagoon, and it was here that I noticed the seas were not quite as vicious, although the reef was very obvious as the seas swept back and forth.

Naomi and YASME III's damaged winch, off Clipperton, after "ferocious winds"

Landing on reefs presents no difficulty to me, and with a reasonable sea, which one can time, I have found one can make successful landings, providing the right time is chosen.

My first attempt became almost a disaster as the dinghy twisted with the surf broadside on, and water boiled over the side, almost capsizing it. Fortunately, I held onto the oars and was able to bring the bow around and rowed back to sea again, where I spent a little time bailing out with an old saucepan. My next attempt was successful, and I swept in on the crest of a wave, which, luckily enough, held enough body to lift me over the reef and onto the sandy beach. Once over the reef it was easy to control the dinghy, and I jumped over the side in the shallow water and dragged it to a higher level.

The last expedition to Clipperton had made their station site at the side of the wrecked lighter, about one half a mile farther south, and had also apparently found the seas a little calmer in that area. But, in my short stay, it was an impossible task at any time to get ashore at that point.

I remained ashore for a while, checking the reef and trying to find a clear opening in it to make my future landings, but there was little to be seen with the continual breaking seas, I chose to mark the point of my first landing with a piece of drift wood, which I stuck into the sand as a guide. I wanted to make further exploration of the area, but I could see squalls building up, and thoughts of Naomi aboard with that fantastic motion and breaking seas around her quite frankly scared me.

Strangely enough, I was able to clear the beach with ridiculous ease, which for the moment made me laugh at my apparent fears. It was the one and only time I actually cleared the beach without capsizing, loss of gear or some damage to myself or the dinghy.

I returned to YASME, and we laboriously hauled the dinghy aboard. I had tried to trail it astern, but the seas picked it up bodily and smashed it under the stern of YASME. Later, I spoke with Dick, KV4AA, and told him my opinion of the place, doing my very best to keep out all words which might shock the innocent. You can imagine that when all words of that kind were removed, it left me little to say!

Radio conditions were bad, but I knew I had to get ashore and do something, even if I made only one QSO. The journey to Clipperton had been the worst in all my years of sailing, and, given the opportunity, I would have told everyone what they could do with the island and the expedition. But I guess, at the back of it all, I just had to make a try at it, even if it was only for personal satisfaction. I told Dick on that second day that I would make the attempt to get the station equipment ashore.

Using the specially built bags, supplied by Hal, K5JLQ, I set off from the YASME loaded down with the entire works, consisting of HT-32, SX-101, Onan generator, tent, a wooden box full of odd junk, a gallon of coffee, and 200 cigarettes. I knew I had to make the attempt, and also that it had to be done first time. There was too much risk involved in going back and forth to YASME, picking up the gear in easy stages. Also, I knew I would not have the strength to fight that sea three or four times in a row. It had to be done, in its entirety the FIRST time.

Slowly I paddled the heavily laden dinghy near the shore and jockeyed for position. I guess about five or six dummy runs were made to get the feel of the surf and wind, until I felt sure in my own mind I was as ready as I ever would be. With head swinging from forward to aft like a wooden doll, to watch each sea, I pointed her bows toward that tiny wooden guide marker stuck in the sand. With the spray it was invisible more often than not. I had been a fool not to have found a longer piece of wood-but it was a bit late to think about that, now!

Gradually I neared the shore, and, at the precise moment, I swung her around with stern to shore and bows to breast the building breakers. In one came. It was a really powerful solid wave which picked us up for the trip in. With both oars working nineteen to the dozen, to control the dinghy, we kept a level heading, still dead in line with my tiny guide stick. It seemed like hours, yet it was probably only seconds. The wave collapsed and I felt that awful grating sound on the bottom. We had hit the reef!

The backward surge of the sea tried hard to drag us back, but we were temporarily held fast on the coral and my oars were flapping around like a frightened bird's wings to hold her in position. Another sea swept in and threw us broadside on, but this time it dragged us onto the "beach". We capsized-but, strangely enough, as the dinghy upset, so the seas receded, giving me a few seconds in which to grab the generator, which was the only piece of gear not protected by K5JLQ's watertight bags, and move it out of the surf line. The other gear was all well covered and could stand complete submersion, but with those seas running I spent little time thinking about anything other than to get the whole issue out of range of the sea.

It was then I discovered a disaster had occurred-my flask of coffee had broken! But I still had cigarettes.

I won't go into the business of erecting the tent and getting the equipment on the air. My antenna was a random piece of wire strung from the tent to a piece of wood hoisted up in the sand. It was a pitiful attempt, but it was an antenna of sorts-and it loaded up and even put out a signal.

Operating FO8AN was one of the most nerve-wracking experiences I have ever had in my life, trying to concentrate on the pile-ups while watching the sky for approaching squalls, watching YASME snatching at her anchor and seeing the occasional sea break over her as she swept into the surf line, and wondering how Naomi was taking it, and whether she had fallen over the side with the fantastic motion of the boat. And wondering if I should operate just one more hour and then get off that beach and back to YASME, knowing I would have to return again, but not knowing if I would make it or not. The night scared me. There was bright moonlight, but it only accentuated the size of the seas and threw shadows which made me tremble as I worked to get clear of the beach and then to land again.

After several runs between YASME and shore, each appearing to be worse than the one before, I sat there on the beach and wondered why on earth I was doing this crazy thing. Have you ever tried to analyze the things you do? Have you come up with the right answer . . . or an answer that makes sense? Nothing made sense to me, yet there I was, leaving my wife aboard a boat she could never handle in an emergency while I sat ashore, frightened out of my life, punching a key or talking into a microphone.

I thought of the thousands of hams throughout the world who had been anxiously awaiting my arrival at Clipperton. For the fortunate few, their waiting was satiated and they made that million dollar contact. I wondered if they thought for one second what had been sacrificed, both mentally and health wise, to give them that one quick contact. Did they really recognize the work and worry which had preceded the expedition long before YASME III was a fact, and did they ever think of the few who had made it all possible . . . ? Perhaps a few who had followed the YASME expeditions over the past years did really appreciate what was going on, but there were also the many who couldn't have cared less as long as they made the contact, and "to hell with who had done what to make it possible."

There are the few, too, who will say 'you volunteered . . . no one asked you to do it. . . etc.' Perhaps this is so, but there will always be the few who will do these things without thought of reward or recognition because they think it is a fine thing for everyone in the fraternity. When I am dead and gone there will always be expeditions. There will always be the fools who will devote their time, money and life to meet the demands of the DXer, but, like many others before them, they will be forgotten in the dust and waves of time.

It was very obvious to me that one anchor could never hold the YASME in her present, dangerous in-shore position. I had other chain aboard, and three other anchors. The chain was stowed in a very difficult spot, but it had to come out. Between

Naomi and myself, we dragged it clear of the bilges and got it on the deck. It weighed about a ton, and was over 3,000 feet long, but it had to be used. Our other 56-pound anchor was shackled on, and, with the aid of the engine, I was able to place it in an advantageous position to help share the load on the other. Although this was holding the boat in position, I also thought that when the next squall came up we would be twice as long retrieving those anchors. It was a mental struggle to decide whether it was wise to have the extra anchor overboard, with a longer delay in getting under way in case of an emergency, or have only one anchor, which would enable us to clear out more quickly but which would double the risk of maintaining our position in the seas.

Our problem was solved for us. A really king-sized sea swept in at the beginning of a squall. YASME was picked up bodily and tossed beachward. There was a sudden tearing sound, and then the whole boat shuddered as the anchor chain strained and snapped. To the northeast I could see the sky darkening. The wind had gusted up, and then the barometer started to fall. Tenth by tenth it fell until the thin needle showed a fall of four-tenths. I had long since started the engine and thrown the clutch in to haul in our remaining anchor, but although we could hear the hydraulic motor pumping, the winch didn't move. Throwing the valves back and forth, topping the reservoir with oil and kicking the chain made no difference. We were stuck there with 60 fathoms of 1/2 inch chain out and a 56-pound anchor embedded in the coral, plus an unknown quantity of spare chain draped over the bows from our lost anchor!

The winch was certainly not going to work. I rigged up a rope block and tackle to attempt to haul the chain in, but it was like trying to move a 50-ton tank with one finger. I didn't even get to first base with my block and tackle, and I realized, too, that if I let this anchor and chain go, we should never be able to stay at the island at all, and our gear ashore would be lost. Some will say: "why not tie a good length of rope to the chain to buoy the anchor and pick it up later?" This works fine in books, but not so well in practice. First, I didn't have any rope on board in sufficient length for that work, and what I did have could never have been strong enough to pick up the weight of chain from the bed of the ocean. I had no alternative but to get the winch working, and pray to God the anchor would hold until I had it fixed.

The first project was to get the chain off the winch, so it could be worked on. This took the best part of four hours, with the aid of many blocks and tackle plus the mast winch. With all this I managed to get about four inches of slack, sufficient to free the chain from the winch. As can also be expected, the nuts and bolts on the winch were fairly rusty, and it took every wrench, hammer and chisel, and lots of angles, to free them all. The entire winch, plus internal pump, had to be drained of hydraulic oil. There were ten gallons in it, and I had to conserve every drop, for we had very little reserve on board. Every drop was pumped out with a small hand pump and into numerous cans. The boat rolled alarmingly, and to pump the winch dry and conserve the oil, plus try to hold on to the cans, and hold oneself on the boat, was an almost impossible task, but it was done. The oil was all sealed up in the cans and stowed very carefully below.

Then came the job of stripping the winch of its million and one cylinders, valves, springs and balls. I eventually nailed a wooden box to the deck to prevent loss of parts, and gradually I got the thing in pieces and discovered the trouble. A main cylinder gasket had blown, causing the hydraulic pump to just circulate the oil without doing any work. Then came the search to find suitable material which would stand up to pressure and also the chemical reaction of hydraulic oil without impairing the normal operation of the winch.

The job took thirty-six hours, non-stop, from night to day and back to night again, working with flashlight . . . trying to keep awake and to balance oneself while working. Naomi had lodged herself on the deck between the fuel drums, holding wrenches and parts and trying to anticipate my needs as I came to each part of the job. Her job was of the worst. She scrambled along that deck . . . climbing over drums and tangled ropes to keep me supplied with coffee. As one cigarette would be blown away by the wind or extinguished by the spray, so she would have another ready for me. She poured hot coffee into me the whole time and never once complained, although I knew it was hitting her pretty hard. So we finally neared the end of the job, and the last and final nut was tightened. All the tools were swept together in a pile and dropped haphazardly down the forward hatch.

Between us we filled the winch and the hydraulic pump very carefully from the dripping cans, clearing the air from the system and insuring nothing was binding.

Then came the great moment. Would it work? I had labored too rapidly on the job, and I realized I could have made a million mistakes. Perhaps a valve was installed the wrong way, or maybe a spring not replaced. My brain was spinning as I thought of all the places where I may have erred. With a prayer on my lips, I went below and started the main engine, and threw the hydraulic pump into gear.

I called to Naomi to find out if it was working, but the scream of the wind made it impossible for me to make myself heard. I scrambled up and along the deck, but there was nothing to see. There was no movement from the winch at all. Then I suddenly remembered I hadn't turned the main valve on. As it screwed down, the winch started to turn with a satisfying grunt as the powerful hydraulic pistons slid up and down turning the massive chain gypsy. We grinned at each other and offered that prayer again. I had held the slack in the anchor chain all this time with my tackle, and within a few minutes we had the chain attached to the winch.

With hardly any change in note from the winch, the chain started to come in. To hear the clatter and clank as it fell into the chain locker was the finest music in the world to us, and although it was blowing 35 to 40 miles an hour we continued hauling in until the anchor dangled over the bow. With engine full ahead, we steamed northward and away from this God-forsaken anchorage into the clear, open sea. Everything, including the radio equipment ashore, was forgotten in our elation to be free of that spot, and that night: although we were both asleep on our feet, we kept sailing farther north until we cleared enough mileage to heave to and rest.

Finally the gale blew itself out. We had felt the worst of it at anchor, and here at sea, with the benefits of a steadying sail, things were much more comfortable. We rested.

We returned to the island again the following day and dropped our big anchor and a smaller one attached to the remains of the chain that had not been lost in our earlier escape, and, once again, I tried to make a few contacts. Again it blew up, and once more we hauled in the anchors and cleared out to sea.

Later, back ashore, I found the tent had blown down, but other than a slight dampness no damage had been done. The generator had well and truly gotten soaked and it took me some time to get a spark from it. By [this] time, as you can well imagine, we both of us had had enough. Physically and mentally we were at our lowest ebb, and I knew that to prolong our stay would be disastrous to ourselves and YASME. The culmination of everything was a final squall that took away our smaller anchor and the remains of our chain. I had reached the end, and although I had promised the DX gang I would be on again later, at 0500 GMT, on the 3rd of December, 1960, F08AN closed down after making only 700-odd contacts.

The weather had now deteriorated to such an extent that it was impossible to remain anchored. The tent was down again, but I knew the equipment was covered and well protected. We stooged around, waiting for things to subside. They never did. I had to take the chance or lose our gear ashore. Naomi had proved herself a fine helmsman, and I knew she could carry out instructions. We got the dinghy over the side, and I instructed her to motor in circles, keeping at least a quarter-mile from the shore. If things got too bad, she was to head north, but to always keep the island in sight.

With these last minute instructions, I climbed into the dinghy and made my way ashore. It would be pointless to go into the gory details of the capsizing, the hole smashed in the stern, the bows broken away. . . . Let it rest that I did get the gear aboard after two attempts, and although I doubt if transmitter, receiver or generator will ever work again they are aboard now. Perhaps when we finally reach Los Angeles we may see them operate again.

It hurt me very much to have to give up at Clipperton, as I hate failures. But with the loss of our ground tackle, two drums of fuel, one drum of water, and lack of food and rest, I thought it better to swallow my pride and get out while we still remained alive and had a boat.

Although we lacked a chart for the voyage from Clipperton to Acapulco, Mexico, the trip was made without any undue occurrences other than headwinds and seas . . . . This could be another story for some other time.

We finally pulled into Acapulco harbor around midnight, December 10th, wondering where the heck to go. We dropped anchor in a favorable spot and hit the sack.

In mid-December Chesser reported that Dave

Tremayne, ZL1AV, had resurfaced, at home in New Zealand. Tremayne said he was "seriously thinking of going on another DXpedition - by myself this time. If I could get a KWM-1 or KWM-2, or something like that, I would be only too glad to pay my own expenses to these places, but the main problem is getting hold of the portable equipment out here for such a venture. I now realize the importance of light equipment after lugging kilowatt gear ashore at various spots with Danny. Believe you me, it's no Sunday picnic by any stretch of the imagination!"

25· The Soul Searching Begins

Thus, to a clanging, clashing, gurgling halt comes the YASME around-the-world expedition. - Don Chesser

Spenceley, having sent Danny's Clipperton story to the DX Magazine for publication, followed up with a letter to Chesser, describing what he called "additional YASME difficulties." With Danny, Naomi, and YASME III safe in the Mexican port of Manzanillo, the soul-searching had begun. Chesser published the report in his 30 December 1960 issue, just a couple of weeks after YASME had left Clipperton (on 3 December).

After five days of rest in Acapulco, Mexico, the YASME III, with Danny and Naomi aboard, sailed from Acapulco bound for San Diego, on 15 December 1960. While crossing the Gulf of Lower California a leak in the stern worsened and presented them with an "emergency situation" on 18 December. YASME's engine greatly aggravated the leak, so that water just gushed in, and the engine had to be shut down. With the electric pump operating, YASME just about held her own to keep from foundering. A call for assistance was sent out.

XF4ICV was contacted, and he immediately informed the Mexican Navy (who, on the trip from Clipperton to Acapulco, had offered the YASME any assistance necessary). Action was immediate, as the Mexican Navy patrol boat "California" was immediately dispatched from Socorro Island and proceeded to YASME's position at 16 knots.

Spenceley said that during the night of 18 December and the next day "many amateur stations monitored our distress frequency, 14307 kc SSB, and kept continuous contact with YASME. Notable were XE1CV, XE1AE, W6LZV, WA6EXM and K6CQU/MM. The last mentioned was Murray, aboard the SS Horace B. Luckenback, headed north toward Danny and Los Angeles. Murray and the Mate, Navigator and Captain of this ship were most helpful. Constant contact was maintained between YASME and Carlos, XE1CV, who relayed data to the Mexican Navy Department and hence to the rescue ship, California."

Danny calculated his exact position at 1900 GMT on December 19th, and four hours later the Mexican Navy Patrol Boat "California" was sighted on the horizon, dead on course for the YASME. A line was secured to YASME at 2355 GMT and the tow to Manzanillo, a Mexican port some 80 miles away, was begun. Speed had to be immediately cut down as it aggravated the leak. At 5 a.m. the next morning the SS Horace B. Luckenbach passed, and Murray transmitted: "And there, big as life, was YASME with the California towing her!"

YASME arrived at Manzanillo the evening of 20 December.

Chesser wrote that "Another milepost in the attempts

Naomi and Danny around the time of Clipperton

of Danny Weil, VP2VB et al, to circumnavigate the world was reached last month-this one possibly marked 'STOP'. After a series of heartbreaking events that would have tried the patience of a saint, Danny and Naomi were towed into San Diego harbor, too tired to be embarrassed, too dejected to be elated over their arrival."

After being towed into the harbor of Manzanillo by a Mexican Navy patrol boat just before Christmas, an inspection of YASME revealed that the leak in the stern was caused by a defective propeller. A new propeller was shipped to Danny by air from the United States and installed by divers. It would have been advisable, Chesser was informed, to have dry-docked the YASME for more extensive repairs, but funds were not available for "this very costly operation."

With the new propeller installed and all leaks cleaned up, YASME III sailed from Manzanillo on 5 January, bound for San Diego, some 1,250 miles to the northwest.

Would this trip, [Spenceley wrote] be free of the hardships and near disasters which had been haunting YASME over the past two months? Silly question!

> Sixty miles out of Manzanillo the exhaust pipe of the Onan diesel generator let go. Because head seas had built up to such proportions that little

progress could be made, it was decided to put into nearby Bahia Chamela to repair the exhaust. As this bay was approached, the seas and surf were of such a size that Danny would not risk entry, and YASME turned about and headed seaward.

On January 8th the wind dropped and things were much better as YASME passed Cabo Corrientes and headed out across the mouth of the Gulf of Lower California. During this time he was able to effect repairs on the diesel exhaust system.

Cabo San Lucas, on the tip of Lower California, was passed on January 10th, and YASME proceeded up the coast. On January 13th, at 1500 GMT, Danny reported his position as 27.12 degrees N. Lat., 114.30 degrees W. Long - or about five miles off San Pablo Bay, with 350 miles to go.

At this point the fates decided that Danny and Naomi were having too easy a voyage, and they decided to cast their evil spell on YASME's main engine. This hardy collection of nuts, bolts and cylinders, which had been ticking over nicely since it had been deloused of the debris picked up in the Guayaquil River, responded immediately with retreating oil pressure, heat, and various and sundry grunts and clanks. It had to be immediately shut down. Lengthy efforts to repair it met with no success. The fault was tentatively diagnosed as a crack in the engine block.

So there YASME lay, in a dead calm, with the contours of Isla de Cedros in the distance. Danny lost no time in advising West Coast SSB amateurs of his predicament, and they, in turn, notified the U.S. Coast Guard.

The next day a Coast Guard aircraft flew down and easily located YASME. The situation was stated to be one of emergency but not distress, as the weather was very calm and food for five days was still aboard. Hope was expressed that a wind would spring up, enabling YASME to continue on her way under sail or to put into Cedros Island, where a cannery exists with some facilities for boat repairs. Naturally, this did not occur.

On January 18th, with the YASME making little progress in any direction, except that she was drifting out of sight of Cedros Island, a Coast Guard plane was again dispatched to drop some requested lubricating oil and to make a reappraisal of the situation in view of the diminishing food supply. On this plane was another Danny, W0RPO, who made SSB contact with YASME some sixty miles out and was able to get a clear and lengthy story of his plight. During this time, the Onan diesel generator, which needed the lubricating oil, apparently gave up the ghost, and Danny was reduced to the power supplied by a small Onan gasoline-driven generator, which had an estimated four hours supply of gas.

With all these facts at hand, a decision was then made to send a U. S. Coast Guard cutter down to tow YASME in. This cutter sailed from San Diego late on January 18th and was due to arrive in YASME's vicinity around 1900 GMT on the 19th. Should difficulties be encountered in locating YASME, a plane would be diverted to aid the search.

Almost continuous communication was maintained with YASME on the 19th by K6BX and W4FVR, who relayed all data to the Coast Guard. At 2000 GMT YASME had still not been located by the cutter, and an aircraft was dispatched. By this time, Danny had managed to get the diesel generator running again, although it ran hot, and he was able to pinpoint his position by giving a bearing on two points of the Lower California coast which had become discernable. With this information, Coast Guard Cutter CG-95318 hove into sight at 2115 GMT and took YASME in tow.

This ended communications with the vessel by radio, but we subsequently learned that YASME arrived in San Diego on January 21st and is tied up at Shelter Island. The San Diego radio gang have taken steps to make Naomi and Danny comfortable after their ordeal and to render whatever assistance they could.

We wish to thank WA6EXN, W3MTG/6, KH6AWS, W6VCA, K6VVA, K6KFY, K8KMS, KL7AWR, and especially K6BX and W4FVR, for the long hours and very efficient communications they maintained with YASME in this emergency.

The January 1961 YASME Newsletter had carried a plea for every Foundation member to dig deep and pledge an extra $10. Spenceley said a 40 to 50 percent return would enable them to carry on. But, the response was only 10.6 per cent, with 52 members pledging a total of "$635." YASME Foundation membership at that time was 490. The call signs of those pledging were listed.

YASME III in Tahiti

Spenceley said "Unless there is a substantial and unexpected change in our fortunes, before the deadline of February 25th, we will have no alternative but to dissolve the Foundation and repair and sell YASME to cover our debts."

Chesser observed "Thus, to a clanging, clashing, gurgling halt comes the YASME around-the-world expedition. Whether it will pick itself up by its own bootstraps depends on Fate - and the DX gang. If you would like to see it continue, contact KV4AA (on 14080 kc daily) for details.

> Possibly no individual operator in the history of amateur radio," Chesser said, "has received greater, wider (or more mixed) acclaim than Danny Weil. His adventures and exploits have been a major topic of conversation at conventions, meetings, in publications and over the air for several years. Whatever the outcome of his present dilemma, he will always remain an impressive phase of our fascinating hobby. If Danny's sea borne adventures do end at San Diego, he will always be remembered by hundreds of DXers for many thrills-and for many new countries.

Chesser also mentioned that Danny had expressed interest in obtaining U.S. citizenship, meaning he would have to give up his British citizenship, which had helped him so greatly in obtaining operating licenses and permission on so many islands of the Caribbean and Pacific areas in past years.

With Danny and Naomi in San Diego and YASME in need of expensive repairs, they rented an apartment, Danny's first place of his own at the age of 42. Danny found a job in a local machine and tool manufacturing company.

Meanwhile, the San Diego DX Club circulated a letter "within the DX fraternity" to gauge support for re-establishing the YASME expedition. At this time, with no publication or individual doing a poll of "needed DXCC countries," the club mentioned the following possibilities for a future YASME voyage: Socorro Island; Clipperton; Marquesas Group; Tahiti; Flint or Vostok Islands; Manihiki Group (Danger Island.); Wallis Island; Tonga; Kermadec; Norfolk and Lord Howe; Sydney, Australia; Willis Island; Timor; Christmas Island; Cocos Keeling; Nicobar-Andaman Islands; Laccadive Islands; Maldives; Chagos; Rodriguez; St. Brandon; Agalega I; Aldabra; Tromelin; St. Helena; Ascension; West Coast of Africa: FQ8, FF7, FF4, CR5, etc.; and Trindade Island.

Spenceley did not publish the YASME Newsletter for April, 1961, and Chesser continued to take up the slack. The foundation was, essentially, dormant during this time, with one exception: it applied for, and was granted, non-profit, 501(c)(3) status by the Internal Revenue Service on 27 Jan 1961.

Otto, K6ENX, paid a visit to Danny and the YASME III in San Diego, in February, discovering that the vessel was being "thoroughly reconditioned in the hopes that the future may hold a resumption of Danny's exploits.

"After talking with him and listening to him," Otto said, "I am convinced that the DX fraternity will be missing a bet by not supporting the Foundation and enabling Danny to continue his travels."

Also at this time Chesser mentioned, from the rumor mill, that "if Danny does resume his world cruise, wife Naomi will not accompany him. In such an event, all efforts will be made to recruit one or two other crew-men and operators to accompany the voyage."

Chesser found more cause for optimism in April, when he wrote "It appears the YASME 'round-the-world cruise might yet pull itself up by its own bootstraps, according to an optimistic letter from Bob, W6ZVQ, this week."

W6ZVQ reported that YASME's engine had been completely overhauled and re-installed, ready for a trial run. With one or two major repair items left, and many small ones, a group of Southern California DX Club members were planning a weekend work party.

The San Diego DX Club's letter was getting some response, most of it favorable, and several have contained good suggestions, which Danny is planning to incorporate, W6ZVQ said. "Not one letter has been downright against the project as yet. The San Diego club will write a note with the general consensus of opinion after receiving the answers, and will mail a copy to each club joining our survey."

W1JNV wrote to Chesser, saying he wondered why some influential West Coast DXers, involved in the movie or TV industry, couldn't get Danny spotted in a few TV adventure segments, such as the "Adventures in Paradise" series.

"Possibly he could get back on a payroll and still continue his round-the-world cruise in style" suggested Al. "A few episodes on TV would sure help put amateur radio into millions of homes that maybe know little or nothing about us." (Danny did appear as a guest on the "Groucho Marx Show").

Spenceley came out with a YASME Newsletter, Number Twelve, for May-June 1961, and noted that this was the first time he had not written a monthly "column" since his first DX column for CQ magazine in 1951. He reported that the YASME III's main diesel engine has been thoroughly overhauled, repaired and reinstalled in the boat. "This was accomplished thanks to the valuable assistance rendered by the Tool and Machine outfit for which Danny works. Billing came to a mere $30.00," he said.

Danny reported the biggest expenditure to date was for a commercial sewing machine which, although it cost $180, had "more than repaid us" in work done on all the sails. To date all had been repaired except the mainsail and deck covers. The quote to repair the sails was said to have been far in excess of the actual cost of the machine. Spenceley particularly thanked members of the Southern California DX Club for their help with the boat. Danny was recaulking and reseaming the decks. John, WA6EXN, donated all the wire needed to rewire the boat, plus all incidental fittings. The dinghy, which was in sad shape, was being repaired properly with fiberglass, for free, said Danny. "Many odds and ends have been acquired by nefarious means," all of which are putting the YASME into good shape at little or no cost to us."

Naomi (a nurse) visits an infirmary in Papeete, Tahiti.

Spenceley informed the membership that YASME III was moored for free, and that "no funds are being used other than those really essential to the actual refit." The biggest expense to be faced would be the overhaul of the Onan diesel generator, which needed the crankshaft reground, new bearings, and reboring of the cylinder and fitting a new piston and rings. Danny was doing all this work.

Danny said that through "scrounging and other means," YASME was getting into far better shape than she had ever been, "and I know full well that she will survive anything when all has been completed. My personal enthusiasm has not waned in the slightest and, as each day passes and YASME improves in condition, so the urge to be off again increases.

"Shortly, I intend to install a rig, take some of the lads for a weekend cruise, and you will be hearing VP2VB/MM on the air again."

It was hoped that YASME III would be ready to set sail around 1 September 1961, and Spenceley appealed for one or two crew members, "preferably hams, not too young, and single," perhaps someone who could sail for two months, then return home by air from Tahiti.

Spenceley said "Regarding the San Diego DX Club's campaign for YASME funds at the radio club level, all replies received have indicated that the YASME trip should be continued and a willingness to support same. Apparently this was all that the Club's initial letter called for and it will be followed up with more definite requests for the green stuff. An offer has been received from W2BIB offering a substantial donation toward YASME and a promise to raise additional funds. If this is followed up the Foundation should have sufficient funds to get Danny on his way and keep him going for a considerable period."

Pledges from 23 percent of the membership totaled $1,237, obligatory upon Danny's sailing. Repairs to YASME in San Diego, to the end of May 1961, totaled $1,123.14.

Spenceley said that from then on the YASME Newsletter would not be published unless YASME was "actually underway."

Danny continued work on YASME through the summer of 1961, having to find time over and above his full-time day job. "In the face of limited funds, new highs in the art of scrounging have been made by Danny to cut financial corners," Spenceley said. A 7-1/2 KW, two cylinder, diesel generator with an original cost of around $3000 was found for just $400 plus the trade-in of YASME's old 3 KW generator. The new generator would supply all of YASME's needs. The old 3 KW machine had been in operation since 1958 on YASME II and had suffered submersion at Union Island in 1959. It had been repaired in Tampa, put on YASME III, and the armature had later been rewired during Danny's stay at VP5VB. Erratic operation dating from the Clipperton Island stop made complete overhaul a must. This overhaul, with replacement parts would have probably cost more than the 7-1/2 KW machine, Danny and Spenceley estimated.

Once the new generator was installed on YASME III, other tasks remaining included installation of fore and aft safety rails; re-installation of all stanchions, caulking of deck seams; fitting new drainage pipes for sinks in the bilge, electrical rewiring, and installation and testing of all radio gear. All of this, plus odds and ends such as buying food and fuel was estimated to take about 25 days of full time work.

Danny had not seen his mother Christine, in England, since April 1958, so he decided to fly to England. He quit his job on 1 September 1961, arriving in England on the 6th. Spenceley noted that all his travel expenses were paid by his mother, and said that after returning to San Diego at the end of September Danny hoped to sail by mid-October. While in England, Spenceley kept in touch with Danny through G2DC, Col. J.M. Drudge-Coates (later a YASME Foundation director).

Following the harrowing voyage of the previous year, in particular the Clipperton landing and various towings, Naomi had hinted she had had enough sailing. However, "her unwillingness to be separated from Danny may cause her to reconsider," Spenceley wryly observed. If Naomi decided not to go, another crew member was essential, and Doug, W6HVN, was by then committed to join Danny.

Spenceley reported that the foundation was "as usual" financially. "It will be just barely possible to get YASME off with our present bankroll. Under present conditions she will certainly sail 'broke' with no extra funds for possible contingencies. Hardly a healthy prognosis. Many members have been more than generous with their contributions and pledges but there have just been not enough of them.

If only a further $1000 could be raised it would place us in a much more favorable position. This would at least cover supplies and fuel at Tahiti and further on."

Spenceley said "Whatever the pro and con opinions of the whole YASME deal, or Danny's personality, or what not - one vital fact has been proven, He does produce. At risk of life and limb he does put rare DX spots on the air. His operating procedures are rapid and efficient. He has given thousands of hams new countries over the past six years when operating from such spots as VP2VB, FO8AN/Tahiti, VR1B, VK9TW/Nauru, VR4AA, VK9TW/Papua, YV0AB, VP2KF, VP2AY, VP2MX, VP2KFA, VP2DW, VP2SW, VP2GDW, VP4TW, VP7VB, VP5VB, HK0AA, HC8VB and FO8AN/Clipperton.

"Some have compared him with Wagner who, as

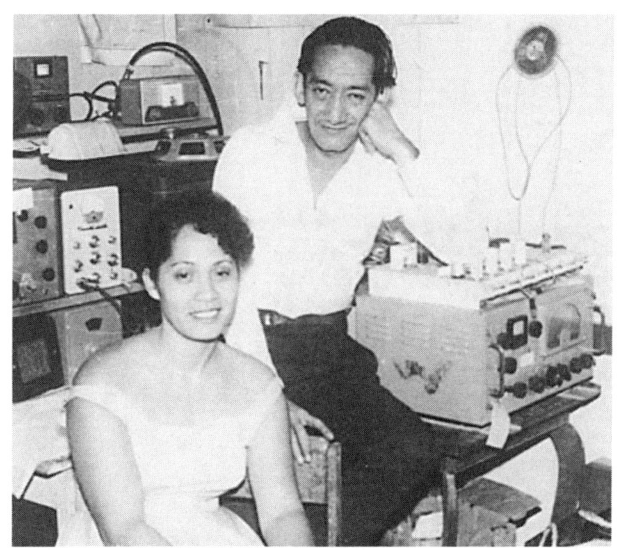

Locals on Tahiti

Danny might say, was a thoroughly rum sort but, for the music he produced, all else could be forgiven!" Spenceley said.

"Can't we help him just a little bit more? If we can, then contributions resulting from activity from his many planned DX stops may make the trip largely self-supporting. To start this off, yours truly is kicking in $50. What say, fellers?"

YASME III Sails Again

Intrepid but trouble-prone Danny Weil is up to his lantern jaw in the miseries again. - Don Chesser

"With sudden decision that apparently caught even his editor and associate, KV4AA, by surprise," Chesser wrote in a special bulletin, "Danny sailed from San Pedro, California on 29 November 1961 and a few days later was reported some 400 miles southwest of Los Angeles, and scheduled to arrive at Nuku Hiva, Marquesas Islands, around Christmas." Chesser reported that new gear aboard YASME III included a Bendix automatic pilot, ship-to-shore radio and direction finder.

A tentative date for departure had been set at 7 November when "this hapless couple," Chesser said, were in an auto accident on 4 November. The car was a total loss. Witnesses said the accident wasn't Danny's fault and, according to KV4AA, "Danny was more angry than hurt as a result of the accident because he had already sold the car and this left him without any transportation at all." An x-ray showed that Naomi had broken a small bone in her neck, and her physician forbade anything but complete rest for a month to six weeks.

So, off they went, with an "able seaman," Bill Bracy, also aboard. Spenceley described Bracy as "a young fellow with a substantial knowledge of sailboats and engines, and while he is not a ham nor a radio operator, he

Locals carry the generator ashore on the Marquesas.

still promises to be a tremendous asset to the trip. Unfortunately, other hams who had been counted on to go apparently changed their minds at the last moment and disappeared."

The game plan was to sail directly to Nuku Hiva in the Marquesas group, where Danny would operate once again as FO8AN (the fact that Danny used this same call sign from the Marquesas, from French Polynesia (Tahiti) and from Clipperton caused no end of confusion over the years). The voyage was expected to take about 50 days, and the expedition was, as Spenceley had warned earlier, "sailing broke" with no funds for any contingencies. "We wish her Godspeed, and trust that DX operations and the generosity of our subscribers will provide the relatively small funds to keep the Expedition underway," Spenceley said.

Clearly, something was needed to inspire DXers to join and contribute to the foundation. The Marquesas did not count as a separate DXCC country from Tahiti and as such it would not inspire DXers to open their wallets.

Spenceley wrote that "Hitherto the Marquesas group has not been accepted by DXCC as a separate country but as part of French Oceania, which covers a considerable Pacific area. It is hoped that the DXCC Committee may be prevailed upon to re-examine this matter in view of what we might call a more liberal, and certainly more popular, approach to these matters, which they have shown in recent years.

"The Marquesa Islands are certainly a separate and distinct island group, and they are separated by some thousand miles of ocean from the parent government in Tahiti," continues Dick. "The same should apply to the Tubuai (Austral) Islands located a good distance to the south of Tahiti. Rapa is an island in this group and has been on the air some years back. Thus, we believe that a division of French Oceania into three countries - the Marquesas to the north, the Society Islands with Tahiti as its hub, and the Tubuai chain to the south, might reasonably come into being under present DXCC criteria. The designation of the Northern Cook group, the Manihiki Islands, as a separate country is a good precedent in this instance. "

Unfortunately, "contingencies" reared their heads almost immediately, and Chesser put it well when he said "Intrepid but trouble-prone Danny Weil, VP2VB/MM et al, our round-the-world DXing voyager, is up to his lantern jaw in the miseries again, according to Dick Spenceley, KV4AA, and Don Wallace, W6AM, this week."

Don Wallace, W6AM (later a YASME director and president) acted as a radio relay between YASME III and Spenceley. All went well until the third day, Dec. 2, when rough seas and 50 m.p.h. winds gave all hands their conditioning dose of seasickness. Then, on the 4th, Danny reported water had apparently got into the ship's main diesel and caused the cylinder block to crack. Simultaneously, an oil line burst on his diesel generator causing partial burnout of a bearing. Danny hoped, however, that it would run until the main engine could be stripped down and the breaks electro-welded.

On Dec. 5th the seas were relatively calm and work proceeded on stripping down the main engine, where three cracks were discovered. Welding was started, and Danny reported that it was going well, with one crack completely sealed, when the diesel generator threw its bearing and wheezed to a stop.

The welding [Spenceley said] had produced a painful eye strain to Danny, and the next day was spent resting up before tackling the generator bearing. Radio contacts were maintained via gasoline generator, of which he has two, but which do not have the power capacity to operate the welder.

> The next two days were spent removing the faulty bearing, scraping off the metal, depositing new metal, machining, and remounting. Danny reported that the crankshaft appeared slightly oval in shape and did not fit the bearing well, but he hoped until the welding on the main engine could be completed. It ran two hours and quit. . . . !
>
> At 1600 GMT Dec. 13th, Danny reported his position at Lat. 13, Long. 128, or about 1350 miles southwest of Los Angeles. Danny is preparing to tackle the generator again, and I am sure he will come up with some workable arrangement. YASME was now in the trade wind area, and steady winds from 8 to 20 m.p.h., plus favorable currents and following seas, are driving her steadi-

ly on to her destination, the Marquises, at an average 100 miles per day with sail. With prevailing winds and without diesel power it would be out of the question to try to head east. Thus he is committed to continue on his route.

YASME left port with 100 gallons of gasoline aboard and, with diesel out, they are using an average of 5 gallons a day for radio and ship's power. Should the generator not be repaired, he will be out of gas about the time the Marquesas are reached. It is hoped gas can be purchased in those islands. FO8AC in Tahiti has been contacted for information on the subject.

And, from Chesser (23 December): At 1600 GMT Dec. 23rd, with W6AM acting as relay station, Danny transmitted:

Traveled about 60 miles since last position toward Nuku Hiva. Things tough here because wind just ain't in right direction for us, I guess. Best we can make without engine is about 60 miles a day, and it looks like we will be at least 14 more days at sea, now. Am getting list of spares needed for diesel power plant, and hope to have ready for next sked. Would appreciate it if you could get hold of a VK station and see if we can get a replacement cylinder block for the Gardner diesel engine in Australia. It is called an L2 model, and the block consists of a two-unit cylinder unit. Gardner has agents there, and maybe we might be able to pick up a used block with pistons. Even if I can stop this one from leaking there's always the chance it might blow up and cause more damage. Gardner also has agents in Canada, but I'm thinking of keeping freight charges down.

On another occasion, relayed by W6AM, Danny transmitted: "I couldn't get enough power to make first class weld. So when I reassembled the cylinder block it worked for a while then leaked water into seams of the welding. Can't fix any more. I've got power plant running on one cylinder, which is just about giving us enough power to run rig, etc."

On another transmission: "I have to fix my ripped mainsail so can move a little faster. We all hoped to arrive Nuka Hiva for Christmas, but with all our troubles it looks like we will be at sea for Christmas and New Years, too. All my time is spent with odd jobs. I had a few hours sleep last night for the first time in two days. Must now QRT, because without mainsail we ain't getting anyplace. Just rolling, throwing me through cabin door and can't even send properly. . ."

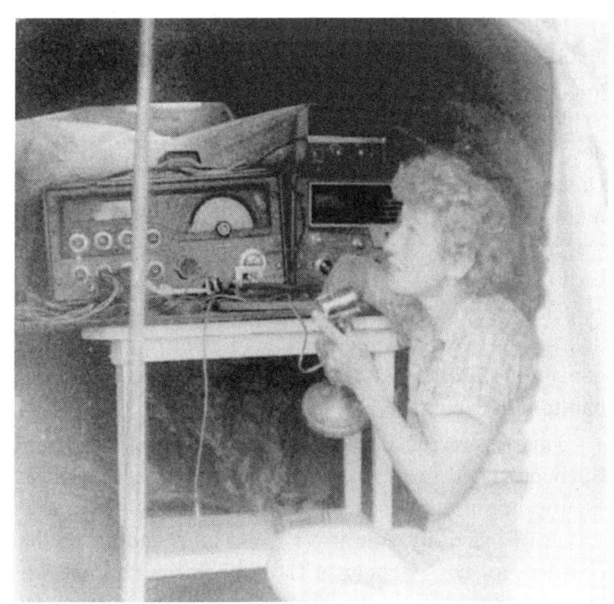

Naomi on Tahiti

The YASME Newsletter for January 1962 (Number Sixteen) reads like a daily log of the voyage to the Marquesas, with the diesel generator running on one cylinder following Danny's welding job. When the winds calmed the main engine was reassembled but the leaks persisted. Christmas 1961 was spent at sea and on December 28 Danny set a one-day record for himself of 140 nautical miles.

The YASME III docked at Nuku Hiva on 2 January 1962, 35 days out from California. After a day of rest, Danny repaired the diesel generator which provided current for the local hospital and then, at the request of the authorities, took YASME out to sea in a five hour search for a missing local fisherman (not found). The next day a QTH was located ashore and it was hoped that the FO8AN/Marquesas operation would begin on January 6th or 7th. Danny found the boom for his Hy-Gain antenna missing but promised to rig up a substitute. He ended up making a boom out of wood.

Meanwhile, working from a list of parts Danny supplied for the Petter diesel generator and the Gardner main engine, K5JLQ and W8EWS cooperated to locate them in the States and ready them for shipping, via Lykes line Steamship, from Houston. They would leave on 18 January for an arrival at Papeete, Tahiti, in mid-February. Danny reported that his main engine was holding up well and should be able to take him from the Marquesas to Papeete (about a thousand miles, expected to take about seven days).

The shipment of parts for the engines, costing around $600, was subsidized by K5JLQ, W8EWS and KV4AA and "it is hoped that contributions arising from Marquise QSLs and further donations or renewals by the Membership will repay this amount."

Danny hoped to operate FO8AN for up to three weeks, with a goal of at least 5000 contacts. He ended up making about 2,500. Log data was transmitted over the air to W8EWS when possible. Radio conditions were described as poor, very few Europeans were worked, and Danny had to lug the gas generator for the radio equipment from YASME to the operating site every day that he was able to operate, that is, when he was not preparing the boat for the next sailing. (Spenceley wrote later that "Openings to Europe caused log transmissions to be delayed in favor of contacts.") "Frequent repairs" to the generator were necessary and gasoline to power it often had to be scrounged.

This was a crucial, turning point operation for Danny, Spenceley, and for the foundation. It was the first of the Pacific Leg in the repaired and refitted YASME III, after a year of inactivity. And it was from an island group that counted, for DXCC, only as Tahiti, from which there was already activity (including even Danny on a previous voyage). In other words, not a contribution-producer.

Spenceley was busy circulating batches of a petition to make the Marquesas a separate DXCC entity, mailing them to some 50 radio clubs. "A glimpse at the map will show the Marquesas as a separate and distinct Island Group removed from the parent Government In Tahiti by some thousand miles of ocean," he said.

In February, Spenceley reported that Danny was enroute to Papeete, and that only one gas generator was working. Danny hoped it would "hold up, along with his main ship's engine, until Tahiti is reached and repairs effected." The repair parts had left Houston on 23 January with an expected arrival at Papeete around 20 February, meaning Danny would stay in Tahiti "far into March." He planned to operate from there (using the same call sign, FO8AN, that he'd used from the Marquesas) with "perhaps an accent on SSB, as this mode is most needed from this QTH."

Here's Danny's recounting of the voyage from Nuku Hiva to Papeete, as published in the YASME Newsletter:

> I had planned to leave Nuku Hiva on Monday, February 5th but, as usual, a few minor events popped up which prevented the departure. The major one was the fact that our crew member had found another skipper to take him along and it took a day or so to arrange all the formalities with the local Immigration Bureau to get him cleared of YASME. By Tuesday, Bill was safely enconced aboard this other yacht. We timed our departure for noon on Wednesday, February 7th, so that we would arrive in the vicinity of the Tuamotu atolls during daylight hours. This group, about 150 miles wide, is far from encouraging and is a meagerly charted area consisting of a mass of coral atolls. Even with an excellent depth finder, one is on edge all the time and hopes like heck that the strong currents haven't swept us off our plotted position. I had noticed a strong EASTERLY current running about 40 miles per day practically all the way from Nuku Hiva. It was rather strange to meet up with such a current in this part of the world as we had normally been fighting a westerly set for many hundreds of miles.
>
> We left with the engine running. My weld job seemed to be holding up and there were no visible leaks externally. A few checks, periodically, on the dipstick showed me that it was holding up internally and had no water in the sump. I breathed a great big sigh of relief as, to get clear of Nuku Hiva and into the true trade wind area, I was counting heavily on the engine. She did hold, for 14 hours, and upon reaching this area, all trade wind sails were hoisted and the auto-pilot "George" took over the helm and we were on our way.
>
> As night fell so did the wind and on went the engine again. After its brief rest it started up with a real big effort and sounded like a bag of bolts being rattled. Oil pressure dropped to zero and on every fourth stroke it gave a wheeze like a cranky old DXer. I shut it down with a mild remark..."how annoying", or maybe it was a little more forceful. I just can't remember as I was too busy cussing the thing out. As it was night time and I was very tired, the last few days at Nuku Hiva had really knocked me out, I decided to leave the engine job until daylight. We did pick up a little wind and I maintained an average 5 knots throughout the night.
>
> At 9 AM I got started on the engine. To strip it down was child's play. I knew every nut and bolt by name and barked my knuckles on precisely the same parts as in past "strippings". During this process the same small parts fell in the bilge and, according to Naomi, I used the same cuss words throughout the whole procedure. A cylinder head gasket had blown and the burning gases had made a very nice groove along the dead flat cylinder head. This groove had to be filled up so now came the fun of getting my old friend (?) the diesel generator going so that I could do a spot of welding... I pushed the starter button and it started! Naomi treated me for shock. This had never happened before! I must have been living right--. Even with

only one lung working, we managed to make a fairly presentable weld on the cylinder head and then it took about two hours of filing to get the thing flat. I found an old cylinder head gasket in the bilge and bunged the whole works together again.

Lack of oil pressure was checked and the oil filter was found to be chock solid with about everything under the sun. I even found a wrench I had lost a month ago and a couple of OK SWL cards! By 6 PM all was complete and she actually started up again with nary a burp.

I had to make a decision, whether to keep the thing running or not. If it ran continuously I could make fast time to Tahiti but would risk a further breakdown. Or, I could make the trip on sail alone and save the engine for possible emergencies. That old iron horse sounded so good that I just didn't have the heart to shut it down. Another big factor was that engine had an AC generator coupled to it which did a number of jobs like charging the autopilot batteries and keeping our deep-freeze running. To get this power from another source we would have had to keep a gas generator running and we only had 20 gallons of gas aboard at that time.

Delay on account of any bad weather would exhaust our gas supply and put us in a rather bad spot. All our cooking was done with 110 AC too. On top of this we had only one gas genny in operating condition and this one sounded like it would give up the ghost at any moment.

Finally we hit the Tuamotus. The area I have always dreaded and hated. On the chart it looks like a slightly befuddled spider had wandered across with inky feet. But, at sea, one doesn't see one darn thing until you are right on top of it. I spent every spare moment shooting stars and then taking a few amplitudes for the compass. I just didn't dare trust my position for more than four hours at a time in view of the strong currents and dearth of information on the charts.

Oddly enough, we came through the whole group and every little island came up exactly where it was supposed to be. Our auto-pilot took the helm all the way and Naomi and I alternated four hour watches just to be on the safe side.

Not one ship was sighted the whole way and, on the night of February 13th, we sighted Venus Point lighthouse shining into the sky. We were right on course and, at 4 AM on the 14th, we stood off Papeete harbor waiting for a pilot. Up to this time we had been subjected to favorable weather but from 4 to 6 AM, when the pilot boat took over, the wind and seas built up to tremendous heights and I was beginning to get a little worried. I dropped all our canvas except a small jib to steady us and then depended on our engine to hold the position. I tried to make contact with Papeete with the Marine Telephone at 5 AM but apparently the guy was either asleep of off duty. At 6 AM (1600 GMT) W6CTO and KV4AA appeared on schedule and, out of the blue, came FO8AC. At this point I sighted the pilot launch plugging through the pass towards me so I hurriedly told all hands that I had to QRT to receive the pilot. The wind had now built up to 45 knots, the seas were breaking over the bow, and me trying to tell everyone what was happening while being tossed around the deckhouse - some fun!

The pilot boat came alongside and threw me a line. Tying it to the winch was a second's job, and we headed towards the pass. The seas, by this time, had built up to fantastic heights and followed us in through the pass. I dared not even look astern as every ounce of concentration was needed to control YASME as the seas threatened to swing her broadside on. The pilot boat kept a steady pull and, at last, we entered into the quiet calm of the harbor. The seas broke and roared on the reef astern as if cheated of their prey but we were inside and safe.

We dropped anchor and brought YASME around, stem to the seawall, and within ten minutes all formalities had been dispensed with and we were free to go ashore. George, FO8AC, was on the road waiting for us and our first move was to wrap ourselves around a fine steak at the nearest cafe. I really enjoyed seeing George again after a six year break. Other very old "acquaintances" were met, some a little embarrassing, as I was no longer a bachelor!

Bad news from the DXCC Desk: Spenceley wrote "In a letter [Spenceley had received] answering petitions which requested a favorable vote on the acceptance of the Marquesa Group as a separate ham country, the DXCC says that no change of status will be made. Their reason: Because France considers French Oceania as ONE overseas territory ---. We hardly think this a valid excuse in view of past precedent in calling the Northern Cook Group separate and in view of the obvious wishes of DX men. We mean to go into this, at some length, in our next Newsletter.

"Unfortunately, DXCC's policy not to commit themselves as to what is, or what isn't a country, until a QSL is received from it, imposes a doubt as to whether even the above will be accepted or not and their stand on the Marquesas precludes any attempt by Danny to cover the nearby Tubuai or Austral Island groups in French Oceania."

YASME's next stop after Tahiti was presumed to be either Flint or Vostok Island, some 600 miles northwest of Tahiti and under British administration, with U.S. claims. "It appears that these Islands are far enough removed from anywhere to qualify for separate country status. From these islands a trip might then be made to Starbuck or Malden Island which should also qualify" Spenceley wrote.

Foundation income news was a "little brighter" with $667.50 received in January, from contributions and renewal of memberships. $100 was cabled to Danny to buy gasoline. "The balance will just about cover purchasing and shipping of repair parts."

After 36 days at Nuku Hiva, Marquesas, YASME III left on 7 February 1962 for Papeete, Tahiti, in the Society Group, some 976 miles to the southwest. Since Bill Bracy had left, Danny and Naomi were going it alone again. The voyage took the expected seven days, with YASME pulling into the Papeete harbor on 14 February 1962.

Danny expected to be in Tahiti until sometime in May, with "plenty of work to be done" repairing all the generators and the ship's engine. The Newsletter advised that he "has not rushed to get on the air from this not too rare stop, but he has been maintaining schedules and should be on for general QSO periods on about March 1."

"Now comes one of those incredible things," Spenceley told the membership, "that could only happen to this expedition - the SS Pioneer Star, with all spare parts aboard, pulled into Papeete on schedule and, through some colossal goof, disgorged all cargo except the stuff earmarked for Danny. It was checked off the manifest but never reached shore! Thus, all these parts are merrily on their way to Sydney, Australia. Arrangements have now been made by the shipping company to have the goods returned to Papeete by French steamer, which should arrive in Tahiti about April 18th. (And I think of all the rushing around we did having these items shipped, by air freight, to Houston in time to catch the sailing of the Pioneer Star on January 22. Ciest la vie, or something!)"

As a result, Danny got on the air sooner, on 27 February, to rack up as many contacts as he could before the parts for the necessary repair work arrived.

Danny's proposed next stops included Vostok and/or Flint islands, then perhaps Malden or Starbuck Island. Spenceley said "These spots should, presumably, qualify as new countries." Danny also was looking around for a Tahitian crewman to join YASME.

26· A Plea to DXCC

"We suggest that the DXCC rules be rewritten in the light of present conditions and possibilities." - Dick Spenceley

Since the beginning of 1962 Spenceley had conducted a campaign for DXCC status for the Marquesa Group of islands. Danny Weil had made some 2500 contacts from there as FO8AN. Spenceley said he thought DXCC status for the Marquesas "would be in accord with the wishes of a large majority of DX-seeking amateurs and that this island group clearly qualifies for such status under existing DXCC rule criteria and precedent."

Spenceley said he had circulated some 9,000 petitions for the Marquesas, through the YASME Foundation mailing list and to DX clubs. The petition read:

"I respectfully urge that the Marquesa Group of islands presently represented by the activity of FO8AN, be considered and accepted as a separate DXCC country. It seems reasonable to assume that this group would qualify for acceptance under interpretation of present DXCC rules as it is a distinct island group some 1000 miles removed from the parent government in Tahiti. Moreover, and especially, I believe that such action on your part would be in overwhelming accord with the wishes of the DXCC membership and the best interests of the DX game."

Spenceley said, in the March YASME Newsletter, that the DXCC Committee "refused to act favorably on this matter." The reply from ARRL's DXCC said:

> Thank you for your letter concerning the proposal to consider the Marquesas group as separate for DXCC purposes. As you may be aware, amateur radio operation has taken place from the Marquesas group in the past, as well as from other groups which make up our French Oceania listing. Such operations have caused us to review [the] question of whether the French Oceania listing should be replaced with four separate listings, each of which would be one of the four groups that make up the French Oceania listing.
>
> In every review that has been made the consensus has been the same in every case. Our listing of French Oceania is based on the very firm fact that the four groups comprising the French Oceania listing make up one overseas territory of France. This fact alone would outweigh any distance or emotional consideration. Should distance be a consideration, it can be shown that in the case of the Marquesas group the factual measurement would be in the order of 300 miles rather than the 1000 miles which has been mentioned. Inasmuch as nothing has changed in the basic consideration of our French Oceania listing, no change will be made.

The letter was signed by the ARRL's DXCC manager R.L. (Bob) White, W1WPO, and it seems that at this point the matter would have been laid to rest. But Spenceley was not to be deterred. Thus, Spenceley's response and arguments, in the March Newsletter are a window into the thinking of DXers. It should be kept in mind, first and foremost, that DXCC was, and is, just one administrative program of the ARRL, and that DXCC was, and is, the premier DX award.

Spenceley had for many years been associated with CQ magazine, the only true competitor to the ARRL's journal, QST. The publishers of CQ had, in the early 1950s, instituted their own award for working "DX Countries," an award based on the ARRL Countries List, but with significant additional "country-entities." Simply put, the CQ award was "looser" than DXCC.

In a long response, Spenceley cited what he thought were DXCC decisions that could be applied to the Marquesas situation, as well as DXCC decisions where "precedent has been flouted and DXCC rules have been 'rubbered' to suit the personal ideas of [DXCC] committee members on what is, or what isn't, a 'Country.'"

> The committee [Spenceley wrote] has given us the impression that a considerable amount of "hard-headedness" exists, sensitivity to any outside criticism (criticism emanating from sources other than within the committee itself) and an immediate rejection of such criticism on a basis of "it can't be any good unless the committee thought of it first." This, we think, is an unhealthy situation. We think that the first consideration of the committee should be to conform to the clearly defined wishes of the DX men they purport to represent and for which they get paid to serve.
>
> We are also quite certain that DX men would not support any petition, calling for the recognition of any area, which, under existing DXCC rules and common sense, would not be considered a reasonable addition. Therefore, the DXCC award as one of the major achievements in ham radio, could never be cheapened by a number of such additions. DXCC is not the private property of the DXCC committee but belongs to its thousands of members whose advice and desires should carry some weight.
>
> Before we go further, let us say that this criticism

deals only with the attitude and decisions of the DXCC committee and not with its otherwise excellent work in administering the award. Nor should it be interpreted as a criticism of any of the many functions of the League. As a member of the ARRL and DXCC we feel that we have the right to criticize, constructively if possible. We feel that no organization should reach the stage where it puts itself above criticism and, of course, we are not saints ourselves.

Our premise is that the addition of countries to the DXCC list, past present and future, has, and will be, a good thing. Every addition will bring zest and thrills of the chase to DX men. Some past additions have been of the "twilight zone" type but all have been easy to accept and live with. On the other hand it is known that there are a few who have deplored the acceptance of many of these, saying that there are enough countries already on the DXCC list etc. etc. But with the additions of so many new ones we feel that these individuals have been hopelessly outdistanced and might as well "throw in the sponge".

Now, regarding the letter in reply to our petitions. The main point of refusal seems to hinge on the emphasized fact that "French Oceania is considered as one overseas territory of France". This seems to be a brand new ruling brought into being by the DXCC committee to suit the occasion. I have never yet noticed DXCC expressing such concern as to how a foreign government rates its possessions. For instance Martinique and Algeria are actually considered as metropolitan areas of France itself but these areas are separate DXCC countries. Here is my answer to this phase of the matter: does the committee know that the French "REF" society considers the Marquesa group as a separate country for their own "DUF" award ?

Our statement, that the Marquesas were some 1000 miles from the parent government in Tahiti, is true. Don't believe us - just look at your map. The committee seems to have brought their microscopes into play and have discovered some outlying rocks which space the groups at the 300 mile figure. This is ok by us as the 300 mile figure is still well over the odd 225 mile criteria which has been chosen by some means to qualify separate status in some cases.

The mention of emotional appeal might mean that the committee deems us emotionally upset in presenting this "hysterical" petition so brashly questioning their decision. I can only say that ham radio, or any other hobby, is for most part, emotional. We are glad it is this way or it wouldn't be much fun.

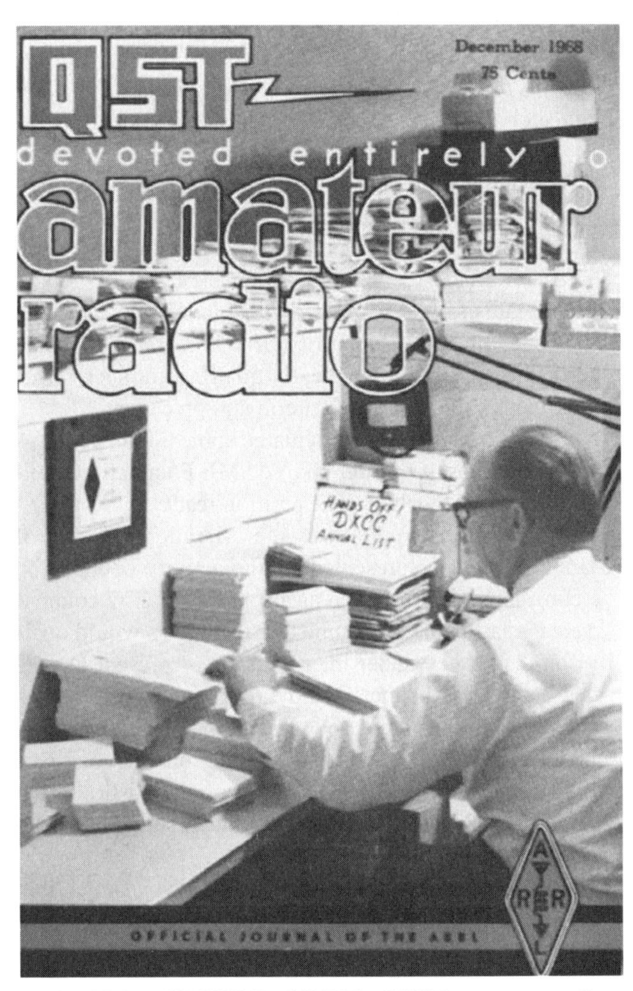

Bob White, W1WPO, ARRL's DXCC manager for more than 20 years, appeared on the cover of December 1968 QST.

The letter [from ARRL/DXCC] continues: 'Inasmuch as nothing has changed in the basic consideration of our French Oceania listing, no change will be made". Well, nothing has changed except the knowledge, via petitions, that DX men want the listing changed. Apparently this is a decidedly small item in committee thinking! -- , the rest, as a well known W0 writes, "from the supreme court of no appeals!"

Now we come to examples of country acceptance, and rejection. These decisions have obviously been arrived at through ultra-flexible interpretation of DXCC rules, or downright ignoring of rules and precedents. Oh yes -- the committee can probably justify their findings, in the fine print, but, well, you read the following and think it over.

Spenceley then listed 22 examples of DXCC rulings, 1959-and-earlier-style. It is beyond the scope of this book to even attempt a history of DXCC rules and "country criteria." In a nutshell, the DXCC rules in 1959 were interpreted and applied at ARRL Headquarters, primarily by Bob White, W1WPO, with advice from others when needed. White was elected to the CQ DX Hall of Fame in the 1990s.

Spenceley may have stated the situation at the time best when he said "We are aware that the word 'country' is a misnomer as applied to many areas on the DXCC listings but a better word for it is hard to find. We might coin the word 'AWSAC' to be used by the committee which would mean 'areas we say are countries."

1. The Cook Islands group
2. The Trust Territory of Ruanda-Urundi
3. Sicily
4. West Indies Federation
5. Finland's Aland Island
6. The Gold Coast (Ghana)
7. Belgian Congo/Congo Republic
8, Nigerian Republic/British Cameroons
9. Malagasy Republic
10. British Cameroons
11. Cameroon Republic
12. Tanganyika and Samoa
13. British bases in Cyprus
14. Alaska's Pribilof Island group
15. Portuguese enclaves of Dao and Damao
16. The Willis (Australia) "Islets"
17. Kure Island
18. Canton Island
19. East and West Germany
20. Neutral Zone of Kuwait
21. Italian islands of Pantelleria, Linosa, and Lampedusa
22. The Sheikdoms of Abu Dhabi and Muscat

In the more than four decades since Spenceley listed those 22 examples, some of them have been placed on the DXCC list, some have dropped off, and a few have done both, as the DXCC rules and their application changed. But, in 1959, Spenceley pushed for only one consideration, the Marquesas, saying "it should be obvious that a remote and distinct island group such as the Marquesas merit separate status when compared with past examples and in acceptances of such rocks like Malpelo and Navassa and sandbars like Aves, Serrana Bank and Baja Nuevo.

Now [Spenceley wrote] do you think that the addition of the few countries, hitherto rejected, would cause any great upheaval or tend to cheapen the DXCC award ? We certainly don't. The committee, with clear conscience, could accept each and every one of then and, thereby, make a lot of people happy.

We suggest that the DXCC rules be rewritten in the light of present conditions and possibilities. We suggest that these rules be amply publicized so that the hams can know what's going on. These rules, once written, should be stuck to. Addenda can always be added to cover unforeseen problems of the future.

Now, we have had our say. Some of our facts may be awry but the overall picture should be clear. We do not think that this effort will cause any change in the 'powers that be' but, if so, it will substantiate some of the contentions we have made above. Remember, it's only a hobby!

Bottom line? The Marquesas did not receive DXCC status, but Spenceley foresaw how DXCC administration would change in the coming years - on 31 March 1998, the Marquesas became a DXCC Country.

The April YASME Newsletter noted that foundation membership was at 350; it did not note that that was down from 490 in February 1961.

While waiting for the shipment of repair parts to arrive, Danny had made some 300 contacts from Tahiti, as FO8AN, through 10 March 1962. It was on the 10th that those parts finally showed up. The shipping line, which had mistakenly let the shipment end up in Sydney, air freighted it to Tahiti, at their expense (the shipping line's traffic manager in New York City was Al, W2WZ). Danny expected to be off the air for about a month while he repaired the generators. He also had received a new filter condenser for the HT-33 amplifier. It seemed a never-ending struggle to keep things running, as Spenceley's description of the work attests:

As we write [said Yasme Newsletter], "the Petter diesel generator has been completely repaired, crankshaft reground etc. and has been running nicely on test. It has now been mounted on the deck, with shock absorbing mounts, rather than inside YASME where its noise and heat made things most uncomfortable."
A marine plywood covering has been made to protect it from the elements. Work on the fuel lines and electrical wiring to the new generator location is going forward. The gasoline generators have been overhauled, new brush holders etc. and all are in operating condition. Work is now proceeding on the main "Gardner" diesel engine and Danny esti-

mates that most work will be complete by April 10th.

Two new tents were expected by ship around the end of April, and a 10 May departure date was set, for Flint Island, some 400 miles northwest of Tahiti.

This island is apparently controlled from Fiji [Spenceley said] and is one of the group comprising Flint, Vostok and Caroline Islands. By any standards and with consideration of the [vagaries] of DXCC decisions, it would seem that this group of islands should qualify for separate country status. Flint Island is the only one having a poor, but possible, anchorage. From Flint, some 300 miles to the north lies the island group consisting of Starbuck and Malden Islands. This group poses problems as no anchorage exists and conditions there are known to be hazardous.

We fully realize the value to the expedition, and to hams, if two new countries can be covered, one after the other, and all means will be taken to cover both. Decent weather is a "must" for Malden or Starbuck operation. Another problem in the inclusion of Starbuck-Malden is the ability of YASME to carry enough fuel, water and supplies. There is no water on Flint or Starbuck and, after these stops, YASME still has a long way to go as outlined below.

The Line Islands group extends 1,500 miles and most belong to Kiribati; Kingman Reef, Palmyra Atoll, and Jarvis Island are unincorporated outlying territories of the United States.

From Flint (or Starbuck), YASME will head northwest or southwest (as the case may be) for Tongareva (Penrhyn) Island, 300 miles away. This island is in the Manihiki group and, we understand, that water and fuel are available there. After a stay at Tongareva, YASME will sail southwest to Pago Pago, US Samoa, about 1000 miles away. From there it's an easy jump to Apia in the new Samoan Republic, should there be a chance that this nation could count as 'separate.' At this time decisions will be made for next stops which could include the Tokelaus, Wallis Island, Tonga, etc.

The first part of the trip is considered the hardest, to Flint and Starbuck, as cross-seas and unfavorable winds/currents may be encountered. After this, in a SW direction, the elements should be more in resonance. All the above is with the assumption that license problems will be resolved.

Hal Sears, K5JLQ, of Houston, was announced as the newest director of the YASME Foundation, in recognition of his "substantial contributions and unceasing and unselfish efforts to further the YASME cause." Sears had worked behind the scenes for some time, was probably one of those who lent money for the purchase of YASME III, and would play a pivotal role in the foundation in the coming years, as we shall see.

27· The Colvins in California, 1957-1964

Colonel Colvin exercised great perseverance, patience, skilled operating technique and technical application in the establishment of his station. - U.S. Army Brigadier General Earle F. Cook (W4FZ)

At an Army Signal Instructor Conference in August 1958, held in Washington, D.C., Lloyd (whose title was Signal Advisor, 49th Infantry Division, California National Guard, Presidio of San Francisco) received a plaque presented by Lt. General J.D. O'Connell, Chief Signal Officer, U.S. Army, which read "In recognition of personal contributions and exceptional operating achievements in various phases of amateur radio and in keeping with the highest traditions of the amateur radio service." Iris joined him on this trip and while in the capitol, the two appeared on a national radio network program to explain the convention, and amateur radio.

Being in Washington at this time allowed Iris and Lloyd to attend the 10th national convention of the ARRL. A military luncheon was part of the convention program, and Army Brigadier General Earle F. Cook, the chief research and development officer for the Signal Corps, addressed the luncheon (which was attended by ARRL President Bailey:

> I am [Cook said] particularly pleased to be here today for two reasons. Mainly because I have a pleasant and rare task to perform. Secondly, I predict that this 10th National Amateur [sic] Radio Relay League Convention will mark a milestone in the history of the ARRL and the amateur service and I for one, will look back in future years to this convention and savor a bit of nostalgic pride in the recollection that I played a part in these ceremonies.
>
> I am here today on behalf of Lieutenant General J. D. O'Connell, Chief Signal Officer United States Army to present the Chief Signal Officer's special award to three individuals who have distinguished themselves among the many individuals engaged in amateur radio and Army MARS activities.
>
> The first of these is W6KG, Lieutenant Colonel Lloyd D. Colvin, an Army Signal Corps Officer who has engaged in the scientific art and fraternalism of amateur radio activities for the past 29 years. Through his determination to establish personal radio communication with other amateur operators throughout the world, he has become the first individual to communicate with 331 foreign and domestic stations of which 301 are confirmed

Lloyd receives a plaque from the U.S. Army for his amateur radio achievements.

> by QSL, each having a different prefix. To obtain this, Colonel Colvin exercised great perseverance, patience, skilled operating technique and technical application in the establishment of his station. Among other accomplishments he has earned many certifications, a few of which are WAC, WAS, WAZ, and DXCC.
>
> In addition, Colonel Colvin was the founder and the first president of the Far East Amateur Radio League and is currently the president of the Northern California DX Club, Incorporated. Colonel Colvin joined the Army Amateur Radio System in 1930. He has been an active member continuously and is presently engaged in establishing many new stations in the present MARS program.
>
> With a communication log containing over 70,000 amateur radio stations worked, including 331 separate call prefixes with approximately 30,000 QSL confirmations, Colonel Colvin is truly an Outstanding Radio Amateur.

In addition to Lloyd, Laddie Smack, W9CYD, of Chicago, was recognized, for his efforts in instructing prospective hams, as was S. Edwin Piller, W2KPQ, for forming a technical training net for MARS members.

Lloyd's plaque read "In recognition of an amateur

radio operator whose personal contributions and exceptional operating achievements in various phases of amateur radio are in keeping with the highest traditions of the amateur radio service."

Lloyd addressed a DX forum at the convention, his first major amateur radio convention appearance, and the topics were his passions: operating, record-keeping, QSLing, and obtaining awards and certificates:

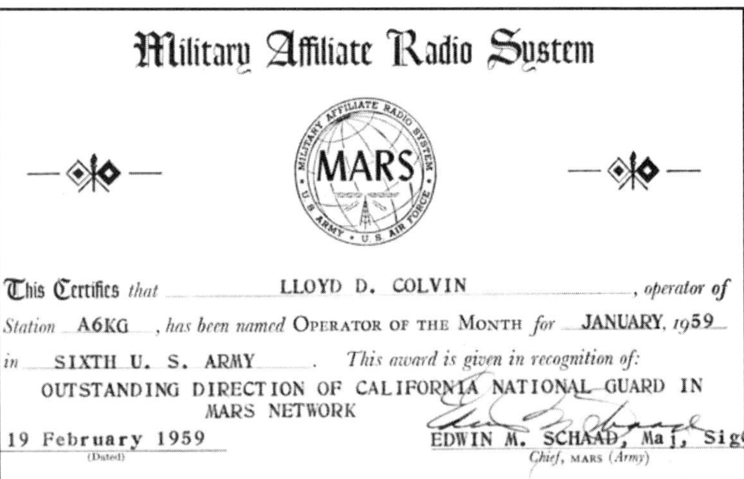
Military Affiliate Radio System

MARS

This Certifies that LLOYD D. COLVIN, operator of Station A6KG, has been named Operator of the Month for JANUARY, 1959 in SIXTH U. S. ARMY. This award is given in recognition of: OUTSTANDING DIRECTION OF CALIFORNIA NATIONAL GUARD IN MARS NETWORK

19 February 1959 (Dated)

EDWIN M. SCHAAD, Maj, Sig
Chief, MARS (Army)

One of many awards Lloyd received for his professional work with M.A.R.S.

I am sure you will agree with me that I am very lucky as both my wife and daughter are licensed radio amateurs. Operating under the twenty two calls that Vic has referred to we have contacted approximately 75,000 amateurs located in 262 countries. On our family QSL cards we have printed the phrase, " 73 from the ham family that has QSOed half the active amateurs of the world". Occasionally I have been accused of exaggerating on that claim.

I have brought with me today proof that amateurs, at least DX amateurs, are not long on the air before I work them. This proof is in the form of a QSL card to me from one of the top DX amateurs of the world. This QSL says "Tnx for enjoyable QSO this AM. Pse send me a card as UR my 1st DX wkd. Hi. Hv had Mi call 4 days, " This card is a hand made card, the QSO referred to was held some 25 years ago, the great DX consisted of working from Phoenix Arizona to Berkeley California and the card I hold in my hand is signed by Victor Clark, W6KFC, now W4KFC, your distinguished chairman. I have made photostat copies of the QSL which I will pass out for you to look at.

My talk is about DX QSLs, DXCC statistics, letters from DXers and other data on DX. This could be a dry talk, however to keep you awake and alert I propose to have you help me. I also want to announce right now that before I sit down I will announce my candidate for the worst QSLer in the room here this morning. I have no idea who that will be, but with your assistance we will find him.

I know that you are all interested in DX and I suspect a large number of you are members of the ARRL's DXCC Club. Will all DXCC members in the room please hold up your hands?

I am pleased we have so many DXCC members as in preparing this talk I made a study of the DXCC members in the first, second, third, fourth, and eighth call areas. I used the DXCC list in last December's QST as my source of information. I found that in the combined districts mentioned there are 1311 DXCC members. In checking my own records I found that I had worked 76 percent of such DXCC members. This is including both phone and CW members, however most of my QSOs were made on CW.

The number of DXCC members in any call area is proportional to the number of amateurs in the call area, the second call area is the largest (in the east) and has the greatest number of DXCC members. K2GFQ has the highest DXCC standing of any K call in the United States, W1NLM not only has a high DXCC standing, but is a blind amateur. In spite of this handicap he QSLs all QSOs 100%.

The top members of DXCC as you know are listed monthly in QST in the DX Honor Roll. I think we should take a few seconds to introduce the Honor Roll members who are present today. Will all members of the DX phone or CW Honor roll who are in the room please stand up? Will you please state your call and present confirmed number of countries worked and whether it is on phone or CW, starting first over on this side of the room please - Please remain standing. Let's give these top DXers a big hand.

All DXers receive letters from time to time from foreign amateurs. Some of these are written by hams who have very little knowledge of English and the results can be amusing. I would like to read one such letter that seems especially appropriate at an ARRL convention. After an enjoyable personal visit to EA6AW, a noted DXer in the

Balearic Islands, my wife and I sent him a gift subscription to QST . The letter was one of appreciation.

Most of us have received another type of letter, one in which the foreign amateur has heard that all Americans are millionaires and he writes asking you for something. The best letter I have ever heard of along these lines was received by famous DXer Reg Tibbetts, W6ITH. This letter is referred to on the West Coast as the Gabardine suit letter. It was sent to W6ITH from UC2AA. After a long build up the letter says "So dear Reg, first class amateur, for our Tanna Tuva trip we need appreciably from you one 75A4 receiver, one KWS-1 transmitter, or equally, one Telrex beam for working all band, 4 pair of shoes sizes 8-1/2, 9, and two 9-1/2, and most importantly 4 Gabardine suits sizes 39, 40 and two 42. Colors please size 39 blue, size 40 brown, and one size 42 brown, other blue."

Before we start on our hunt for the worst QSLer I would like to briefly describe a few common types of QSLers and I will demonstrate what I mean by naming actual DX calls.

First we have the 100% QSLer, who appears to send a card to everyone worked. Top DXers in this category include W1ADM, W2WZ, W3ALB, W4DHZ and W8BTI. I have a half dozen or more cards from each of these amateurs.

Next there is the type who, when he finally gets time to QSL, sends his cards out without including all essential data. By essential data I mean the date, time, frequency and RST. I am sure Bob White over here can give you a long list of foreign offenders in this category. Some of you will be surprised to know that some of our top American DXers also send out incomplete QSLs. Some such American DX hams are: WlBFT, W1GOF, W8DUS and W8UPN.

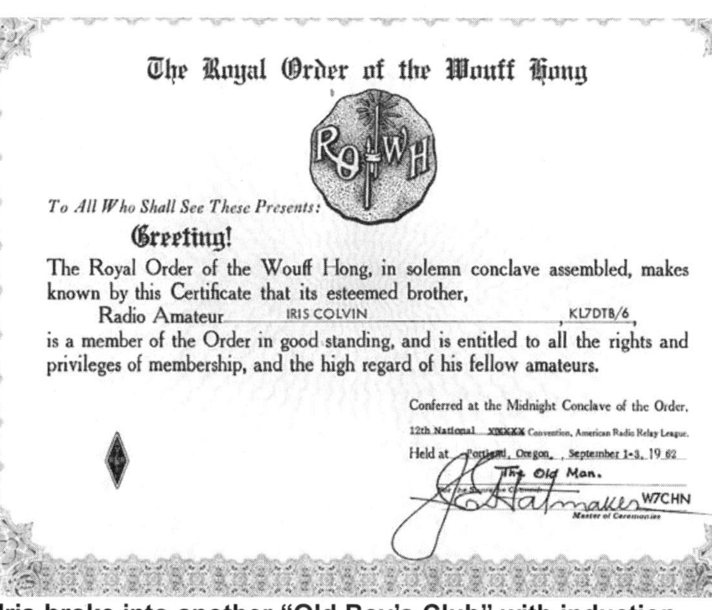
The Royal Order of the Wouff Hong

To All Who Shall See These Presents:

Greeting!

The Royal Order of the Wouff Hong, in solemn conclave assembled, makes known by this Certificate that its esteemed brother,
Radio Amateur IRIS COLVIN, KL7DTB/6,
is a member of the Order in good standing, and is entitled to all the rights and privileges of membership, and the high regard of his fellow amateurs.

Conferred at the Midnight Conclave of the Order,
12th National XXXXX Convention, American Radio Relay League.
Held at Portland, Oregon, September 1-3, 19 62

The Old Man.
W7CHN
Master of Ceremonies

Iris broke into another "Old Boy's Club" with induction into the Royal Order of the Wouff Hong in 1962.

Then there are the ones who when you ask them over the air about long overdue QSLs will tell you they cannot understand it as they mailed you a card and it must have been lost in the mails. There is one well known DXer who uses this excuse to the point that it is impossible that he is correct. This particular amateur I have worked from four continents, have had 22 QSOs, have sent him a QSL for each QSO and am still waiting for the first card from him. I am talking about Bert Brown, W4FU.

Next there is the amateur who sincerely tries to keep up with his QSLing but is just a plain poor book-keeper. The amateur I picked as representing this type is a well known European DXer. After this talk was prepared I received the sad news that the amateur is now a silent key. I am referring to Joaquim, CT1JS, who very recently passed away, Two years ago my family and I visited Joaquim in Port Portugal. CT1JS met us at the ship, showed us around town, wined and dined us, and in general made us feel very much at home.

While at his house I mentioned that he had been the first CT1 worked by me immediately after World War 2, when I was in Japan. I wondered why it was that I had not received his QSL card as I knew he did QSL at times as I had seen his card pass through the QSL bureau. He said that he could not understand it as he always QSLed. A few minutes later we were taking about W6 stations. Joquim said he had worked several W6s a few nights ago and suggested we look them up in the log. He went to his operating room and on his operating table was a large pile of papers of all sizes and shapes. CT1JS started looking through the big pile of papers. This was his log. It was not bound and was just a bunch of scribbled notes. The reason he had overlooked sending me a QSL card years before was apparent. In any case we will never forget CT1JS and I sincerely hope that those of you who worked him will not forget him either.

A famous last type of QSLer is the ham who QSLs all the new countries worked but doesn't bother answering cards from any amateur who is not a new one. I am not going to read a list of amateurs I consider in this category as I expect we will uncover a few when we attempt to find the worst QSLer in the room.

I think without further delay I can now proceed to find this worst QSLer. To prepare for this I have made up a list of all DXCC members in the eastern United States whom I have worked but who did not QSL. I have arranged them with the worst offender first and so on down the list. The top offenders are all stations whom I have worked from several continents and who have been sent several QSL cards without replying. Now this is just a game and all in fun. If your call is listed please hold up your hand promptly. If anyone fails to hold his hand up will the amateurs sitting nearby please point him out? All right here goes:

I want to apologize to W____ for the way in which I have exposed him and to tell him that I will immediately take him off my black list as soon as I receive his QSL, which I hope will be as soon as I sit down.

Speaking of sitting down, before I do I have one more thing I want to say. My wife, the lady with the blond hair sitting over there, has with her the QSL cards from all of you DXCC members whom we have worked under any of our calls. If you will see her she will be glad to show (you) your card after it has been around the world a couple of times and in addition she has a special QSL card to give you in memory of our personal meeting here in Washington, D.C. Thank you.

Joe Reisert recalled running into Lloyd again at the ARRL Convention in Portland, Oregon, in September 1962. "I was just returning to San Jose from Vancouver, BC, with my new bride of two weeks. Lloyd was very gracious and surprised to see me married. Later he asked if I had QSLed all our contacts. I said "'yes, but why are you asking me that?' He replied, 'at the DX breakfast tomorrow I will reveal who is the worst QSLer in the group.' He already was QSLing all contacts and filing them, and keeping statistics."

28· Life in the Home Front Military

Unfortunately, a person has only one life to live and cannot do all the things he would like to do - Lloyd Colvin

The October 1958 issue of The National Guardsman reported on the convention, noting that Lloyd's citation "praised Col Colvin for his 'personal contributions and exceptional operating achievements in ... amateur radio."

Because radio amateurs were present in large numbers both in the active military and as MARS members, the August 1958 issue of SIGNAL, Journal of the Armed Forces Communications and Electronics Association (AFCEA), published an article to educate its readers, written by Army Col. Fred J. Elser. The article was titled "Amateur Radio in the Services":

> No one can be certain who the first service radio "ham" was, but probably one of the first was Maj. Gen. Joseph O. Mauborgne, USA Ret. When he was ordered to the Signal School at Fort Leavenworth, Kansas, in 1909, General Mauborgne took with him his spark set so that he could "pound brass" thru the rigors of a winter on the Kansas plains. This early interest must have had something to do with the fact that in the Philippines, General Mauborgne was instrumental in the installation of a powerful Telefunken quenched spark transmitter at Fort Mills, Corregidor Island in Manila Bay. He also collaborated with the late General "Hap" Arnold in pioneering the first wireless communication between aeroplanes and the ground.
>
> The modern counterpart of this early pioneer of "Wireless" in the military services is well evidenced by the activity of such "hams" as General Curtis LeMay [W6EZV], Vice Chief of Staff of the Air Forces (whose recent exploits in Africa with another "ham," Arthur Godfrey (K4LIB) received considerable notice); Rear Admiral Henry C. Bruton, former Director of Naval Communications, George Bailey, Executive Secretary of IRE and Director of AFCEA, who communicates around the world nightly, and Brig. Gen. Earle F. Cook [W4FZ], Chief, Research and Development Division, OCSigO, whose article in SIGNAL on "Efficient Spectrum Utilization" was published in November 1957.
>
> Although individual amateurs in the services were active many, many years ago, it was not until World War I and after, that the services began to appreciate the value of organized amateur radio as a source of personnel, equipment and frequencies in time of need. In fact, it is reported that the commanding officer of the Navel Communication Center in Washington at the close of World War I was of an entirely opposite opinion. He was not convinced until his aide, an amateur operator, mustered for his inspection the "hams" in the comm-center; a matter of over 90% of the men there at the time!
>
> After the bridging of the Atlantic and Pacific oceans on short waves by "ham" operators in the

MARS Is Expanding with Growth of Alaska

Messages from outer space? most, but not quite.

The Military Affiliate Radio stem (MARS) is primarily ncerned with training and perimenting in radio communications, coordinating amateur d military radio operations, d providing additional radio mmunications circuits and rsonnel in event of a local or tional emergency.

One of the services of MARS sending and receiving mesges for soldiers to and from ends anywhere in the United ates.

This group also establishes d supports clubs concerned th radio communication, ofs instruction in radio theory d procedures, and assists embers in the assembly of ectronic equipment.

Beginning as the Army Amaur Radio System (AARS) in 25, at Fort Monmouth, N. J., became a joint military ornization known as Military nateur Radio System, in 48. At this time membership is restricted to military only. of civilian radio amateurs, the age limit for membership was lowered from 21 to 16 years of age.

Selected extension courses of the Army Signal Corps School are available to qualified MARS members to enable them to further their training in communications and electronics.

MARS headquarters for USARAL is at Fort Richardson. It is directed by Lt. Col. Lloyd D. Colvin, who has been a radio amateur and member of MARS for more than 30 years.

Members of the USARAL MARS Station, AL7USA, this summer started a school for amateur radio operators, with instruction broken down to 13 hours of code and 13 hours of theory instruction.

"The class was a huge success," Col. Colvin reports, "with all graduates expressing a desire to become MARS members after receiving their Federal Communication Commission novice-class license."

terested in becoming a MARS member can write or contact the USARAL MARS Director, USARAL Support Command, Fort Richardson. Complete information regarding membership will be mailed on request according to the station director.

The Army National Guard publication *The Pioneer* published this article in 1960, while Lloyd was temporarily stationed in Anchorage.

early 20's had added long distance to prior organization for continental message-relaying and the handling of local disaster traffic to the accomplishments of "ham" brasspounders, the services took concrete steps to affiliate amateur operators more closely into the twin services. About a third of a century ago, both the Navy, with its Naval Communications Reserve (N.C.R.), and the Army with its A.A. R.S. (Army-Amateur Radio System) I initiated organizations, which, with minor charges, still function today.

In the summer of 1925, when the Pacific Fleet made a cruise to New Zealand and Australian waters, to complement the high-power long wave radio apparatus aboard, the Navy brought to active duty Lt. Fred Schnell, an amateur of Hartford, Connecticut, who had been the first U. S. amateur operator to communicate across the Atlantic on short waves in 1923. Schnell's low powered short wave transmitter aboard the Seattle soon made the call NRL famous on the amateur frequency bands, and his musical 500 cycle note became familiar to amateurs all over the world.

A newly established short wave station at Fort Wm. McKinley in the Philippines erected by Lt. (now Col. Ret.) H. P. Roberts of the Signal Corps, was one of the first transpacific contacts made by Schnell as he started across the ocean, and was the last to be contacted by the Seattle before the "hook" was dropped in San Diego after the cruise was over.

At Fort Monmouth, meanwhile, Capt. (now Brig-Gen. Ret.) Tom Rives was one of the "sparkplugs" in the newly organized A.A.R.S. Soon the organization was in full force: 2CXL communicating periodically with stations at the Hq's of the nine Corps Areas, which in turn relayed messages, information and tests back and forth through the state Hq stations and so on down to the individual members. The systems grew and flourished: soon they were to be put to the real test.

The few thousand amateurs employing their talents in the services during World War I expanded to many thousands in World War 2. The onset of that emergency found not only amateur men and women in uniform but many in civilian attire as instructors or maintenance/production personnel. The release of millions of dollars of scarce equipment such as radio receivers, and indicating meters eased shortages in the early stages of the war, and the conversion of vacated amateur frequency bands gave sorely needed kilocycles not occupied by other stations.

At the end of World War 2 the Navy revived the N.C.R. The Army and the Air Force jointly established the MARS (Military Amateur Radio System, later changed to Military Affiliate Radio System). Membership in these organizations was made widespread. At present any U. S. citizen 16 years of age or over who is a radio amateur can join MARS. Certain distinct advantages accrue to the individual. For example, in MARS, active members become eligible to receive surplus equipment made available from time to time. They may also receive correspondence course material on a variety of subjects from television repair to Sideband theory. Naval Reservists have similar advantages, including the opportunity of taking training cruises on Naval ships.

In addition to the organized opportunities, many individual service amateurs still carry on the old traditions of individual advancement: both for themselves and for their services in modern electronics fields as new as transistors or operation "Vanguard" [an American rocket program].

Nor are all the service amateurs uniformed. Civilian engineers are avid "hams" too. In 1909 when the late George Eltz became the first Vice President of the newly formed New York Radio Club (later to become the Radio Club of America and Institute of Radio Engineers) he little thought that in 1958 he would be a division director in the U. S. Army Signal Corps Engineering Laboratories, but still active "on the air" with single sideband from his home in Belmar, N. J.

Perhaps Leo C. Young (W3WV) of Washington, D. C., had a better idea of his future part in the Naval service, since he has had an important part in the engineering of Naval radio facilities since years before the Fleet changed over to short wave.

When a radio car was proposed for President Roosevelt's train to keep him in touch with Washington during his continental trips, the job was given to Col. Bill Beasley (W9FRC). His ingenious solution of the antenna problem, in particular, enabled the Signal Corps radio teletype operators to "get the message through" from all parts of North America during and after World War 2. Although now retired from the Army, Bill still keeps up with activities of the radio car and service amateurs like George Hart, whose single sideband transmissions from the train were widely received, even from inside tunnels.

Another mobile "sidebander" is General Griswold of the Air Force. He became active right after World War 2 from Guam and lately has flown all over the world with various single sideband equipments aboard his plane.

Some service amateurs are extremely active. One Signal Corps officer, Lt. Col. Lloyd Colvin, was recently featured on the cover of a radio magazine for his feat in first communicating with several hundred radio amateurs having different "prefixes" all over the world. Then there is Cdr. Ray Meyers, USNR, who is organizing the AFCEA Net. Ray has pounded the key for more years than he wants to remember.

The spirit of the radio pioneer is not dead. The scope of amateur radio has widened, and the number of the practitioners of the art has greatly multiplied. Complexities have grown, but so have the tools for coping with them. The essential spirit of inquiry, worthwhile accomplishment and "camaraderie" of the frequency bands is as alive in '58 as it was in '02 when that first amateur, Marconi, heard the first signal across the Atlantic.

On 27 January 1959 the U.S. House of Representatives passed bill H.R. 3506, restating and expanding MARS, "To authorize the establishment, maintenance, and operation of auxiliary communication networks composed of licensed amateur radio operators for military radio communications:

Be it enacted by the Senate and House of Representatives of the United States of America in Congress assembled, That in order to promote and stimulate interests of individuals or groups in military radio communications, and for practical training of such individuals in skills which may be utilized in the operation of military communication networks, also increasing the number of persons technically qualified to operate and maintain military communications, the Secretaries in the Army, Navy, and Air Force, jointly or severally, subject to approval of the Secretary of Defense.

The operation of this radio network shall be conducted on radio frequencies assigned from those allocated to the participating military departments.

Late in 1958 Lloyd was nominated for the 1958 Edison Award, sponsored by General Electric. Although not a winner, the award committee did issue a letter of commendation on 23 January 1959, to "Lt. Col. Lloyd D. Colvin, W6KG, a candidate for the 1958 Edison Radio Amateur Award, [who] is hereby commended by the official judges thereof for rendering devoted service to the public while pursuing the hobby of amateur radio.

"Documentary proof of this service has been submitted to the Edison Award Council and reviewed by the undersigned judges on January 23, 1959.

"The above candidate has demonstrated to a high degree the Federal Communications Commission's official justification for amateur radio in the United States: 'In the Public Interest, Convenience, or Necessity.'"

Committee members were E. Roland Harriman, Chairman, American National Red Cross; Rosel H. Hyde, Commissioner, Federal Communications Commission; and Goodwin L. Dosland, President, American Radio Relay League.

Lloyd's nomination was from Col. Bernard J. Kitt, chief of staff of the California National Guard, saying

> Colonel Colvin is the Regular Army Signal Advisor to the 49th Infantry Division, California Army National Guard, and is an enthusiastic and well known radio amateur. Under his guidance during 1958 a network of some 60 amateur radio stations has been formed consisting of stations located at National Guard Armories in principal communities of Northern California.
>
> Colonel Colvin has done much of the organizing of these stations during off-duty hours while operating from his own amateur radio station W6KG. He has spent many hours of his own time and energy assisting individual stations to obtain their licenses and successfully build and operate their stations. He has prepared and forwarded weekly information letters to all stations, distributed FCC forms, offered technical assistance when needed, and personally designed and installed many stations and antenna systems. He encouraged and helped the stations and operators to operate on the amateur radio bands, become members of MARS, and operate in a daily radio network.
>
> During a disaster period of high winds and floods in Northern California in the Spring of 1958 many of those stations relayed important information and proved their ability to render valuable emergency communication service.
>
> The contributions made by Colonel Colvin in the

> field of amateur radio during 1958 are beyond that expected in the normal performance of his duty and have contributed directly to the establishment of some 60 amateur radio stations strategically located in National Guard Armories located throughout Northern California in areas where they have, and will continue to be, of great assistance to the public in event of emergency or disaster.

Lloyd designed and put into action a National Guard radio network covering much of northern California. In January 1959 an article appeared in The National Guardsman, Official Publication of the National Guard Association of the United States, to encourage other Guard areas to establish similar networks. The article was compiled from information Lloyd supplied as the Signal Corps adviser to the 49 Infantry Division of the California Army National Guard. The article identified the network, dubbed "Radio Wisdom," as Lloyd's "brainchild" and it covered 70,000 square miles of northern California. Lloyd said the secrets of the network's long range capability were "a good antenna system using coaxial cable and the selenium rectifier, which provides indoor power for the vehicular type radios."

Covering California from the Oregon line to Bakersfield (600 miles) and from the Pacific Ocean to Nevada (300 miles), the network was praised for offering "unexcelled training," rapid transmission of official messages, and emergency links to other services in the event of a large-scale disaster.

The net grew piece by piece and eventually comprised smaller, local nets feeding the umbrella net, Radio Wisdom, which then affiliated with MARS. The structure was very similar to the ARRL's National Traffic System on the amateur radio bands.

The National Guard pointed out that Radio Wisdom's MARS membership also gave the National Guard units access to equipment not normally available to them, tactical infantry equipment with which Lloyd was intimately familiar.

This equipment (hand-held and portable radios) was mostly designed for short haul radio work of 200 or so square miles.

By supplementing Army-issue radio gear with higher-powered shortwave equipment - ham radio HF-type gear - Lloyd's brainchild became a MARS analogue of the ARRL National Traffic System's "regional nets."

In other words, local National Guard units would use, at their armories, selenium rectifiers to operate their vehicular radios from 110 VAC and coaxial cable fixed antennas to connect by radio with other armories in the network. If necessary, their mobile radios were always available for their original intended use. This doesn't sound like Rocket Science - it sounds more like ham radio Field Day. Lloyd knew about Field Day, and before long he and Iris would take off on the world's longest Field Day.

The National Guardsman praised the network,saying that "men charged directly with communications responsibilities are learning 'on the air' to handle a steady flow of important information rapidly and efficiently. They're all urged to take a turn at operating the stations, and the problem of training radio operators has lessened considerably as many of the operators stop in voluntarily, to rehearse correct radio procedures and to help fabricate new equipment.

The Guardsman said that the Radio Wisdom stations were operating regularly on the ham radio bands and working stations all over the world "on equipment designed for a mere 15 mile range."

Not long after its organization, the system proved its value during floods in Northern California, when "operators in scores of armories scattered over the area funneled up-to-the-minute flood information into Division Headquarters, including regular reports on the activities of Division units ordered to emergency duty."

49th Division Signal Officer Maj. Bernard A. Sword offered hints for local 'Guard units:

> Set up your stations near the Hq of each unit. It won't be used unless its close by. Install fixed station type antennas (see diagram) for range, but keep the design simple.
>
> Stations which, for training purposes, must conduct a heavy interchange of radio traffic, should schedule their drills on the same evening. Go to your local radio amateurs for advice on which frequencies will offer the most reliable communications.
>
> ...keep the stations on a daily operating basis, a practice which not only will insure that your equipment is in good operating shape but will help sustain interest.
>
> Most important, use the network for official Division business at every possible opportunity, for training purposes and to keep every man in the organization communications-conscious.'
>
> [That's] the surest way to get 'radio wisdom!

Time Magazine Hops on the Bandwagon

Time magazine (16 February 1959) jumped on the cold war communications bandwagon with an article about single sideband, which was rapidly taking over from AM (amplitude modulation) as the amateurs' phone transmission mode of choice. The article, cleverly entitled "Power on the Side," was significant because it stressed that radio amateurs had pioneered the use of SSB for long-distance communication, and quoted Gen. Butch Griswold, a licensed amateur.

Despite some of the hilarious technical descriptions in the article, it was a watershed. Imagine, technical terms like SSB in a mass-circulation magazine! It publicized the role hams played in pushing new (and sometimes not-so-new) technology past hide-bound bureaucrats, just what Lt. Col. Lloyd Colvin had done in the California National Guard. *Time* said:

> When a howling mob of Venezuelans besieged Vice President Richard Nixon in Caracas last spring, the most urgent problem was to get the word to Washington -- fast. But how? Newsmen had tied up just about every telephone line leading out of Caracas; the U.S. embassy's own radiotelephone required a link through a Venezuelan switchboard.
>
> Air Force Colonel Tommy Collins, pilot of Nixon's plane, solved the problem. A dedicated 'ham' (amateur radio operator), he had brought along the ham's newest kind of equipment, a portable single sideband transceiver (transmitter-receiver). He flipped open his suitcase, pulled out a breadbox-size radio set, dropped an aerial out the window, in a matter of minutes was talking to ham operators in Washington. The hams called Administration officials on their telephones, hooked their stations into the phone line through a phone-patch, and soon Vice President Nixon's party was talking directly to the White House.
>
> For U.S. hams, such communication has become routine. Since 1957 thousands of them have installed the same 175-watt single sideband (SSB) transceivers in their homes and cars, have become accustomed to chatting with fellow hams in Australia, Alaska or South Africa as they bowl along superhighways on their way to work.
>
> "Key & Carrier." SSB makes a transmitter at least four times as potent as an ordinary amplitude modulated (AM) rig of the same power. AM transmitters operate over a wide belt of frequencies, a carrier frequency plus a band of voice frequencies on either side. Radio engineers long ago realized that either of those voice sidebands contained all the necessary frequencies for intelligible conversation. The carrier wave might be likened to the stream of air that is pumped through an organ. By pressing various keys, the air stream is modulated to produce separate notes. In effect, SSB eliminates the basic air stream, sends out only the signals that activate the keys. At the other end, the receiver resupplies the carrier wave, which, combined with the 'key' signals, recreates intelligible speech.
>
> Since each sideband uses only one quarter the power of the original carrier, putting only one sideband on the air saves the transmitter a considerable amount of electrical work. SSB's one narrow sideband takes up less space than AM in the already overcrowded radio frequency spectrum, where the carrier waves of AM stations often beat against each other in whistling confusion. Getting rid of the carrier and one sideband before they reach the transmitting antenna also means that the SSB receiver may be designed for extremely sharp tuning.
>
> [The] result is that SSB is far more potent than AM, and even as efficient as CW (continuous wave), the old workhorse system of communication that requires all messages to be sent in International Morse Code. A further asset is the ease with which SSB can be "patched" into any telephone line at both transmitting and receiving ends. Thus it enables any householder to telephone, with the help of an SSB operator, almost anywhere on earth.
>
> Patchwork. SSB was used by commercial radiotelephones as far back as 1923. But it showed serious defects when engineers tried to use it for the higher frequencies (3 to 30 mc), which cover greater distances by bouncing off the ionosphere. Finally, radio engineers learned how to manufacture the stable oscillators that high-frequency SSB called for. During World War 2 the Army used SSB between Algiers [where Lloyd Colvin was stationed] and Washington, D.C. In 1944 the U.S. Navy established a limited point-to-point SSB circuit between Washington and Pearl Harbor.
>
> But as late as 1954, SSB rigs were considered too bulky and too fragile to serve in the Air Force's far-ranging planes. Goaded on by hams, Collins Radio Co. of Cedar Rapids (no kin to Nixon's pilot Collins) kept refining and simplifying its equipment, developed a rugged, lightweight mechanical sideband filter that gave SSB sets the needed versatility. Among the most enthusiastic hams were General Curtis LeMay (known as "Curt" to hams

all over the world) and his deputy, Lieut. General Francis Griswold (known on the ham air as "Butch") - the two top commanders of' the Strategic Air Command. In 1956 Griswold mounted a standard Collins SSB ham rig on his personal plane and took off on an inspection flight to the Far East. No matter where he flew, General Griswold was able to keep in touch with LeMay by talking to U.S. hams and having them arrange a phone-patch with SAC headquarters. Sometimes Butch could talk with Curt directly. The Air Force was convinced.

Just Testing. We thought of SSB first for what we call our Positive Control System, says General Griswold. This is the "Fail-Safe" system by which planes on an atomic bombing mission can be sent to a prearranged line of departure but may go no farther without specific orders from SAC. "Around the North Pole there is what is called an auroral absorption zone, which gives us trouble in radio communication. And, of course, in our business the North Pole is a very important area. If any radio system can pierce the auroral absorption zone, SSB can. If it can't, we can always use it to relay our signals around the trouble."

The simplicity of relaying SSB signals was demonstrated by SAC's Airborne Electronics chief, Lieut. Colonel Joseph Beler, on a recent polar flight. While headed from Greenland to Alaska over the top of the world, Beler called a SAC base in England by SSB, asked for a telephone connection to headquarters in Omaha, had Omaha patch the call onto SSB again, had it relayed by SSB to North Africa, then relayed by SSB once more to Dhahran on the Persian Gulf. I could have talked direct to Dhahran," Beler admits, "but I was just testing."

Today almost every SAC plane has its SSB. Service bases all over the world are equipped not only with official SSB equipment, but service-operated SSB ham stations that keep their people in touch with their homes via ham relays and phone patch. SSB's most emphatic admirers are SAC's high brass, who have seen it turn their telephones into instruments for exercising worldwide command. "Communications is synonymous with command," says SAC's commander, General Thomas S. Power. "If I don't have communications, the only weapon I have is my desk - and I can't throw it very far, and it's not very lethal."

When Lloyd and Iris began their world travels in 1965, their equipment of choice was the latest Collins S-Line. And thousands of young hams sent to Southeast Asia during the Vietnam War remember warehouses full of Collins HF gear, the very same 75S3 receivers, 32S3 transmitters, and KWM-2 transceivers that were, for some two decades, the rigs of choice for hams who could afford them.

Pondering Army Retirement

Lloyd marked a milestone in June 1959 and faced the decision of whether to retire from the military. His official letter said that "...you will complete 20 years of active Federal service on 10 August 1960. If you desire voluntary retirement ... you would be entitled to receive retired pay based upon 23 years, 3 months and 24 days of service.

"On date of appointment in the Regular Army (28 June 1946) you were credited with 6 years, 2 months and 4 days of service. You accepted this appointment on 22 July 1946. Therefore, your constructive service credit on date of appointment is increased by your Regular Army commissioned service from 22 July 1946. You will complete 20 years of constructive service on 17 May 1960."

Lloyd had kept close track of his Army personnel file over those years and in September 1958 wrote to the Pentagon for help in removing "certain unwarranted correspondence from my official AG 201 files," a negative comment from a superior officer, which Lloyd referred to in his request as an admonition. "I feel strongly that it is not fair to me for the correspondence to remain in my files." The correspondence was subsequently removed per a 13 November 1959 letter. Lloyd, Iris and Joy were finally settled in the Bay area and Lloyd, expecting to be reassigned sometime in the second half of 1960, asked the Army to "please hold off as long as possible on my reassignment. I would like to stay here as long as possible."

Joy said "Lloyd had a number of choices and traveling around the world promoting amateur radio was the one he chose! It was not an easy decision. He thought long and hard over many days about what his best choice would be. He loved radio, and I believe that he felt he could positively affect the lives of more people by working in the field of amateur radio than continuing in the military."

Perhaps Lloyd had had enough of the Army's entrenched bureaucracy and snail's pace, as well, a pace once described by General Creighton W. Abrams, when he was Army chief of staff, who said "If I can move the Army one degree, I will have considered my tenure a success."

Iris and Lloyd made their first major, structured presentation regarding their travels at a meeting of the Southern California DX Club on 10 September 1959. They wrote a script for their talk. But it wasn't long before they began speaking off the cuff, from few notes, or none at all.

As some of you may know [the script went] Iris and I have an ambition to visit DXCC in person. We are getting close to our goal. The QSL cards on display represent 93 of the countries we have visited so far. In most cases we have visited the hams whose cards are on display and in all cases the QSLs are for QSOs made from one of our 23 calls. The 23 calls we have held are also on display.

The purpose of our talk tonight is to tell you about, and show you movies of, our Caribbean trip which we made last month." [The Colvins saved thousands of snapshots over the years, but most of the films they made were lost - ed].

In many respects our trip was a duplicate of a similar trip made earlier this year by Elvin Feige, W6TT, and his wife. In fact Elvin will not be too pleased over the way we arranged for the trip. We contacted the travel agent, a lady, who arranged for Elvin's trip and told her we wanted a similar trip. After many hours of research she came up with the exact itinerary that we eventually followed. Her price however for the trip included hotel rooms at a figure between 50 (and) 100 dollars per day. We told her this was too high and offered to buy just our airline tickets from her. She did not want to do this so we ended up buying our tickets direct from the airlines.

We ran into an unusual situation in buying our airline tickets for travel in the Caribbean area. The airline ticket agent in Oakland called their SF office to confirm flight reservations. We had told the agent that we wanted tourist class tickets, however on the phone the agent, another lady, kept talking about lst class tickets. We kept trying to catch her attention and said no, we want tourist tickets. Finally she turned from the phone in an exasperated manner and said 'You can have tourist tickets if you want but it will cost you more money.' We asked her what she meant and she explained that there was a special 17 day rate on lst class tickets over the route we wanted and that tourist class tickets would cost us almost $200 more than lst class tickets. We took the 1st class tickets!

Incidentally we never had so much to eat and drink on a plane before in all our travels. PAA now advertises that on their lst class flights you can have all you want to eat and drink. They are not fooling. When we arrived in Puerto Rico after a 4 hour trip from Miami we could hardly get up out of our seats.

We made out OK on hotels at all stops without reservations as this is the off season for tourists. Everyone normally likes to visit the Caribbean when it is winter in the USA. We found satisfactory accommodations at far less prices than $100 a day, however the very best hotels in all countries visited do charge around $100 a day,

I think the greatest difference noticed on this trip from others made in previous years is the fact that all the hams we visited had rotary beams. The most common and most widely used beam was the Mosley tri-bander. I would guess that we saw nearly a dozen of these beams in use during this trip. Iris will now tell you some of the hams and countries visited.

I will continue [Iris said] with a description of our Puerto Rico visits. We visited KP4AMR, Carlos, and the two very famous DXers KP4CC, Juan, and KP4KD, Ev.

Surprisingly we found all three stations to be very similar. They all run about 200 watts, use old HRO receivers, and have 3-element beams. We especially [enjoyed seeing] KP4AMR as a year ago when he was WP4AMR he was our one and only WP4 contact and we of course used his QSL for WPX credit. KP4AMR, Carlos, is the general manager for the local commercial telephone and cable system. He lives in a nice house and has a large room for amateur radio.

We planned on visiting KP4AMR and then going over to visit Ev, KP4KD. We had previously made a date to visit KP4KD at 10:30 that evening. We were having a nice time at KP4AMR's house when I noticed it was past 10:00. We told Carlos we would have to leave and he suggested we phone KP4KD and tell him we were on our way. We were certainly glad he made the suggestion as, when Ev answered the phone, he said he had already gone to bed. I asked him if we should cancel our visit, or still come. He said to come over and that he and his wife would get dressed again.

We arrived at his house about 11 p.m. We apologized for being a half hour late. He answered that it was 12:30 in the morning and we were over an hour and a half late. What had happened was that we had failed to change our watches on our arrival in Puerto Rico and we were still on the time used in Miami and the Virgin Islands.

After having got up and dressed again they did not seem anxious to go right back to bed, so we stayed and had a nice visit. This is KP4KD's 41st year in amateur radio. Any history of amateur radio in the future I am sure will include Ev. He was the first amateur in the world to operate from the Caribbean area. He was very active in the very first days of amateur radio, and is still active. He has won innumerable operating contests. He is the QSL manager for KP4 and has helped many KP4 hams obtain their licenses.

While visiting his station we noticed one outstanding thing that impressed us as to his activity. For nearly twenty years he has used the same old HRO receiver. You all know the big dial on the HRO, well he has tuned his receiver for countless number of hours and he always holds the dial with his right hand and lets his little finger drag against the front panel of the receiver. It is almost unbelievable but he has done this so often that where his little finger touches the panel all the paint is gone and he has actually worn a deep impression half way through the thick metal front panel.

KP4CC lives in a rather poor section of San Juan. His transmitter, and much of his other gear, is home made. While there he greatly impressed us by pulling out an old log book to show us that we were the first Japan station he had worked after World War 2. His one ambition in hamdom appears to be to catch up with, or surpass, the many accomplishments of KP4KD.

While in Haiti we visited HH2AB and HH2CL. It would be hard to find two stations with more different equipment. HH2AB is a colonel in the American Air Force. He has the latest line of Collins SSB equipment plus several spare 75A4s and 32V transmitters. HH2CL is a native and has a small homemade 10 watt transmitter. In spite of all this we noted more QSLs confirming on the air QSOs at HH2CL than we did at HH2AB's elaborate station.

In addition to his regular job Claude, HH2CL, has a side job and hobby of making coffee. He was very enthusiastic in describing how he makes the coffee. He sells a considerable amount of it locally in Port-au Prince. The coffee in Haiti is very strong and bitter. Neither Iris nor I like it. Well, as you can imagine, Claude and his wife offered us some of their special home made coffee. We just couldn't say no, so we held our breath and drank it. We told them it was wonderful but we noticed a doubtful expression on their faces and I am not sure we hid our true feelings too well.

We had previously worked many stations in Jamaica and had expected to visit several of them. Most of them we wanted to see turned out to be visiting in either the US or Canada. We did visit the QTH of VP5AA and had a nice personal visit with VP5BF, who is general manager of the combined communication system that serves all airlines in Jamaica. He showed us the communication facilities at the airport and we took a picture there which is one of those passed out for you to look at.

Late in 1959 Lloyd learned that he had been selected to be Chief of Army MARS, at the Pentagon in Washington. This selection seems to have been "floated" via a telephone conversation as no written record of it exists. Lloyd declined, in a letter dated 27 November 1959, to Col. George P. Sampson, Chief, Army Communications Service Division:

> I am writing this letter as a follow-up to our recent telephone conversation. My immediate reaction when I was notified that I had been selected to be assigned under you as Chief MARS was one of great joy and happiness. I have had a secret desire for years to be Chief MARS.
>
> When I saw you, and many other Signal Officers I have previously worked with in the Pentagon last summer I kept thinking to myself how fine it would be to be assigned to duty in the offices of the Chief Signal Officer. As I think you know, I have long been a faithful admirer of you and your ability. There are very few people in our government who have an understanding of the over-all technical problems of Signal Communication that you have. I am certain you will be a Brigadier General in the near future. I would consider it a great privilege to work for you again.
>
> Unfortunately a person has only one life to live and cannot do all the things he would like to do. I have made a careful study and evaluation of my standing and future prospects in the U.S. Army. At the same time I have studied my future prospects in civilian life. I am forced to arrive at the conclusion that the future opportunities in civilian life appear to offer more to me. I am still young

enough that I can make a new career for myself in civilian life. I have decided to apply for retirement after completion of 20 years active duty, which will be in 1960.

I thank you for considering me for the Washington assignment.

The Colvins visited the U.S. Virgin Islands in the summer of 1959, and described, in a talk to the Oakland Radio Club, their visit with the famous Dick Spenceley, KV4AA (who, at the time, was in the process of resigning as DX manager for CQ magazine in order to devote his time to the fledgling YASME Foundation).

Danny Weil was not there but Dave Tremayne, ZL1AV, who was to accompany Danny briefly on the next voyage of YASME, was.

The Colvins did not mention that Spenceley and his wife had nine children; that one of those nine was severely disabled and whom, according to others who visited the Spenceleys, was accorded particular love and affection; nor did they mention Spenceley's wife by name:

One of the stations we visited on our recent visit to the Caribbean was the well-known KV4AA in the Virgin Islands. Dick invited us over that evening for dinner. When we asked for instructions on how to get to his house, he said "Ask Anyone." He was right! The first taxi cab driver that we called knew where he lived even without an address.

We were somewhat late in arriving because it took so much longer than we had expected to get from our downtown hotel to Dick's house. It is located on a narrow street of rather decrepit -looking stone walls. We had no trouble spotting the beam, but we spent another half-hour trying to find the right entrance into the house.

Eventually Dick saw us or we might still be on the other side of the wall. Once inside, we found a nice home, patio and swimming pool. We met Dick' s charming native wife who served us drinks and an island-dinner of fish. We also met Dick Jr., a daughter and granddaughter and other guests including Dave, ZL1AV. We saw the remains of YASME II and equipment that had been salvaged [from the wreck off Union Island in the Grenadines], plus the extensive accumulation of expedition QSLs belonging to Danny Weil.

After dinner, Dick said he always worked the rig for an hour, so we retired with him to the hamshack. He has a Johnson Kilowatt final with several different exciters including sideband; A 75A4 receiver and Hallicrafters receiver, a good homemade 20-meter beam [and] wire-type antennas for the other bands. He is planning to buy a tri-band beam. In the next 45 minutes, Dick worked 31 stations plus relaying a message to our daughter Joy, here in California. And there was no contest on! It is easy to see how Dick is 6th top DX man of the world!

When we were ready to leave, Dick said he would WALK back to our hotel with us. We insisted on calling a cab but were overruled. Were we surprised when we were a half-block from home - but down a one-way street.

29. Treading in Danny Weil's Footsteps

We decided to locate the spot where Danny Weil, VP2VB, operated from. - Iris Colvin

Iris and Lloyd also visited the British Virgin Islands, home of Danny Weil's amateur radio license:

The trip across was a most interesting one, on a boat named the Flying Fish, which actually did almost fly. It is a hydrofoil ship, made in Germany. It has large wheels on either side, both back and front, that look something like those used on river-boats, with the addition of propeller-like blades. In action, these cause the boat itself to rise up out of the water and skim over the surface at about 45 miles per hour. Thus, when we say that our transportation on our Caribbean trip was by air, you can see that this is almost literally true.

When we landed at the dock in the British Virgin Islands we saw the single main street with a few huts clustered together - this was the sum total of the town of Tortola. Most of the population appeared to be on the dock; not to meet us, however, but to buy fish. The fish market itself had just arrived. That is, a fishing boat with its hold filled with water and live fish. The shoppers picked out the fish they wanted to purchase and the merchant promptly caught it in a net, knocked it over the head, weighed it, and the happy buyer went away with his "fresh fish."

We then walked around the town, which takes only 10 minutes to see everything. The supply of fresh water here is obtained, as in Bermuda, from rain. In our movies you will see one of the hills that has been covered over to collect the rainwater and drain it into storage basins.

Of course, there are no hams at present to visit here, but rather than spoil our record, we decided to locate the spot where Danny Weil, VP2VB, operated from. This was not difficult. We asked some of the natives. Their faces brightened up with interest. Yes, they remembered very well the ham who had set up a radio station. They pointed out a large white house on a hill at the end of town. He had set up his station behind this house. Doubtless this was the biggest event that had ever happened to this sleepy little town with only a hundred or so natives.

Iris spoke to the club of visits to a couple of European hams she and Lloyd had made during their time in Germany five years earlier.

I have also chosen EA3IT to talk about, since I worked 'Alfonso.' My luck is probably due to the fact that Spanish Don Juans love to work lady-hams!

On arriving in Barcelona, I immediately called Alfonso. It was about 9 in the morning. He was anxious to have us see his station and said 'come over this afternoon' and suggested about 11 o'clock. Thinking that he wanted to have the visit during the day time, I said '11 o'clock this morning?' I know better now! - 11 o'clock at night is afternoon Barcelona.

Somehow, Alfonso had managed to round up all the local hams, so that when we arrived at EA3IT, we were greeted by a regular ham-fest. It was truly an expression of exuberant hospitality. The language difficulty of our not knowing Spanish only added to the international flavor of the situation. One ham knew German very well and most of them understood French. We "made-out" with some difficulty and lots of fun.

Lloyd then takes up the narrative:

[Another] visit I would like to mention is Tibi, HA5BB, who also was worked by me, from Germany. The contact was made when the "behind-the-iron-curtain" hams were being heard but not WORKED. I was surprised and pleased when he answered my CQ. He called me 4ZC - with NO prefix. We had a nice QSO, but I was sure that he didn't know WHERE or WHO I was - but I DID get a QSL and found that the omission of a prefix was merely to protect himself from the constant and strict monitoring by the Hungarian government. Every ham was required to keep a carbon copy of his log - and send it in to the government, monthly.

When we were fortunate enough to visit Budapest, we also had the pleasure of seeing Tibi, his wife and small son - his home and his own rig. Both receiver and transmitter were completely home-made, using relatively obsolete parts according to our standards. Most of the other hams that we saw there were at the club station, where they did most of their operating. Equipment is just not available for good private stations. It is also difficult to get a call without political pull. So many of the SWLs are fine radio men and operate code up to 30 words per minute.

No place in the world are visitors more welcome than in Hungary - they hate to have you leave. During our stay here, we were invited on a continual round of dinners, dances, and entertainment. We met many important government officials, royalty, artists, actors, and musicians, as well as ALL the hams. We had a government car (with chauffer and interpreter) constantly at our disposal. 'We wondered whether we were receiving the red-carpet treatment or just being kept under surveillance.

When we left, we found that we had not spent all of our Hungarian money, so we put it in an envelope and mailed it to Tibi, thinking that he could use it. We didn't hear from him for several months. When a letter finally came, we found out the reason: he had been in jail, and because of us. Our letter to him had been opened, the money discovered and poor Tibi held for extensive questioning. To bring this story up to date, Tibi was one of the few who were fortunate enough to escape after the slaughter in the Budapest public square. The last we heard from him and his family was from France, where they were all together.

Iris and Lloyd made another club presentation that fall, in November 1959, this time to another general interest club (not necessarily DXers), the Menlo Park Radio Club:

Thanks to Uncle Sam, Iris, Joy and I have traveled extensively. We have visited 93 countries, held the 23 calls shown on the display, worked some 80,000 amateurs located in 272 countries, and have received over 30,000 QSLs. The QSLs on display represent all of the 94 countries we have visited in person except HV, the Vatican City, which we have never worked. Most of the QSLs on display are from amateurs whom we have visited in person as well as working them from one of our calls.

Our last overseas assignment was in Germany. While there we visited all countries not behind the Iron Curtain and we even managed to visit at least one country behind the Iron Curtain [Hungary].

Most all European hams use American receivers or build their own. There are not many receivers in Europe [made in Europe] that will compare with ours, and the few there are are fantastically expensive. This means that European hams must try to obtain their receivers through American surplus markets in Europe, and through American hams and friends. Only in the rarest of cases are European hams permitted to import American receivers, or other communication equipment, without paying prohibitive taxes. Relatively few of the European hams use rotary beams made of aluminum tubing. One reason for this is that aluminum tubing in Europe is as expensive, or more expensive, than here. Figuring the average European's income, this means that a beam there may easily cost three times what we pay here, on a relative basis.

And, a visit to Trinidad (still a British colony, VP4) in those days:

Iris and I could tell you some kind of a personal story about each of the 93 countries and amateurs represented in the QSL display. This would take too long. Instead we will attempt to give you a brief story an a few of the more interesting hams visited. Iris will now take over and tell you about some of our visits.

I am going to start off by telling you of a visit that concerns a problem that occurs very seldom in amateur radio. It is generally accepted that one of the fine things about amateur radio is that no matter what your occupation, creed, or color, you are accepted on a common ground in amateur radio. When we arrived in Port of Spain, Trinidad, the first amateur we contacted was VP4TI. He is a British banker and we visited him in his downtown offices during the morning. He invited us to visit him at his home that afternoon and we gladly accepted.

After leavings his office we decided to attempt to locate VP4LW, Bill, whom we had worked often from Europe. Bill lives some distance from Port of Spain but we finally found his telephone number and called him by telephone. He said he would like to meet us and I told him we were to visit VP4TI that afternoon and suggested that Bill meet us at VP4TI's house. Bill said that would be fine and that he would see us there.

We arrived at VP4TI's house that afternoon as planned. We saw his station, his nice house, and were introduced to several relatives. We were all in the living room having a drink when the door bell rang. VP4TI went to the door and said in a loud voice "You know better than to come to the front door. If you want to come in this house go around to the back to the servant's entrance." It was our friend, Bill, VP4LW, at the door and he was a Negro. Bill went around to the servant's entrance and Iris and I were forced to go back to the rear door to talk to Bill as VP4TI would not let him enter his front room. You can imagine how embarrassed we were as we had asked Bill to come there. I am happy to say that this is the only incident of this kind we have ever encountered in our travels" [up until then, 1959].

Lloyd, the building contractor, found Bermuda interesting:

Last summer Iris and I made a trip to Bermuda. We are going to show you a 2 minute movie of that trip. We have several hours of movies we could show but more or less by accident we showed this 2 minute movie to a group of amateurs on our return from Bermuda and it seemed to be better received than our long movies. I guess this must prove that the shorter a movie can be made the better it is.

Bermuda is a British possession. It is possibly the only British possession that will still allow American amateurs to obtain a local station license. There are no income taxes in Bermuda. This sounds wonderful at first, then you find out there is almost no industry or prospects for same. The film is very short so I would like to mention in advance three things of particular interest. The weather is hot so most cars are open with a picturesque fringed awning on top. There are no fresh water wells in Bermuda and fresh water is obtained from rain. Wherever a suitable hill can be found they pave it over and let the rain water run down the paved area to a reservoir.

In addition almost all houses have arranged to collect the water that falls on the roofs. Most roofs are covered with coral slabs. These are white thus reflecting the heat, and remain pure thus being a sanitary way to collect water. At the very end of the film we show some scenes of a new house being built in this manner.

Both Iris and Lloyd enjoyed speaking to groups of non-hams, giving them a chance to preach the gospel. Iris spoke to a meeting of the "Officers Wives Club" in October 1959:

I am going to tell you a little about our recent visit to the Caribbean Islands, and then show you some movies that we took. We visited both the American and British Virgin Islands, Puerto Rico, Dominican Republic, Haiti, and Jamaica. We did not visit some of the other islands such as Cuba and the Bahamas, that we had already seen on previous trips.

The pictures you are going to see are of special interest to Amateur Radio operators, and you will see quite a number of antennas, as well as hams that we visited enroute. This seems like a good opportunity, however, to show you that Amateur Radio is a wonderful hobby, as well as a big advantage when one is traveling. The people we visited are all "good friends" over the air. We always have a lot in common with them. They almost always speak English, an advantage in foreign countries, and we usually invite them out, and ourselves gain the advantage of being shown the sights by a native.

One most pleasurable part of our trip was in the company of a ham friend in Haiti. He has the call sign HH2CL. Claude speaks good English, but his family speaks French only. We got some practice on our French. Haiti is really the most interesting, as far as native customs and entertainment is concerned, of any country we visited. I saw my first cock fight and also my first voodoo dance. The audience at both was primarily native (not tourist) and they enter into the excited frenzy which characterizes the interesting, even if somewhat gruesome display, especially the latter which is enacted in dim candle light.

But they did burn a chicken alive, using the blood to cast a spell upon the witch who then proceeded to burn herself. Maybe there is something to it, otherwise I cannot see how the bewitched woman escaped from being burned.

You can see how fortunate we were to have someone explain the intricate details of the ceremony. Claude manufactures coffee, so then he insisted that we have some of his special brew - very strong, syrupy espresso type coffee which neither Lloyd nor I like. Being good guests we tried hard to show him how much we appreciated it. Maybe we overdid it a little, because he insisted that we bring some home with us. Anyone here like Haitian coffee?

I might just add a few words about the interesting country of the Dominican Republic, since unfortunately we did not get enough pictures of the fabulous, all-new buildings and public squares which seem to dominate the city of Ciudad Trujillo. The city is named after the Trujillo brothers, rulers of the country. In a few short years they have managed to build up an unbelievable economy from practically nothing. Their pictures are displayed everywhere, from government buildings to bars. Lloyd nicknamed them the Gold Dust Twins.

In exact contrast to Haiti, the entertainment here is expressly for tourists. To prove this, we attended the "Agua Luce" [Agua y Luz], which means dancing waters. The show, composed of intricately manipulated fountains and waterfalls, combined with varied lights and music to produce a true dramatic effect, was to begin at 8 o'clock. When we arrived we were amazed at the vastness of the auditorium, but even more amazed at being the only people present, besides the large staff of waiters, technicians and doormen. Unlike American business practice, however, the show was put on - just for us."

The Colvins told a meeting of the Northern California DX Club, and others, of running into a celebrity ham in the Dominican Republic:

I would like to tell you something about our visit to the Dominican Republic. Ciudad Trujillo is, I believe, one of the most intriguing places we have visited. Amazing because one constantly wonders how the two Trujillo brothers could have built such an array of fabulous new buildings in a short time and out of a poor, broken economy.

The fine public buildings, squares and statues are tangible evidence of prosperity even though one feels that this prosperity is entirely false. In our movies you will see the new homes that are replacing the slum areas - yet the money to finance this false front must have been obtained by exploiting the poor. The people appear to have confidence in their government and feel that their country is impregnable.

Here, we visited John Dulles - John B. Dulles, that is, not John Foster Dulles. HI8JBD. His home, like most of the houses and buildings in the city, is new. His radio equipment is new, a Collins 75A4 receiver and 32V3 transmitter, a commercial steel tower with Mosley tribander and commercial rotator. At one side of the radio room we saw a display of silver cups and trophies. Thinking that these might be awards for radio proficiency, I read the inscription on the largest one. It had been presented to JBD by Wheaties breakfast cereals. This, like the others, was an athletic award. We found out that John is a rather famous golf player.

When we left the Dominican Republic, we felt as if we had seen everything that we were supposed to see, without actually seeing anything.

Back in the real world, Lloyd continued to monitor his Army records, and late in 1959 these comments appeared in his "Overall Efficiency Index"

"A well qualified officer from a technical standpoint in field of signal communication who is definitely a staff officer rather than a commander. His tendency is to do all details of job himself instead of delegating authority to his subordinates. Physically he is of medium size and build, trim and soldierly. So far as I know he has demonstrated no mental, moral or physical weakness."

Another superior said "Has tendency to do all detailed work himself rather than fully employ the subordinates at his disposal. Characteristics fit him better for staff assignments than for command."

The following spring Lloyd began looking into retiring, specifically about perhaps remaining in some capacity in the National Guard. Lloyd wanted to know how remaining in the Guard, and possible promotions, might affect his retirement pay from the regular Army. The answers he received were all negative. The only way he could work a deal was if he had a "minor disability" that necessitated retirement from the regular Army but allowed him to be a Guardsman.

In February 1960 Wayne Green left as editor of CQ magazine. Lloyd expressed a business interest in CQ but the publisher was not interested in selling. October QST's "How's DX" said "W6KG signs on as DX editor for a prospective ham-type periodical engineered by W2NSD." This leak, or rumor, turned out not to be true (Wayne Green would found 73 magazine in 1961 but Lloyd never wrote for it).

Lloyd and Iris, the inveterate QSLers and always looking for ways to economize, corresponded with the U.S. Post Office Department regarding post cards, looking for ways to allow them to be mailed for less than first-class postage, but are rebuffed. At the time, the first-class rate for post cards was four cents each, third-class cost two cents.

Lloyd received a teletype message dated 20 May 1960 saying that he had been selected for assignment to USA Alaska. He continued correspondence with the Pentagon in an effort to clear some material from his personnel file, and a friend there told him that the "Efficiency" report in question, from his time in Germany, might not be "as damaging" as he fears. Lloyd wrote to the Army colonel who was his rating officer there, asking for a re-rating, but is denied, then writes to the Efficiency Reports Review Board at the Pentagon, asking that "every consideration be given to removing" the report from his record.

Lloyd hoped for a promotion from lieutenant colonel

to full colonel, and in July 1960 the commanding general of the California Army National Guard wrote to the Army's Adjutant General at the Pentagon, saying:

> Lt. Colonel Lloyd D. Colvin, Signal Corps, has served for the last four years as a regular army advisor to the 49th Infantry Division, stationed at the Division Headquarters at Alameda, California.
>
> During this period he has held positions as Division Signal Advisor, Advisor to Division Trains, and Executive Officer of the Division Advisor Group. In performing these assignments he has demonstrated outstanding leadership, personal relations, knowledge, skill and ability to organize. He has consistently displayed a commendable attitude toward duty and enthusiasm for assigned tasks...
>
> Largely through his efforts and professional skill during his tour of duty, this division has developed an area wide radio network that functions daily during the armory training year. During a necessary absence of the Division Signal Officer from field training in 1958, he also assisted as acting Division Signal Officer in a superior manner in addition to his regular duties as an advisor. He is a distinct credit to the Signal Corps and the United States Army.
>
> I consider Lt. Colonel Colvin to possess the military and professional qualifications for promotion to Colonel, Signal Corps, and recommend he be promoted to such grade. I will appreciate it if this letter is placed in the records to be reviewed by the board which considers Lt. Colonel Colvin for promotion to Colonel.

Lloyd pulled out all the stops and had the following "CERTIFICATE" placed in his records:

> To the best of my knowledge the following statements are true and accurate.
>
> On my arrival in Europe in October 1953 I was assigned as the first Battalion Commander of the 315 Signal Battalion, Micro-Wave and Radio Relay (later redesignated as the 102 Signal Battalion, Microwave and Radio Relay), The Battalion Headquarters was at Karlsruhe, Germany. The Battalion was part of the 4th Signal Group whose Headquarters was at Heidelberg Germany.
>
> The Battalion was confronted with a large number of problems. We operated nearly 60 different radio sites located in Germany and France. We were assigned nearly $14,000,000 worth of highly complicated and delicate radio equipment. Most sites were on mountain tops in isolated areas. Our problems in transportation, messing, housing, training, and operations were far greater than encountered in most Signal Battalions. In spite of the many challenging problems I considered it a splendid assignment and I thoroughly enjoyed being the unit's first Battalion Commander,
>
> From the beginning it was decided that technical control of the huge radio-telephone system (the largest military radio-telephone system in the history of the world) would be located with 4th Signal Group at the Headquarters of the US Army Europe (USAREUR) in Heidelberg. The officer assigned to exercise this technical and engineering control was designated OIC USAIREUR Radio-Telephone System. All command functions in the Battalion remained with the Battalion Commander and his Company Commanders.
>
> In 1954, Colonel Philip Rose was assigned as CO 4th Signal Group. Shortly thereafter I was transferred from the Battalion to the Headquarters 4th Signal Group where I was designated OIC USAREUR Radio Telephone System. This assignment was carried out successfully and during this period I received several letters of commendation, which are in my official TAG files.
>
> In late 1955, Colonel George F. Moynahan, Jr, was assigned to Europe and relieved Colonel Rose as 4th Signal Group CO. Immediately upon assuming command (8 December 1955) Colonel Moynahan called a staff conference which included all key officers of 4th Signal Group, the 102 Signal Bn and the USAREUR Radio-Telephone System. Colonel Moynahan stated that he felt that the Micro-Wave and Radio Relay Battalion had been poorly coordinated in the past and that the OIC of the USAREUR Radio-Telephone System had failed to fully employ the subordinates at his disposal.
>
> These same words were later used in the efficiency reports rendered by Colonel Moynahan on me covering periods of a later date when I had no subordinates at my disposal. During this meeting, on the same day he became my rating officer, Colonel Moynahan relieved me of my duties as OIC USAREUR Radio-Telephone System and transferred all such duties to the S-3 officer of the 102

Signal Battalion. He said that in his opinion the technical and engineering control of the Micro-Wave and Radio Relay system should always have been assigned to the S-3 officer and that the Command of the Signal Battalion had been poor due to improper organization in the first place.

I disagreed with this sweeping statement and attempted to point out to Colonel Moynahan that conditions during the difficult build-up of the Micro-Wave System were different than he presently found them and that while I agreed that certain changes were now in order I did not feel it was fair to say or imply that past organization or command had been all bad. Colonel Moynahan replied that I was to leave for the United States soon and he would appreciate it if I got "out of the picture" as soon as possible as a new and better organization was to be formed.

The next day without contacting me in person, Colonel Moynahan had the furniture in my office moved to his office and I was transferred to a small room located on a top floor end as far from the Group Commander's office as possible. All of the personnel of the USAREUR Radio-Telephone System started reporting to the S-3 of the 102 Signal Battalion with the only person remaining at "my disposal" was my German civilian secretary (later she was transferred to the S-3 office Headquarters 4th Signal Group). On 16 January 1956 formal orders, confirming the earlier verbal orders, were issued relieving me as OIC USAREUR Radio-Telephone System and assigning me as Assistant to the CO 4th Signal Group. No such TO&E position existed and it was common knowledge in the 4th Signal Group that I was in effect "marking-time" until my orders for return to the United States were issued.

In spite of this unusual and difficult position I made every effort to assist the Commander 4th Signal Group in any way I could. My duties consisted almost entirely of minor technical assignments and the conducting of visiting dignitaries through the technical operations of 4th Signal Group. On several occasions Colonel Moynahan congratulated me for the efficient manner in which the few jobs assigned to me were accomplished. On 26 January 1956, Brigadier General Wesley T. Guest, Signal Officer USAREUR, personally thanked me for a job well done and presented me with a letter of appreciation from General Desfemmes of the French Army (copy enclosed).

On another occasion I received thanks from Colonel William D. Hamlin (now Major General William D. Hamlin) for the manner in which I had escorted him and a group of officers from Washington, D.C. on a month long visit and inspection of the technical operations of 4th Signal Group.

For a short time, just before my departure for the United States, I was assigned the job of coordinating the activities of an aviation flight control section consisting of 4 aviators. I had no command control of these officers.

I met Colonel Liedenheimer, who was the [e]ndorsing officer on my efficiency reports during this period, only a few times in his office and at occasional parties. Under these circumstances he was not in a position to do anything but go along with the ratings given by Colonel Moynahan.

I did not see the two subject reports in Europe nor were they discussed with me. I first learned of the reports much later in the United States.

But ... the Army wheels were turning, and on 1 September 1960 Lloyd was reassigned, and he and Iris moved to Alaska. These were the days of the famous cold war "Dew Line" across Canada. Through February 1961 Lloyd described his duties as "Signal Officer US Army, Fort Richardson, Alaska. Directed all the US Army Signal Corps operations and installations in Alaska. Was in charge of much of the equipment and personnel operating 'White Alice' and other electronic radar, and protective equipment installed in the far north for communications and for the protection of the United States in the event of a sneak enemy attack over the North Pole."

Lloyd also was involved in Army MARS operations in Alaska. An Anchorage newspaper, The Pioneer, featured Lloyd and MARS station AL7USA, where Lloyd was teaching students Morse code in preparation for their Novice Class ham tickets. The article noted that the MARS system's most noteworthy task was "sending and receiving messages for soldiers to and from friends anywhere in the United States," in addition to conducting classes and supporting amateur radio clubs.

Lloyd told the newspaper that MARS had begun in 1925, at Fort Monmouth, N. J., as the Army Amateur Radio System, of which Lloyd had been a member in the 1930s. When the name was changed in 1948, the new system, MARS, welcomed civilians to join military personnel as members.

At the same time, the age requirement was lowered from 21 to 16.

Lloyd was director of the MARS headquarters for USARAL (U.S. Army Reserve Alaska) at Fort Richardson. The Army MARS system in Alaska had members in Anchorage, Fairbanks, Fort Greely, Wildwood Station, Homer, Delta Junction and Kodiak, and Lloyd told the Pioneer he hoped the system would expand to "every city and village in Alaska.'

A letter from Edward K. Blasdel, M.D., in Berkeley (18 November 1960) attested that "Mrs. Edna White was examined by me this date, found to have severe osteoarthritis of the spine. She states that because of this disability, she needs the help of her son. Lt. Col. Lloyd Colvin, presently on active duty in Alaska. I believe she is justified in her request to have her son released from active duty, to assist in her care and support."

During this time Lloyd, with Iris's help, prepared applications for many amateur radio awards, drawing on the 30,000 QSLs they had on file.

30. Return to Civilian Life - 1961

I am convinced that passage of the proposed [reciprocal licensing] bill can go a long way toward mutual trust and understanding between our country and all others. - Lloyd Colvin

Lloyd's retirement from the regular Army became official in February 1961. In a resume he would write more than 10 years later, he said

> Retired from the Regular US Army Signal Corps, rank of Lt Col, after more than 20 years service. Immediately formed Drake Builders, a general contracting company, and started various construction projects in Northern California. Buildings were erected in the cities of Alameda, Berkeley, Richmond, Oakland, San Francisco, Stockton, Daly City, El Cerrito, and San Jose, California.

In 1962 the US Army Signal Corps sent a questionnaire to all retired personnel in order to determine how successful they were in civilian life, what jobs they had, and what salaries they were receiving. The results of this survey showed Lloyd Colvin as receiving more money for his services in civilian life than anyone else answering the survey.

> From 1961 to the present approximately $15,000,000 of new construction was built by Drake Builders. This included residential subdivisions, commercial buildings, hospitals, and apartments. Several millions of dollars of construction was built for the government under the Housing Urban Development Program. The Government programs included the requirement for an affirmative action program, including plans for the use of minority workers on the projects. Many conferences were held with various HUD, FHA, and other government agencies in connection with such government programs. Appeared in person before the City Planning Boards and the Board of Supervisors in large cities such as San Francisco and Oakland, California, to present details of proposed construction.
>
> During the period 1966 to 1969, Lloyd Colvin and Iris Colvin took an extended leave of absence from Drake Builders and made a trip around the world, carrying with them an amateur radio station from which they communicated to other amateur radio stations throughout the world.
>
> With the contacts previously made by army signal corps travels and amateur radio contacts, they were able to visit the personnel in charge of government communications in most countries visited. As a result, they were invited to visit major communications and broadcasting facilities throughout the world. About 50 countries were visited on this tour, and the control, regulations and success of communication, both government and private, were studied first hand.
>
> During this same period, a successful book, entitled "How We Started Out Building Our Own Home In Our Spare Time And Went On To Make A Million Dollars In The Construction Business" was co-authored by Lloyd Colvin and Iris Colvin, his wife. This book was published by Prentice Hall. Over 100,000 copies have been sold to date.

In 1961 the Colvins decided to sponsor an amateur radio award of their own, "The Colvin Award," and immediately became involved with Clif Evans, K6BX, the founder of CQ magazine's "Certificate Hunter's Club" and a cantankerous sort, but a blind man who worked tirelessly for amateur radio. The Colvin award was for "working pairs of land based stations licensed to the same family (immediate) under the following conditions:

"3 pairs of stations each pair located in a different continent, or 4 pairs of stations each pair located in a different country (ARRL List), or 10 pairs of stations each pair located in a different US State."

Lloyd wasn't happy with the way the rules were presented in Evans's Awards Directory, and Evans replied that "Upon receipt of your first awards rules I wrote you that such rules did not meet minimum standards for Directory listing in that it was directed as self glorification of self and family, gave an award for working one person, and promoted the individual rather than Amateur Radio."

"However," Evans wrote, "I told you that if you modified the rules as I later listed it in the Directory and made it to ANY family (not one person) then it would barely slide by lowest Directory standards.

"Then, after agreeing to such listing you went ahead and embarrassed me considerably by giving the original version to the magazines and contradicting the Directory listing ... and now you request that the original version replace what is in the Directory. The flat answer is NO.

"I have had scores of letters protesting your award does not meet Directory lowest standards and it has even reached the point that many CHCers have written asking that it be thrown out by K6BX and if I will not agree to throw it out, then it should be put before the awards board."

Eventually, this was worked out, the rules revised, and new awards were (re)issued to the first 16. (#1 was Tom

Stuart, W4ML.) The last known award issued was #346, to VK3ABA, on 20 August 1969. Lloyd noted at the time that only three of the first 16 awards issued had been to amateurs who'd worked the Colvin Family. "The other 13 were for working amateurs who have placed remote countries of the world on the air, such as Danny Weil, Bob Roberts, etc.," he said.

By November 1961 the Colvins were back in California, with Iris laboring under the call sign KL7DTB/6. She wrote to the FCC, saying "I received my General Class Operator's License in 1945 during World War 2. I have been continuously licensed since that time. I would like very much to have a call that would indicate that I am not a newcomer to the amateur radio ranks. I know that it is legal for you to issue three letter calls in California which start with W6. All such calls of course must be re-issued calls that have become available by radio amateurs dying or letting their radio amateur station licenses expire. Will you please issue me a call starting with W6 rather than WA6, K6, etc?" This was a special request; although first licensed in 1945, Iris had never held a W6 call sign!

Lloyd wrote on 8 December 1961 supporting the "reciprocal licensing" bill being considered by the U.S. Senate, to Senator Claire Engle, a member of the Committee on Interstate and Foreign Commerce,

"I am writing this letter to urge favorable action on bill S.2361, which I understand is now being considered by you and the Committee on Interstate and Foreign Commerce:

> I have been a radio amateur for over 30 years. As an officer in the US Army I have traveled in nearly 100 countries and personally know thousands of radio amateur operators the world over. I am convinced that passage of the proposed bill can go a long way toward mutual trust and understanding between our country and all others. There is no practical security problem in permitting reciprocal licensing as there are always hundreds of radio amateurs listening to each and every transmission on the amateur radio frequencies. Any unauthorized transmission would be immediately monitored and reported.
>
> If we are to avoid world war we must bring the people of the world closer together in a spirit of friendliness. Passage of bill S.2361 will help to accomplish this. Please give it your support.

In 1962, among the first of Drake Builders's projects is a new home for the Colvins, at 5200 Panama Ave, in Richmond, California. They would live there until their deaths.

One of Lloyd's ham neighbors, in Berkeley, was caught up in Cold War fever, as the following 24 January 1962 story in the Oakland Tribune attests:

> **"Anti-Red Crusade By Berkeley 'Ham'"**
>
> BERKELEY - A ham radio operator whose favorite epithet is 'left-winger' has started an anti-communist movement among amateur radio operators across the nation.
>
> "So far," says Fred Huntley, W6RNC, a 44-year-old bachelor, 'more than 50 ham operators (have) expressed interest in the Anti-Communist Amateur Radio Network (ACARN).' His adherents include a ham operator in Chile, a newspaper executive in Chicago, and 'several military men in Southern California."
>
> The purpose of the 3-month campaign is 'to educate the amateur radio operator on the dangers that threaten the existence of the hobby of amateur radio - as well as the existence of the United States as a nation.' The 'education' thus far consists of bulletins and study outlines distributed by Huntley. They quote prominent persons, suggest membership in patriotic organizations, and recommend anti-Communist literature.
>
> Huntley says he feels that every citizen can do his part to fight Communism and that proper education is the place to start. "Education now, action later," he says of ACARN. In the meantime, he hopes to 'alert America' via a group of radio amateurs united by a spirit of dedication.
>
> Huntley works as an electronics mechanic at Alameda Naval Air Station and broadcasts his anti-Communist messages from a $2,000 radio set in his five-room home atop Grizzly Peak Blvd.
>
> He says his anti-Communist activity has been increasing through the years, but that the idea for ACARN came from correspondence with Roscoe Morse, state commander of the American Legion. Although his movement does not endorse any specific organizations, Huntley has investigated the John Birch Society and personally thinks "it is very fine."

Lloyd wrote to A.S. "Arnie" Trossman, the editor of CQ magazine, in June 1962, to remind him that "As you know I was the second radio amateur to qualify for the "Arne Trossman CHC-200 Top Honors Plaque." It has not yet been presented to me. I am writing this letter to suggest that a good time and place to present it might be at the ARRL National Convention in Portland, Oregon in September. I am scheduled to present a talk on "DX QSLs" in the DX portion of the convention. My talk is to be at 1030 AM on 2 September. You will know best as to how this will fit into your plans and desires." His effort was rewarded at the convention.

The Colvins were back on the club-speaking tour in 1962, at least in California, as their fame spread. Lloyd hammered away on the themes of travel, QSLing, record-keeping, and awards chasing:

> I recently retired from the United States Army Signal Corps. During my 21 years in the Army my family and I were fortunate in being able to travel in 76 different countries and operate under some 23 different calls, as listed on these poster boards. Since 1929 we have now made over 100,000 QSOs and have approximately 50,000 QSL cards filed alphabetically at our QTH.
>
> With some 50,000 QSLs to choose from it has been fairly easy to qualify for a large number of awards, which are based on proof of contact by producing a QSL. As this is an ARRL convention I would like to state my personal belief that in spite of the many hundreds of awards now available, the award which is still considered number one by most amateurs is DXCC. I would like to ask all members of DXCC in the room to please raise your hands.
>
> Great! It looks like most all of you are DXCCers. In such a gathering I think it highly appropriate to pay honor to the several very top DXers who are with us. Will all of you who have a DXCC standing of 300 or more please stand up? Now, starting over on this side, please give your call and DXCC standing using the all-time high listing.
>
> Let's give a hand to these terrific DXers. They have really reached a fantastic DX standing and their calls are known by many, many thousands of amateurs throughout the world.
>
> I have had some unusual experiences associated with DX and QSLs and would like to tell you of a few of them. I am sure all of you have worked someone at some time (who) had an unusually poor fist and was difficult to copy. One day I worked a W4 (who) was almost impossible to copy. The sending was so bad I called my daughter and wife, who are both amateurs, to the radio room to listen to the worst sending I had ever heard. A few days after this QSO I received a QSL from the station I worked and here it is...
>
> In the case of handmade cards I would like to show a couple that must have taken the owners of the cards hours to make.
>
> Speaking of women on QSLs, I received this QSL from ZK1AR two years ago. You will notice the pretty girl on the back. After a recent QSO, this year, I received (another) ZK1AR QSL. What happened to the girl? I bet there is a story behind this. Did he get mad at women? Did he marry the girl, or what?
>
> Last year, while I was still on active duty in the Army, we were in Alaska, where I was the Signal Officer at Fort Richardson. One morning I was informed that the commanding general had TVI, which he thought was being caused by an enlisted family that lived nearby in the enlisted mens' housing area. I went to the quarters that afternoon in my uniform. A lady answered the door and said she was WL7DPE and that her husband, who wasn't home, was also a licensed novice operator.
>
> She took me to the rig, which was in the basement. Things were a real mess and I volunteered to try to correct some of the more obvious troubles. For example, the antenna relay was connected so that when they transmitted the receiver remained connected to the antenna. While I was working she brought all of her children to the basement and told them "See, children, a general is fixing Mama's rig."
>
> Just as I was about to tell her I was a colonel, not a general, I realized she wasn't referring to my army rank, she was referring to the fact that I had a "general" amateur class license.
>
> This doesn't quite complete the story. A few weeks ago I heard a nice, fast operator on 14 megacycle CW with a KL7 call. I called the station and found it was our old friend Louise, WL7DPE. She now had her own general license and is an excellent operator. Unfortunately, her OM has no license now as he has failed the technical part of his general license (exam) four times.
>
> I have taken (the list of) DXCC members in QST and found that I've worked approximately 90 percent of DX members of the world. I pulled (my) QSLs for VE7 and the U.S. Sixth and Seventh Call Areas and they are displayed on the wall over there. If you are a DXCC member and your card is not there, we either have not QSOed or I do not have your QSL.

> While I [composed] this display I made a list of DXCC members who were worked several times but who have not answered my several QSLs. In most cases these amateurs worked me from different QTHs. Ladies and gentlemen, the time has come. I now will announce the worst QSLer in the room. Of course, this is all in fun, and if your call is in my list please do not be too offended, and please admit you are present - do not make your neighbor point you out.
>
> I will start with the worst offenders and continue down the list until we find the worst QSLer in the room....

Not long after his separation from the regular Army, Lloyd had sought a disability allowance, saying that he was sterile as the result of having contracted the mumps while in the Army in 1953 (during the Korean War period). The request was disallowed, but Lloyd pursued it, had sperm count tests done in 1962, and in July, 1962 was awarded a disability payment of $47 a month beginning in October 1962. He was 38 years old at the time; Iris, 39.

Archival records are slim in the early 1960s, but Lloyd continued to apply for more amateur radio awards, going back to his DL4ZC days in Germany 10 years before. One thing the Colvins saved was an issue of Gus Browning's DXers Magazine, in which Gus reprints a long article about his DXpedition that appeared in "Signal," a publication of the Collins Radio Company. Among other things, the article says "As of last October, he had made more than 150,000 contacts from 46 transmitting locations and is still going strong." The seed was planted in the Colvin's minds by now!

In 1961 the first of the Baby Boom generation turned 15 years old, and new amateur radio licenses issued began to swell, aided by the FCC's adoption of a Novice class license a decade earlier. Most of these new licensees, it must be said, were boys. Cold war fears and the now discredited "Missile Gap" sent many of these youngsters on to engineering school, and newspapers were ripe for stories about hams. One about the Colvins appeared in the Berkeley Gazette on 4 June 1963.

Newspaper reporters and editors are notoriously ignorant of all things scientific and technical, and, to make things worse, they sometimes like to portray people who pursue scientific hobbies - like amateur radio - as aliens, or nerds. This article, headlined "Hilltop 'Ham Operators' Courted in Dots and Dashes." was a pretty good example.

The article's writer invented the word "hamsters," which should have made any ham reading the article want to throw things.

These objections aside, the article did document Iris and Lloyd's story through the years of World War 2, especially how amateur radio led Lloyd to the Army Signal Corps, and followed the family history through Alaska, Japan, and Germany. "Subsequent Army posts," the article said, "have led the Colvins to almost 100 different countries. No matter where they go they always have a friend either through Army or ham radio contacts."

With more than 200 awards under his belt, by 1964 Lloyd was scraping the bottom of the barrel. Among his collection were Illinois Counties Certificate; Orange County Award; CIM (Montreal); Pea Picker Certificate (Walla Walla); WAVCKA (Australia); Trans-Canada; Worked Birmingham (Ala); Lincoln Award (Ill); St. Louis; Lone 4; Presidential Award.

In the spring of 1964, Lloyd took out a life insurance policy and made the ARRL the beneficiary. The minutes of the July 1964 ARRL board of directors meeting record, in Minute 27, "On motion of Mr. Compton, unanimously voted that the Board extends its special thanks to Lloyd Colvin, W6KG, for his sponsorship of a future annual monetary award, known as the Colvin Award, to a deserving radio amateur." Lloyd would pay the annual premiums on this policy and, at his death in 1993, the ARRL would receive the benefits - $150,000.

Ever the record-keeper and QSLer, Lloyd wrote to ARRL after the 1964 November Sweepstakes contest, lamenting that the contest exchange no longer included a signal report (Lloyd was the contest award winner for the East Bay section in the CW Sweepstakes contest in 1963 and 1964, and on both CW and phone in 1965):

> I have no objections to the shortened time for SS operating. In fact a change now and then is good for all. I wish however to complain over one point. Please, please restore the RST exchange in the SS contest. I have talked to about 25 active SS operators both over the air and in person. Every one of them would like to see the exchange of reports required in SS contests. I offer the following points in favor of this:
>
> It is of little use to exchange QSLs for a contest without RST. The ARRL and many other organizations require minimum RST reports to be shown on QSLs submitted for certain awards.
>
> During the last two years' SS contests I have had an average of one in eighteen stations worked either give or ask for an RST report during the contest QSO. This takes up time and would not occur if ARRL would put the RST back in the (exchange).
>
> Operating would be greatly improved during the SS if you know how well the other station is receiving you.

Of those 1961 days, Joe Reisert remembered "IBM transferred me to San Jose, California in late 1960. In 1961, I joined the Northern California DX Club. In those days, they met in Oakland and were, as I soon found out, a male only club. A few months after I joined, in popped Lloyd and Iris at a meeting. A month later, Lloyd put in a nomination for membership for Iris. Lloyd and Iris briefly left the meeting when the vote was taken. It wasn't 100% in favor but Iris did get accepted as the first female club member. Immediately after the voting was counted at this meeting, several of the members immediately left and never returned to the club or its meetings!"

31 · Dick McKercher, W0MLY's Africa Operations

Suddenly there was a pounding at my door and, upon opening same, I was informed - and saw plenty of evidence - that the joint was on fire! - Dick McKercher

In May, 1962, the YASME Foundation ceased to be a one-man (Danny Weil) show, with the appearance of Dick McKercher, W0MLY. McKercher, from a small town in Iowa and in his early forties, was an active DXer and a YASME contributor.

Unfortunately, no archival material exists to show just how McKercher cooked up a scheme to operate from Africa under the auspices of the foundation. With no warning, Spenceley's June issue of the YASME Newsletter (mailed in mid-May 1962), simply led by saying "Our object in getting the June NEWSLETTER to you a bit earlier than usual is timed with the expected arrival of W0MLY in GABON on May 26th. Operating frequencies will be found on reverse side."

Talk about lack of fanfare! In fact, Spenceley seems hardly to have known McKercher, as he refers to him as "George" when W0MLY's on-air name had always been "Dick."

Spenceley said that McKercher had "firm permission" to operate as W0MLY/TR8 for up to three weeks, and that after that he "plans to operate" from the Central African Republic TL8, Cameroons TJ9, Dahomey TY2, Togo 5V4, Mauritania 5T5, Mali TZ "and any other exotic DX spots within reach," where, it was thought, he could probably get a license "by personal appearance." There were no plans to operate from several nearby countries that were not as rare, namely, Tchad, Congo, Upper Volta, Guinea and Ghana. "George also eyes Annabon Island but difficulties are imposing," Spenceley said.

McKercher planned at least three months in Africa, using an HT-37A and SX-115 "courtesy of" Hallicrafters. Hy-Gain had provided a beam and vertical antennas.

Spenceley wrote that "this African venture has been made possible through the financial support of a dedicated DXer and is conducted under the auspices of the YASME FOUNDATION."

What "dedicated DXer?" McKercher himself, it turned out, who fronted the five thousand or so dollars for the trip. "Contributions are requested, with QSLs, from ALL hands so that partial costs may be defrayed. Should this work out satisfactorily, future expeditions, in like manner, are distinct possibilities. QSLing will be in the same manner as with the YASME expedition and KV4AA will be QSL manager. Toward fast confirmations attempts will be made to take W0MLY's log over the air but only if such action does not interfere, to any large degree, with general contacts."

Meanwhile, in Tahiti, Danny Weil reported having made some 3200 contacts, and that the YASME's main diesel engine was repaired and working perfectly. Danny had been denied permission to sail directly to Tongareva (Penryhn) Island from Flint or Starbuck, due to a law that all boats must call in at Rarotonga, the capital of the Cook group, before going to any other islands in the Cook or Manihiki grouping. "This means." Spenceley said, "that YASME would have to sail some 1000 extra miles, first down to Rarotonga and then back north to the Manihikis. In any case, and we are still working to have this ruling set aside, Danny will make all attempts to operate from the Manihikis as this is a most needed area for a good many of us. "

A convention photo of Dick McKercher, published by Chesser in October 1962.

Bill Bracy, YASME's former crew member who had transferred to another yacht in the Marquesas, "returned to haunt us upon this other yacht's arrival in Papeete," Spenceley said. "Seems they want to get rid of him too and the immigration authorities hold that he is still Danny's responsibility. At worst, this only involves the cost of plane fare from Tahiti to Honolulu but this we can ill afford should immigration persist in their views." First it was Dave Tremayne, now Bill Bracy, crew members who caused headaches.

YASME membership, just as W0MLY was opening up from Africa, was 389. "Special answering frequencies for YASME Foundation Members" were listed in the Newsletter.

W0MLY/TR8, Gabon, May 30-Jun 6 1962, 2750 QSOs.

W0MLY/TL8, Central African Republic, Jun 13-18 1962, 2200 QSOs.

W0MLY/TN8, Congo Republic, Jun 18-21 1962, 1100 QSOs.

W0MLY/TT8, Chad, Jun 22-27 1962, 1700 QSOs.

W0MLY/TJ8, Cameroon, Jun 29-Jul 5 1962, 2000 QSOs.

TY2MY, Dahomey, Jul 9-16 1962, 1624 QSOs.

5V4MY, Togo, Jul 17-22 1962, 1646 QSOs.

W0MLY/TZ2, Mali, Jul 24-26 1962, 700 QSOs.

The YASME Newsletter Number Twenty-Two, July 1962, reported these operations. W0MLY was on the air almost continuously since arriving in Africa on 30 May 1962. He began in Gabon as W0MLY/TR8, then crossed into the Congo on 8 June for 41 CW contacts as W0MLY/TN8. As the chart [in the newsletter] illustrates, he moved quickly from one country to the next, set up, and got on the air. The July issue of the YASME Newsletter came out just as the operations were beginning, with Spenceley outlining the itinerary as best he could, saying that after Dahomey McKercher would "attempt to get on the air from Togo, 5V4-land. Following this, if Dick's nerves and endurance have not completely collapsed, we are working out plans for him to go to Yemen," where a gasoline generator would be necessary and feelers were put out to the Cyprus and Aden Radio (RAF).

McKercher also had eyed the new DXCC countries of Rwanda and Burundi but discretion was the better part of valor, political conditions there being extremely unsettled (and the likelihood of getting operating permission probably nil, anyway).

Some questioned McKercher's signing from first the Central African Republic and then from Bayanga, a town in the northwest part of the Congo, on the same day (GMT) but this was later explained to most everyone's satisfaction. He simply packed up his station, crossed the border by land, and set up again. He spent an extra, third, day operating SSB from Congo (where the need was greatest) when a swollen river prevented his departure on the 20 June.

Most of the eight countries McKercher operated from were "new" (for DXCC, back to 1960), but not Cameroon, which Spenceley said he thought should be as "there have been substantial territorial changes made with the absorption of the old British Cameroons. Old FE8 should join the deleted countries."

The June Newsletter had suggested special YASME answering frequencies but they were soon abandoned. It was observed that it "takes no mastermind" for non-members to figure out what the special frequencies were. McKercher was operating fast and furiously and there was no time in the heat of the pileups to consult a list of YASME members, which, at any rate, would not have included members who had joined since his departure.

McKercher's pileup control was brutal but he produced QSOs. Spenceley wrote that "All this operating, travel, heat, poor accommodations and what not, tend to make one somewhat touchy and this has been reflected, to some extend, by the blasts that W0MLY has directed at stations not conforming to somewhat rigid DX procedures. We suggest that the recipients of these reflect on the above and not feel too bad about it. The DX station must exert complete control for any semblance of order in such an operation and if your call is in the log you will get a QSL."

A controversy erupted over McKercher's transmitting SSB on 14001 kc. He did it as a matter of expediency and probably made many more 2-way SSB and SSB-to-CW contacts than if he had moved above 14,100. And while SSB on 14.001 was legal from the countries from which he operated, it violated a gentlemen's agreement among most hams.

Spenceley observed that "No complaints about this have been received from SSB men but a number of CW men are quite bitter. It was not the original plan for such use of SSB and I wrote and, via QSOs, pleaded with Dick to change. Finally, in Tchad, he stated that he had tried other SSB frequencies many times and was always clobbered. Also, if he couldn't use SSB on 14001 he would 'immediately pack up and come home.' The matter was then referred to W8EWS who placated him and told him to continue as he thought best. I do not condone this use of phone but hold that it is now beyond my control without an actual scuttling of the expedition which would invoke greater condemnation. It does seem to me, however, that those so vehemently against W0MLY SSB use might initiate their complaints against an overwhelmingly greater source of phone QRM on the CW bands stemming from daily use by South and Central American phone stations who consistently and uncaringly make use of the CW portions of the band for their AM phone transmissions."

Enough people were angered over this that the ARRL weighed in with an Official Bulletin from W1AW:

> AFTER 0001 GMT JULY 13TH DXCC CREDIT WILL BE DENIED FOR ANY AM/SSB PHONE CONTACT WHERE ONE OF THE PARTICIPANTS IS USING SUCH MODE OF TRANSMISSION BETWEEN THE FREQUENCIES OF 14000 AND 14100 KC.

Spenceley wrote that "While not going into the question of their right to do this, and as a CW man, I for one, am all for it! Others may feel differently, such as Mike,

W3AYD, and I quote hi(s) comments which have been transmitted to quite a few hams:

> (From W3AYD, Nightletter to Hartford) "The ARRL 'CW Forever' boys are at it again, this time trying to negate foreign countries' licensing authority by coercion. DX hams with no phone restrictions in their licenses will go to 14000/14100 just to demonstrate their independence from ARRL threats and will probably refuse to work US hams. Will the great Hartford fathers also refuse recognition for CW contacts above 14100 kc?

"After a few kilo QSOs as W0MLY/TJ8, Cameroons, Mac [was it Mac, or Dick, or George? - who knows?] tried his Dahomey DX luck as TY2MY. Poor choice of phone working frequencies created difficulties," said QST's "How's DX" in September 1962.

W0MLY's operations triggered a healthy increase in YASME Foundation membership, up from 389 on 21 May to 619 on 7 July. These funds were to be apportioned 40% to defray W0MLY expenses and 60% to the YASME III and Newsletter sides of the ledger. The June Newsletter asked for additional small contributions, to be enclosed with QSLs for McKercher's various operations, to help pay for the trip. The funds in the YASME treasury "wouldn't have taken Dick as far as Coshocton, Ohio," as Spenceley put it. Long-standing members were understanding and cooperative, but small notices sent to non-members (inviting them to join) along with W0MLY QSLs did raise some questions. Sometimes a brand-new member accidentally got an invitation to membership slip.

Meanwhile, switching gears, and continents again, Danny Weil was in Tahiti, ready to leave for Flint Island, with all engines and gas generators repaired and working.

A McKercher setup somewhere in Africa.

K5JLQ had located a 6- KW Onan diesel generator in the States to replace the Petter and it was hoped to get this shipped to Danny at his next big port, probably Rarotonga, Cook Islands. Information about getting operating permission and call signs for Flint and Starbuck had apparently been incorrect, and same was expected shortly. Unfortunately, word from DXCC was it was "doubtful" that Flint Island would be accorded "country" status.

Spenceley said he was "quite unable to follow this line of reasoning," that Flint and Vostok and Caroline Island are known as the "Southern Line" Islands and are hundreds of miles from any other group as were Starbuck and Malden Islands, which Danny and Spenceley also were eyeing. Spenceley said that Flint had been picked as " being one of the most certain to qualify for separate status!"

All this was "solved" when, finally, the authorities at Honiara, Solomon Islands, said they had no authority to grant VR7 call signs for any of the above islands. "If we cannot untangle," Spenceley said, "Danny will probably have to sail for Rarotonga and then try to beat his was back to the Manihikis."

The McKercher operations took place across Spenceley's monthly publishing schedule, so Don Chesser's DX Magazine picked up between issues with the following sad tale:

"DXpedition Ends in Fire"

"The time was 0545 GMT on July 26th I was operating from the second story of the house (in Bamako, Mali Republic, operating as W0MLY/TZ2) and for some time had been aware of some sort of hubbub going on downstairs. As such noises are fairly common around here I had not paid too much attention to it. Suddenly there was a pounding at my door and, upon opening same, I was informed - and saw plenty of evidence - that the joint was on fire!"

Thus, in a note to his QSL manager, Dick Spenceley, KV4AA, W0MLY described the sudden end to a fabulous DXpedition that had permitted him to operate from eight rare, new African republics since May 30th: Gabon, Central African Rep., Congo, Tchad, Cameroon, Dahomey, Togo and Mali. He had planned to add a ninth country, Mauritania, 5T5, before terminating the tour, reports Dick, but the premature end was sudden and final with the destruction of all the Hallicrafters equipment.

"[I] dashed back to the rig," continues the account, "where I had been in QSO with MP4BBW and told Ian that I had to QRT on account of fire. Grabbing my suitcase, I hurriedly stuffed in all the few belongings within reach, plus logs, and exited through a window onto an adjoining roof. Crossing this roof, I clambered to the ground. My intention was to return and save the gear but by this time the whole premises were ablaze and that idea was out of the question. Luckily my plane ticket was at the airline office, being altered for the Mauritania trip, along with my passport. The house burned to the ground and took two neighboring shops with it. The fire had started under the staircase during some cooking chores."

McKercher somewhere in Africa.

Dick, W0MLY, subsequently boarded a plane and returned to the United States, arriving in New York on July 29th. After a dinner with members of the North Jersey DX Association he visited W8EWS on August 2nd, the Hallicrafters plant in Chicago on August 3rd, and was due to arrive at his home in Iowa on the 5th.

Although prematurely ended, his DXpedition had covered nearly two months of almost continuous operating, with an average of 1,600 contacts from each of eight countries, wrote KV4AA in the August issue of the YASME Newsletter.

During the African trip, excellent conditions prevailed up to and including the TJ8 stint, but deteriorated badly during his stay at TY2MY," reports KV4AA. "They picked up a bit at 5V4MY but hit bottom during Dick's short TZ2 stand. Everything considered, W0MLY did a tremendous job and gave thousands of DXers one or more new countries.

“The TZ2 logs have not arrived at KV4AA as yet. The delay may be due to the prevailing censorship in Mali. If they haven't arrived by August 5th, duplicate logs will be airmailed from W0MLY, stateside.”

The YASME Newsletter Number Twenty-Three for August 1962 took up the tale, noting that McKercher's nearly two months of continuous operating netted an average of 1600 QSOs from each spot. The fire in Bamako, Mali, had prematurely ended things, as McKercher had planned to end his trip with one more stop, in Mauritania.

Danny Weil left Papeete, Fiji, on 20 July 1962, headed for Rarotonga, after thumbs-down on Flint Island from both the licensing authorities and DXCC. The weather was pretty lousy and on 26 July "Danny reported that he was five miles southeast of Rarotonga in full gale winds and high seas." Port was made on the 27th.

Danny obtained the call sign ZK1BY thanks to some help from resident ham Bill Scarborough, ZKlBS, and hoped this call could also be used on his next stop, the Manihiki group. Danny hoped to be on the air from Rarotonga for about two weeks, beginning 5 August 1962; Special answering frequencies for Foundation Members were listed.

While in Rarotonga YASME was due to be careened for the repair of a few small leaks while awaiting arrival of the new 6-kw generator from the states. Next stop, the nearest island in the Manihiki group, then either Danger Island or the new Republic of Samoa, ex ZM6.

Spenceley responded to a proposal going around to extend the U S. phone band down to 14150, saying it "would surely chase all the Latin phones down to the CW segment below 14100 and this condition is already bad enough. We feel that no further change in U.S. 14 mc phone band frequencies is practical until moves are made, and cooperation received from ALL Latin American Radio Clubs whereby they would agree to keep their phone stations above 14100, and police same. The I.A.R.U. has a signed agreement to this effect concerning European Phone operation, i.e., European phones have agreed to stay above 14100 kc.

"It would seem to me that an unending and vigorous campaign should be initiated by ARRL, under Mr. Hoover's more aggressive leadership, to have the segment 14350 to 14400 returned to us. This would indeed be an exhilarating change to the past policy of apparent passive acceptance of frequency cuts."

YASME Foundation membership had further climbed

on the heels of the W0MLY Africa operations, to 759. A note went out to non-members seeking QSLs:

An outdoor view of the McKercher setup in the previous photo. The tent can be seen at the base of the antenna mast.

> Dear OM, We have received and are holding a QSL card (or cards) from you confirming contact/s with the YASME AFRICAN DX EXPEDITION whose operator, W0MLY, was active from many of the new African Republics during June and July, 1962.
>
> This African Expedition, while conducted under the auspices of the YASME FOUNDATION, was financed, under separate account, by a dedicated DXer, and our expenses have been a good many thousands of dollars. The YASME FOUNDATION also sponsors the YASME III voyage by Danny Weil who is now in the Pacific attempting to put as many rare countries on the air from this area, as possible.
>
> The African Expedition is an experiment to see whether expenditures may be recovered by means of small and relatively painless contributions sent to us by each amateur, with his QSL, after contacting the expedition. If this proves successful such expeditions may be continued thereby giving you new countries and adding much zest to the DX game.
>
> While no exact amount is specified as a contribution, the usual amount being received is one dollar per country contacted. If all amateurs would subscribe to this amount all expenses would be easily repaid and future expeditions, of like type, would be distinct possibilities.
>
> It might be pointed out that, normally, after contacting a rare country, sending him a QSL via air mail and including International Reply Coupons, your costs would be near to a dollar anyway. Also, you might have to wait months for your return QSL or maybe not get it at all! QSLing for our expeditions have been extremely rapid and, in the case of the African expedition, many have received their QSL within ten days after contact.
>
> It is not our intention to impose financial hardship upon any amateur and we feel, in most cases, both American and foreign amateurs can afford the amount necessary to enable the expedition to break even on expenses. Those who cannot, or will not, contribute must be satisfied to receive a rubber stamped confirmation of contact on their own QSL card which will be quite legal for DXCC credit. Some delay will occur in the sending out of such confirmations as, in all fairness, we feel duty bound to answer all contributors' cards first. Therefore, ALL stations are assured of a confirmation even if NOTHING is contributed.
>
> No one connected with the YASME FOUNDATION receives one penny for their considerable work and service to you in planning, financing and answering QSLs for the expeditions. So, we solicit your help in the form of small contributions in ratio to the expeditions countries that you have contacted. Foreign amateurs may send 9 International Reply Coupons as the equivalent of one dollar, or, the equivalent of a dollar in their own currency.
>
> Many may wish to join the YASME FOUNDATION. The membership fee is $5.00 per year. With this you will receive the monthly YASME NEWSLETTER, by airmail, which contains news and plans of the YASME expeditions plus other current DX information. Members of the FOUNDATION (which now number about 800) receive QSLs by airmail without further contributions (although many still contribute with QSLs) and all efforts will be made by the expeditions to contact FOUNDATION members. - KV4AA.

The West Gulf DX Bulletin praised the W0MLY operations in its issue of 31 July: "Two months and eight countries have gone by since Dick, W0MLY, left the States on his YASME Foundation sponsored DX safari. Having arrived in New York on Sunday, 29 July, Dick is now homeward bound for a well deserved rest. To say that his trip was a success would be somewhat of an understatement. It would be impossible to make a meaningful estimate of the number of new countries and/or mode countries he dispensed other than to say that it must be astronomical. To Dick, then, a much deserved well done for a superb performance. To Dick, KV4AA, our sincere appreciation for a difficult job masterfully done."

At the time the September 1962 Yasme Newsletter was published, YASME Foundation membership had climbed to 852. Danny had made some 1800 contacts from Rarotonga while waiting for the new generator for YASME III which was due to arrive there on 30 September, with Danny planning to leave about 10 days later, for Penrhyn (Tongareva) Island. A new exhaust also had been installed on YASME's main diesel engine; the old exhaust, near water level, sucked in salt water to the engine during "any kind of rough weather.

[But] YASME's troubles are not limited to the open seas," Spenceley wrote. "On August 20th and 21st a 75 MPH gale hit Rarotonga and lasted over 48 hours. Danny had a rough time, even though in the harbor, both anchors were out and other lines tied YASME to everything available." One anchor chain snapped but the other chain and lines held preventing YASME from being swept ashore. This anchor chain was supposed to bear a ten ton strain and Danny hoped that local divers could recover it. The tent he operated from ashore was flooded but all gear was removed to a safe place in the nick of time.

> Penrhyn Island is the farthest northeast of the Manihiki group and about 600 miles due North from Rarotonga. It will be a difficult voyage as YASME will be bucking adverse winds and currents for the entire trip. Penrhyn has an adequate entrance into its lagoon where a safe anchorage is assured. Danger Island, the most Westerly of the group would be somewhat more easily reached and would save hundreds of miles on YASME's route but has no safe anchorage nor deep enough entrance to its lagoon for YASME.
>
> Another solution for the Manihiki operation exists. This is from Suvorov (Suwarrow) Island some 350 miles northwest of Rarotonga. This island is inhabited, offers a good anchorage, and is most simple to reach. Suvorov is some 200 miles from Danger Island and 50 miles nearer to Nassau Island, both known to be definitely included in DXCC's definition on the Manihiki group. Well, you might say, why doesn't Danny go to Suvorov? There's one little hitch: DXCC will NOT commit itself as to what group Suvorov will be counted as.
>
> At the Portland Convention Hal, K5JLQ, and Golden, W8EWS, put this question directly to Bob White, W1WPO. Mr. White stated that the burden of proof is upon us. We should ascertain what status the Cook Island authorities attach to Suvorov and inform him. As we have said before, during our abortive attempts to have the Marquesa group accorded separate status, we hold that the acceptance of the Manihiki group was something "pulled out of the hat" by the DXCC Committee. The Cook Island authorities know of Manihiki Island but of no "Manihiki Group" as such. All the Cook Islands, north of a radius of 150 miles from Rarotonga, are known to them as the Northern Cook Islands.
>
> Suvorov is in this group. In accepting the "Manihiki Group" DXCC stated that this group had been known as the "Northern Cook Islands." Obviously Suvorov should then count with the Manihikis. A hypothetical question was put to WlWPO: Suppose QSLs were received at DXCC covering contacts made with Suvorov. What would DXCC count it as ? - no commitment.
>
> We stated in our March Newsletter that DXCC would NOT commit itself as to what is, or isn't, a "country" in advance of an [expedition] there. This was denied by DXCC but here we have the same thing again! We will continue our efforts to establish the status of Suvorov in hopes that Danny may be saved several hundred miles of hazardous voyage. Of course, if Suvorov is NOT with the Manihikis it might be called "The Central Cook Islands" and a brand new country. -- Ha
>
> After the Manihiki operation YASME will call in at the new Republic of Samoa (ex-ZM6), which DXCC advises is NOT a new country, and then probably to Niue, ZK2, and Tonga, VR5. Itinerary is subject to decision upon arrival, at Samoa. Nauru Island is recognized as one of the rarer spots in this area and is a definite YASME possibility.

Chesser's DX Magazine reprinted these comments on 6 October 1962.

While it seemed that Spenceley was constantly at war with DXCC, his sense of humor never failed him. He reprinted an article from the South Jersey Radio Association's newsletter, saying the article "lampoons the recent W0MLY African trip and we were quite tickled by it. It's a good cure for any one taking this stuff, or himself, too seriously." (Incidentally, the article's author, Dale Kentner, W2ZX, was a YASME Foundation member.)

"ANYONE FOR DX?"

THE 'NEVER A DULL MOMENT' DEPT: As most DXers know by now, R. McKercher, W0MLY, #2 boy on the string of KV4AA the Kingfish's stable of so-called YASME DXpeditioners, has been operating both phone (SSB) and cw on a frequency of 14001 kc from various of the new African republics (most especially Gabon which the Kingfish needed). To fully provide appreciation of a situation, it must be stated that McKercher is best described in polite terms as a premiere prima dona of the air waves, less politely as a thin-skinned boy from Iowa, drunk with the power which accrues when many Ws pleadingly howl at him, functioning as a self-appointed guillotine operator to Ws who he decrees are calling in violation of his sometimes weird set of rules.

He sits on his African throne(s) and rules with a heavy hand; some very well known old-time DXers feeling the whiplash of his judgement. They in turn, completely frustrated, run to the Kingfish for solace, pardon and release from the Black List. The Kingfish listens sympathetically because he has learned from experience that very few dollars come in from people who are on said lists. One W8 actually choked up and found it difficult to talk when the Kingfish, who had interceded with MLY, passed down the verdict to him (the W8) that MLY was apparently unrelenting.

Thus the high-riding McKercher deals out his favors to the hoi polloi from 14001 kc, listening for W SSB in the W band and also accepting many non-W SSB calls on and around 14010 kc. Quite frequently around 2100 GMT the lower end of the cw band takes on the nauseating aspects of the Citizens Band at its gory worst; all of which of course is not calculated to warm the cockles of the hearts of the CW man. Actually the warming was taking place under the collars of a great many CW devotees. Humorously enough, of all people, the Kingfish, an old CW man, was caught in the squeeze, and when he found it pertinent to ask MLY to move his SSB frequency, he got told off by MLY that either he would operate the way he wanted to or he was catching the next plane home. The Kingfish, known more for his practicality than his idealism, had no desire for nightmares featuring winged dollar bills flying NORTH, so he let the matter drop.

This was l'etat d'affaires on July 9 when two things ripped open the calm still of the night; McKercher opened up from Dahomey as TY2MY (his 6th country-stop) and a cavalry column flying the Star Spangled Banner in the van, rode out of West Hartford in all directions at once, as the pit band played "Stars and Stripes Forever", double fortissimo. The rescue column, in true tradition, having unsealed orders to put a double whammy on the unorthodox MLY fone clambakes, pulled up in formation and let go with both barrels with [Special Official Bulletin] #604 dated July 6th, reading in part as follows: "'Therefore ARRL announces that effective 0001 GMT on 13 July 1962, DXCC credit will not be given for contacts where either station is operating by telephony, using frequencies between 14 and 14.1 mc." So once again, praise be, young innocent amateurs may walk the green pastures of 20 CW without fear of meeting a fate worse than death!

Thus we have our pleasant and relaxing spare time hobby of amateur radio. But the tale goes on. McKercher, defiant to the end, continued untiringly to knock off SSB contacts from TY2MY (which, to give credit where credit is due, he does with considerable finesse when he keeps his temper down) right up until 0000 GMT on the 13th using his 14001 frequency. As this is written, he has just issued a palace edict that as of 0001 Friday he will operate SSB/CW from 14101 kc. In any event, except for a little wind knocked out of the McKercher sails, everyone will now be happy; his QSLs will count for DXCC, dollars will continue flying south, MLY will probably next go to Togo, and another chapter in ham radio will have been written.

Jack Agee, W8BGU, agreed:

Quite by chance I read the article "Anyone for DX?" - I happened to be amongst those calling W0MLY/TT8 when I heard him tell W2XZ that he had already worked him from TT8 and, therefore, was striking him from the log for both contacts. I fully appreciated his one contact per station per country practice as it gave that many more stations a chance to work him and because of his fine operating, there was never any doubt in your mind when you did work him. When I heard W0MLY slap down some of the big (?) boys I felt like shouting "Hurrah and Hallelujah! I had plenty of chance to note his operating technique which far exceeded that of any SSB DX operation that I have ever listened to; I worked him once in each one of the eight countries -- no sweat, FB!

$5000 had been contributed toward the W0MLY African trip, and total expenses were $4,727.97. Of this amount $4,305.50 was recovered via contributions and division of membership fees (40% W0MLY, 60% Regular Foundation). This division was made in cases where it was apparent that W0MLY activity was a major factor. This left a balance of $422.05 which was further reduced (Spenceley wrote) by the conversion of some foreign currency on hand here plus the value that can be recovered from a good many IRC'S [International Reply Coupons, exchangable for postage] on hand.

"More funds are expected to dribble in and the chances are good that the entire amount may be recovered.

"Should amounts over the expenses be recovered we should like to earmark them in partial repayment for the Hallicrafters gear loaned to us and destroyed in the Mali fire. The Hallicrafters Company accepted this loss with good grace and we are indeed indebted to them for all their help. (Hallicrafters was back on board, at least for this expedition.)

"We consider." Spenceley wrote, "total recovery of funds expended will only be due to the fact that W0MLY covered EIGHT fairly rare countries.

"Going to one or two countries wouldn't have gotten us very far financially. Only some 500 W/K VE stations did not contribute and they have been sent form letters as printed in the August NEWSLETTER.

This idea has already borne fruit. Response from overseas stations was, as expected, very poor. Overseas economies are generally booming so I do not see too much reason why such stations could not help to a greater degree.

In the final analysis it is estimated that if each station had kicked in 30 cents per contact per country we would have had it made. As is, the W's bear the financial brunt."

32· Danny and Naomi in the Pacific

I believe the YASME expedition will go down in the annals of ham radio once the whole story can be told. Meanwhile, let's enjoy it while it lasts. - Hal Sears, K5JLQ

On 8 October 1962 Danny and Naomi sailed from Rarotonga for Suvorov Islands, a voyage of some 600 miles. The good word from DXCC was that this would count for the Manihiki Group, that is, the North Cook Islands (which joined the DXCC list on 1 March 1959, for contacts back to 1945). The same call sign, ZK1BY, would be used that had been used on Rarotonga (YASME Newsletter Number Twenty-Five October 1962).

Danny Weil, left, and the author, in San Antonio, Septermber 2001

Suvorov featured, according to Spenceley, a good anchorage and a passage into its lagoon. A stay of 10 days to two weeks was foreseen, "or until a satisfactory number of contacts have been made." After that, a 500-mile westward voyage to American Samoa and then to the new Samoan Republic. Hal Sears, K5JLQ, was on his way to Australia from Houston and planned to drop in on Danny and Naomi on Samoa.

At a major convention in Portland, Oregon, a resolution had been made to have SSB stations operate at the low end of the 14 Mc. phone band, 14200 to 14275. "The idea behind this," Spenceley wrote, "was to facilitate contacts with DX SSB stations, more and more of which, it was claimed, were now operating below14200. It was the opinion that AM phone was now mostly used for local rag-chewing and they could easily move up to occupy the 14275-14350 segment of the bands. Whether this will work out we don't know. We haven't noticed much of an exodus to this end to date."

Western Samoa

Danny intended to give this new "band plan" a try and to ask SSB stations to reply between 14200 and 14215 kcs. "Opposition is to be expected from die-hard AM stations (they are die-hards or they would have SSB by now) but," Spenceley said, "if we persist, we hope to override it. We do hope that SSB with its obvious advantages in economy, band width and carrying power, will take over completely in the next few years and AM signals will become a rarity!"

YASME Foundation director Golden Fuller, W8EWS, continued to act as QSL manager for Danny's operations, and worked up some statistics from his FO8AN operations (both Marquesas and Tahiti): Foundation members contributed 60% of the total income and made 16% of the QSOs. Spenceley observed that "This recap might be looked upon as a moderate return in view of the fact that the countries operated from were not very 'rare' (DXCC refused to consider the isolated Marquesas Group as a separate one). It adds up to 31% who are keeping this thing going and 69% riding along in the rumble seat!"

On the other hand, W0MLY's African expenses were fully recovered "and the original $5000 has been returned to the sponsor" (who was never identified - it could have been McKercher himself). "Total expenses were $4,736.91. We count this quite an achievement and take this opportunity to thank all hands for their generous support. We always stand ready to promote like expeditions if the spot is sufficiently rare and a good part of the expenses stand a chance of being recovered. We would welcome any suggestions from the membership to this end. One good spot for such a trip would be San Felix Island, off the coast of Chile. This would be a brand new one and we mean to contact CE3AG re possibilities."

Danny dropped anchor at Suvorov Island on 11 October 1962 (YASME Newsletter Number Twenty-Six November 1962). Record daily runs from Rarotonga peaked on 9 October when YASME covered 190 miles. Danny said that Suvorov was "a typical Southseas dream island with lush foliage and a lagoon with water so clear that you can see down a hundred feet."

"Upon arrival," Spenceley wrote, "Danny and Naomi received an enthusiastic welcome from its lone inhabitant. This character, a 60 year old New Zealander, came up from Rarotonga three years ago for a rest cure. He was a storekeeper in Rarotonga and has a wife and three kids there. He will return in a year or so. After running out of canned goods he found it no problem to exist on Suvorov which abounds with fruit and has a lagoon full of fish just begging to be caught. Danny sold him some canned goods and used one of his thatched huts to store some of his gear. (Yep, the bloke had money with him, and we suppose Danny made some attempt to make him a member!). Government boats call in at Suvorov once per year and other boats drop in now and again."

Danny began his Suvorov, ZKIBY/Manihiki operation on 13 October, and the power transformer in the HT-33 linear amplifier promptly went up in smoke. Conditions on 20 meters for the duration of the stay were very poor and Danny was encouraged to try 40 meters more. Operating up to 17 hours a day, Danny produced 2800 contacts in 80 countries in 16 days.

For what it's worth, in 2002, given a good operator, decent band conditions, and modern equipment, 2800 contacts would be a reasonable total for 17 hours. In 1962 the sunspot cycle was near minimum and the Pacific is a long way from the what were then the high concentrations of hams.

At Western Samoa

Hallicrafters "has again extended their usual helping hand," Spenceley wrote, to us with a discount deal on two HT-41 linear amplifiers which have already been shipped to Danny at Pago Pago. This should take care of our PA [power amplifier] problems for some time to come.

YASME III left Suvorov on 28 October for the 500-mile trip to Pago Pago, American Samoa. Western Samoa in 1962 became the first nation in the Pacific Islands to become an independent state, after New Zealand. The name was changed to just Samoa in 1997. The boat was in good shape save for a leaky oil seal in the main engine which leaked about a quart of oil per hour. This oil was caught and thrown back in the engine. Danny and Naomi arrived on 1 November and were met by KS6AN, with whom he had been in contact. Several stations were active from American Samoa so Danny did not plan much operating (even if a license was possible). Hal Sears, K5JLQ, was due there around 6 November for a week's stay.

Danny came on from the new Republic of Western Samoa, signing ZM6AW, on 28 November 1962. His signals were reported to be weak (still no linear amplifier) but, as W8TTN wrote to Chesser, "I noticed someone remarked in your magazine that they never heard Danny at ZKIBY. If so, they never listened. Between Oct. 17 and Oct. 28 1 heard him every day for several hours on 15 meters from 1730 GMT until the band went out. He spent much of the time calling CQ since few stations were calling him. He was very easy to work even though his signals were weak." (Chesser 8 December 1962.)

(By this time Spenceley and Chesser had cooperated for several years; Spenceley wrote reports for Chesser to publish between issues of the YASME Newsletter; the same reports, verbatim when appropriate, showed up in the Newsletter a week or two later.)

The YASME Newsletter Number Twenty-Seven for December 1962 recapped Danny's stay in Apia, Republic of Samoa. He had made 2,450 contacts and would stay until after Christmas and possibly a few days into 1963.

Spenceley's war with DXCC continued, and he wrote that the reason for Danny staying "is due to the fact that a new prefix will go into effect for the independent state of Samoa on January 1st. Unfortunately, current DXCC policy is such that they will not accept Samoa as a new country, new prefix or no, although in reality it is exactly that. This follows the DXCC stand as in the case of Nigeria, Congo, Malagasy Republic and other nations which have acquired independence. DXCC' s scheme of double totals in the Honor Roll listings tends to balance Honor Roll possibilities between old timers and newcomers in as much as the first total represents countries which are always currently available. A pity that they could not follow this up by deleting ALL old countries and adding ALL the new ones that have received independence. Such action would keep things current in the country department and at the same time add zest to the DX operations of the old time DXer, who would have to work them to keep in the running, and who is now just running out of countries to conquer."

Hal Sears had been on the air in American Samoa beginning 6 November, as K5JLQ/KS6, using Danny's gear for a few contacts. On November 9th Danny "figured that he was just marking time" there, and YASME took off, with Hal aboard, for the 80 mile trip to Apia, arriving there the same day.

> Every courtesy was extended to Danny, [Spenceley wrote] by the top brass in Apia who have evinced a strong interest in ham radio. He was offered a choice of QTHs, free electricity, and the call of ZM6AW." Several days of rough weather prevented Danny from setting up his gear ashore but the job was finally accomplished and ZM6AW appeared on the air on November 15th. K5JLQ flew back to Pago Pago on November 12th to catch a plane for Australia. Danny received and installed a new power transformer in the HT-33, and late in December the two HT-41 amplifiers donated by Hallicrafters.
>
> We have a situation in the Republic of Samoa, which Danny would like to put before the Membership. There are four licensed hams, and several other fellows, who are greatly interested in ham radio but there is no ham station there nor do the fellows have the cash to get the necessary gear. On Danny's instigation the Apia Amateur Radio Club has been formed and it holds its first meeting today (Nov. 30th). We call upon the Membership to give the matter some thought and, if possible, to donate that old rig or receiver so that a permanent club station may be set up in Apia.
>
> With Xmas just around the corner and with a thought towards what happiness such a station would bring to the lads there it seems that this project could be accomplished without too much difficulty. To this end W6GN has contacted Ted Henry in Los Angeles and he has agreed to box and ship any gear so donated. The Foundation will pay the shipping expenses from LA to Apia. Should you be able, or know of anyone able to donate such gear please drop KV4AA a line so that things may be sorted out and duplication of effort avoided. What say gang!

The previous newsletter had asked for suggestions for future stops. The Kermadecs, Campbell Island, British Phoenix, Nauru, Wallis, Willis, and British North Borneo led the lists.

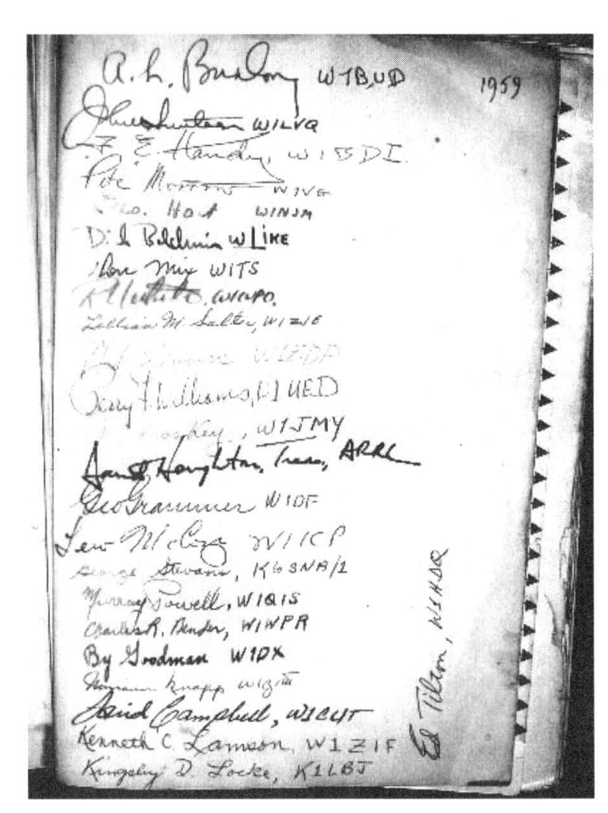

A prized possession of Danny's is an ARRL Handbook, signed in 1959 by members of the Headquarters staff (author photo)

The YASME Newsletter Number Twenty-Eight, January 1963, signaled the beginning of the end. While on Samoa and especially during the visit of Hal Sears, Danny and Naomi began some serious soul-searching. YASME was tentatively scheduled to depart 10 January 1963, for Wallis Island.

On 10 December, during Sears's visit, Danny mailed the following letter to Dick Spenceley, published in the Newsletter:

> This letter is the result of long discussions with Hal [K5JLQ] and Naomi and I feel it is not entirely without purpose [Danny Weil wrote].
>
> The future of the expedition, its financial position and general opinion of the Directors were the main topics of the conversation, but being a three sided affair at this end, I hope this letter will clarify the situation and give you an opportunity to voice your opinion. There appears to be three main facets in the present situation: Personal, financial and finally, what is advisable for the future.
>
> PERSONAL - Naomi is unable to carry on further owing to her health. To date, she has done a wonderful job under rough conditions, but it had to end

somewhere along the line. She will return to the USA on January 10th and I will hire a deck hand to continue the voyage with me.

FINANCIAL - I am under the impression that the expedition is still working at a loss and is dependent entirely on being underwritten by the Directors. Apparently, incoming funds are insufficient to maintain the voyage. While I, as an individual, have always been prepared to continue, I do see the logic of the situation and can fully appreciate that, however keen one may be, the cash situation must be considered first and foremost. We cannot obviously continue at a loss regardless of enthusiasm.

GENERAL - It is an accepted fact that Gus, W4BPD, will have covered the entire Indian ocean rare spots by the time the Pacific tour is over with me. To head into areas already covered by Gus would be repetitious and certainly not advantageous financially. Therefore, I suggest that I endeavor to cover all immediate areas in the Pacific and then return to the United States. I think it would take about a year to cover worthwhile spots in the Pacific such as Wallis, Fiji, New Hebrides, New Caledonia, Willis, Solomons, Nauru, Canton etc. I am aware that all of these spots have had activity in the past but are still needed by many. Other spots can be covered upon advice.

YASME must be hauled in Fiji for hull overhaul. This will be our largest expense as we are fairly well equipped in all other fields both marine and radiowise. You are aware of my personal feelings in the expedition, but even I can see that to continue at a loss is impracticable. Naomi's return to the States will naturally affect my personal feelings but, rest assured, the expedition will be carried on just as efficiently as before altho I may not have the best of cooking available. I feel that the lads who have helped to support this trip should be given the opportunity to work as many spots as are available here in the Pacific before we call it a day. Such decision could be subject to change depending on the circumstances which might exist as the Pacific tour nears its end. - Danny, VP2VB etc.

In the same newsletter, Spenceley said "The YASME trip would be self supporting should all equipment and boat function for extended periods without major repair or replacements. Recent shipments to Danny of a new diesel generator, two linear amplifiers and other gear plus a much needed Addressing machine for the NEWSLETTER has imposed expenses upon us amounting to around $3000. We are now $500 in the red and this amount was just given us, on a short time loan basis, by one of the Directors. Pending cost of hull repairs Danny could carry on, under present status, for a couple of months more during which funds would be coming in as result of operations."

Danny Weil, a man of many interests

Also published in the newsletter was this letter from Hal Sears, K5JLQ:

> Since some individuals may regard Danny's letter as merely propaganda intended to enlist sympathy and financial support, I would like to comment on just what is behind it .... something which I believe the thousands of wonderful hams who have supported the enterprise are entitled to know .... from one who has just returned from visiting Danny and getting a first hand view.
>
> You can be sure that Danny's letter was not easy for him to write. Seldom if ever expressed, he has never, up to now, wavered from his original objective despite the adversities, which would have overwhelmed other than a most extraordinary person. This goal, to be the first Englishman to circumnavigate the globe singlehandedly, might well have been accomplished had he not become involved in ham radio resulting in the loss of YASME I on a far away reef in pursuit of a rare DX spot. Without bitterness, Danny plunged enthusiastically ahead on a somewhat altered course, never really daunted nor doubting eventual success.
>
> Arriving in Pago Pago, I found both Danny and Naomi really down, principally due to sheer exhaustion having had little rest or sleep since leaving Rarotonga and even there the endless work

and duties preparatory to leaving. Doing the job of at least three hands was enough to exhaust a lesser man. Handling the responsibilities of pilot and navigator, plus deck hand, plus engineer, means little or no sleep while underway as I later observed during the voyage from PagoPago to Apia. Thereafter, the long vigils at Suvorov, fighting adverse band conditions, rig trouble, etc., plus the ever present duties around the YASME including hull repairs with a Scuba outfit where one ear drum popped, all this plus his usual excellent radio job, turning in some 2500 contacts.

Then up anchor, through dangerous reefs, fighting heavy seas, and finally arriving at Pago Pago shortly before my arrival but with still zip enough left to get together with numerous hams, including Mike, KS6AN, to work out details of ZM6 operations. This was a prelude to our conversations and an attempt to explain why Danny at first said that he had just about "had it" and was strongly seconded by Naomi. After a good night's rest and a couple of shore meals spirits were lifted and we talked at length regarding the expedition's future.

Altho it deserves much more, at least a brief word about Naomi's contribution to the expedition. Some day I hope it may be fully told just how, despite the hardships and isolation which must have meant real loneliness, she went steadily ahead as wife and companion, cook and nursemaid, deck hand, assistant navigator and secretary with never more than a droll and resigned remonstrance. I can vouch that Danny will miss her greatly for they worked together as a team, each understanding and appreciating the other. Somehow I just can't believe that Naomi's sea going days are over for she fits too well into the scheme of things not to be a natural part.

We discussed the possibility of carrying on as planned. To do this would certainly call for some kind of a rational existence with at least the minimum comforts and enough help to do the job in a normal way. Altho the considerations are complex, the most glaring obstacle is cost and there is no room in the budget for employ of necessary deck hands etc. Especially it leaves no room, under present income, for other than the bare necessities of the participants. Danny is highly sensitive to this and while it might be suggested that he could occasionally indulge himself, he could never be able to see it that way, especially realizing shortage of funds and being intent on putting every penny of this to the maximum use.

While at Apia, I insisted that they live ashore for a few days if for no other reason than to help perspective. Relieved of the responsibility of having to take from YASME's limited funds, both Danny and Naomi relaxed and their appreciation and enjoyment was extremely heart warming. Of vital concern to Danny was the attitude of the ham fraternity in general. I believe the YASME expedition will go down in the annals of ham radio once the whole story can be told. Meanwhile, let's enjoy it while it lasts. - Hal Sears, K5JLQ.

Wallis Island Ends the Danny Weil Era

All of you have been very generous in your thoughts and faith in me. I will not betray this faith by making the wrong decisions at this time and I feel it is time to stop now. - Danny Weil

YASME Newsletter Number Twenty-Nine for February 1963 carried the news: Naomi flew home to Tampa, via Hawaii, in early January 1963. Danny left Western Samoa on 20 January 1963, and arrived at Wallis Island on the 22nd. Danny was sailing alone, back in the Pacific where he'd been seven years before.

Danny immediately cabled New Caledonia and received permission to operate as FW8DW. While listening on YASME Danny noticed interference from the government radio station, so he sailed to the far south side of the lagoon and set up shop ashore. In the next several weeks he made 3,000 FW8DW contacts, one of his best efforts from anywhere.

Spenceley said that Danny planned to leave around 7 February for Suva, Fiji Islands, a 650-mile, five-day trip. There, some YASME leaks would be fixed, and "a decision will be made whether to carry on with the expedition through 1963 or terminate it. This will be Danny's decision and, whatever it may be, it will have the backing of the FOUNDATION Directorate."

DX Magazine in its 9 March 1963 issue said "Scheduled on now: Fiji Islands, VR2EO, operated by Danny Weil, VP2VB etc., for about a week, probably on all bands, CW and SSB. This will probably be Danny's swan song from his life of adventure and DXpeditioneering, for after VR2EO he will sail the YASME to the United States to be sold, after which he will retire to a more prosaic livelihood. Thus ends one of the most colorful, exciting and eventful sustained DXpeditions in amateur radio history. (K8KFP)

The YASME Newsletter Number Thirty, March 1963, was the last published by Dick Spenceley, KV4AA, who stuck with Danny Weil from the beginning, in 1955, to the end. Spenceley's poignant comments follow:

Danny pulled the big switch at Wallis Island,

FW8DW, at 0026 GMT on February 8th, after slightly over 3000 contacts, and sailed for Suva, Fiji Islands later in the day. The 650 mile voyage was fairly uneventful and YASME III pulled into Suva at 0045 GMT on February 12th. The night of February 9th was spent at Wailing Island when Danny pulled in there to grab a little sleep and also to effect repairs on a slow but persistent leak in the stern. From there to Suva, reefs abounded and careful watch had to be maintained.

At Suva Danny met Barry, VR2BZ, a DXer of some note, who showed him around and was most helpful in having Danny given the call VR2EO. YASME went up on the slip on February 18th for substantial repairs to her hull, rigging etc. Word was received today (Feb. 28th) that YASME was now back in the water but small leaks were still noted which may mean a return to the slip

VR2EO should open up for general QSOs on the weekend of March 2nd. Danny plans to make about 1500 contacts, with SSB favored, before leaving Fiji.

We regret to state that Danny has decided to terminate the expedition after his stay in Fiji. He will sail YASME back to the USA, either to the West Coast or to Corpus Christi, Texas. He will return substantially via the same route by which he sailed West or, might go via KH6. We considered this trip difficult, apparently against prevailing winds and currents, but Danny says it can be done OK and that's the way he wants to do it. The trip to Texas would take an estimated 100 days.

Danny has spent the last 8 years of his life on this venture, has survived two shipwrecks and has operated from the following DX spots in this order:

FO8AN/Tahiti
VR1B/BritishPhoenix
VK9TW/Nauru
VR4AA/Solomons
VK9TW/Papua
YV0AB/Aves Island
VP2VB/British Virgin Islands
VP2KF/St.Kitts
VP2AY/Antigua
VP2MX/Montserrat
VP2KFA/Anguilla
VP2DW/Dominica
VP2LW/St. Lucia
VP2SW/St. Vincent
VP2GDW/Grenada
VP4DW/Trinidad
VP7VB/Bahamas
VP5VB/Jamaica
HK0AA/Baja Nuevo
HC8VB/Galapagos
FO8AN/Clipperton Island
FO8AN/Marquesas
ZK1BY/Cook Islands
Manihiki Islands
ZM6AW/W Samoa
FW8DW/Wallis Island
VR2EO/Fiji.

Danny takes up a new interest, 1967

Spenceley published a letter from Danny setting forth his decision, written from Suva, Fiji., on 15 February 1963:

> The stack of mail which greeted me upon arrival, at Suva has taken considerable time to read and digest [Danny Weil wrote]. I did set out to evaluate each letter and answer each and every statement made by each individual but found, in the final analysis, that all letters had the same basic ideas.
>
> I have decided to turn around and come home to the USA directly. YASME is fit for sea and after a few VR2 contacts made with the boys. All of you have been very generous in your thoughts and faith in me. I will not betray this faith by making the wrong decisions at this time and I feel it is time to stop now. I know that you do not require my rea-

sons for this decision but it makes me feel better to give my personal thoughts along this line,

It is an impossible situation for me to carry on without Naomi. We have been a team in more ways than one, and to carry on with the job feeling as I do is pointless and will only cause much unhappiness to us both. To proceed to the remaining rare spots left in the Pacific would be too costly to be covered by contributions we have been receiving. The distances involved, plus the risk, just don't make the effort worth it.

YASME cannot be sold here nor would she bring even a fair price in Australia where prices are lower than in the USA. There- [The rest of this newsletter does not survive].

It was now time to close out accounts. Charles Biddle, W6GN, continued as the foundation's treasurer, and wrote on April 7, 1963:

I am enclosing the photo copy of the reports sent to me by Danny and Naomi covering the period August 4, 1961 to January 17, 1963 [Charles Biddle wrote]. The balance indicated at the beginning of this period of $18.87 was the closing balance on the last report that Danny sent to both you and Golden. This must have been before Hal was a director. The photo copy of the report that I recently sent you covering the period November 9, 1962 to January 17, 1963, when added to these copies will complete the detailed data covering the period set forth in the recapitulation attached and covering the period August 4, 1961 to January 17,1963. I am sending copies of this letter and of these reports to Hal and Golden.

Dinner celebrating Danny's U.S. citizenship

Each of these reports, when submitted by Danny or Naomi, have been accompanied by the detailed receipts for the various expenditures. These receipts were grouped as received and reported and are in my files for any I.R.S. examination or, of course, for the information of the directors.

In order to simplify the record keeping, any money that has been sent by you to Danny has been charged off to Ship Expense. The statements of account rendered by Danny and Naomi have only been used to obtain information with reference to reasonably large capital items purchased or sold by Danny, the loans and payments between Naomi and Danny and the Foundation and donations made direct to Danny. The information so obtained has been used to change the accounts to reflect what happened based on this information. Other than the changes so brought about, all cash sent direct to Danny by you or by me or any one else and reimbursed by you is still charged to Ship Expense.

If you, Golden or Hal have any questions that I can answer in connection with these reports, please write to me. Neither Naomi nor Danny are accountants and so I guess we can't be too critical when the figures get a little sketchy. I have done the best I could to make the books come out even in spite of some of the sketchy figures sent to me.

I am enclosing a bill for $7.38 from Henry Radio covering a B&W800 choke just sent to Danny at Tahiti. This is for his HT33A. He apparently operated it on an antenna that was shorted out and blew the choke. He is now using a HT41 final but would like to get the HT33A in working condition. Will you please send Henry Radio a check for this amount.

I have been in contact with Danny daily. He seems to be coming along very well. No problems at all to date. I hope that it keeps up like this. - Charles Biddle, W6GN.

By July, YASME III had wended her way, Danny at the helm, through the Panama Canal and was docked at Freeport, Texas, near Galveston. The closest YASME Foundation director was Houston's Hal Sears, K5JLQ, who wrote, on 8 July 1963, to the Yasme directors (Spenceley, Fuller, Biddle, and Drudge-Coats):

Now that the YASME is tied up safely at Freeport, Texas and what was once speculative is now "fait accompli," [Hal Sears wrote] we are faced with the happy responsibility, "happy" that is from a standpoint of a safe conclusion, of deciding our future course of action with respect to final disposition of the vessel and associated property.

Because of the Foundation's economic position, our thinking is confined much more to how than what. It seems obvious that our obligations should be cleared up now that the expedition is finally and definitely concluded and the only possible way is by liquidation. Fortunately, the YASME arrived in fairly good condition even though a little worse for the wear and at least without expending tremendous sums, is a salable commodity.

Danny and I have talked this over at length and the following is a summary of our thinking. Firstly, we have a strong desire to return the vessel but only if we can do so without laying out excessive amounts. Although I personally do not have much time to devote to its use, with Danny around as is definitely planned, we think this would work out.

DANNY WEIL, THE FLOATING CALL SIGN
To 3,500 Radio Hams He Is VP2VB
—Post Photo by Ed Valdez

VAGABOND VOYAGER

***Houston Post* clipping, 1963.**

Obviously, I would have to put up the wherewithal on some basis.

Now this will only be possible if after exploring the market and making our best efforts to get the highest price we are still in a competitive position. You can rest assured that we will do everything possible to realize the maximum amount on the vessel's disposal even if that cuts us out from personal ownership. With this in mind, I have asked Danny to contact the agents which sold us the vessel to solicit buyers on a non-exclusive basis. There is little probability that there would be a local buyer but I shall also explore this possibility. We hope that this meets with your approval.

Now that the Foundation is static and there are no more contributions, all future financing hinges on the YASME's disposal. Because the project is in my own back yard, I am agreeable to providing the necessary funds for whatever work is required to make the boat attractive and effect basic repairs. As you know, it hit a log on the return trip and the bow was damaged following which temporary repairs were made. The hole created in the bow was leaded up very effectively so that the boat is now very tight and not taking water but it does represent a defect which might seem important to a prospective owner.

Of course, the boat needs a great deal of cleaning up, painting, and so forth, and the engine, although running okay at the present, has numerous temporary repairs so that it might be a good idea to put it in first class condition. We believe that it would be a good idea to pull it out to do this work whether or not it is sold to others or acquired by ourselves. One or two thousand dollars spent in this way would greatly enhance its desirability. Therefore, unless there are objections or something unpredicted intervenes, we plan to follow this course.

As you no doubt know, there is considerable equipment and so forth even including the sails, with exception of the last one (mainsail) which are residue from Danny's previous operations and which he feels are rightfully his own. Most of this came off of YASME II which he salvaged, this vessel having been purchased by his efforts making lectures, etc. He has a list of this which I have not studied but recommend that we do not split any hairs in this respect since the sale of the vessel plus equipment which we have provided seems more than adequate to cover outstanding indebtedness. Rather strict accounting has been kept on this and if you desire, when he returns, we will circulate the information. I feel that W6GN has all this in order.

We propose that Denny spend necessary time handling not only repairs but working with the disposition and that he is entitled to reimbursement for this service. In order to come up with something definite, I proposed a moderate compensation of $500 per month during this period which of course will come out of the sale of the YASME and agreed to advance these payments as well as to stand back of the other costs involved in the meantime. I have offered him the free use of my house in Freeport which is conveniently located just across the canal from Bridge Harbor Marina where the boat is docked.

Also, he intends to set up a kind of small shop at my house to handle the major part of the work and thereby keep costs to a minimum. The length of time required cannot be exactly predicted but he believed that we should allow two months which I think is conservative. He will probably require at least one helper which can be ordinary labor and will not represent too much cost.

At this time it is not possible to predict the extent of the work which will be done since of course a buyer might come along who would take the vessel at an attractive figure as is. Or, he might request that minimum work be done. On the other hand, if the disposal project drags on, I personally would favor doing the whole job. Since this is a preliminary letter, several factors could easily change rather quickly. Anyhow, any observations and ideas will be gladly received.

There seems little problem in Danny being gainfully employed as soon as the YASME matter is cleared up. It might be suggested that he handle the YASME matter on a part time basis but I believe this would be a mistake, partly because Freeport is a considerable drive from Houston (about 75 miles from my house) and also once he has gotten into his shore activity, split time would not be practical. I believe that money expended even though it might run to two or three thousand dollars is a safe investment and back this up by taking a personal risk if necessary.

I think it would be a good Idea if W6GN could get up the final accounting and at this time all those having claims should present them to him as a kind of final house cleaning on any past vagaries or inaccuracies. I believe everything is a matter of

Danny and Naomi Weil in their Texas clock shop, 1982

record and I refer only to this type of claim since should we ever get into all the ifs, ands, and buts created by the emergency of the moment, etc., etc., we would be opening a veritable Pandora's Box.

Personally, I think that we should all rejoice that this whole project has come out to a much more happy conclusion than might have been imagined at one time. We are winding it up with the popularity of the Foundation, Danny, and other principals at an all time high and in a fashion it is going out in a blaze of glory. If Danny and I retain the YASME, there remains the possibility of some future fun although at the present time I wouldn't venture a guess as to what form it might take. Certainly I am sure it will never be another round the world expedition.

With regard to the Foundation itself, it is my own hope that it will be retained rather than dropped if that is possible. It is dubious if there will be much, if any, residue in funds after the final settlement but considering everything which has been contributed by all concerned, I think this should either remain in the treasury or, if possible, use same to compensate Danny in some small way for his unpaid services. I know that he is not asking for this but it has been suggested, so I merely restate it. Of course, this can very well be presupposing since I have no way of estimating the sale price of the vessel, tax regulations having made charge-off almost impossible, and pleasure craft of this kind may be a drag on the market. Therefore, instead of residue there may be a deficiency but should this come about, it would mean that the sale price would probably be attractive to me in which case I could absorb same in the purchase. Anyway, I am sure it will work itself out one way or the other.

I hope the foregoing will give you some idea of what is going on and will look forward to hearing from you.

By the way, enclosed is a clipping of a story appearing in a local newspaper. It contains the usual inaccuracies but appeared on the front page and wasn't too bad. A local TV station taped an interview but I don't know if they ever presented it. Later, the reporter who wrote the news item plans a feature story and possibly same may be attempted by TV. As you know, Danny will give a talk at Lake Placid to the Florida gang July 13, 14, and also there may be something planned for him in New Orleans on the return. We plan to get the Houston gang together for a dinner party with Danny. - H.A. Sears, K5JLQ, PA0HAS.

The Houston Post published a story on the new arrival in the area, titled "Vagabond Voyager Retiring Seabag After 8-year Cruise," by Bernard Murphy:

All over the world the news went Monday that Danny Weil of London had arrived Safely in Freeport. His arrival in Texas, after a single-handed 10,000-mile voyage in a 52-foot sailing boat from the Fiji Islands in the South Pacific, was big news to radio hams - even in Russia.

For Danny, a tall, lean former watch repairman and one-time airman, has been serving radio amateurs for more than eight years as a one-man floating contact. Radio hams in 32 countries have supported Danny, and their voluntary contributions have helped him to sail, almost always alone, about 20,000 miles around the earth.

But when his travel-grimed ketch, YASME 3 was safely tied up in Freeport, Danny sought not his radio but a telephone to call his XYL (wife in ham language) in Tampa, Fla., to tell her that after more than a year he was home. Mrs. Naomi Weil, the former Miss Naomi Kay of Tampa, has been working as a secretary in Florida while Danny completed his voyage.

"I met her in Tampa about three years ago," Danny said in Houston Monday. "It was just a platonic friendship but when I sailed away alone again I realized there must have been something more to it. So I called her up by radio through a ham friend. "I courted her by radio and plugged the question en route to Panama.

"Her final answer came to me in Panama - 'Yes" - so she flew down and we were married in the Canal Zone and she joined me on the boat. She sailed with me for more than a year but became so ill in Tahiti that she had to fly home and leave me to conclude my cruise alone.

Danny Weil in his home workshop, 2001 (author photo)

"Now I feel the time has come to stop isolating myself from the world except for radio links, and I plan to settle down. I'm flying to Tampa to Naomi and then I'd like to settle in Houston and work here."

Danny's host in Houston is radio ham K5JLQ, otherwise knows as Hal Sears, chairman of the board of Hal Sears, Inc, and Geo-Space Electronics, at 2424 Branard Ave. Both Danny and Hal are directors of the YASME Foundation, Inc, an organization supported mainly by radio enthusiasts - known as DXers.

Danny said that DXers are radio hams who collect certificates for contacting different foreign countries. "I have been going around the world to uninhabited islands," Danny said. "By talking to me they earn new certificates. I've talked to hams in Africa, America, Russia, all over the world. "Every new country I called from meant talking to about 3,500 amateurs. I carried 1,000 gallons of oil to work my generators on the islands. It took me 10 to 14 days in each port to call them all."

Weil said that after finishing nine years in the Royal Air Force, he decided to get away from the smoke of London by sailing round the world. He spent four years building his first YASME ("Freedom" in Japanese).

Before his three-month trip to [Freeport from Fiji], Weil sailed every ocean in the world except the Indian and wrecked his first two boats along the way. In the Virgin Islands, he met a well-known American ham, Dick Spenceley, who persuaded him to become a floating country for hams to call.

"My mother, who only made one trip with me and wouldn't go again unless wild horses dragged her, said 'Don't be a bloody fool, my boy,'" Danny said.

On 15 December 1963 the YASME Foundation sold YASME III for $15,000, checks then going to Biddle, Spenceley, Sears, Fuller ($450 each); and a check to Danny for $414.

Biddle wrote to Spenceley on Feb. 16, 1965:

I would like to reduce the amount of work to be done by any of us in connection with this I.R.S. audit to a minimum [Charles Biddle wrote]. If we turn our records over to the agent he can do the work to dig out the information that he wants. In connection with the names of the donors and the amounts of donations, if we can give the agent the call letters of all donors during the period under examination and the amount donated by each and also give him a call book, he can look up the names and addresses of the donors.

All we should be able to do is to reconcile the call letters and amounts given under the misc. donations amount given for each month. That is the amounts per month that I gave

Danny Weil with 2nd Operator Spunky, and two telegraph keys given to him during his travels (author photo)

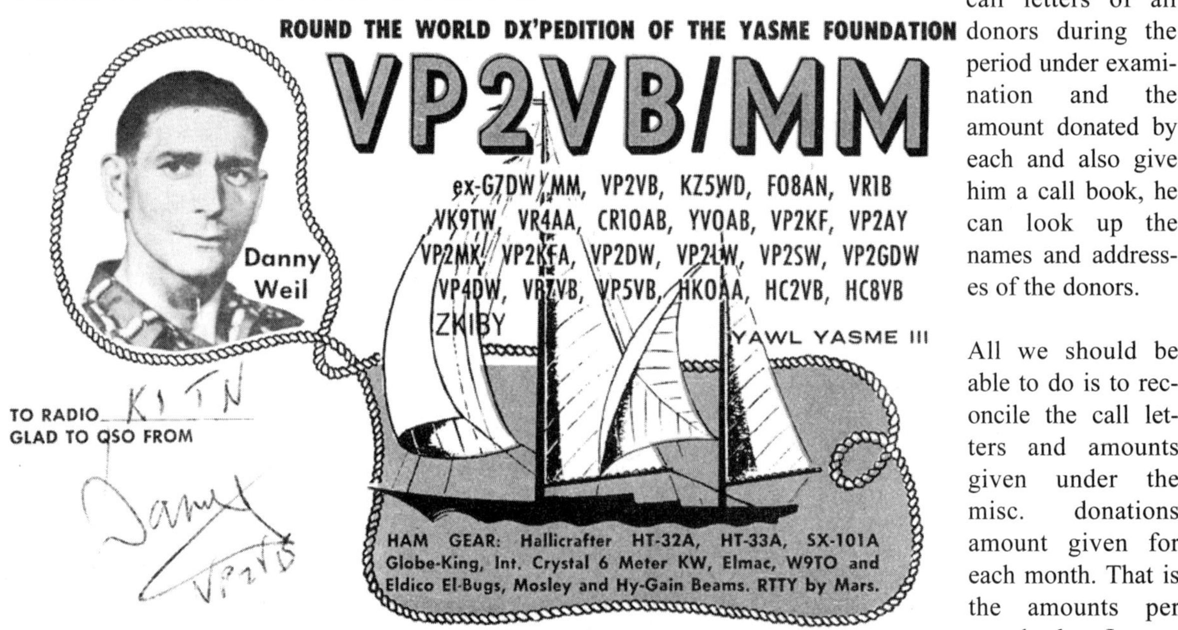

The author's YASME QSL card for an "eyeball QSO" in September 2001

you in my letter of February 6th. If your deposit slips are complete and add up to the amount of the month's deposit and each deposit slip has on it or attached to it a list of call letters of all donors and the amount of each we can give the agent the deposit slips and the call book and let him work it out.

Don't get involved in making a big job of this yourself as most of the actual work of putting this together can be done by the I.R.S. agent providing we can give him the complete data to work with. If you can send Mr. Vawter the data as requested in my letter of February 6th, that is the bank statements and checks listed and any record you have that ties in the donors by call letters to the amounts of donation and the monthly totals given in my letter, we can put the thing together in such shape that the agent will do most of the work.

There is no hurry about this. Mr. Vawter is to call the agent as soon as he has all of the data for him. The agent knows we have to get the dope from you at St. Thomas and so I am sure that he will be patient. I would rather flood him with more data than he requests-that is the kind of data that he can dig at to his hearts content-so that he can get a full picture of our operation as soon as possible. It will probably be easier on you to get as much of this data to Mr. Vawter or me so that this matter can be cleaned up here. I certainly would prefer that they do not have to get their St. Thomas I.R.S. agent to take any part of all of this examination down your way. I feel sure that we can get sufficient dope from you so that this will not happen.

Again let me plead with you, Dick, don't do any work in this tax matter that you can turn over to the agent and let him do. As long as we can give complete records or work sheets, he can get all the dope he wants. - Charles J. Biddle, W6GN.

JLY 5, 1963 NO. 5220 PRICE 3d

CHRISTCHURCH TIMES

Also incorporating Highcliffe, New Milton and Ringwood Times

Danny's Made It

DANNY Weil, the Christchurch lone voyager, has made it. He has completed his round the world expedition in nine-years in his 60 ft. ketch "Yasme 111" and has dropped anchor at Freeport, a Texas seaport, his last port of call.

And now, his mother, who lives in 6, Dennistoun Avenue, Christchurch, is planning the biggest voyage of her life. She has booked a passage on the Queen Mary for October and will stay six months with her son, to catch up on all the experiences of the past nine years.

The last part of Danny's trip—9,000 miles from Fiji Islands—took him nearly three months, but there was a big welcome awaiting him from sailing and amateur radio enthusiasts at Freeport.

Danny, who is aged 44, is to settle in America, and has been offered several "good" jobs. He will rejoin his American wife, Naomi, at Tampa, and will start to put his adventures into a book from daily logs kept on the voyage.

LECTURE TOUR

Before Danny finally settles down he will be making a big lecture tour of conventions in the States talking about his experiences. He is already a public figure in America and has appeared on television and broadcast on radio networks when he made a previous visit.

Danny left Christchurch in 1954 to sail single-handed round the world, an incident packed trip which he volunteered to undertake on behalf of radio "hams". He has set up radio stations on remote islands, and the trip has been logged by some 100,000 amateur radio enthusiasts throughout the world.

He has met many famous people, and dined with many including the Governor of the British West Indies at Tortola, who with his wife went aboard "Yasme". He has also been called on to help with sea rescues off many coasts he has visited.

There has been plenty of excitement during the past nine years, battling through storms and high seas. In the Bay of Biscay he met with waves 40ft. high, and on one occasion sharks followed his boat for days.

On another occasion he ran into a plague of dragon-like locusts. Thousands covered his decks during the Gibraltar-Canary Islands run.

The trip had its humorous moments too. A former watchmaker, Danny offered to repair the town clock on a small island in the West Indies group. It had not received attention for 100 years, and he found the trouble was not mechanical but a lizard hanging on the back of the pendulum. After carefully removing the reptile Danny had no trouble in getting the clock going again.

Danny, leaving Poole in 1958 at the start of his voyage. Inset shows him with his mother, and the then Mayor of Poole.

The Christchurch TIMES heralded Danny Weil's completion of his nine-year sailing journey in this 5 July 1963 article, reporting Danny, 44, had dropped anchor in Freeport, Texas. The large photo is of YASME II as it left the English port of Poole in 1958. The inset photo is of Danny's mother, the mayor of Poole (in 1958), and Danny.

Danny Weil was inducted into the CQ DX Hall of Fame in 1969. CQ's DX columnist John Attaway, K4IIF calling him the pioneer of the worldwide DXpedition. "He had," CQ said, "more narrow escapes from death than 100 average men would expect while pursuing peaceful occupations. His exploits will live on as long as little groups of amateurs gather to reminisce about the "good-ole days" of DXing."

In 1969, at the time he was inducted into the CQ DX Hall of Fame, Danny was living on a five acre farm in Seguin, Texas, near San Antonio, where he was employed by General Dynamics. Later, he and Naomi ran a clock shop.

Danny gradually disengaged from amateur radio, after serving as a director and, for a time, as president of the YASME Foundation. "The great adventurer settled far from the sea, but the sound of the crashing breakers must still be in his ears," Spenceley said.

Danny Weil became a U.S. citizen. Naomi died in 1994. Danny still lives in San Antonio.

33. The Colvins, and a Changing of the Guard

Written application has been made to operate in approximately 150 countries. Iris and Lloyd say they will go anywhere. - YASME Foundation press release

In 1964, the Colvins approached Wayne Green, publisher of the new 73 magazine, because he had formed an "Institute of Amateur Radio," for possible sponsorship or collaboration. Green wrote:

> Good to hear from you, and that your trip is well along in the planning. I would be very interested to hear which countries have given you permission to operate so far...this in itself would make an interesting article. F [French] calls probably can be facilitated by Bernard Malandain F9MH, who is the French Consulate in New York. 934 5th Avenue. He helped me get FO8AS and FL8NSD. Now that reciprocation has gone through and is in the works with France I expect he will be helpful.
>
> One scheme that I have been considering ... re DXpeditions ... it would be nice if the Institute of Amateur Radio could accept donations of ham gear from hams and commercial manufacturers, which equipment would be turned over to the Consulate for use by a ham club at home ... with the DXpeditioner setting up the gear and instructing locals in its use and upkeep.
>
> If we are to hold our frequencies at Geneva we have a gigantic sales campaign ahead of us. We have to convince all of the small countries that ham radio is of great value to them...and it is. Hardly any country is so small that it does not benefit from the opening of electronic industry and from an expansion of radio communications and applications of electronics. Without a basis of hams there is virtually no staff for such companies. The more hams the more electronic industry in any country. Plus emergency communications etc.
>
> Financing. I'm trying to find out the name of a good tax lawyer that we can consult on this. You might check with Ed Simmons W6CLW who set up the Ed Simmons Foundation as recipient for his royalties, etc, This outfit, run entirely by Ed, pays his bills, digs oil wells, runs a fishing fleet off Mexico, etc. The Institute cannot be a tax-free organization since it is engaged in lobbying in Congress for amateur radio. You can't do both.
>
> Would it be of any help if I sent you a formal letter agreeing to buy your articles, thus making you a professional writer? I should think that then you could charge your expenses off as a writer. I could make it seem reasonable for your expenses to be more than the pay for articles by giving an option for a possible future book with rather good royalties if published. Again the fellow to see is a tax lawyer...this would seem like a very good time to spend a few hundred dollars on expert advice.
>
> If I find out anything more I'll let you know. If there is any way I can help. I'm available.

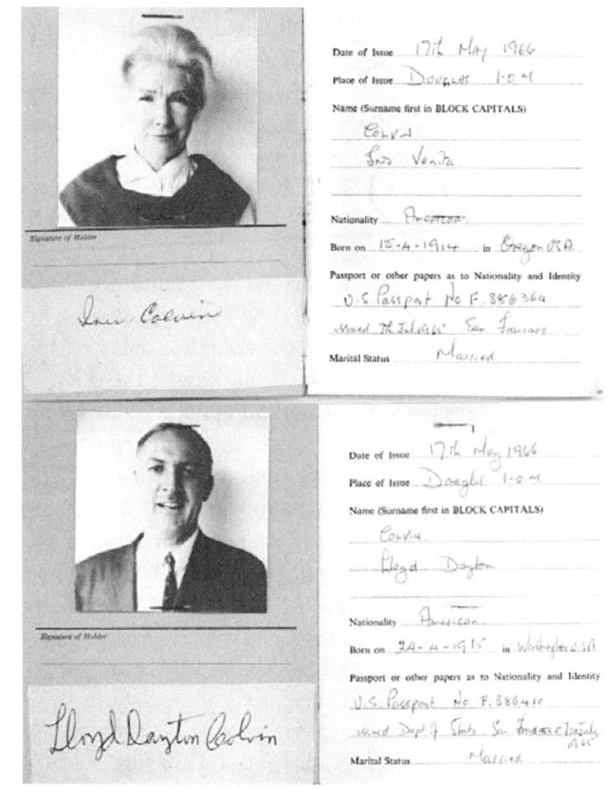

British (and French) licenses and visas would be very useful around the world and the Colvins maintained both.

Lloyd wrote to the National Geographic Society in May 1965 that "I recently retired from the U.S. Army. My wife and I have been active radio amateurs for many years.

"We are planning to travel to remote sections of the world for the next ten years and operate our amateur radio station in each country visited. Our immediate itinerary is to visit several countries in the Indian Ocean and after that to visit all countries in Africa.

"We are being sponsored partially by the 'YASME Foundation,' a non-profit organization. Enclosed is a statement of Nature of Activities of the YASME Foundation as filed with the Internal Revenue Department

[sic] with their last tax returns. The Board of Directors of YASME have suggested that I write to you to inquire about the possibility of obtaining a National Geographic Society Grant in turn for which we will provide pictures and articles about each country visited. We do not ask for any money in connection with the grant but feel sure that some kind of commission in the form of a written document from the National Geographic Society would assist us in traveling to remote sections of the countries we are to visit,"

A photo of Iris, Lloyd and Joy was enclosed. Nothing came of this.

Iris told 73 magazine in a 1981 interview that "We had always thought about going out on a DXpedition, and in 1964 I said to Lloyd, 'Why don't we go?' We contacted YASME because we knew it was still active, although no one was out at the time. We asked if they'd sponsor us, and they said yes.

"They had no money in the Foundation at that time, but we paid our own way, anyhow, so it didn't matter. They provide a big service for us in the way of QSLs. The donations they receive help with that. They also helped with licensing in countries where we had trouble."

Chris Brown, KA1D, the 73 staff member who wrote the interview story, said "Tanned and healthy, the Colvins approach retirement age with energy and enthusiasm instead of dread. Part of the reason is their lifestyle. Financially secure and unfettered by family or business constraints, they are free to roam the world together in an ongoing DX odyssey."

In the summer of 1965 Iris and Lloyd sent a letter to the last known directors of the YASME Foundation, saying they planned a 10-year DXpedition to visit and operate from "almost all [the] rare DX countries in the world," starting in August 1965. They would "stay until the pile-ups ended," and travel mostly by air. They proposed reviving the YASME Foundation to sponsor the trip. "We will pay sufficient funds into YASME to keep it going (minor contributions will be solicited to maintain the tax-free status of YASME)."

The Colvins had already hired a tax lawyer, who OK'd the arrangement, and they had "suitable radio equipment and antennas" for the DXpeditions. They also had begun writing to various countries to get operating permission.

"It is proposed," they said, "that the past efforts of YASME, its directors, and the operators it sponsored be briefly described on the QSL cards of the new YASME DXpeditions, in order to, in some measure, honor the past operations and operators and keep their efforts in the minds of the ham-public.

"Lastly, and of the most importance, will you indicate in your letter whether or not you are willing to continue as a Director of YASME? We certainly hope that you will as we need your help."

Shortly thereafter, an announcement came from the foundation with the familiar "The YASME sails again!

"Not actually, but at least in spirit! [the announcement said] The YASME Foundation (non-profit, all contributions tax deductible) has been reactivated under its old by-laws and organization with some changes and additions to its officers. Its immediate objective is to sponsor the world-wide DXpedition activities of Iris and Lloyd Colvin. This News Letter is sent as a one time effort to publicize the reactivation of the YASME Foundation. For further YASME news read the West Gulf DX Bulletin. Frank Campbell, W5IGJ, is Editor of the West Gulf DX Bulletin. He is also Publicity Director for YASME."

The officers and directors of the YASME Foundation were: President, Danny Weil, VP2VB; Vice-President, Hal Sears, K5JLQ; Secretary and QSL Manager, Bob Vallio, W6RGG; Treasurer, Ed Peck, W6LDD; directors: Golden Fuller, W8EWS, Dick Spenceley, KV4AA, Charles Biddle, W6GN, Jack Drudge-Coates, G2DC, and Frank Campbell, W5IGJ.

The announcement of the new, upcoming YASME DXpeditions, said

> Iris Colvin, KL7DTB/6, and Lloyd Colvin, W6KG, are departing on an extended DXpedition that is hoped to include operation from most of the rare and semi-rare countries of the world. Travel will be primarily by commercial aircraft. The gear includes a Collins 75S-3 receiver, 32S-3 exciter, 30L-1 amplifier and in most cases a Hy Gain triband beam (A 14AVS vertical is available as an alternate). Iris and Lloyd have already visited 96 countries and held 23 different calls and need no further introduction to the DX fraternity.
>
> Written application has been made to operate in approximately 150 countries. Iris and Lloyd say they will go anywhere. If permission to operate can be obtained and you can provide transportation here is your chance to send them to work you from that country you still need - hi. The present operating plan calls for operation from Pacific areas, then the Middle East, then Africa.
>
> Iris and Lloyd leave the USA in August. First operation will be in the Pacific Areas in September. To make it easy to remember, Iris and Lloyd will operate on the same announced frequencies as the Don Miller-Chuck Swain DXpedition which are: 7000-10, 14045-55, 21045-55 KC CW, and for SSB 7090-100, 14100-110. 21400 KC, listening as directed. Only one QSO per band, per Mode is requested. QSL address top of page. Please SASE.

Contributions as you see fit. Time GMT. All QSLs answered.

"All QSLs answered," and they meant it. Iris and Lloyd told 73 in the 1981 interview that "We use a log system that YASME has used ever since we've been connected with it: We can thank the infamous Mr. Don Miller W9WNV for designing it. In fact, he gave us the first copy of the log we use. It's a good system and we have used it for as long as we've been operating. It consists of a small form with attached carbon, about five by seven inches in size. Each sheet holds 100 entries and when a sheet is full, we send the carbon to YASME for filing and verification."

Iris later said "When we decided to go on our first DXpedition in 1965, we sold our house, closed out the five construction contracts we had going, advertised for and found a good home for our dog, and then left. Our intention was to go out indefinitely. After three and a half years we got tired and went back to California for a rest. Now we have a different house; it's a place we can go back to as a home base and it even has a permanent amateur radio station."

In May and June 1965 the Colvins sent enquiries to Guinea, Afghanistan, Burma, Sudan, Bechuanaland Protectorate (Botswana), Yemen, Mongolia, Nepal, Maldives, Iraq, Brunei, Saudi Arabia, Jordan, Kuwait, Wallis, New Hebrides (now Vanuatu), Niger, Basutoland (later Lesotho), Malawi, Ivory Coast (now Burkina Faso), Swaziland (approved provisionally). Luxembourg, Sri Lanka, and San Marino rejected the requests. The rest did not reply.

Here's the letter that went out and which they would use for some time to come:

> I hereby apply for permission for myself and my wife, Iris A. Colvin, to operate our amateur radio station in [name of country] for a period of not more than two months between [starting date] and [ending date].
>
> The following points in support of the request are submitted, and it is respectfully requested that you consider them carefully:
>
> a) I have been an active radio amateur in the United States since the early days of radio. I obtained my first license and operated my amateur station in 1929. I have taken examinations and passed them successfully for all classes of radio operating licenses issued by the United States Government. I am technically qualified to install, operate and maintain an amateur radio station in accordance with the rules and regulations of your Country. Iris Colvin has been a licensed radio operator since 1945.
>
> b) My wife and I have traveled all over the world and have already operated our amateur radio station in several foreign countries in a manner that has been a credit to the radio amateurs of the world, and has served to further good relations between the Countries operated in and the United States.
>
> c) The requested operation is desired. In order to permit a maximum number of radio operators throughout the world to communicate with [your country]. While there are presently active amateur stations in [your country], there are many amateur stations throughout the world who are still waiting anxiously to establish their first communication with [your country]. Such a contact by amateur radio will permit them to receive various awards for excellence of operation.
>
> d) We have small compact radio equipment of the very latest technical design that we will take with us for the operation. Briefly, it will consist of a 100-watt station operating on both CW and SSB. The complete station, including antenna, will weigh only about 150 pounds and can be packed in one large suitcase.
>
> e) We have made arrangements for our travel to [your country] and places to stay while there.
>
> f) We have sufficient funds for the trip and will in no way require any Government assistance, except the requested permission to operate our amateur radio station.
>
> g) It is especially appropriate at this time that favorable action on our request be taken, since the President of the United States on 28 May 1964 signed into law a Senate bill number 920 amending the Communications Act of 1934 so as to permit reciprocal operating agreements between the United States and other countries for radio amateurs. This means that legally the United States may now consider an agreement to let radio amateurs of your country operate their amateur radio stations in the United States. Favorable action by you on our request for permission to operate in [your country] would establish a favorable precedent should radio amateurs from [your country] desire to request permission to operate their amateur radio stations in the United States.

h) The requested operation will comply in full with the present amateur radio rules and regulations of [your country]. If desired, we will be happy, at the end of the operation, to submit a copy of our log showing all contacts made, frequency, time of contact, etc.

The color Colvin family photo was always included in these requests.

The Colvins corresponded with several Indian amateurs regarding the Laccadives (Lakshadweep) Islands, a group of 27 islands, only 10 of which are inhabited, hoping to operate in August 1965. There were no resident amateurs on the Laccadives. It was noted that the well-known Gus Browning had problems with red tape, and that even getting a license just for India would be difficult or impossible, with no reciprocal agreement. And there's no power on the Laccadives.

A letter from K.R. Phadke of the Indian department of communications in July said "I am directed to refer to your letter on the subject mentioned above (OPERATION OF AMATEUR W/T STATION IN THE LACCADIVE ISLANDS) and to regret that it is not possible to consider your request for operation from the Laccadive Islands. However, if you are interested in operating your Amateur Station from some other part of India, you may please fill up the enclosed application form and send to this Department for considering the case on its merits."

A September 1965 letter from the India Department of Communications, Wireless Planning and Coordination, said the Colvin's licensing request was being considered but "without action."

They also sought help from the U.S. embassy in New Delhi, who advised "It is our understanding that (licensing in India) is presently impossible. The Embassy is exploring with the Government of India the possibility of entering a bilateral agreement under Public Law 88-313 which would permit reciprocal licensing. However, we have not yet reached the stage of formal negotiations. While we would have no objection to your receiving mail at the Embassy, we would consider it inappropriate to have the radio equipment licensed at the Embassy address."

A request in 1965 to Cyprus, where a 1964 U.N. resolution calling for a cease-fire had ended, in August 1964, fighting between Greek and Turkish Cypriots, brought the following terse reply "I have to refer to your letter of the 16th June, 1965, applying for permission to operate your amateur radio station in Cyprus for a period of not more than two months during the next year, and to inform you that your request cannot be granted as the Government has revoked all amateur licences for the time being."

In September 1965 Reunion Island (France) demurred, citing the lack of reciprocity, and the same month Zambia said "As yet, there is no reciprocal operating agreement between Zambia and any other country to cover the case, and it is therefore necessary for special authority to be obtained from the Minister of Information and Postal Services. This authority is being sought on your behalf, and you will be advised of the outcome as soon as possible - General Post Office, Telecommunications Division"

Also in the spring of 1965, permission was sought to enter China, most-needed DXCC country at the time. The Colvins sent this modest letter:

The requested operation is desired in order to permit a maximum number of radio operators throughout the world to communicate with the People's Republic of China. While there are presently active amateur stations in the People's Republic of China, there are many amateur stations throughout the world who are still waiting anxiously to establish their first communication with the People's Republic of China. Such a contact by amateur radio will permit them to receive various awards for excellence of operation.

Although formal permission of the U. S. Government for us to visit The Peoples Republic of China has not been issued, it is anticipated that no difficulties in obtaining such permission will be encountered if you reply favorably to this letter. A room in a tourist hotel is all that we will require in the nature of accommodations during our stay in the People's Republic of China.

An inquiry to Iran was turned down and one to the Federation of South Arabia (Aden) revealed that "the regulations of this Administration only permit the operation of an amateur radio station by a person of British nationality or by a British protected person."

Nearly 40 years later Joe Reisert, W1JR, said "One thing that always impressed me about Lloyd and Iris was that they could never be called 'Ugly Americans.' They were always friendly and courteous to a fault. When they went to any DX location to operate, they had proper authorization, or they didn't operate, and they met the natives and hams, socialized and genuinely tried to accept the local population. On more than one occasion, Lloyd told me of all their letter writings and correspondence to get proper authorization and licenses. He mentioned many failures when they even went personally to the country but were later refused."

In 1981 Lloyd told 73 magazine that "In all such matters you've got to be part politician to make any progress. You both must be quite adept at working your way through bureaucracies after all your dealings with foreign governments".

Iris added "That's one thing that you've got to do. I remember an instance in Africa back in 1971 when many [new] countries were being formed. In one place we visited, we felt that they were willing to let us operate but we sensed that there was another problem. They simply didn't know how to go about granting an amateur license. They had never had to do it before, having had no amateur service of their own. They didn't even know what a ham license should look like.

"We once spent two and a half months in a rare country, with our gear, and couldn't get a license. In spite of that, we never fell for the temptation of getting on the air illegally. We've always had legitimate licenses on all our trips. We also have made it a special point not to commit any type of minor crime, such as operating without proper permission or not paying all our bills."

Goal: One From Every Country

3roadcasting Team on Tour

ΝΚ KRYSTYNIAK
Post Reporter

olvin and his wife one long, continual

30, has been a ham usiast for 37 years. wife have just com- roadcasting tour of Pacific.

IAVE SET the goal asting from every the world in the ears.

vas a signal officer . Army for 23 years. as a lieutenant col- in 1965 a group 'ASME agreed to is travels.

is an organization adio operators inter-

BETWEEN YASME I and III, there was YASME II and another boat.

YASME II sank when it hit an uncharted reef off the coast of New Guinea. The other boat, still unnamed, blew up in an English harbor.

Colvin and his wife take their 400 pounds of radio gear by air. It is quicker, he said, but more expensive than by boat. However, not all of their travels have been by air.

THERE WAS the time, for instance, when they were taken by boat to the Island of Ebon, in the South Marshalls. A boat was supposed to pick them up in two weeks, but didn't make it back for two

insistence on this individuality in dress, the Colvins said.

THE NATIVES are usually friendly, Colvin said, although they are usually hard to convince that there is nothing military or spy-like about the broadcasts.

The Colvins broadcast in an international code. They have communicated with a Saudi Arabian king, an Austrian archduke and an American presidential candidate, Barry Goldwater.

The Colvins are now touring the United States, where they give lectures on their travels and meet with ham radio clubs. They will go to England in May, where they end their tour.

THE FIRST broadcasting point on the new travels, in case any of you half million hams are interested, will be the Isle of Man.

From there they plan to go to Africa and possibly Albania and Turkey.

The last two countries are extremely difficult to get into, however, and Colvin said he will not be too disappointed if he does not make it.

LLOYD COLVIN
Globe-Trotting Ham

IRIS COLVIN
Liked Ebon Diet

DANNY WEIL
Started It All

Iris and Lloyd toured the US in the spring of 1966 and a stop in Houston, where Danny Weil lived, resulted in this article in the Houston Post (7 April 1966). The article recounted Danny's travels, and the Colvins' recent operations from Pacific islands. Of their time on Ebon Atoll, which was extended two weeks while they waited for a boat to take them off the island, Iris said there were no supermarkets on the island, so [we] ate breadfruit and Papayas. "It was very good for the waistline," Iris said. The Post said the Colvins were getting ready to end their tour in England, then operate from the Isle of Man. The photo captions are "Lloyd Colvin, Globe-Trotting Ham," "Iris Colvin, Liked Ebon Diet," and "Danny Weil, Started it All."

34· Enter Don Miller

In general, the less correspondence and the more actual seeing-people you do, the more licenses you will get without any trouble. - Don Miller

Don Miller, W9WNV, was sought out for his advice concerning both equipment and licensing and on 8 July 1965 he wrote from San Bernardino, Calif. {the ellipses ... are Miller's):

> Dear Lloyd & Iris: Just got back here from this rather hectic trip. Hope you got in touch with Hal Sears. Here [are] a few points we talked about, which may be of some help to you.
>
> Take the KWM2. You may be able to get a used one. First of all it is a good back up rcvr and back up xmtr, weighs less than 20 Lbs., and runs off the pwr sply for the 32S xmtr. If you can get also the portable pwr sply for the KWM2, it plugs into the back and fits into the suitcase with it, and with this unit you can then run either the 32S xmtr or the kwm2 on 220 as well as 110V. (30L1 also runs off 220 so that you can have a complete stn for 110 or 220 V). Get a waters [Waters] keyer ... forgot, you like the bug).
>
> Hy-Gain 14AVS or equivalent is all the antenna you need. You can get it up 20 feet or more on a piece of bamboo or pole. Use at least 4 radials for each band .... 15, 20 and 40 is probably all you will use. Be careful assembling it for the first time because you will kink the aluminum elements with the new type clamps they use and it is then tough to disassemble with the kinks in it. Kink it as little as possible.
>
> Take a supply of new & used callbooks and handbooks. Most hams in isolated areas are tickled pink to get these. Enclosed are permits and info on KG6R and KG6D. Please return these to me as I don't have time to make a copy right now.
>
> Enclosed is correspondence from India. Again, please return to me ASAP...my only copy. Regarding VK license, please write to KV2QJ ... vk2qj (this typewriter not too good) ... but do not mention you got his call from me or mention me in any way ... he did me a favor and I don't want to extend his accommodation ... he has done this for a number of W's so I'm sure he will help you in getting a VK ticket ... don't mention anything about a DXpedition .... they hate this ... just mention you are taking a trip around the world with your ham set, etc ... don't ask whether this ticket is good for the Is. ... the regs say that it is good for any Austral. Terr., so just pick up the ticket in Sydney after John (qj) has made the arrangements and forwarded the forms, and take the VK permit with you and you will have no trouble getting local authorities on the various islands to grant you a local permit once you show them the VK license .... (this is confidential...)
>
> Enclosed a few log sheets ... let me know if you like them & I'll ship you a couple of thousand ... write on one to try it out .... Under "mcs" you can write the freq. & mode when you change, give sig. Strength under "S" (unless other than R5 or T9, then you can give entire rpt), and check for qsl sent, etc. ... quite compact, maybe too small for you and Iris ... let me know."
>
> Anything you need or want to know, I'll help you if I can and if I don't know the answer, I'll refer you to someone who does. I'll give my services as a qsl manager & organizer after I return from my own trip, if you need one... 73 & best of luck on your exciting adventure!

The Colvins wrote to New Zealand regarding operating from the Tokelau Islands, and a June 1965 letter from the New Zealand Tokelau Islands Administration advised that "before granting permission for you to operate the station I should be grateful if you would give me further particulars about these arrangements please. You may not be aware, that there are only three visits made to the group by trading vessel each year, there are no hotels or boarding establishments as such, and there is no electricity in the group."

In October 1965 F. Gillies wrote

> The Official Representative Cocos (Keeling) Islands has referred to this office your letter of 18th June, 1965, seeking permission for yourself and your wife to enter the Territory and operate an amateur radio station there for two months in 1966.
>
> The Authority responsible for the issue of amateur Radio licences in Australia and its Territories is the Controller, Radio Branch, Postmaster General's Department, Melbourne. Before a licence in respect of one of the Territories is issued it is necessary to obtain the concurrence of this Department or the relevant Territory Administration.

I am afraid that so far as the Territory of Cocos (Keeling) Islands is concerned there is no point in your planning such a visit. The only accommodation which could be used by visitors, if available is that provided by Qantas Empire Airways. This Company caters only for aircraft passengers in overnight transit and for single persons in regular employment or on other official duties. This accommodation is heavily taxed, and the Manager states that it would not be available for purposes other than those already stated.

I understand that you have also submitted applications to operate amateur stations in the Territories of Papua and New Guinea and Norfolk Island. Advice regarding the outcome of these applications will be conveyed to you by the Controller Radio Branch or the respective Territory Administration.

The Cooks Islands looked a little more promising, Rarotonga writing that "Before we can consider your (18 June 1965) application for a licence to operate an amateur radio station in the Cook Islands it will be necessary for you to obtain a permit to enter the Cook Islands. To obtain this permit you should write to the Clerk of the Executive Committee, Government of the Cook Islands, Rarotonga, giving details of your intended visit, length of stay, mode of travel to and from the Cook Islands, accommodation which has been arranged and finance available during your visit.

"When you have obtained a permit to enter the Cook Islands please advise me accordingly and further consideration can be given to this matter."

An inquiry to Mauritius in May 1965 was finally answered in March 1966, director of telecommunications F.W. Lovell saying "I must apologise for the delay in finalising matters, but you will be pleased to hear that authority has now been given for the issue of an Amateur's Licence for a period of two months. I enclosed an application form. Will you please forward a certified copy of your present Amateur Licence and also give the name and address of your representative or contact in Mauritius, if any."

Bayene Desta, "Manager Radio Division" for the Imperial Board of Telecommunications of Ethiopia, responded that the Colvins request was under consideration. This is mentioned mostly because in most cases those countries not disposed to amateur radio operations by outsiders never responded at all.

The Republic of Nauru wrote to the Colvins during their stay in England in the summer of 1965 that

Your letter dated 18th June, 1965, addressed to the Department of Radio Communications in this Territory has been referred to me. There is no such Department in Nauru. (At the time Nauru was a trust territory of Australia, using VK9 call signs.)

Your desire to come to this Territory and to establish an amateur radio station has been noted as has your advice that you have already made arrangements for accommodation upon arrival in this Territory.

There are no hotels or lodging establishments here and permission to enter the Island of Nauru is given by the Administrator only where he is satisfied that proper arrangements for the accommodation of the visitor have been made. This usually involves sponsorship of the visitor by a personal friend. Perhaps you would be good enough to let me have details of the arrangements which you say have been made for the accommodation of your wife and yourself. Your enquiry will then be further considered. -R.E. Vizard, Official Secretary"

Canberra weighed in:

The grant of a licence to operate an amateur station on Nauru is the responsibility of the Administrator and he has informed me that he advised you by telegram on 8th February that a licence would be granted subject to no interference with local communications and no inconvenience to neighbouring homes. The Administrator added that nothing further had been heard from you in the matter.

Since authority to issue licences lies with the Administrator, and not this office, further enquiry in the matter should be directed to the Administrator. The travellers cheque for ten dollars and letter of 4th May, 1966, from the Postmaster-General's Department which accompanied your letter are returned herewith. - F.D. Gillies, for Secretary.

The Colvins' ten dollars was returned and Lloyd wrote to a friend in Australia in January 1966:

I am writing this from Majuro just before we get on the ship for Nauru. A few minutes ago I received the radiogram from the controller radio branch Melbourne. The message reads:

"Regret unable consider authorizing operation your amateur station Nauru stop decision dependent upon results examination all relevant aspects your proposal by administrator nauru and department of territories stop you will be advised further when enquiries complete stop returning your checque by airmail."

The worst part of the above radiogram is that they are returning my money. I am sure they can never issue a license without the payment of the required fee. Tony will you please immediately go down to the RI, see Mr. Buckley, and pay him the required fee for my application? I am inclosing a travelers check to reimburse you for the fee charges. Will you please also give Mr. Buckley sufficient money to pay for sending me a radiogram at Nauru with the information on the call sign assigned to me? Perhaps you may have to personally send me the radiogram with the call sign info Tony. In any case if there is any money left over - keep it for your trouble. If you require more money I will send it at once.

"Thanks for all your help - in haste."

The Solomon Islands looked more friendly, with C. C. Wright, Comptroller of Posts and Telecommunications writing

Such requests take a considerable time to process as they are normally referred to authority in England (technically, the Solomons were administered by Australia, until their independence in July 1978). I am not quite certain from your letter as to what the actual date of your arrival would be so have taken this course of action which would circumvent the delay in obtaining the authority from England. It is possible to use the call sign of one of the local amateurs for the period in question and I would be pleased to allow the use of my own call sign as I am at present inactive. This call sign is VR4CW and I would be grateful for your comment as to whether this course of action is satisfactory.

While still visiting G2DC in England in the summer of 1965 another letter came from Don Miller, who was on Fiji (Miller wrote in stream of consciousness style; the ellipses … are his):

Please excuse the typing ... I'm at a second-class hotel in Fiji and using their typewriter; To get to the point, I'm writing to try to be of some help to you & Iris and possibly give you some idea of the way people react around here. Today when I went in to see the Radio Director of the Post Office here in Suva in Fiji Is. I was going to request a VR2 [Fiji] permit not because it's rare (a few Euro stns still need VR2) … It's not particularly rare, mainly because of Owen's (VR2DK) continuous activity and his W2CTN QSL service, but I wanted the ticket to help open the door for other licensing in this area and possibly to get on for a few contacts.

Well, the first thing they showed me was a file of correspondence from others who had been denied calls, and right on top was that photo of you & Iris in your station, you know, the one you sent me also … they asked me if I knew you, and I said I did but I was reluctant to say that we were good friends or anything because I felt that they would be afraid to issue me a ticket as a special favor because I might tell you and you would raise a stink about it because they had denied yours thus far.

What I did here was this. I got a specially-arranged audition with the Postmaster General, a fellow by the name of Crushank {spelling?) and spent about 4 hours with him discussing amateur radio and I informed him in detail that the United Kingdom had passed the reciprocal licensing agreement and what it meant to us. Finally (and I mean finally), he agreed to give me a call without writing to London or any other formalities, and he issued me VR2EW, good until December of this year at which time I can renew it if all is in order. I don't intend to use it, except possibly for a few QSO's from Owen's tomorrow night, but also it will help me get my VR5 call and a couple of others.

What I suggest you do is to not correspond further with these chaps, you know, this type in particular, their "no" gets more and more definite and finally they got to the point where they refuse to reverse their original decision even if they know they are wrong (English method) ... I think you know what I mean .... Well the point is, I am absolutely certain, that when you and Iris get here, go in and have a word with him and he will give you your license without any trouble ... especially now that he's set a precedent with mine (There is one American licensed here but he owns Fiji property and spends about half his time here. W6AL, perhaps you know him). There is a small licensing fee.

Just finished with 5W1 ... what an effort it was to get the ZM7 permit!!! I couldn't begin to explain to you here what I went through and the scheming involved, etc etc etc ... also got a few other juicy ones for use in a few weeks .... maybe our paths will cross somewhere, After YJ and possibly one other I am going to meet Chuck for Indonesia ops and possibly help with BY and go to XZ and then

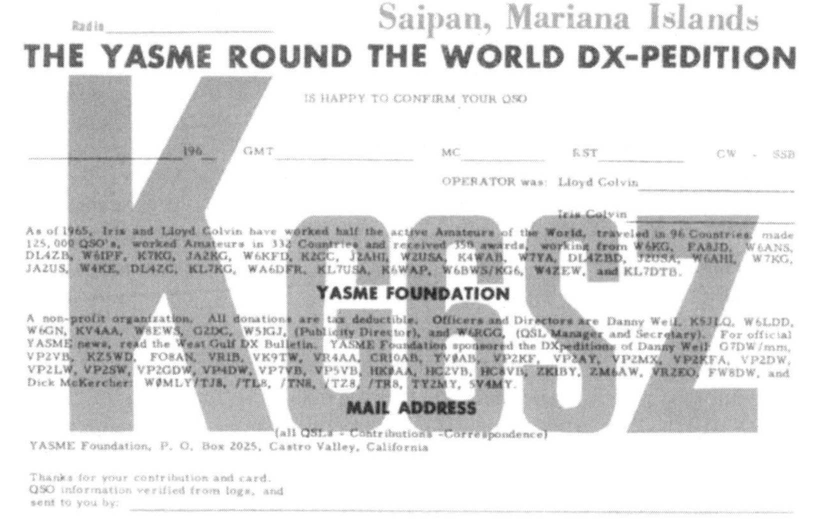

Radio ______ Saipan, Mariana Islands

THE YASME ROUND THE WORLD DX-PEDITION

KG6SZ

IS HAPPY TO CONFIRM YOUR QSO

______ 196__ GMT ______ MC ______ RST ______ CW - SSB

OPERATOR was: Lloyd Colvin ______

Iris Colvin ______

As of 1965, Iris and Lloyd Colvin have worked half the active Amateurs of the World, traveled in 96 Countries, made 125,000 QSO's, worked Amateurs in 332 Countries and received 358 awards, working from W6KG, FA8JD, W6ANS, DL4ZB, W6IPF, K7KG, JA2KG, W6KFD, K2CC, J2AHI, W2USA, K4WAB, W7YA, DL4ZBD, J2USA, W6AHL, W7KG, JA2US, W4KE, DL4ZC, KL7KG, WA6DFR, KL7USA, K6WAP, W6BWS/KG6, W4ZEW, and KL7DTB.

YASME FOUNDATION

A non-profit organization. All donations are tax deductible. Officers and Directors are Danny Weil, K5JLQ, W6LDD, W6GN, KV4AA, W8EWS, G2DC, W5IGJ, (Publicity Director), and W6RGG, (QSL Manager and Secretary). For official YASME news, read the West Gulf DX Bulletin. YASME Foundation sponsored the DXpeditions of Danny Weil: G7DW/mm, VP2VB, KZ5WD, FO8AN, VR1B, VK9TW, VR4AA, CR10AB, TVØAB, VP2KF, VP2AY, VP2MX, VP2KFA, VP2DW, VP2LW, VP2SW, VP2GDW, VP4DW, VP7VB, VP5VB, HKØAA, HC2VB, HC8VB, ZK1BY, ZM6AW, VR2EO, FW8DW, and Dick McKercher: WØMLY/TJ8, /TL8, /TN8, /TZ8, /TR8, TY2MY, 5V4MY.

MAIL ADDRESS

(all QSLs - Contributions - Correspondence)

YASME Foundation, P. O. Box 2025, Castro Valley, California

Thanks for your contribution and card.
QSO information verified from logs, and
sent to you by: ______

back through here for VR5, ZM7 (that trip will cost about three thousand bucks), FW8, and possibly ZL1 (Manihiki) and ZK2, plus three new ones we found that will count as new countries, and then try to get to Heard, which will also be expensive, but for which we have found a way.

There are other less rare spots planned, such as VK9 Cocos, and we are still trying to get the HS authorities to give us at least a temporary permit, etc ... Should be back in the States in Nov or Dec.

Anyway, if you stop the correspondence and simply go in to see the Postmaster General when you get here, with your wife, you'll got your VR call and, in general, the less correspondence and the more actual seeing-people you do, the more licenses you will get without any trouble....hope this is of some service. "This is CONFIDENTIAL! -Don Miller.

The Colvins wrote to the "Office of the Official Representative, Christmas Island," and in July 1965 received word from J.W. Stokes that

To reach here you must first have some local resident to accommodate you -- there are no tourist facilities here. I cannot help you in this regard. Having that, you must then have the permission of the British Phosphate Commissioners, 515 Collins St., Melbourne, Victoria, Australia, to travel to and from here in one of their ships - there is no other way of reaching here. Then, you must have an entry visa stamped in your passport. This can be obtained from an Australian Consul if you have a valid travel document, are in good health and can produce evidence of having had a clear chest X-ray during the three months before arrival. Normally, the visa would be valid for a period of no longer than three months.

Having reached here, there would be no legal objection to your operating your amateur radio station. Quite a lot of amateur radio operation takes place here. The logs of the many U.S.A. contacts made by Christmas Island stations VK9XI and VK9DR can be inspected at station W2GHK, New York, if desired.

The first announcement in 1965 of upcoming operations by the Colvins under the YASME banner said they "hoped to include operation from most of the rare and semi-rare countries of the world. Travel will be primarily by commercial aircraft. The gear includes a Collins 75S-3 receiver, 32S-3 exciter, 30L-1 amplifier and in most cases a Hy Gain triband beam (A 14AVS vertical is available as an alternate). Iris and Lloyd have already visited 96 countries and held 23 different calls and need no further introduction to the DX fraternity.

Operations were to begin in the Pacific in September 1965, on the same announced frequencies as the Don Miller-Chuck Swain DXpeditions. "Please SASE. Contributions as you see fit."

In early September 1965, with their business affairs handed off and the logistical backing of the resurrected YASME Foundation arranged, the Colvins left for Saipan in the Marianas Island group. They did not have a concrete itinerary (as they would on later trips). Operating permission was easily obtained as the Marianas were under United States jurisdiction. They did not travel light in those early days - hauling a complete Collins S-Line, including a 30L1 linear amplifier. Thanks to Lloyd's status as a retired Army officer they were able to island-hop using, when available, military transport.

The Marianas, (KG6SZ) operation marked their first of what would be more than 100 countries, their baptism. They opened up from the U.S.-administered Saipan on 12 September 1965, working JA1BNW, and closed with WA6KNC on 25 September. The KG6SZ log shows about 3,000 contacts. They used the Don Miller-design carbon paper log sheets, from which the cut-up strips of the carbon copies would be glued to their QSLs. For the next 25 years DXers would become familiar with these "automated" QSLs, in the days before personal computers.

KG6SZ was the first "YASME ROUND THE WORLD DX-PEDITION" QSL with a look that never changed. It takes a magnifying glass to read the text on this one-sided card:

"As of 1965, Iris and Lloyd Colvin have worked half

the active Amateurs of the World, traveled in 96 Countries, made 125,000 QSO's, worked Amateurs in 332 Countries, and received 350 awards, working from W6KG, FA8JD, W6ANS, DL4ZB, W6IPF, K7KG, JA2KG, W6KFD, K2CC, J2AHI, W2USA, K4WAB, W7YA, DL4ZBD, J2USA, W6AHI, W7KG, JA2US, W4KE, DL4ZC, KL7KG, WA6DFR, KL7USA, K6WAP, W6BWS/KG6, W4ZEW, and KL7DTB."

Those reading more fine print on this QSL would see "YASME FOUNDATION. A non-profit organization. All donations are tax deductible." The foundation officials and directors are listed, as well as the call signs of the Danny Weil and Dick McKercher operations. Readers were directed to the West Gulf DX Bulletin for foundation and DXpedition news.

Next stop was Truk Island, in the East Carolines (now Micronesia), and 700 contacts as KG6SZ/KC6. Later in October they moved to Yap Island, in the West Carolines, making 3,600 KC6SZ contacts, including an appearance in the CQ Worldwide Phone contest. (KC6SZ, DXCC #10,407, 3 March 1969.)

For the rest of their travels the Colvins would make an appearance in any major contest that coincided with their operations (CQ and ARRL).

In these first operations the thought of making sure they achieved DXCC from each stop hadn't yet occurred to Iris or Lloyd. It was some years before this would become their consuming passion. Nevertheless, the Yap operation did yield 100 countries, and a DXCC certificate, #10,407. This was issued 3 March 1969, during a break in traveling.

Iris and Lloyd did not even tally DXCC countries worked while they operated, generating records and applications only later (as QSLs poured into YASME's Castro Valley, California post office box).

The next stop was Majuro Island, in the Marshall Islands, which had become a trusteeship of the U.S. in 1947 (and later, in 1979, self-governing), and another DXCC, for KX6SZ, #10,357 on 31 January 1969. The first of 1100 contacts was with ZL2BAV, the last with W9UX. The Colvins would return here 6-20 January 1966.

Gregg Greenwood, WB6FZH, has memories of these earliest of Colvin operations that are typical of those of the many teen-aged hams in the 1960s (including this writer):

> I wanted to send you these fond remembrances - of the Colvins - of a 15 year old ham in the early 60s. I lived in San Rafael, California, across the bay from their "home" QTH. I had a tower and 2 element quad that allowed me to follow them throughout the Pacific on their treks, on the low end of 20CW, usually. It was a special treat when I would just send a quick call when the band was open, and get a greeting by my first name, while many listened hoping to hear their callsign returned. Later at a Pacific DX location [of my own], I had the fun of doing the same thing for a few years.
>
> Those black and reddish pink (fading a bit from age) QSLs are still some of my favorites, because they verified that I could follow Lloyd around the world with a homebrew 807 CW rig or my trusty Viking Ranger, and a Drake 2B (They are all in my collection and in working order.)
>
> I met him only one time in person as I recall, at one of the "greater Bay Area Hamfests," or perhaps another bay area club gathering. It was good to associate a "face with the fist." I even remember thinking of Lloyd and the sailboat, YASME, when I lived in Kaneohe, Oahu, Hawaii at the edge of Kaneohe Bay (93-99), where, of course, I became more aware of the Pacific, sailing, and my new neighbors with the familiar place names that I first saw on QSLs from W6KG/somewhere new.
>
> I am 53 now, and I keep at least one of those QSLs in a picture album with other favorites that I take to radio club meetings to show new or prospective hams what DXing, and QSL collecting, is about, and how sometimes you get to meet those you contact on the radio again in person.

35· Ebon Atoll - The "Country" That Wasn't

Under the circumstances, please withhold new-country status of the K7LMU/HC8E operation until we can advise you of our findings on Ebon Atoll. - Iris and Lloyd Colvin, to the ARRL

QST for January 1966 announced three additions to the ARRL DXCC Countries List: Spratly Islands, Ebon Atoll and Cormoran Reef. "Ebon Atoll is located in the Marshall Islands group at 167 degrees East and 4 degrees North. Confirmations for operation from Ebon Atoll made under permission from either Ecuador or the U.N. Trust Territory will be accepted for credit." QSLs could be submitted beginning March 1, 1966.

In the summer of 1965, if not earlier, word was circulating that Ebon Atoll was a likely candidate for DXCC country status. At the time, there was no DXCC requirement that an amateur radio operation must take place before a location could be added to the Countries List. So, the race was on, with both Dr. Don Miller, W9WNV, and his traveling second-operator Chuck Swain, K7LMU, as well as Iris and Lloyd Colvin planning to operate from the South Pacific that fall.

Miller and Swain got to Ebon Atoll first, at the end of October 1965. Or so they claimed. QST later said "In the summer of 1965, Charles W. Swain, K7LMU, and Donald A. Miller, W9WNV, among others, were planning a DXpedition to the South Pacific. [They noted] that some maps (e.g., Rand McNally 1963) showed Cormoran Reef as under the jurisdiction of Costa Rica and Ebon Atoll as under the jurisdiction of Ecuador. On July 25, 1965, in a letter to Swain, the Consul of Costa Rica in Los Angeles authorized operation on Cormoran using the call K7LMU/TI9C, provided the amateur radio rules of Costa Rica were observed. On July 26, 1965, by a similar letter, the Consul of Ecuador in Los Angeles authorized operation on Ebon using the call K7LMU/HC8E, adding, 'Since this is sovereign territory of Ecuador, you will abide by the regulations of amateur radio in Ecuador during your operation.'"

The Colvins, planning their South Pacific trip and aware of the DXCC allure of Ebon Atoll, made inquiries.

Lloyd wrote on 25 August 1965 to the Department of Radio Communications, Radio Amateur Licensing Division, Government of Ecuador that he had written them just nine days before, asking for permission to "operate my amateur radio station for a few days on Ebon Atoll (167 deg E. 4 deg N.)."

> I have since discovered [Lloyd said] that for such a short period of operation, the proper procedure would probably be for me to use my present amateur call W6KG signing W6KG/HC during the period on Ebon Atoll (discussed with Ecuador Consulate - Los Angeles and Federal Communications Commission in Gettysburg, Penn., U.S.A.) I am leaving this date to start the trip to Ebon Atoll," Lloyd said. "Unless you advise to the contrary, I will assume that you have no objections to the contemplated operation.

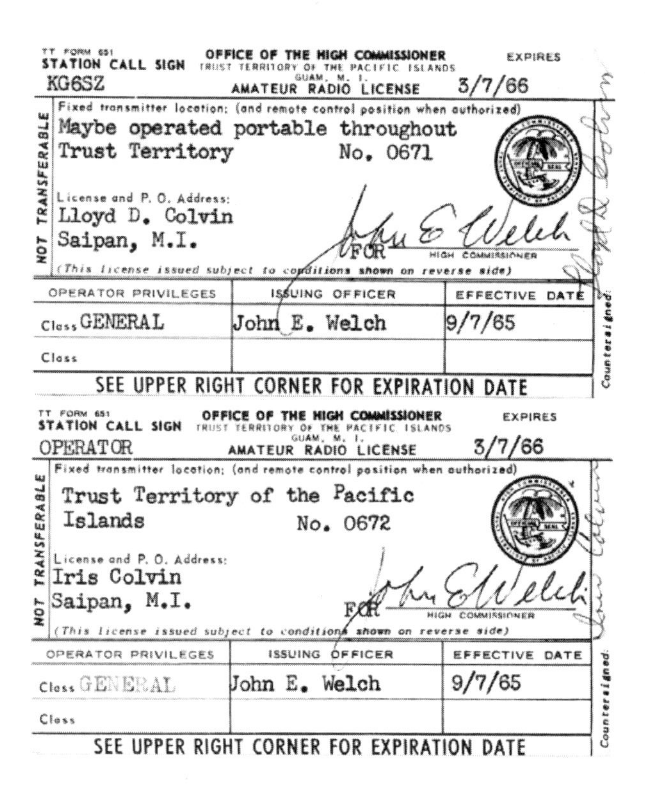

TT FORM 651 STATION CALL SIGN KG6SZ
OFFICE OF THE HIGH COMMISSIONER TRUST TERRITORY OF THE PACIFIC ISLANDS GUAM, M. I.
AMATEUR RADIO LICENSE
EXPIRES 3/7/66
Fixed transmitter location: (and remote control position when authorized)
Maybe operated portable throughout Trust Territory No. 0671
License and P. O. Address: Lloyd D. Colvin Saipan, M.I.
John E. Welch FOR HIGH COMMISSIONER
(This license issued subject to conditions shown on reverse side)
NOT TRANSFERABLE

OPERATOR PRIVILEGES	ISSUING OFFICER	EFFECTIVE DATE
Class GENERAL	John E. Welch	9/7/65
Class		

SEE UPPER RIGHT CORNER FOR EXPIRATION DATE

Countersigned:

TT FORM 651 STATION CALL SIGN OPERATOR
OFFICE OF THE HIGH COMMISSIONER TRUST TERRITORY OF THE PACIFIC ISLANDS GUAM, M. I.
AMATEUR RADIO LICENSE
EXPIRES 3/7/66
Fixed transmitter location: (and remote control position when authorized)
Trust Territory of the Pacific Islands No. 0672
License and P. O. Address: Iris Colvin Saipan, M.I.
John E. Welch FOR HIGH COMMISSIONER
(This license issued subject to conditions shown on reverse side)
NOT TRANSFERABLE

OPERATOR PRIVILEGES	ISSUING OFFICER	EFFECTIVE DATE
Class GENERAL	John E. Welch	9/7/65
Class		

SEE UPPER RIGHT CORNER FOR EXPIRATION DATE

Countersigned:

At the same time, Lloyd told the FCC's San Francisco, and Gettysburg, Pa. offices that he intended to "operate my radio amateur station signing W6KG/HC in Ebon Atoll for a period of not more than one week, sometime within the next 90 days.

"This operation has been cleared," Lloyd said, "with the Government of Ecuador (Consulate of Ecuador - Los Angeles) who has no objection to such operation.

"I am leaving the United States by aircraft this date to start the journey to Ebon Atoll. Unless advised to the contrary, I will assume that the FCC has no objections to the contemplated operation."

If it seems odd that Lloyd would inform the FCC about his intended operation, keep in mind that in those days U.S. amateur regulations required signing /portable when operating anywhere other than the license address ... even a move down the block within one's own town required it.

The Colvins wrote a letter to the ARRL on 22 September 1965, from Saipan, the first stop on their South Pacific trip, regarding Ebon Atoll. This letter no longer exists, but a reply from Bob White, W1WPO, ARRL's DXCC administrator, dated 5 October 1965, does. White said:

> The book on the Trust Territory of the Pacific is quite interesting, particularly with respect to the population and local control of Ebon. From the information in the book it would appear that a new face appearing on the scene would be quite noticeable. It would certainly seem that the local authority on Ebon would be rather well informed as to what goes on and due to the size of Ebon could, if they wished, keep a rather strict control over any activity.
>
> It is quite obvious that Ecuador does feel inclined to believe they have a claim to Ebon, as evidenced by their issuing you the permission to operate under their authority. The FCC is also aware of the permission that has been given. Should operation take place from Ebon, under the Ecuadorian authority, we would have no alternative but to recognize that operation.
>
> The matter of our listings for the Trust Territory of the Pacific Islands is one that we go through periodically. Some say that inasmuch as the Trust Territory is one identity, logically, it should be only one listing. However, the fact that this Trust Territory does consist of an area covering some three million square miles it makes for a rather bulky item and breaking it down into smaller pieces is simply a practical compromise. The time to make changes in the area concerned with the Trust Territory is when the areas concerned become independent countries in their own right.

On 30 October 1965 Lloyd wrote to the ARRL from Guam:

> By direct radio communication this date, I informed Bob White that the operation now in progress by Chuck Swain, K7LMU/HC8E, was surrounded by facts and questions that are being investigated by several persons and sources. He asked that all available information be sent to the ARRL. [In 1965 Bob White was active on the air and may have had a radio contact with Lloyd, or received a relay by radio of Lloyd's message, although such communication with ARRL headquarters employees has historically been scrupulously avoided.]
>
> There is doubt as to the location of the station. Beam bearings taken from Saipan, Guam, Wake, Hawaii, Majuro, Japan, and other places in the Pacific indicate that the station is not on Ebon Atoll. We are departing by ship for Ebon Atoll within 48 hours. The trip will take five days. Upon our arrival in Ebon Atoll, the recent presence or non-presence of Chuck Swain on Ebon Atoll can be verified. The Trust Territory is extremely upset over the Chuck Swain operation, and are doing everything they can to locate him and put him off the air.
>
> The Trust Territory officials claim that if Chuck Swain is operating at Ebon Atoll, he has and is committing the following serious violations: He entered the Trust Territory without authority, he is using an illegal call sign in the Trust Territory, and he is operating on radio frequencies not authorized for use by amateurs in the Trust Territory.
>
> Whether it is called Ebon Atoll, Ebon Reef, or Ebon Islands, The Trust Territory officials claim absolute authority to issue all amateur licenses and control all amateur operation there, in the same manner as all other areas of the Trust Territory.
>
> The Trust Territory does not recognize any claim to Ebon by Ecuador or any other country. All operations in the Trust Territory are by United Nations Directive, the highest authority possible. Operation in the Trust Territory using a call issued by Ecuador is not authorized.
>
> We have been specifically directed that operation by us in Ebon using any authority or call issued or approved by Ecuador will not be done (physical restraint and possible imprisonment to follow if we should violate this directive).
>
> Before we left the USA, we requested permission of Ecuador to operate on Ebon Atoll, 167 deg E - 4 deg N. In reply, we received permission from the Ministry of Communications authorizing us to operate in "all Ecuadorian Territories." If the Chuck Swain authority to operate is similarly worded, it obviously cannot be used legally in the Trust Territory. If it specifically authorizes operation as K7LMU/HC8E on Ebon Atoll 167 deg E - 4 deg N, the Trust Territory wants to know this so that an official complaint can be made through proper channels. Will you please advise us by return letter on this last item?
>
> Under the circumstances, please withhold new-country status of the K7LMU/HC8E operation until we can advise you of our findings on Ebon Atoll.

The investigation of the Chuck Swain operation at K7LMU/HC8E did not originate with us. We were made aware of the operation and investigation by the Trust Territory officials contacting us and giving us most of the information contained in this letter.

This letter was written while the K7LMU/HC8E operation was in progress. But it still appeared that Ebon Atoll would be a new DXCC country, and the Colvins left for there, arriving in mid-November.

In 1965 one did not just pick up a cell phone and call a number on another continent. Sometimes amateur radio could be used for instant communication, always keeping in mind third-party restrictions. On 11 November 1965 Lloyd sent a letter, from Ebon Atoll, to Bob White, saying that at the request of the Director of Communications of the Trust Territory of the Pacific, the ARRL, the District Administrator of Majuro, and "several amateur radio organizations and several individual radio amateur operators, we have conducted an investigation on Ebon Atoll of the supposed operation there, starting on or about 28 October and extending on into early November, of Chuck Swain, signing K7LMU/HC8E.

"Ebon Atoll consists," Lloyd wrote to the ARRL, of a number of small islands shaped in a circle with a radius of three miles. If one stands on the shore of any one island one can look out across the inner lagoon and see all the other islands of Ebon. All of the islands appear to have been once connected together like the top of a large volcano. Even today the islands are partially connected together by rock formations, which in some cases can be walked across during low tides. All the islands on the outer sea side have reefs (partially submerged) a hundred feet or so off shore. All of the reefs are a hazard to navigation and must be avoided by ships. All of the islands that are large enough to live on, are inhabited.

Ebon Atoll has no regular radio communication to the outside world; there is no aircraft service, no electricity; and ships from the outside world arrive only a few times a year. The arrival of any ship is big news and word travels from native to native like fire in dry brush with a strong wind. These people are completely dependent upon the arrival of a ship for everything they use or need except the few items grown on the islands (like coconuts). All eyes are constantly searching for a ship. No ship can land without the natives immediately swarming to it.

There is only one safe passageway into the lagoon (or center) of this small circular island group. The Sailing Directions for the Pacific Islands, issued by the Hydrographic Office, U.S. Navy, advises entrance to the lagoon through Ebon Channel and then going ashore from within the lagoon. No boat can enter through Ebon Channel, night or day without its being seen by responsible natives.

All ships and responsible ship owners in the Pacific and South Pacific are aware of the fact that no one can come into the Trust Territory without formal approval (for specific period, with authorized parties specified by name). Our investigation consisted of talking personally with all persons who speak English. The local Magistrate, in turn, had enquiries made throughout all of Ebon Atoll, with the following results:

No ship that could have carried Chuck Swain entered or left through Ebon Channel during the period of 27 October to this date. No person, foreign to Ebon Atoll, is known to have been on its shores, or operated an amateur radio station during the stated period. It is the opinion of all local natives that no one could have come ashore and operated an amateur radio station during the period in question without being seen by someone.

Authority, by name and date, for Chuck Swain to enter Ebon Atoll was not granted by the Trust Territory. On or about dusk of 27 October, a strange ship (approximately the size of an LST) was sighted out at sea near Ebon Island. The natives assumed that it would lay off shore and enter Ebon Channel the next morning. The next morning, the ship was gone, and has not been seen since. This is a very unusual occurrence, since Ebon Atoll is not on any regular shipping lane.

It is our considered conclusion that Chuck Swain did not operate his radio station, K7LMU/HC8E, from Ebon Atoll, or the reef thereof, or any island thereof, at any time during the period 27 October 1965 to the present date, 11 November 1965.

It is interesting to note that although this was a Don Miller operation, the supposed licensee was Chuck Swain. Since the Colvin's letters at the time do not refer to Miller, it is not known if they knew that Miller also was involved (although probably they did).

On 24 November 1965 the Ecuadorian director of communications, Gabriel Jarrin A., wrote to Lloyd (on Guam) that his office had not given "any kind of permission" to Swain. He also mentioned that Lloyd's provisional permit given on 26 August had expired, since it was valid only for the period 1 to 30 September 1965.

"In any case," the director said, "you should have operated only when you were located within our territorial waters, in other words within 92 degrees west, 2 degrees north and 4 degrees south, but in any case the permit did not cover Ebon Atoll, 167 degrees East, 4 degrees North, which does not belong to the Republic of Ecuador."

When the Colvins operated from Ebon Atoll, of course, they did so under their U.S. amateur licenses.

Don Miller had long had paperwork, it seemed, to operate from the Trust Territory, as a letter dated January 30, 1963 suggests. It's from John M. Spivey, acting deputy high commissioner, the United States Department of Interior, Trust Territory of the Pacific Islands, Office of the High Commissioner, Saipan, Mariana Islands:

> Dear Captain Miller:
>
> We are pleased to comply with your January 3rd request to our Director of Communications for a Trust Territory Amateur Radio Station permit. This letter will serve as authority for you to operate W9WNV (Portable) in the Trust Territory except Kwajalein Island, subject to the Trust Territory "Rules Governing Amateur Radio Services" and permission from the local official in charge. Many Trust Territory locations have limited electrical power, hence our requirement for local permission.
>
> Your offer of extra equipment is appreciated. I am sure local amateurs would make good use of anything surplus to your needs. A copy of our "Rules Governing Amateur Radio Services" is enclosed.

Iris and Lloyd Colvin began operating from Ebon Atoll on 15 November 1965, eventually making 4,500 QSOs between 15 November and 24 December 1965 (first QSO WA6IQM, last QSO K5MWZ.)

Warren Davis, W6IBD, had telephoned Bob White, and relayed to the Colvins (who were using the call sign KX6SZ/Ebon) that White had suggested that "some official proof, other than our own statement, of the non-presence of Chuck Swain on Ebon Atoll be forwarded to the ARRL." In response, the Colvins managed to get a signed statement from the two highest officials on Ebon Atoll and sent it along to the ARRL:

Those officials, Amram Alik and Naisher Lojmen, secretary and magistrate, respectively, of the "Ebon Council," said that they had investigated "the reported presence of an American, Mr. Charles Swain," on Ebon Atoll during the period 27 October to 4 November 1965. "During this period," they said, "no ships entered Ebon Lagoon and no American was seen ashore. This has been verified by inquiry throughout Ebon Atoll. In our opinion Mr. Charles Swain was not on any of the islands of Ebon Atoll during said period."

In talks they would later give to radio club meetings, Iris and Lloyd said

> Ebon Atoll had been given new country status by the ARRL and the lure of operating from a new country took us to this southern island in the Marshall Islands. We took a copra ship, the Micho Queen, with our generator, radio equipment, and supplies for two weeks. At the end of the two weeks the Micho Queen did not return, but, fortunately, one of the missionaries on Majuro had given us a dictionary and the people understood us no matter how we pronounced the words.
>
> Ebon is a matriarchy and the women are most important. We were not truly accepted there until one night we heard chanting above our radio signals. It was the women of the island singing for us and bringing gifts of food, woven baskets, and flowers. We were there for Thanksgiving and Christmas. Our fuel was running low so we didn't operate the radio except to get messages and information. Over time we learned to share our food and our lives with our neighbors, and many of the island's residents became our friends. Our ship finally returned, on New Year's Eve.

The Colvins continued to write letters, and a reply dated 26 November 1965 reflects the situation. It's from John E. Welch, Director of Communications, Trust Territory of the Pacific Islands, Office of the High Commissioner, Saipan, Mariana Islands:

"Thank you for your letters of November 11 and 17. Before proceeding further in the matter of the Ebon operation, I would appreciate your comments concerning the attached copy of a letter from Dr. Donald A. Miller.

"He has been advised," Welch said, "of your letter concerning the K7LMU/HC8E operation and the K7LMU/TI9C operation, where a Donald Miller claimed to be operating from Cormoran Reef in the Western Carolines."

The letter from Miller, dated 26 November 1965, was on the letterhead of the "WORLD RADIO PROPAGATION STUDY ASSOCIATION" (Trustees E.C. Akerson, W4ECI, Gus Browning, W4BPD, Robert Crane, W4ARR, James Long, W4ZRZ, and Walter Wendelken, W3RIS).

Miller wrote to Welch:

> Being active in Amateur Radio, and being on a current trip through the Pacific and Southeast Asia, operating my ham station from many small Islands

and unusual places, I am curious about a recent Amateur Operation from the Trust Territories. If this operation was, indeed licensed and authorized by you, and received your sanction, I would like a similar license and for the same authorization to be extended to me. If it was not such an operation, I would also like to know about it.

The operation to which I am referring is the recent operation of KX6SZ/portable HC8E/Ebon Atoll, in the Marshall Islands. Mr. Colvin has been signing this call-sign on the air and says that the Trust Territory has authorized him to use this call-sign. If so, could I have the same permission?

My F.C.C. call-sign is W9WNV. I can be reached during this trip at the following address: c/o Mr. M.H. Meyers, Communications Dep't., Qantas Airlines, Qantas House, 70 Hunter Street, Sydney, Australia (NSW). Incidentally, I am a US Citizen, a Medical Doctor, and 29 years of age, home address is Box 3278, San Bernardino, Calif., 92404 USA. - /s/ Donald A. Miller, MD.

EBON COUNCIL
EBON ATOLL
MARSHALL ISLANDS

15 NOVEMBER XXXX 1965

To Whom it May Concern,

We have been asked to investigate the reported presence of an American, Mr. Charles Swain, on Ebon Atoll during the period 27 October to 4 November 1965.

During this period no ships entered Ebon Lagoon, and no American was seen ashore. This has been verified by enquiry throughout Ebon Atoll.

In our opinion, Mr. Charles Swain was not on any of the Islands of Ebon Atoll during said period.

Amram Alik

Amram Alik , Secretary Ebon Council

Naiser

Naisher Lojmen, Magistrate Ebon Council

Miller wrote this letter after the claimed K7LMU/HC8E operation from Ebon Atoll. If Miller and Swain already had operated from Ebon, why would Miller now be asking for "a similar license?"

Lloyd wrote to Welch that "You have asked for our comments on the statement of Mr. Don Miller of 26 November, in which he states that we have operated in Ebon, using the all KX6SZ/HC8 Ebon. This is absolutely not true. We have never at any time, not even once, used the call KX6SZ/HC8 Ebon."

In a letter to Miller dated 10 December 1965, Welch said

Attached is a copy of a letter from Mr. Colvin. This should answer some of the questions posed in your letter of November 26. Mr. Colvin was issued our special event call, KX6SZ, for the Ebon operation. We most certainly would not issue a portable call for another country. A copy of your letter will be sent to Mr. Colvin with a request for comments.

We have received several reports concerning illegal operations on Ebon and Cormoran Reef in the Western Carolines. The Cormoran Reef operation used the call K7LMU portable TI9C. This is Mr. Swain's call, however, the operator was identified as Donald Miller. We intend to bring both operations to F.C.C. and AARL [sic] attention. We would appreciate your early comments before doing this.

Iris and Lloyd concluded their Ebon Atoll operation on 24 December 1965 with a QSO with K5MWZ. They had made 4,500 contacts, including operating in the CQ Worldwide CW contest. In a letter dated on the last day of their operation, to ARRL general manager John Huntoon, W1LVQ [later W1RW], and to Bob White, W1WPO [later W1CW], the Colvins said

We wish to bring a very serious matter to your attention. During the past few years and especially during this year Charles Swain, K7LMU, and Don Miller, W9WNV, have conducted a number of DXpeditions to rare parts of Oceania and Asia. They are excellent operators and ardent DXers. They have identified themselves as a team, and if one of them does a good job of DX operating they have both taken credit for the operation, due to joint planning, travels, etc.

It is therefore logical that if one of them does a poor or unethical job of operating, they should both take credit for the same. With this explanation, and for the sake of simplicity, in the rest of this letter, we refer to the separate operations of Chuck Swain, the separate operations of Don Miller, and the combined operations of Chuck and Don as, simply 'Swain-Miller.' In their great zeal to operate from new countries, it would appear that Swain-Miller have gone beyond the moral and legal limits governing amateur radio.

Based on the facts and documents in our possession, and also in the possession of the ARRL, we present the following serious allegations:

Swain-Miller did make false representations to the ARRL, to Iris and Lloyd Colvin, and to the radio amateurs of the world, that Ebon Atoll was sovereign territory of Ecuador (denied by Government of Ecuador in letter 24 November 1965).

Swain-Miller did make false representations to the ARRL, to Iris and Lloyd Colvin, and to the radio amateurs of the world, that Ecuador had authority to issue permits for amateur radio operation on Ebon Atoll (denied in letter from Government of Ecuador in letter of 24 November 1965).

Swain-Miller did make false representations to the ARRL, to Iris and Lloyd Colvin, and to the radio amateurs of the world, that they had permission from the Government of Ecuador to operate as K7LMU/HC8E on Ebon Atoll (denied in letter from Government of Ecuador on 24 Nov 1965).
Swain-Miller operated for a number of days, in late October and early November 1965, working approximately 3,000 amateurs throughout the world, signing K7LMU/HC8E, giving, falsely, their QTH as Ebon, Longitude 167 deg E, Latitude 4 deg N, (station and operators were not on Ebon, nor on any island of Ebon Atoll, per conclusion of investigation by Iris and Lloyd Colvin in letter of 11 November 1965, and of investigation by Ebon Council in letter of 15 November 1965).

Swain-Miller have caused to be sent out, to radio amateurs throughout the world, QSL cards confirming QSOs with K7LMU/HC8E, Ebon Atoll, thus constituting written proof of either illegal operation (no proper permission from Ecuador or the Trust Territory), or a complete hoax (not actually on Ebon Atoll).

As longtime members of the ARRL (Lloyd continuous ARRL membership for 37 years), as strong supporters of the ARRL (See Minutes 1964 ARRL Board of Directors Meeting), as active members of DXCC (Lloyd over 300 countries confirmed as W6KG, and Iris over 200 countries confirmed as KL7DTB/6), and active DXpedition operators (now on YASME world-wide DXpedition), we urge and request that the ARRL investigate the five allegations specified.

If any one of the allegations is determined by the ARRL as being a true statement, strong disciplinary action should be taken against Swain-Miller; their past and future operations should be carefully reviewed and investigated and appropriate announcements be made in QST.

Although the ARRL announcement of DXCC Country status for Ebon Atoll had said QSLs would be accepted beginning 1 March 1966, the matter was in limbo through the end of the year. December QST finally announced that Ebon Atoll and Cormoran Reef - both "Miller countries" - were withdrawn as DXCC countries.

36· Exit Don Miller

"Iris and I simply cannot be associated with an organization that also sponsors Don Miller in any way. - Lloyd Colvin

QST's "DXCC NOTES" said

> Reference is made to the DXCC Note, page 90, January 1966 QST.
>
> Announcement was made therein of the addition to the countries list of Ebon Atoll (HC) and Cormoran Reef (TI). The action was based primarily on authorizations issued by recognized representatives of the Ecuadorian and Costa Rican governments. Information developed subsequently indicates that the named countries claim no jurisdiction and that the authorizations were invalid. Accordingly, there was no choice but to revoke the earlier action and delete the two locations from the countries list.
>
> Because of the unusual circumstances surrounding this activity, a more detailed explanation is in order.
>
> Copies of the letters were mailed by the amateurs involved to the Federal Communications Commission with the request that, if acceptable, they be forwarded to the ARRL Awards Committee. Upon receipt of the letters from FCC, the Committee considered all information then available, including the maps and the copies of authorizations from the named consuls. The Committee particularly noted the statement that the consulates had contacted Quito and San Jose, respectively, to obtain operating permission. The Committee thereupon acted to add the two islands to the countries list, and announced the action in January 1966 QST. Based on this action, the DXpedition plans were completed and several thousands of amateurs were provided with new-country contacts.
>
> Subsequently, it was determined that the maps relied upon were inaccurate, that the consuls had mistakenly issued the authorizations, that they had not checked with their home administrations as claimed, that the two areas are actually within the United Nations Trust territory and under United States Administration. After learning of these most unfortunate errors, the Awards Committee reluctantly concluded that Cormoran Reef and Ebon Atoll do not meet the criteria for separate status recognition and that they must be withdrawn from the list. To rule otherwise would have resulted in lowering the stature of the countries list. Hq. deeply regrets no credit can be given to those who spent many hours attempting contacts on the basis of the original announcement.
>
> The ARRL Executive Committee, at its late-September meeting, was in receipt of a request from W9WNV to review the action of the Awards Committee. After extensive discussion, the Executive Committee found no basis on which to remand the question to the Awards Committee. In view of numerous requests for clarification on what country confirmations from KX6SZ/Ebon might be claimed for, and inasmuch as Ebon has been shown to be part of the Marshall Islands (KX6), DXCC credits for KX6SZ/Ebon confirmations may be claimed for Marshall Islands only.

The two important upshots of this were, first, that K7LMU/HC8E QSLs would not count for anything, since it appeared that Miller-Swain had never set foot on Ebon Atoll. And, second, the stage was set for revision of the DXCC countries criteria, particularly in the area of "separate political/administration.

In the next two years, Chuck Swain died at sea, and Miller found himself the target of more and more questions concerning his operations. ARRL actions in disallowing some of Miller's operations culminated in a lawsuit filed by Miller against the League, which was settled in 1968.

A lengthy story in October 1968 QST noted that Ebon Atoll was a "major difficulty":

> The second item which caused major difficulty was operation from Ebon Atoll (HC8E) and Cormoran Reef (TI9C). These countries have since been deleted from the [DXCC countries] list, and a detailed explanation provided in QST (December 1966, p. 95) which will not be repeated here. We do wish to note, however, that Dr. Miller misrepresented certain foreign consulate activities to the Awards Committee, resulting in a decision which later had to be reversed. Additionally, a letter from the Office of the High Commissioner, Trust Territory of The Pacific Islands, which has jurisdiction over the areas involved, supported charges that the installations actually were not physically present on either island. However, disqualification eventually resulted from an entirely separate factor, license

invalidity. Yet the point is an unavoidable consideration in the over-all appraisal of complaints concerning Dr. Miller's activities

After their operation from Ebon, Iris and Lloyd finished their first DXpedition trip with an operation from Tarawa in the Gilbert Islands (now West Kiribati). They made 2,500 contacts signing VR1Z (first QSO W6EUF on 4 February 1966 and last QSO K0HHU on 13 February 1966). Then it was back to California, for a post mortem of the expedition and to take care of the business affairs of Drake Builders.

In the spring of 1966 the Colvins flew to England, where they obtained call signs G5ACH (Lloyd) and G5ACI (Iris). These British licenses would be kept current in years to come, as they were often crucial in easily obtaining operating permission in British territories (and many former territories). Iris and Lloyd stayed with Col. Jack Drudge-Coates, G2DC, a longtime DXer and a member of the YASME Foundation board of directors, using it as a home base for operations from the Isle of Man, Jersey, and Sark Island.

The Colvins were at sea concerning future operations, especially after the Ebon experience. They wanted to activate rare countries and seemed, at this time, not to have any goals for visiting as many different countries as possible. It should be remembered that in 1966 there was no 5-Band DXCC, no CW DXCC or single-band DXCCs, no "WARC" bands, no amateur satellite bands, no moonbounce. All these future challenges would change the definition of a "rare" country; in 1966 putting a country on the air on one band, both CW and SSB, was enough to satisfy demand.

While in England the Colvins corresponded with ARRL concerning DXCC country status for Sark and Rockall Islands in the British Isles. The best way to make a rare DXCC country was to invent one!

On-going questions about Don Miller operations had the DX world in some chaos. Many DXers held QSLs that had been accepted for DXCC credit but were in danger of being disallowed, as well as holding QSLs that might never be accepted. Miller was rumored to have black-listed some DXers, keeping a mental list of them and refusing to work them, and as a result some disillusioned, longtime DXers had dropped out of the game in disgust, some never to return.

In mid-1966, Miller was still traveling, and operating, while ARRL pondered the Ebon situation. On June 13, 1966, the Colvins wrote to Hal Sears K5JLQ, of Houston, a YASME Foundation director and, at that time, arguably the most active and influential foundation director, after Spenceley. It was only a year since the Colvins had affiliated themselves with the foundation and, not only was the foundation not aiding them financially in any way, but the reverse was true: the Colvins would pump money into the foundation when necessary to keep the paperwork - QSLs - flowing. Other than help with QSL chores, the foundation provided the Colvins with only one other benefit - an umbrella "scientific organization" that could be cited in requests for operating permission (and, probably for I.R.S. tax purposes as well).

The letter to Sears was confidential, for obvious reasons. Lloyd (and Iris) telephoned Sears from England (a big deal in those days) and then wrote:

> In the Spring of 1965 I attended a meeting of the NC [Northern California] DX Club. A member of the Club reported that he had just come back from Los Angeles where he had attended a meeting of the SCDX Southern California Club. He said that almost all of the meeting was taken up with a discussion of whether or not to expel a member of the Club. The member under discussion was Don Miller. The reason for the proposed expulsion was the unethical, and perhaps illegal, operation of Don Miller at W6AM during a recent (ARRL) international DX contest. According to the testimony of several members of the Club, Don Miller was heard calling and working DX stations that no one else in the area could hear at all.
>
> It developed later that the whole matter was taken up with the ARRL by various amateurs, and, as a result of positive investigation by the ARRL, the-operation of Don Miller at W6AM during this contest was disqualified and an announcement to this effect was published in QST.
>
> Hal, this meant very little to me at the time, and I almost forgot it. In August 1965, at an ARRL Convention in San Jose, California, WA6SBO, Iris, and I were in charge of a DX display room. Don Miller came in the room and almost immediately Bill, WA6SBO, asked Don 'What is the truth about your calling DX stations during the DX contest that were not even on the air?' Don Miller answered with a smile 'Surely there is nothing wrong with calling DX stations during a DX contest. Perhaps I had a schedule with them - if I thought I heard them when they were not on, it could be blamed on all the QRM on the bands.'
>
> At the same San Jose meeting, while Iris and I were talking to Don alone, he told us of various ways we might be able to operate in foreign countries even though the official position of the Government was to forbid operation by US hams in that country.
>
> Hal, I have said very little to you, or anyone else,

about the Ebon controversy. After the death of Charles Swain [Swain drowned in a boat accident in the South Pacific after his and Miller's Ebon "operation"] I thought it best to drop the whole thing in deference to a dead man. I want to briefly tell you the facts. First of all the real guiding light in the K7LMU/HC8E operation was Don Miller. When Iris and I got on the air in the South Pacific, we contacted Don and told him that we might go to Ebon. He tried being nice, he tried being tough, and, in short, he did everything he could to talk us out of going. We obtained our licenses to operate on Ebon from the only legal authority, the Trust Territory of the Pacific (separate licenses for both Iris and myself) and traveled to Majuro (head administrative office for that area).

While there, we heard K7LMU/HC8E on the air saying they were on Ebon. Every time that we called them they went off the air completely. I am sure that they heard us and must have been afraid to talk to us. A week later we arrived in Ebon. A complete official investigation was made, by the local officials. K7LMU/HC8E was never on Ebon Atoll and never operated from there.

It is an impossibility for anyone to have operated on Ebon Atoll and not been seen or heard [there]. We told Don Miller this over the air; Don immediately wrote a letter to the High Commissioner of the Trust Territory and asked them to revoke our amateur licenses as we were operating on Ebon in an illegal manner. The Trust Territory answered Don's letter denying anything was illegal with our operation but stating that they had information that Don Miller had been monitored illegally operating within the Trust Territory saying that he was on Cormoran Reef.

Hal, during this controversy, I had a QSO with Ack, W4ECI. Ack told me that he had seen the documents that Don and Chuck had from the Government of Ecuador giving them permission to operate on Ebon. I wrote to the Government of Ecuador and asked:

1) Does Ecuador claim Ebon Atoll as sovereign territory of Ecuador?

2) Does Ecuador feel that it has authority to issue permits for amateur radio operation on Ebon Atoll?

3) Did Ecuador issue a permit to Chuck Swain or Don Miller to operate an amateur radio station on Ebon Atoll using the call K7LMU/HC8E?

Hal, I received a letter back from the head of the communications department in Ecuador saying "No" to all three of these questions. I have received communications from several amateurs in the Pacific Areas who claim that their beam headings were wrong on several of the QSOs with both Don and Chuck.

All of this above information, and more, was transmitted by various sources to the ARRL. As a result the Ebon Atoll operation by Chuck and the Cormoran Reef operation by Don are being declared illegal and will not count for DXCC. The ARRL officials told me of additional complaints that they have had on Don Miller's operation, while I was at the ARRL Hqs recently.

As I told you over the phone, please read again the editorial in the June 1966 issue of QST. Hal, I think you will agree that the incidents just listed paint a picture of a ham who is so anxious to work DX that he oversteps the limits recognized by the normal law-abiding radio amateur.

Iris and I simply cannot be associated with an organization that also sponsors Don Miller in any way.

Just before we left VRlZ for the USA, Don Miller contacted us over the air and the possibility of our selling him our Collins gear (after Ted and Chuck were lost) came up. A sked was set for a day later. We found out that it was impossible to fly our radio gear out of VRl (we eventually sent it to England by boat). During our sked the next day Don Miller made us feel like we had done him the greatest injustice in the world - that we were not anxious to help out the DXer - that we were going back on an implied promise to him - etc, etc. I tell you this story Hal, because I am sure you now find yourself in the same position,

Anyone who even thinks of helping Don and then does not, is made to feel that he is mean and uncooperative. Iris and I know that all you have tried to do for Danny, us, or Don has been in the best possible spirit, and we are very sorry that you are once again put in the spot of being in the middle of a controversy not of your choosing.

Hal, I am sure that you will agree that it is wonderful for YASME to be in the black for once. Let's keep it that way, and when the money is spent, we must do it so that all the one-dollar contributors will be happy. A possible program that would help all DXers and still not get us involved in the deep way that YASME and Ack have been in the past is for YASME to loan the basic equipment for one or two DXpedition stations, with rough rules as follows:

1) Loan is for limited time - for specific DXpedition - with supporting information supplied and desired by YASME,

2) All QSLs for DXpedition will credit YASME for the equipment it supplied.

3. The persons in the DXpedition will sign promise to return the equipment in as good condition as received.

What do you think of such a plan? Unless we hear from you to the contrary, Hal, we will assume that YASME is not going to have donations for Don Miller made through YASME, and we sure hope that you can understand and appreciate why we feel this must be so. - Lloyd Colvin, Iris Colvin.

The June 1966 QST editorial mentioned in the above letter was ARRL's way of laying the groundwork for the announcements to come:

"DXPEDITIONS-A CAUTION"

In recent years expeditions to remote areas have provided lively interest for the DX-minded among us. Properly organized and conducted, they have been and can continue to be high spots in the challenging game of DX.

We must express a warning, however, concerning a more recent trend of laxity among some DXpeditioners as to the validity of their exotic-area licensing documents. There have been several incidents in which amateurs (mostly U.S., but some others as well) were too eager to accept the word of a local or minor official as to operating authorization, without bothering to determine whether such people had the power to grant privileges. Such incidents are hardly a contribution to international relations, since they incense the legitimate amateurs and national societies in parent countries who play the game by the full rules. In administering DXCC affairs, the League cannot undertake individual and detailed investigation of the validity of each and every operation authorization; normally we must assume that such operations are legitimate, unless there is a warranted challenge of legality. In a few of the latter instances, credits have had to be withdrawn.

Perhaps in most cases, the DXpeditioners were acting in good faith, knowing of the haphazard procedures which (it must be admitted) exist in some remote areas, and so specific instances are best left unstated here. We can better understand the feeling of our foreign brethren by a fictitious but parallel example in the reverse direction: How would we - or our government! - view the operation of a Lower Slobbovian amateur in the Virgin Islands on the basis of a Harbormaster's okay - without reference to FCC?

Organizing a DXpedition - or any plans for operation on foreign soil-should include meticulous attention to the authenticity of licensing documents. Failure on the part of voyaging amateurs to assure themselves on this point can destroy much of the healthy international relationships most DX activity helps create, and might ultimately cause a reappraisal of the granting of new-country status to DXpeditions.

Miller also claimed to have operated from the Laccadives Islands. The Colvins, while in Senegal, wrote on 23 June 1967 to K.R. Phadke, the assistant wireless adviser to the government of India,

An American radio amateur, Mr. Don Miller, recently operated from the Laccadive Islands using the call VU2WNV. The American Radio Relay League later questioned certain aspects of this operation and a meeting of the Board of Directors was held by the ARRL with Don Miller present.

On 13 May 1967, after the above mentioned ARRL Board of Directors meeting, Mr. Don Miller sent a newsletter to some 3,000 radio amateurs throughout the world saying: 'Even with the aid of letters from the Indian Government, the (ARRL) Committee was unable to demonstrate that VU2WNV was illegal or unauthorized from the Laccadives. This call was issued specifically in response to a request for a permit for the Laccadive Islands.'

On the 16th of June 1965, we request permission to operate our amateur radio station in the Laccadives. On 9 July 1965 (file No. W22 (6) 65)

you said that it was not possible for us to operate in the Laccadives, but that operation from some other part of India would be considered. On 20 July 1965, we made application to operate in India proper. All necessary forms were submitted together with application fee. By letter 24 September 1965 (file No. W 22 (5) 65) you stated that our application had been received and that we would be informed of the decision in due course. No further correspondence from you has been received.

In view of Mr. Don Miller's world-wide newsletter, we hereby request that our pending application for permit to operate in India be changed to read "Permit to operate in the Laccadives." Please assign our Laccadive call-sign in advance, as was done for Mr. Don Miller, and rush it to us by air mail, since we are ready to leave for the Laccadive Islands as soon as the call is received". - Lloyd Colvin W6KG, Iris Colvin W6DOD.

In November 1967 the ARRL issued a white paper on DXCC that would lay the groundwork for actions to come, including disallowing some Don Miller operations as well as tightening DXCC rules for DXpeditions. The white paper was from the ARRL Awards Committee, seven active hams on the ARRL Headquarters staff who would study and make recommendations on controversial matters such as DXCC and contest disqualifications, particularly regarding gray areas in the rules. The ARRL said its particular concern was that DXpeditions "be conducted on the highest plane of ethics and good sportsmanship, as well as legality, since they represent the image of U.S. amateur radio to the rest of the world."

Directly inferred in the white paper was that any questionable DXpedition would reflect poorly on DXCC, which reflected poorly on the ARRL, which would in turn be a black mark on the entire worldwide amateur radio community and perhaps even endanger continued as well as future frequency allocations to the Amateur Radio Service.

Miller had been accused by some DXers as having "selective hearing," in ignoring calls from operators who had not contributed sufficiently to his operations. The League said it saw nothing unethical in preferential mailing of QSLs following an operation, but that "selective hearing" might very well "violate not only the rules of DXCC but probably FCC Rule 97.111, which says "An amateur station shall not be used to transmit or receive messages for hire, nor for communication for material compensation, direct or indirect, paid or promised."

"Only the Federal Communications Commission, of course," the ARRL said, "may take disciplinary action as it may find necessary in the administration of its rules. ARRL's function is the proper enforcement of its own DXCC rules."

The Awards Committee said that it had received "many complaints" about Miller operations and that although evidence was often circumstantial, the volume of complaints had come to a head. Two particular Miller operations were cited; the first was 1S9WNV from the Spratly Islands in October 1965, in which the main concern had been solicitation of funds and selective hearing. The committee said there was "indeed evidence that Dr. Miller had refused to work certain stations, most of them on the [DXCC] Honor Roll, at least in the early stages of the Spratly activity. As stated earlier, refusal to work a particular station or stations is an operator's privilege, and the evidence that this 'list' substantially comprised non-contributors was inconclusive."

Miller's claimed operation from Ebon Atoll (and Cormoran Reef) was the second that caused "major concern," even though both had been deleted from the DXCC Country list almost immediately. The committee said Miller had "misrepresented certain foreign consulate activities" to the Awards Committee, and that the letter from the Office of the High Commissioner, Trust Territory of The Pacific Islands, which has jurisdiction over the areas involved, supported charges that Miller was never on either island. Although Miller's Ebon Atoll operation had been disqualified on the basis of license invalidity, his never having actually set foot on Ebon Atoll still was an "unavoidable consideration."

Other charges against Miller were accumulating (a letter from the U.S. Coast Guard said no permission had been granted for Miller to land on Navassa Island, and a letter from India said that the government was at that time not permitting any amateur radio operation from the Laccadive Islands), but Ebon Atoll stood out. When the Miller saga played out - when Miller sued the League -- even though Ebon Atoll was no longer on the DXCC list (for reasons having nothing to do with Miller; it simply did not meet the criteria for a DXCC country), Ebon was not a factor. But it was a great factor to Lloyd and Iris Colvin, who actually had been there, had operated in good faith and under proper (U.S.) jurisdiction.

The Awards Committee white paper announced that Miller's membership in the DX Century Club was suspended, and that "no credits would be given for future contacts with Dr. Miller operating from DX locations until further notice."

Setting the tone for future DXpeditions, the committee said it hoped to be able to "prescribe a method, such as photocopies of logs and accompanying affidavits, which will permit the reinstatement of credit from locations where the DXpedition was properly authorized."

In the years of his DXpedition operations, Don Miller appeared often in the U.S. at club meetings and conventions. He was intelligent, poised, well-spoken, and could be charming. On the air, he is remembered nearly 40 years later as one of the all-time great CW operators. But, as the ARRL white paper noted, Miller's operations had polarized DXers - many believed him. Miller made life miserable for several dedicated League staff members, and helped drive a wedge between the ARRL and CQ magazine, which supported Miller right up to the end, a division that would take years to heal.

On 9 June 1967, Miller had appeared at a meeting of the Northern California DX Club. Hugh Cassidy, WA6AUD, remembered 27 years later that it was perhaps the largest attendance ever at a NCDXC meeting, with about 90 DXers there. As Cassidy later said, 1967 "was a time of change, although this is probably realized more in retrospect than it was at the time. DXing was going through a wrenching internal soul-search, the ARRL being driven into a major study of its corporate navel; things never would be the same again. It was Don Miller who brought about the changes - he and no one else. He was a superb operator ... just ask any Old Timer about Don, but don't stand close: some are still angry, and they intend to stay that way. They were there and they are the ones who bear the scars of those years. It was a terrible time and even today some of the elders have difficulty in talking about it."

QST for October 1968 closed the Miller story. Miller had first threatened to sue the ARRL in early 1967, when the Awards Committee began investigating charges against him. He finally did file a lawsuit in February 1968, against both the League and personally against its general manager, John Huntoon, W1LVQ. The suit was settled out of court on 15 June 1968.

37· Colvins' First DXpedition Concludes

We are not in favor of [a proposed no-code U.S. license]. Though many say CW is a thing of the past and we don't need it, it never seems to disappear completely. - Lloyd Colvin

In December 1965 Iris and Lloyd were back on Majuro, in the Marshall Islands Trust Territory, and trying to get permission to operate from the Territory of Papua and New Guinea. But a letter from the Australian Postmaster General's department of posts and telegraphs to the authorities at Port Moresby said their application of the previous summer had been "far from complete" and that they would have to apply again (radio licences for Papua New Guinea were issued from Melbourne). It's too bad - the Colvins never did visit or operate from Papua, the site of the destruction of Danny Weil's YASME I.

Instead, they went to Tarawa, in Gilbert and Ellis Islands (now W Kiribati) where their operation included 700 contacts in the ARRL DX Competition, 'phone. Their first QSO was K6EBH, the last K6RZX, and they received DXCC #18,716, on 19 April 1979. VR1Z ran from 4 to 13 February 1966

Iris and Lloyd lugged a Collins S-Line on this first expedition, along with the famous TH-3 tribander, doublets, and a special telescoping pole they had bought in England. They said that "22 other amateurs from various countries had applied for permission to operate, but we were the only lucky ones to be granted such authority. " In the 1970s, when the Ellice Islands became the separate country of Tuvalu, the Colvins would credit their 1965 visit as "a big factor" in convincing the authorities to let us operate on the first day it became a new country.

Back home in the spring of 1966, the Colvins tended to their family business and thought about their

Benin

Azores

DXpedition future. They were very interested in putting rare DXCC countries on the air - visiting as many countries as possible and making DXCC from all of them was a goal that would come to them much, much later.

"Rare" DXCC countries at the time included so-called U.S. Department of State "restricted travel areas." In 1966 these included Albania, "Communist China," Cuba, North Korea, and North Vietnam.

The State Department made exceptions for journalists and scholars, i.e., for "persons who are professionally engaged in obtaining and disseminating information to the public and who can demonstrate that the purpose of their proposed trip is legitimately related to these activities." Such a person could get passport validation (a visa) by providing proof of one's credentials and intent from an editor, publisher, or the like. A visitor also would need an entry visa from the country in question, something State could not help them in getting.

Lloyd and Iris drafted a letter (5 April 1966) to the State Department; it is not clear if it was actually sent, as there is no reply in their files.

> We are actively engaged, [they said], in world-wide radio, scientific, and educational development, under sponsorship of the YASME Foundation, a non-profit amateur radio organization. We started our full-time world-wide travels, under YASME sponsorship, last August and have traveled in eleven countries in the Pacific area since then. Prior to this, we have traveled in approximately 90 countries on scientific and journalistic travels throughout our lifetime.
>
> We request that this application be considered under both the current State Department policy for travel of journalists and scholars, and, the current

State Department policy for travel of Doctors or Scientists.

In the countries that we visit, we establish our amateur radio station and teach the people of that country basic principles of radio communication and the great value of talking via amateur radio to other peoples of the world, thus greatly improving world-wide understanding and appreciation of the problems of people throughout the world. The value of amateur radio during public health emergencies is stressed. We are regular contributors to amateur radio publications throughout the world and as journalists have done much to inform the general public of conditions existing in the many countries visited.

We are both graduates of the University of California and are scholars in the field of radio and communications. We have already done much to improve international relations and feel sure that our travels in Albania and Communist China will similarly result in improved relationships and understanding between these countries and the USA.

The Colvins were planning to go to England, where they would use the home of Jack Drudge-Coates, G2DC, as a base of operations. They were still hopeful of putting on the air rare, or new DXCC countries - in this case, Rockall, a hunk of rock sticking out of the Atlantic Ocean off England's coast.

A letter from J.M. Hern, G3NAC, in April 1966 discussed a Rockall operation. Hern, a flight lieutenant, described himself as "Royal Air Force Amateur Radio Society Expedition Officer," and was plotting a Rockall

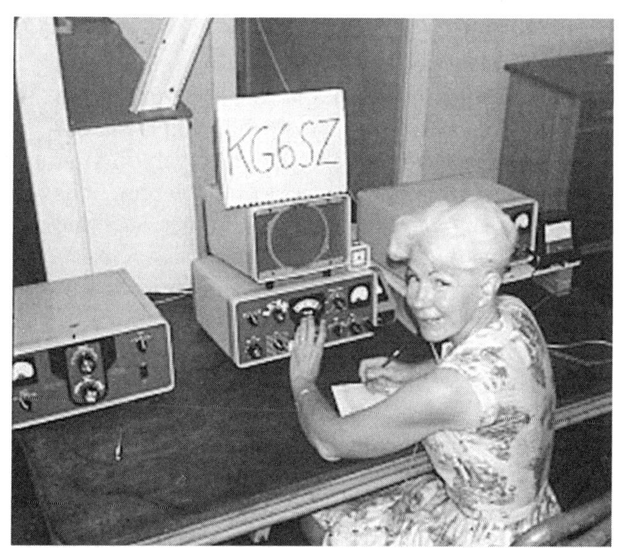

Marianas

operation. He said his group would be going to Rockall on "her Majesty's ships" and he could get permission for Lloyd to accompany them, but not Iris.

Hern said Rockall was "a very dangerous proposition." The rock is only 70 feet above sea level, only two people and equipment can get on the top at once, and the waters are treacherous. Hern only expected to be able to operate about 10 hours, and put their chances of even getting there at 30 per cent.

The Rockall operation never did happen. A year later Hern wrote to Lloyd, who was in Senegal at the time, and was not encouraging. He cited 'powerful storms" and said there was only one place to stand, a ledge 66 feet high. As for equipment - "How will we support a mast? I can get the generators (and spares) but suggest [a] 14AVQ in a bucket of cement is the answer. Please let me have your comments as soon as possible."

Using their new British amateur radio licenses (G5ACH for Lloyd and G5ACI for Iris), from 6 to 30 May 1966 the Colvins operated from the Isle of Man. Used both GD5ACH/W6KG and GD5ACI/WB6QEP, and this is the last time they used both their call signs from a country. Lloyd made DXCC and about 3,000 QSOs. (GD5ACH, DXCC 10,237, 25 November 1968.) (GC5ACI, DXCC 10,247, 17 March 1969.)

Next operation was GJ5ACH, Jersey Island, and 3,000 contacts (first QSO HA0HC, last HB9KB). GJ5ACH DXCC was #10,259, 4 December 1968.

A "DXCC YL" listing in QST had failed to include Iris, but the magazine made up for it in June 1966, saying "The DXCC YL listing printed last October should certainly have included KL7DTB/6, Iris Colvin, who had [DXCC] endorsements for over 200 countries under that call. Since September, 1965, Iris and her OM, Lloyd, who have been on a worldwide DXpedition sponsored by the YASME Foundation, have worked DXCC from several locations (confirmations yet to be checked) as they visited such places as many islands of the Trust Territory of the Pacific - Saipan, Yap, Truk, Majuro and Ebon. Their next DXpedition will find Iris also trying to work DXCC YL [women operators in 100 different DXCC countries] which she feels will be a most interesting but difficult thing to do."

Iris and Lloyd went to tiny Sark Island for an operation 3 to 26 June 1966, racking up 3,600 GC5ACH/W6KG QSOs, the first with LA1KI and the last with GW6YQ.

The Colvins then sought DXCC status for Sark from the ARRL following their departure, based on the "distinctively separate administration" criteria, saying that "Le Dame de Sercq rules this Island like a dictator." They noted that de Sercq "has the authority to, and does issue orders that few other rulers in the world could enforce," and that she had banned automobiles; female dogs (except her own); and houses visible from the sea shore.

In 1974 the Colvins again sought DXCC status for Sark, citing the same arguments. ARRL General Manager John Huntoon, W1RW, replied that "We still can't get away from the fact ... that insofar as we can determine, there are only two separate administrators, one for Jersey and one for Guernsey & Dependencies. If there is an additional one for Sark, we have yet to learn of it.

"In other words, as we see it, the Dame (or successor) is responsible to the Privy Council, which includes the aforementioned administrators."

The Colvins also were looking at Sudan as a destination. John H. Garrett, OD5EE, wrote to them from Beirut to say that the only licensed amateur in Sudan was one Jim Collins, ST2BSS, a call allocated to the Boy Scouts of Sudan, but not active on the bands. The Colvins were instructed to contact Collins, who could make the proper overtures to the authorities. Collins said Lloyd and Iris were welcome to stay with them and to use the equipment there: an S-Line. Garrett suggest they "mention in [their] letter the free publicity the Boy Scout movement would receive by the use of Amateur Radio to publicize their activities, etc."

Gambia

Azores

Iris was issued the call sign WB6QEP in 18 July 1966, finally enabling her to shed the KL7DTB/6 call.

Before leaving for home, Iris and Lloyd operated as ZB2AX from Gibraltar, from 9 August to 4 September 1966, making 4,500 QSO, and making DXCC #10,585, issued on 12 June 1969. First QSO was DJ9NX and last was G3FMN.

In the summer of 1966 the Colvins still were looking to operate from rare DXCC countries. They made another inquiry to Albania, citing U.S. Senate Bill 920 (28 May 1964), and also followed-up on inquires to Bechuanaland, Uganda, Gabon, the Voltaic Republic (Burkina Faso), Guinea, and Somalia. The response from A.S. Arraleh, director general, "Republica Samala, Ministero delle Communicazioni e Transporti," is typical of the bureaucratic doublespeak they often faced:

> With reference to your Application dated 4th August, 1966 in which you sought permission to operate your Radio Amateur while you are in our country, and to convey the approval of your request. This is however subject to the provisions of the ITU convention 1959 and other rules and regulations covering Radio operations.
>
> This temporary licence is liable to withdrawal at any time if considered necessary in the national interest. Moreover, you are requested to submit all in formations or data in connection with your Radio Amateur with which you want to operate in Somalia before commencing operation.
>
> Finally, formal approval will be given to you when the requisite data of your Amateur Radio is

received by this office.

After repeated inquiries to Spain in 1965 and 1966 for "request for permission to operate our small amateur radio station while visiting Spanish provinces in Africa," Lloyd traveled to Madrid "for the sole reason of visiting the principal Spanish offices involved in the granting of permission to operate an amateur radio station in Spanish areas of Africa." Two telecomm officials said no.

The Colvins also had hopes of putting Indonesia on the air. In 1966 the country was in great turmoil, only Indonesian citizens were granted amateur radio licenses, and few were active on the air. A letter from the American vice consul in Djakarta, Robert C. Stebbins, told the Colvins that an operation by them would be "illegal." He cited Don Miller and Chuck Swain, who claimed to have operated from Indonesia, and said that his office had been "unable to verify any such operation, or even the fact that they ever were in Indonesia. It would appear that any queries we might be able to make to the Indonesian government at this time [on the Colvins' behalf] would be counter productive.

"The political situation has not yet completely stabilized and foreigners entering Indonesia to carry on activities of this type are automatically subject to suspicion. Last year we were only able after much effort over a period of seven or eight months to obtain the release from jail of a missionary who was accused of carrying on unauthorized radio transmissions, in addition to other charges made against him."

Iris and Lloyd operated from Madeira Island from 15 September to 11 October 1966 as CT3AU, making 5,600 QSOs (first CT3AS, last VK3XM). Their DXCC was #10,184, 25 October 1968.

From 17 October to 12 November 1966 the Colvins operated CT2YA from Azores, for 5,300 QSOs (first UA1DZ, last WA9DQS) and earned a certificate for the CQ WW Phone Contest and DXCC #10,284, dated 20 December 1968.)

In January 1967 Iris and Lloyd traveled to Virginia and a meeting of the Virginia Century Club, to receive the club's annual VCC Award. Previous recipients of the award, which began in 1962, were Gus Browning, W4BPD; Ack Ackerson, W4ECI; Stuart Meyer, W2GHK; and Don Miller, W9WNV, together with the late Chuck Swain, K7LMU.

The Colvins were back in the U.S. not only to get caught up on Drake Builders business, but also to fine-tune the YASME Foundation and their relationship with it. Lloyd and foundation treasurer Ed Peck held a meeting and vote was taken, by mail, "in order to present the proper picture should the tax exempt nature of YASME Foundation be examined by the Internal Revenue Service." The IRS filing for the year ending 30 June 1965 showed no activity; the filing in June 1966 said the foundation's income was $1,728.75. A few hundred dollars had been spent for printing QSL cards. $132.80 was spent for radio equipment which was sent to Lloyd - replacement tubes.

Those voting by mail included President, Danny Weil; Vice-President, Harold A. Sears, K5JLQ; Secretary, Robert B. Vallio, W6RGG; Treasurer, Edward F. Peck, W6LDD; and directors Jack M. Drudge-Coates, G2DC, Charles J. Biddle, W6GN; Golden W. Fuller, W8EWS; Richard C. Spenceley, Sr., KV4AA; and Francis K. Campbell, W5IGJ. They approved the following:

(a) IT IS RESOLVED that the Treasurer be authorized to purchase radio equipment from cash on hand and that said equipment will be loaned to amateurs approved by the Foundation who are going on DXpeditions, and who agree that the confirmation cards, or QSLs, issued for contacts with said DXpeditions will give printed

The Colvins' home in Richmond, California. (Photo by Leon Fletcher)

acknowledgment to YASME FOUNDATION for supplying the equipment used, and further providing that said amateurs using said equipment will first sign a written agreement agreeing to return said equipment immediately upon completion of the DXpedition in good condition, reasonable wear and tear excepted.

(b) IT IS RESOLVED that YASME FOUNDATION continue to sponsor the DXpedition of Lloyd and Iris Colvin.

After two months in the U.S. the Colvins began a swing through Africa, from Senegal, making 4,400 6W8CD QSOs from 22 February to 14 March 1967. (First HI8LAL, last W9JT). 6W8CD DXCC #10,351, 29 January 1968. While there the Colvins made inquiries to Togo and the Republic of Biafra, which had seceded from Nigeria early in 1967 (and remained a secessionist state until 1970. They wrote to Biafra that "At this young and formative period for Biafra, the free world-wide and highly favorable publicity which we can and will give is a thing to be highly desired....we are ready to depart immediately for Biafra." A further inquiry in late September, while the Colvins were in Accra, Ghana, also went unanswered (Biafra was never a DXCC Country).

From 21 March to 9 April 1967 it was 5T5KG, Mauritania, and 5,600 QSOs (first DJ5IO, last G3KOR). They earned a certificate for 1,111,350 points in the CQ WPX Contest and DXCC #10,263 (5 December 1968).

The Colvins then moved to the Gambia for a ZD3I operation from 8 May to 2 June 1967; 6,400 QSOs (first WA9MJE, last EL2D) and earned DXCC #10,243, 29 November 1968.

The Colvins returned to Dakar, Senegal in June and July, and from there made inquiries to Guinea and Cameroon.

They then went to Sierra Leone for a 9L1KG operation from 29 July to 20 August 1967, for 4,600 QSOs (first WB2KTB, last K6CCM), and a WAE CW Certificate for 114,005 pts and 3rd place multi-op non-Europe. Their 9L1KG DXCC #10,266 was issued on 5 December 1968.

Ghana

How We Started Out Building Our Home in Our Spare Time and Went On to Make a Million Dollars in the Construction Business

Peter Dodd, G3LDO, was in Sierra Leone at the time, working as a communications and instrument engineer for Sherbro Minerals, a U.S. company. He remembers:

> The Colvins were operating from the Brookfield Hotel, Brookfield, Freetown. I received a message from Freetown, an S.O.S. from Lloyd that his power supply had blown up. I don't think that there were any hams living in Freetown so there was no one else that he could contact. As luck would have it I was due to go down to Freetown on one of my regular maintenance checks of the radio equipment in Freetown, so I was able to take the [power supply] from my home-brew SSB transceiver, which Lloyd was able to use to fix to his Collins equipment, which showed he was more than just an appliance operator. [They were using a Collins 516-F2 power supply.]
>
> I took a picture of the [the Colvins'] QTH with thebeam in place and I felt that the photo was important so when I went to Dayton a few years ago I made a copy, hoping to give it to someone who might find it interesting. I was intrigued with the way they carried the beam around - it was in a cylindrical container about 5 feet long when broken down, as I recall.
>
> The place where Lloyd and Iris operated from was originally called the Government Rest House and was a relic from the colonial days. I stayed there myself for a couple of months before being allocated a bungalow.
>
> I originally went to Sierra Leone as the police force communication officer and I finished up with a bungalow just round the corner from Brookfields Hotel. The AC supply was very bad at the hotel and the voltage used to surge up and down, which

caused the electrolytics in my PSU (the same one I lent to Lloyd) to all fire off in rapid succession one Sunday morning. The repaired and modified PSU was relatively surge-proof. In fact the AC supply was so bad that the fluorescent lights in the dining room used to sometimes go out during the low part of the voltage cycle.

Lloyd and Iris treated me to a meal at a local restaurant/nightclub. I got the feeling that Iris enjoyed the mini break from ham radio.

The next operation, 5L2KG from Liberia (23 August to 19 September 1967) netted 5,300 QSOs and a certificate in the WAE Phone Contest. The 5L2KG DXCC #10,240 was dated 27 November 1968.

Iris later told a club meeting that their equipment had been shipped separately to Monrovia and they arrived at the airport with just their suitcases. The customs officer asked if they had any radios. Iris, hoping to leave the problem of clearing their ham gear to when it showed up, said they had "Only a small receiver, 4 by 5 inches.

"Show it to me!" the Customs officer said, in an authoritative voice. Iris said "I searched through my suitcase and could not find the receiver. I looked in Lloyd's suitcase - no radio. The officer still insisted that he must see the radio. Finally, I explained that it must have been packed with our other luggage sent by air freight. He insisted he must see everything, but until he had looked at all the radio gear he didn't know what to do. Finally, he said "Pass, take everything through."

Iris also remembered some "unusual interference" during that operation. "Loud music and chanting became louder and louder as people slowly approached our QTH. Not being able to operate, we went out to watch and follow along with a funeral procession. About 10 people clad

Ghana

Liberia

in black robes carried the coffin and many people followed along, making loud and mournful wailing. Reaching the cemetery, they continued their prayers and final goodbyes to the deceased.

"There seemed to be some question as to who should be the one to pay the final respects. As we watched the pallbearers began to argue and soon the mournful occasion ended in a fist fight."

DM2BON in East Germany was the first QSO from TU2CA in Ivory Coast (29 September to 17 October 1967); JA2KKG was the last. DXCC was achieved.

From Ghana, operating 9G1KG 21 October to 3 November 1967, Iris and Lloyd made 1,700 QSOs, plus many more in the CQ Worldwide Phone Contest (900,726 points and a certificate). First QSO was W6EUF, last was W9VNE. DXCC #10,342 was dated 18 March 1969.

Their inquiry to Togo paid off with operating permission and the Colvins opened up as 5V1KG for a 7 to 16 November 1967 operation. First QSO was 5N2AAX, last was W6SR, out of 3,000 total. DXCC #10,311 was dated 6 January 1969.

The last stop on this west Africa tour was TY2KG, Dahomey (now Benin), running 1 to 26 November 1967. Aside from a 721,994-point effort in the CQ WW CW Contest, they made another 1,400 QSOs, the first with OK2KBH, and the last with WA6ZQU.

Time Out For a Book

When it comes to books, "write what you know," the saying goes. Iris and Lloyd knew construction, and, somehow, they found time to write a book. It was published by Prentice-Hall, in 1967, with the breathtaking title "How We Started Out Building Our Own Home in Our Spare Time and Went on to Make a Million Dollars in the

Construction Business.

The Colvins had returned to the San Francisco Bay area in 1961 (from Alaska) and in the next 10 years they built a number of apartment houses and at least two hospitals, including the Carlson Convalescent Hospital. Daughter Joy had graduated from U.C. Berkeley in 1962 and moved to New York, where she married Richard Gilcrease in 1969. Iris and Lloyd were presented with their first grandchild, Justine, born on July 8, 1971. Shortly thereafter, the Gilcreases moved to California to work with the Colvins in construction and property management. Grandchild Number Two, Vanessa Gilcrease, was born in November 1972.

Iris and Lloyd lived in a three-story triplex, that Lloyd built, at 5200 Panama, in Richmond. This was the scene of most of the famous annual Fourth of July parties that Lloyd and Iris gave during the 70s, 80s and 90s. Although this house was sold after Iris died in 1998, as of 2002 the large free-standing pole and yagi antennas are still there.

Joy says "Richard and I lived a street over on Sacramento, and the back yards were adjacent. The Colvins also owned several adjacent structures on both streets - a small hillside kingdom."

Although Prentice-Hall no longer has records, it was reported that upward of 100,000 copies of the "Million Dollars" book were printed/sold. This writer had no trouble locating copies of the book, via the Internet, from used book sellers. Despite having what may be the longest title in book publishing history, Million Dollars is a good piece of work, and just about everything it said in 1967 applies in 2002. Iris and Lloyd wrote a foreword, titled "What this book will do for you," and here it is:

> This book will tell you how to do what we did, mostly in our spare time - we built our own home, and went on to make a million dollars in the construction business. The first half of this book will show you how to build your own home, at a price

Ivory Coast

Mauritania

> which will allow you to sell it at a nice profit. The last half of this book will tell you how to organize and run a medium-to-large construction company, in a manner that should assure you a million dollar - or more - profit.
>
> Our success in the building business was attained during the twenty-five year period from 1940 through 1965. This was a period in which inflation prevailed and the value of the dollar declined. This is the atmosphere in which the building business thrives. With considerable assurance, we feel confident in predicting that the next twenty five years will see continued overall inflation and further decline in the true value of the dollar. It has been reported that there is only one major problem facing the United States Government on which all members of Congress are in agreement; no matter what political party is in power, Congress is in unanimous agreement that some further inflation and decline in the value of the dollar seems inevitable.
>
> [In 1940, we were stationed in Anchorage, Alaska; Iris was employed by the Government accounting department, and Lloyd was First Lieutenant in the United States Army. Our total savings amounted to $1,000 (equivalent to about $2,500 today). Neither of us had any special training in construction. We were renting a small apartment at fairly high rent. We were both 25 years old; we had a child, a one-year-old daughter. We decided to build our own first home; we took all of our savings and bought a lot. In order to build our home, it was necessary to borrow money from the bank, to pay for materi-

als and to do much of the work of actual construction.

We hired one qualified, older carpenter, and the three of us did most of the carpentry and painting of the small, two-bedroom house that we built from plans copied by us from a magazine article. Much of the construction was done in evening hours and on weekends. For the first time, we learned what such things as floor joists, studs, headers, and plates are. Also, we encountered our first experience with subcontractors (plumbing, electrical wiring, grading) and immediately we found out, that the prices for sub-contract work vary greatly.

Soon after this first house was completed and we were nicely settled in it, World War 2 started, and we were transferred, by the Army, to another part of the world. We sold the house that we had built ourselves, at a nice profit (although we received only a relatively small down-payment, and took a note and mortgage for the rest of the payment). We repeated this same building and selling process on our own homes twice more during the war, but scarcity of building materials and workmen did not permit us to build at several places where we lived.

A few years after World War 2 ended, we were transferred to New Jersey. We immediately bought a lot and built our own small home. After our very first home in Anchorage, we no longer did much, if any, of the actual construction work; we sub'ed everything out to sub-contractors. Immediately after finishing this house and moving into it (less than three months from purchase of house until

Togo

occupancy), we discovered that we could sell it easily for a very nice profit, if we wanted to. This house was in an area surrounded by subdivisions, being built by large construction companies on a mass-production basis. For the first time we realized that somehow or other we had built a home, as a single-house enterprise, that could be sold on a competitive basis with the so-called big home builders who built medium-priced, tract homes for sale to the masses. You can do the same, now or anytime. In the first half of this book, we will tell you why this is so and how you can do it.

Flushed with the knowledge that we could compete with the big builders, we entered on our first subdivision project. It was very small, to be sure, consisting of only six houses in a 12-house subdivision, of which six other houses were being built by another small builder. We found out that there is a considerable difference between building one house and six houses all at once, as a subdivision. We encountered our first real problems with unions, lending institutions and government agencies. These problems, and how to lick them, are partially explained in the first half of this book, and in greater detail in the latter half. It should be stated here, however, that it is because of these problems, and the attendant extra expenses that go along with being a "big builder," that you can build one home of your own, even your very first home, at a price as low as that of the big subdivision builder. You must, however, follow the simple rules and recommendations given in this book.

To continue with our own history, from 1950 to 1961, we continued to build our own homes wherever the Army sent us. In addition, in several locations we built small-to-large subdivisions in the nearby areas. Most of the actual supervision of construction was done by Iris, with Lloyd spending his spare time from the Army in assisting with plant, site locations, financing and sales.

In 1961, Lloyd retired from the Army as a Lt. Colonel. From 1940 to 1961, approximately $300,000 had been saved, principally from our construction activities, which had been run, more or less, on a spare-time basis.

In 1961, for the first time, we decided to devote our full [time] to the construction business. A survey was made of areas where construction was booming; California headed the list. From 1961 to 1965, we built several millions of dollars worth of apartment houses, hospitals, and other large buildings. In 1965, at age 50, we retired for the second time, with a net worth of more than $1,000,000. The last half of this book tells you how you can do likewise, and how you can keep much of your profit, without giving it all to the government in taxes.

We would like to stress that, even though the construction business is considered one of the most risky businesses in the world (very high failure rate), if you follow closely the recommendations in this book, you will most likely not fail financially; you may not make a million dollars, and you may make more than a million dollars; it is our considered opinion that most likely the worst that can happen to you, on any single construction project, is to break even.

Isle of Man

Sark

We wish you careful reading and good building. - Lloyd Colvin and Iris Colvin.

Following their travels and operations in 1965, '66, and '67, the Colvins would keep the home fires burning for the next eight years. During that time the YASME Foundation was dormant, as well. But an article in the January 1969 issue of 73 magazine, "The YASME World-Wide DXpeditions," served as a reminder of what had transpired and hinted of things to come. Why was this major article published in 73, and not in QST or CQ? Iris and Lloyd were longtime, loyal ARRL supporters and, indeed, their interest in DX depended on the ARRL's DXCC award. In 1969, QST did not pay for articles (73 and CQ did), but payment would have been inconsequential to millionaires, at any rate.

"The YASME World-Wide DXpeditions," the article began, "have given many a ham his first contact with a rare country during the last decade and a half of amateur radio. To some amateurs YASME means a ship; to others it stands for a mysterious group of rich amateurs; but to the majority of the amateurs of the world, it stands for DX. Everyone who has had anything to do with YASME has been interested in DX."

The article was a press release for the YASME Foundation, but it also was a primer on DX. 73 editor and publisher Wayne Green, W2NSD, had published the Danny Weil stories in 1956, in CQ, and the 73 article recounted those days, as well as the genesis of the YASME Foundation, "a group of influential hams interested in DX," the article said.

The YASME DXpeditions so far mentioned [through Dick McKercher's 1963 eight-country Africa trip] were conducted on the basis that amateurs worked were urged to make a small contribution with their QSLs to help defray a portion of the costs. This was, and still is, a highly controversial subject. Some amateurs feel no one in amateur radio should ever send money to help a DXpedition, while others feel a small donation is more than justified. In any case, the Directors of YASME found that in order to keep DXpeditions such as Danny Weil's going they had to put a lot of their own money into the Foundation. This became a discouraging and expensive procedure and the Directors decided to let the YASME Foundation become dormant unless someone wanted to pay their own way on a DXpedition and let the YASME Foundation help in other ways than direct financial responsibility.

The article then introduced the Colvins as the white knights, saying that "During the DXpeditions made by Lloyd and Iris, some donations have been received by the YASME Foundation. About half of the money so received has been used on QSLing expenses. The balance is held by the YASME Foundation."

Lloyd Colvin had toyed with the idea of making at least some of his and Iris's travel costs tax-deductible by becoming a professional writer - by publishing DXpedition stories. QST would never have considered an on-going series but Wayne Green already had a track record of publishing such work. Lloyd and Green did discuss such an arrangement, both before and after the Colvins joined forces with the YASME Foundation, but nothing came of those talks. For the Colvins, it would have been a joint effort - Iris, the artist, was the better writer of the two. It just never happened.

Another Colvin venture was begun in 1969 - the "World QSL Bureau." It was in Lloyd's genes to be a methodical keeper of records, undoubtedly honed by two decades in the Army, one of the greatest of bureaucracies.

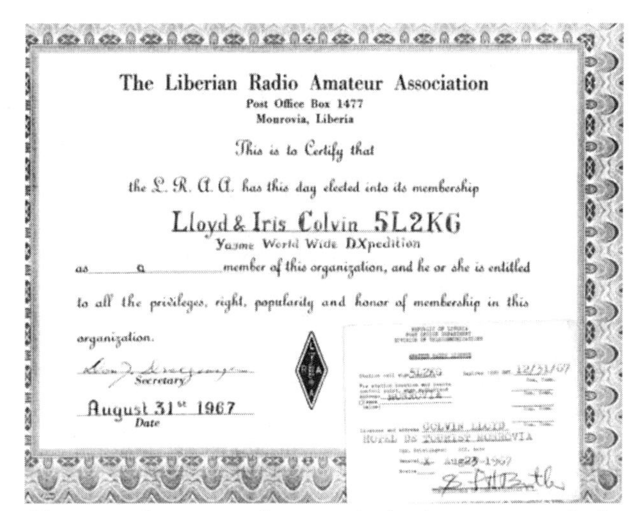
The Liberian Radio Amateur Association
Post Office Box 1477
Monrovia, Liberia

This is to Certify that

the L. R. A. A. has this day elected into its membership

Lloyd & Iris Colvin 5L2KG
Yasme World Wide DXpedition

as a member of this organization, and he or she is entitled to all the privileges, right, popularity and honor of membership in this organization.

Secretary

August 31st 1967
Date

Almost to the very end, every single slot on every single line of his log book was filled in (which was necessary because the logs were subsequently "cut up" and pasted on QSL cards). And of course QSLs ruled; the Colvins sent them and they collected them.

Their scheme for the World QSL Bureau was to act as an out-going agent for QSLs, anywhere in the world. Send them to Bureau and they would be forwarded, at six cents each. "We distribute QSLs by combination of direct, other bureaus, and clubs. The majority of QSLs are sorted, processed and mailed out within 48 hours of receipt. All QSLs are checked for possible QSL managers and are sent direct to such QSL manager with an SASE included for reply via the ARRL QSL Bureaus. We utilize every possible aid to operate the World QSL Bureau in an accurate and efficient manner, including the use of computers, postal stamping machines and mechanical equipment. We are presently handling more than 1,000,000 QSLs a year. Our goal is to handle all the QSLs in the world."

Although the Bureau was advertised for about a year, no records exist of its operations. Around the same time an outfit calling itself the "Continental QSL Bureau" operated as a clearinghouse for QSLs within the U.S., but it ran afoul of U.S. postal regulations.

Both Colvins became very active on the air from their new home station in California. Lloyd, a die-hard CW man, even applied for 'phone Worked All Zones in November 1969. They liked the ARRL Sweepstakes Contest, too (no DX involved) and, when the ARRL removed signal reports from the contest exchange of information, Lloyd mounted a campaign that lasted several years to get signal reports reinstated saying "it is of little use to exchange QSLs for a contact without RST."

Lloyd was still occasionally involved in Army National Guard activities, serving as a communications adviser. A letter of appreciation in 1971 from the commanding officer of the California Army National Guard 249th Signal Company thanked him for his help in the phone patch program during their annual training at Ft.

Irwin, California, for his "time, equipment, and personal expense."

In 1972, Iris and Lloyd fed their need to make radio contacts by operating in both weekends of the ARRL International DX Competition, 1.2 million points and Division Leader on 'phone and 836K points on CW. In 1973 W6KG racked up 269K multi-single in the CQ WPX SSB Contest, 807K m/s in the ARRL Phone DX Contest, and 1 million points m/s in the ARRL CW DX Contest.

In November 1974 Iris and Lloyd did a W6KG multiop, 517K points, in the CQ WW CW Contest; in 1975 they teamed up for 600K in the ARRL phone DX Contest and another 486K in the ARRL CW DX Contest.

On 1 January 1973 Iris took over as the ARRL's W6 QSL Bureau Manager, which was in some disarray at the time. She served for one year.

One of the highlights of the annual DX convention in Fresno was a bus ride there for DXers in the San Francisco Bay area. The biographical handout for bus riders in April 1973 described Iris and roasted Lloyd:

Senegal

"W6DOD - Iris is one of Not-Very-Many YLs to make the (DXCC) Honor Roll. She is past president of the NCDXC, did a great job as manager of the Six-Land QSL Bureau. Her husband, what's his name, is active in most of the contests. Gives the Fresno Bus a touch of class by riding with us every year.

"W6KG - Lloyd is Iris's husband."

The West Coast DX Bulletin (High Cassidy, WA6AUD, editor and publisher) published its poll of Most Needed Countries in May 1973. The poll was conducted by John Irwin, K6SE/2, a professor in the Department of Earth and Planetary Science at Newark State College in Union, New Jersey. 123 WCDXB readers responded. Below are listed the 50 most-needed DXCC countries. Note that four of the top 10 are not, in 2002, on the list (deleted years before) - Tibet, Saudi Arabia/Iraq Neutral Zone, Kamaran Islands, and Sikkim. The number following the country name is how many poll respondents out of 123 "needed" it.

Sierra Leone

1 Tibet 87
2 S. Sandwich 85
3 Iraq 85
4 Clipperton 84
5 Bouvet 81
6 China 78
7 Saudi/Iraq Neutral 73
8 Kamaran 70
9 Mt Athos 70
10 Sikkim 69
11 Burma 60
12 South Yemen 49
13 Malpelo 44
14 Spratly 44
15 Somali 41
16 Tromelin 38
17 Heard 32
18 Geyser Reef 31
19 Albania 31
20 Cocos Keeling 29
21 Blenheim 28
22 Bangladesh 27
23 Palmyra 27

24 Congo Republic 27
25 Glorioso 27
26 Crozet 26,
27 Farquhar 25
28 Bhutan 24
29 Formosa 24
30 South Shetland 24
31 Cambodia 24
32 Qatar 23
33 Laccadives 22
34 Port. Timor 21
35 Macao 20
36 South Georgia 20
37 Abu Ail 19
38 Rio do Oro 19
39 Trindade 19
40 Juan de Nova 18
41 Chad 17
42 Ceuta/Melilla 15
43 Des Roches14
44 Republic Guinea 14
45 Agalega 13
46 Tokelaus 13
47 Brunei 12
48 Egypt 12
49 Crete 12
50 Central African Rep. 12

In the early 1970s amateurs in the Bay Area often shopped at Amrad Supply, in Oakland. A handout for the general public that Amrad kept on their counter gives a pretty good picture of the state of ham radio public relations in that era:

> The field of radio is a division of the much larger field of electronics. Radio itself is a broad study that is still further broken down into a number of smaller fields, of which "shortwave" or high-frequency radio is a small part.
>
> The largest group of persons interested in the subject of high-frequency communications is the more than 350,000 radio amateurs located in nearly all countries of the world. Strictly speaking, a radio amateur is anyone interested in radio non-commercially, but the term is ordinarily applied only to those hobbyists possessing transmitting equipment and are licensed by the government.
>
> Amateur radio is a scientific hobby, a means of gaining personal skill in the fascinating art of electronics and an opportunity to communicate with fellow citizens by private short-wave radio. Amateur radio is a fascinating hobby with many phases. So strong is the fascination offered by this hobby that many executives, engineers, and military and commercial radio operators enjoy amateur radio as an avocation even though they are also engaged in the radio field commercially. It captures and holds the interest of many people of all walks of life, and in all countries of the world where amateur activities are permitted by law.
>
> Amateurs have rendered much public service through furnishing communications to and from the outside world in cases where disaster has isolated an area by severing all wire communications. Amateurs have a proud record of heroism and service in such occasions. Many expeditions to remote places have been kept in touch with home by communication with amateur stations on the high frequencies. The amateur's fine record of performance with the "wireless" equipment of World War I has been surpassed by his outstanding service in World War 2.
>
> By the time peace came in the Pacific in the summer of 1945, many thousand amateur operators were serving in the allied armed forces. They had supplied the Army, Navy, Marines, Coast Guard, Merchant Marine, Civil Service, war plants, and civilian defense organizations with trained personnel for radio, radar, wire, and visual communications and for teaching.
>
> Basically, to become a "ham" licensed by the U.S. government requires learning the International Morse Code and taking a test in receiving and transmitting and learning Amateur Radio Regulations and basic electronics theory.

Mauritania

The thrill of sitting in one's own home and being able to talk to people in all parts of the world, including the Communist world, without the use of wires, and using equipment that in some cases is home built, is a feeling that is hard to beat.

A Break from DXpeditions

We always have [our radio equipment] out in full view. That way airport security guards, airline crews, even custom agents rarely bother us. They can see we're not trying to sneak aboard anything that's illegal. - Lloyd Colvin

The ARRL Foundation was incorporated on 21 Sept 1973 and Lloyd was one of the original directors, elected by the ARRL board of directors to a two-year term as one of four "public" directors (the other five seats filled by members of the ARRL [board of directors] as provided in the Foundation by-laws.

The foundation's charter described its mission, "To operate exclusively for charitable, educational and scientific purposes entitling the corporation to exemption under the provisions of section 501c(3), and more specifically, to study and contribute to the development of amateur satellite programs and other innovative programs related to the purposes of The American Radio Relay League, Inc." It existed to raise money for special projects.

Lloyd's resume for the foundation said "1961-Present: President and General Manager of Drake Builders, a California Corporation operating as general contractors building subdivisions, government buildings, office buildings and hospitals. Have directly supervised the building of approximately $100,000,000 of new construction since graduating from University of California, including construction in New Jersey, North Carolina, California, Alaska, Canada and several foreign countries.

Also in 1973, Iris and Lloyd bought John McLean of California, a United States entity of John McLean of London, England, and about a year later changed the name to Drake Builders. In the coming years Drake Builders would build several million dollars worth of buildings, many of them Housing and Urban Development low-cost apartments.

Lloyd decided to apply for appointment as a commissioner of the Federal Communications Commission, sending a letter to the White House on 15 January 1974, saying "I feel that I am fully qualified to hold such position and believe that I can bring to the Commission a wide field of experience that will benefit the Federal Communications Commission and the nation. Lloyd cited his background in both military and civilian communications and extensive world travel. "I have had a reputation,

Liberia

all of my life, for hard work and the ability to successfully complete all projects undertaken," he wrote.

Lloyd's letters of reference were from a local lawyer, an Oakland estate and business planner, who wrote "I understand there are no Electrical Engineers on the Commission at this time. Mr. Colvin is an Electrical Engineer and could bring technical proficiency to the Commission," and from YASME Foundation treasurer Ed Peck, who had been a communications officer in World War 2 and was a lieutenant colonel in the reserves in 1974.

An appointment (which did not come) to the FCC would have required a move to Washington and the job certainly would have interfered with the Colvins' travel, so perhaps it's just as well.

Iris and Lloyd made enquiries to licensing authorities in Burma, Iraq, and China. In the case of the latter, not long after President Nixon had visited China (following back channel negotiations by Henry Kissinger), Peking University told Iris and Lloyd that "We have received your letter of 31 Oct 1975 to Mr. Wong-Lin-Lung. I am sorry that we cannot satisfy your request under present conditions. We believe as long as American and Chinese relationships will carry on as stated in the Shanghai Declaration the friendship between the peoples of these two countries will develop further in the future."

In the interview in the October 1981 73 magazine, when China was Number One on everyone's DXCC "needed list," Lloyd said he had sent Kissinger a letter about ham radio and asking if he could help them get into "Red China." Eventually someone in the state department replied with the names of several people to write to in China. The one response received was the one mentioned above.

Lloyd told 73 that "Though [the Chinese] they are not

permitted to contact the outside world yet, it is my opinion that the day is coming.

Iris said "The Chinese don't want their people to contact the outside world. That's the stumbling block. If there were some way we could overcome that... we might get somewhere. Right now, they are just afraid to have their people have any contact with the rest of the world."

On 16 November 1975 the YASME Foundation surfaced after several years of obscurity, A press release once again announced that "YASME Sails Again!" Iris and Lloyd would hit the trail, in the Pacific, and back to the Ellice Islands, where they had operated in the mid-1960s as VR1Z, to open the brand-new country of Tuvalu when it became independent, as Tuvalu, on 1 January 1976.

They did not announce any further itinerary after Tuvalu.

"In the interest of saving valuable operating time," Lloyd and Iris asked for not more than one contact on the same mode on any band, and that callers not ask

"What is your call?"

"Where are you going next?"

"When will you switch from CW to SSB (or vice versa)?")

"When are you going to operate on some other specific band?"

and "Why don't you call by districts?"

The YASME Foundation had just reconstituted itself, naming the following: President, Don Wallace, W6AM; Vice President, Danny Weil; Secretary/Treasurer, Bob Vallio, W6RGG; Directors, Tom Taormina; WA5LES, Dick Spenceley, KV4AA, Dick McKercher, W0MLY, Charles "Rusty" Epps, W6OAT, Nobuyasu "Nob" Itoh, JA1KSO, Martin "Martti" Laine, OH2BH, Lloyd Colvin, W6KG, Iris Colvin, W6DOD, and Dave Duff, VK2EO.

On the Air, Colvin-Style

When it came to on-the-air operating, Iris and Lloyd were slow, steady, and methodical. With three possible exceptions - Tuvalu, Abu Ail, and Mozambique - they operated from DXCC countries that were not particularly rare, enabling them to crank out their standard 6,000 to 9,000 contacts over two or three weeks at a pace well below frenetic. In the early days of their travels Iris helped with CW and Lloyd helped with 'phone operating; later they each would stick to their favorite mode. Joe Reisert observed "one thing for sure, like working Gus Browning, you were sure you were the station they were working as they signed calls both coming and going!"

In the years 1965 to 1993 DXing changed dramatically in several ways, including more and more use of split frequency operation, and so-called "list" and "net" operations. The Colvins' operations just caught the beginning, in the early 1990s, of the explosion of Packet Cluster spotting of stations. These techniques all put more pressure on DX stations to work faster and faster. Lloyd was asked in 1987 about lists and nets. "They are important in some situations," he said. "I'm not against them. We prefer to work on our own, but if conditions make lists or nets necessary, we do them. "

New DXers often had to be educated in DXing techniques, something the Colvins did plenty of. They had infinite patience when it came to giving out a new country to a neophyte DXer ("if you don't have it, it's rare").

Lloyd said he could work CW at "a little over 40" words-per-minute, but that he usually kept it around 25 wpm for the sake of the newcomers.

Sometimes it's hard to remember what country you're in, and what your frequency limits are. Lloyd told 73 magazine that "It's not so much keeping up but rather dealing with the annoyance of being in a country where you can't operate on a frequency that you did the previous month in another country.

For instance, there are some countries now that allow their amateurs only a 100-kHz segment of 80 meters. During our operations, it is a great temptation to go where the Americans are operating on 80 meters, but often we are unable to do this, and we don't."

Asked in 1987 about computers, Iris said "We use a computer for accounting in our business, but I don't think we'll ever use computers in our hamming." One reason: they did not want to add still more weight and gear to their travels - even with a laptop (still a real luxury in 1987) they would have had to carry plenty of paper logs as back-up.

73 asked Lloyd in 1981 about the new no-code license, and he said "We are not in favor of it. It seems to me that we must not lower the standards of amateur radio to those of CB. We have a little something special with the code requirement. Though many say CW is a thing of the past and we don't need it, it never seems to disappear completely.

73 also asked "Do you feel that in this era of liberalization it is difficult to maintain standards in amateur radio?" and Lloyd replied "To a degree. However, I don't go along with those who say that the conditions on the bands are worse than in years gone by. As I look back at my early days in the hobby, and I started in 1929, I'd have to say that it was just as bad or worse back then.

"People have just forgotten.

"In 1929. when two people in the same city got on the air they QRMed each other so badly that neither could hear anything, even if they were on opposite ends of the band. It's simply not true to say that things are worse today than in the old days.

"As far as politeness goes," Lloyd said, "we are in no worse shape today than we were in 1929. When I first became a radio amateur, we had 35,000 hams in this country, that's all. We have ten times that number today, and while there are more troublemakers around, the ratio is just about the same as it was in 1929.

"I don't really think that conditions today are any worse than when I started."

In 1987 it was still politically correct to ask about women hams as a group. Iris said "I think that women are becoming better operators. I usually try to contact the YLs, and I've noticed recently that the YLs are much more self-confident in pileups. They talk slowly, seem morc poised, and in general are capable of making contacts in good style."

38· Pacific Islands Operations, 1976

Lloyd and Iris have friends in every part of the world, having met and talked with them on the air. - Fiji Islands newspaper article

The Colvins for a time called their operation from Tuvalu their "greatest DXpedition." They were the first to put the new country on the air, and from 1 to 18 January 1976 they made 6,600 QSOs (the first person there to work them on New Year's day was none other than Dick McKercher, W0MLY; the last QSO was with K5JVF). They were the only outsiders on the island during the three-day celebration of the island's independence; the population had increased from 500 to perhaps 1,000 for the celebration, with visitors and dignitaries from surrounding islands. They did not apply for DXCC from VR8B.

The Tuvalu Committee decided that at the stroke of midnight on the first their GEIC (Gilbert and Ellice Islands Corporation) call signs would change from VR1Y and VR1Z to VR8B and VR8C.

While they waited, Iris and Lloyd operated as VR1Z from 16 to 31 December, making 3,700 QSOs and DXCC #18,716, dated 19 April 1979. First QSO was K6EBH and last was K6RXZ.

They had been allowed to enter the island only after showing that they had reserved accommodations. Their hotel on Fanafuti, the only one on the island, had five rooms, the others occupied by visiting dignitaries. The weather being so warm year around, the windows were all permanently open and sounds, especially Iris's voice when she operated 'phone, carried across the island.

At a club meeting in California

Boarding on Western Samoa

This would not do; the Colvins were putting a new country on the air - how could they operate all night without interfering with the various conferences in progress in the hotel? They solved the problem by using the only space that was completely closed, the bathroom. There was hardly enough room for the equipment, let alone the operator.

They had traded their Collins S-Line for smaller and lighter equipment: the solid-state Kenwood "pair," the R599 receiver and T599 transmitter, and a Heath SB-230 linear amplifier with solid-state final output. This equipment worked perfectly for them through their entire swing through the Pacific.

Next stop was Suva, Fiji Islands, where the Colvins were able to operate in both the ARRL CW and Phone DX Competition weekends (about 1,000 QSOs in each). Signing 3D2KG from 28 January to 23 February 1976 they logged a total of 7,500 contacts in 113 countries, and this seems to have been the first operation where they kept a running countries-worked total. The first QSO was with WA8JUN and the last QSO was with UA1OBY. Their 3D2KG DXCC was #18,079, 18 September 1978.

Openings into Europe were "fantastically good" and produced more than 1000 contacts. They loved the long path opening to Europe from 0630 to 0830 GMT, which was followed occasionally by a short path opening from 0900 to 1100 GMT. During the contest weekends they worked lots of W and VE stations on both 10 meters and 80 meters, but not many outside of the contest periods. Contests work! They made a Worked All Continents in 28 minutes on 20 Feb, on 20 meter phone: TU2GA, I0IJ, ZP5LX, JA7SGV, AH3FF, and W7TX.

The local paper published an article about the Colvins' visit, saying "Visitors to the Fiji Islands during February 1976 include two world-known amateur radio operators, Lloyd Colvin and Iris Colvin, from San Francisco, USA.

The Colvins are devoting their time to their lifelong hobby of amateur radio. They have traveled in 122 countries and carry with them a small amateur radio station which they set up and converse, by both Morse Code and voice, to many of the million amateur radio stations owned by private citizens throughout the world. In Fiji, the Colvins operate under the call 3D2KG, which has been issued to them. There are several local radio amateurs, called "hams," including 3D2RM, 3D2EY, 3D2CM, 3D2AJ, and 3D2ER.

Lloyd and Iris are staying at the Metropole Hotel, have erected a light-weight aluminium beam on the roof, and, with their small radio receiver and transmitter communicate with other "hams" in every part of the world. During the few weeks in Fiji, the Colvins have talked to some 4,000 other "hams" and made contact with 60 different countries.

The Colvins express the hope that more Fijians will become amateur operators. A technical and operating knowledge of radio, as well as an ability to send and receive the dots and dashes of Morse Code, is required in order to obtain an amateur license. No commercial traffic is allowed, but the hobby develops technical skill and spreads good will to other countries. Lloyd and Iris have friends in every part of the world, having met and talked with them "on the air."

Lloyd and Iris plan to leave here soon and operate from another island of the Pacific. They will meet in person the local hams which they have previously met via radio and get acquainted with people of the islands.

Certificate of Appreciation

Presented to LLOYD COLVIN

in recognition of your address

before the Kiwanis Club of Port Vila, New Hebrides

Your contribution to our club is deeply appreciated.
We hope this certificate will serve as a lasting memento
of this pleasant occasion.

President — Program chairman — Date: 14th May 1976

The Colvins, seasoned world travelers by now, were fascinated by the island nation of Nauru. Phosphate-rich Nauru reportedly had the highest per-capita income in the world in 1976 and would be rich until the mineral deposits ran out. The government was trying to promote tourism, to protect against that day. The Colvins said each Nauru citizen was getting about $10,000 a year, to do nothing. In fact, they said, "it is the family, rather than the individual, that is credited with the money, and then only if the phosphate being mined comes from land belonging to that family. The government is investing the peoples' money for them. Although they insist on being called a Republic, It is actually one of the most socialist states [we] have ever seen," they said.

Iris and Lloyd had had permission to operate from Nauru when they were on the Marshall Islands 10 years before, and had chartered a boat to get there, but it was not allowed to dock because of "conditions in the "harbor". This time, Dave, C21DC, had agreed to sponsor them and said there was plenty of room in the hotel to live and operate. If the Colvins could not be issued a call sign, they would still be able to use C21NI, the club station call.

So they flew from Fiji to Nauru on Nauruan Airlines, in the most luxuriously appointed aircraft they had ever seen. The flight was cheap, too, one way the island was working to encourage visitors.

The Colvins were expected to stay at the hotel, another government venture to promote tourist income for the island. On the plateau behind the hotel, a large swimming pool and recreation area was under construction. Dave offered to let the Colvins stay with him, but that would have interfered with his own radio operating, so they declined. They could find no "ideal" spot for their antennas, so after two nights in the hotel they located a place on the north end of the island where they could live with a native family. "This gave us a much better opportunity to know the people and see the island."

They met a Nauruan who offered to let them use one of his eight cars and they did, but insisted on paying him for it. He insisted - showing us the 7 other cars that he had for his own use. We accepted his kind offer but also insisted upon paying for the rental of the car.

The Colvins gave up on getting their own call sign and instead operated C21NI from 1 to 25 March 1976. Once again they were able to work both the ARRL CW and DX Phone Competition weekends (in those days, that contest was two weekends for each mode). They made about 1,500 contacts on each weekend, and totaled 7,500 C21NI contacts in all, the first with JA6DHI and the last with UB5JBY. They worked 116 DXCC countries and their DXCC was #17,887, 28 June 1978. They reported more "beautiful openings to Europe, including one day when bands were open continuous[ly] to Europe for 10 hours."

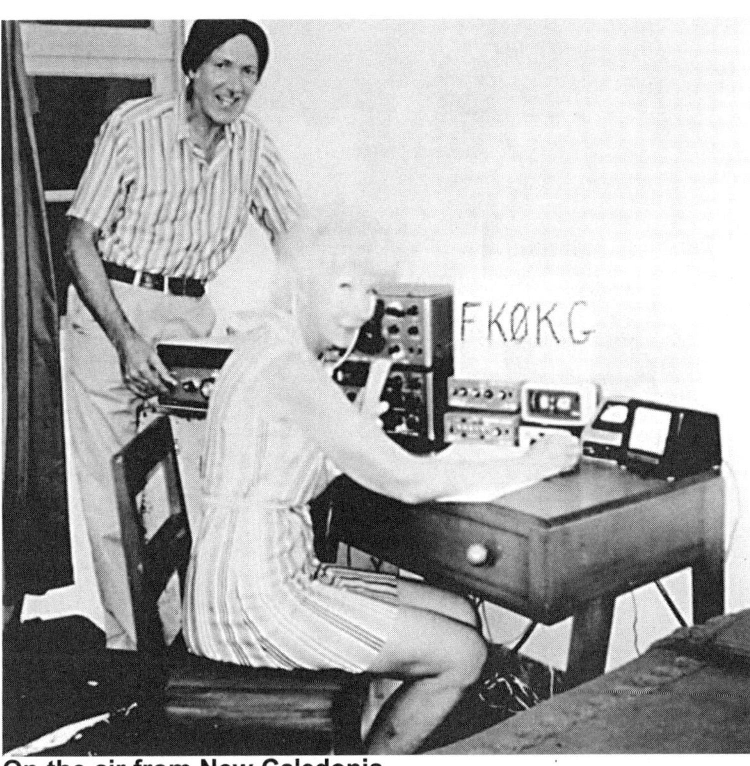

On the air from New Caledonia.

Iris and Lloyd met three Japanese hams who were making a one-day stop on Nauru: JA0CUV/1, Tack, JA3KWJ, Kazy, and JA2PJC, Hotsui. They were on their way to operate from Tuvalu and New Hebrides. The Colvins gave them a tour of the island. The Japanese told them about an operation, C21DX, a couple years before, and why the island's president now refused to issue any more amateur calls to visitors.

It seems that a "very long and descriptive account" of the C21DX operation had appeared in a Japanese ham magazine. Unfortunately, the article also mentioned the multiple wives of one of the ministers on the president's staff. The president wasn't happy about this and decided not to issue any more call signs to visiting hams.

Another local, named Bill, was operating at C21NI and was trying to get a call sign of his own, but even that seemed to be in limbo.

The only other legitimate ham, besides C21DC, was Ken, C21KM, who operated "maritime mobile" from the Nauruan ship the Enna G, a government project to establish a luxury cruise business for tourists.

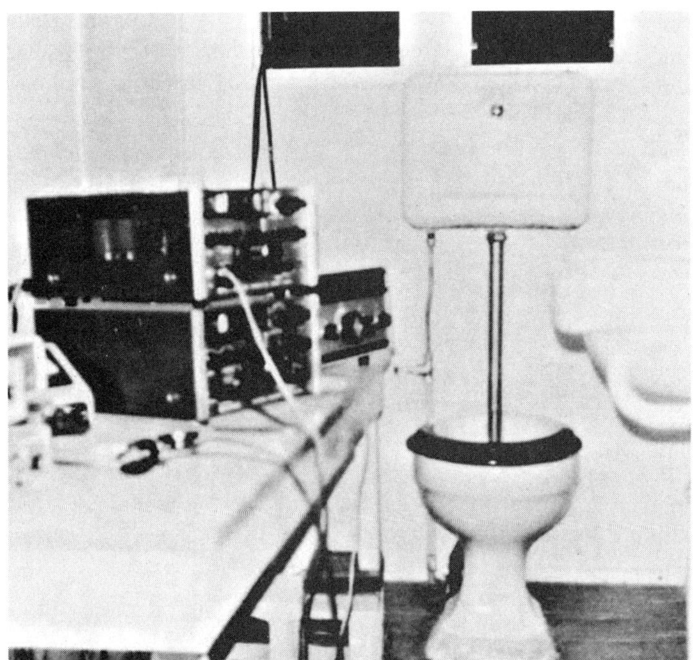
The seat of operations from Tuvalu.

Lloyd and Iris found, it seemed to them, that most of the administrators, planners, school teachers - and even laborers and workers - were foreigners. "The Nauruans are lazy and they," Iris wrote, "themselves, freely admit this. They have imported workers from other countries. There are large sections occupied by the foreigners, the Chinese, the Gilbert-Ellice, and the Filipinos, who live in their own separate area.

"When the British Phosphate Company became the Nauru Phosphate Co (when Nauru became a Republic) the British administrators remained to keep the operation going. Eventually - perhaps - the Republic can operate its own company, but this does not seem likely in the very near future."

Lloyd and Iris, veterans of World War 2, were intrigued by the remains of the Japanese occupation during the war: concrete pill boxes along the 12-mile stretch of beach, at regular intervals of a few hundred feet, and gun posts and look-out stations "built into the coral precipice which forms the interior plateau of the island."

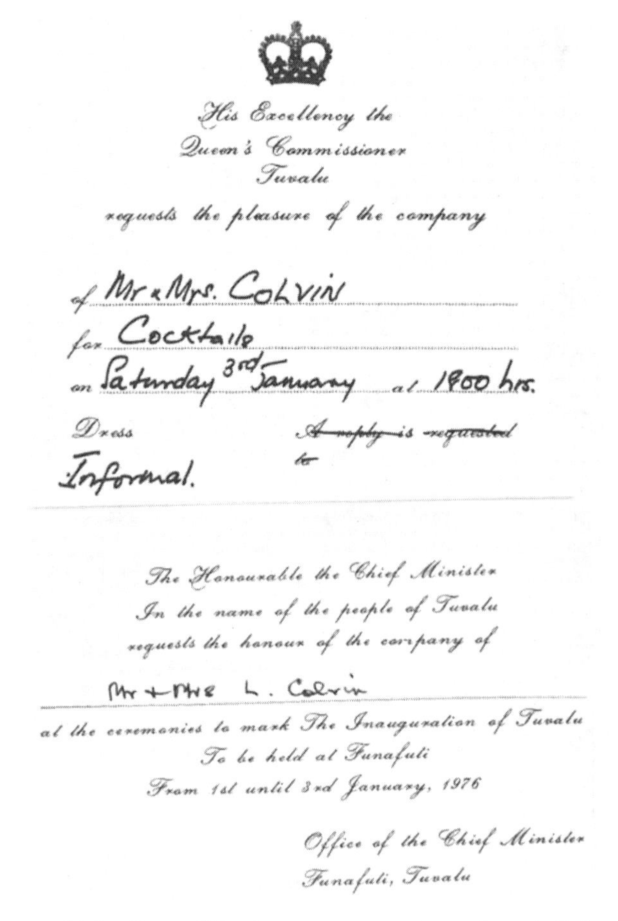
His Excellency the
Queen's Commissioner
Tuvalu
requests the pleasure of the company

of Mr & Mrs. Colvin
for Cocktails
on Saturday 3rd January at 1900 hrs.
Dress ~~A reply is requested~~
Informal. ~~to~~

The Honourable the Chief Minister
In the name of the people of Tuvalu
requests the honour of the company of
Mr + Mrs L. Colvin
at the ceremonies to mark The Inauguration of Tuvalu
To be held at Funafuti
From 1st until 3rd January, 1976

Office of the Chief Minister
Funafuti, Tuvalu

At the time of the Japanese occupation almost all the native Nauruans were deported to other islands. In 1976 they were returning, to establish their claim to their portion of the phosphate income. "The wife of one Nauruan official," Iris said, "became extremely indignant when I asked her nationality. Her skin coloring and features were distinctly European. She spoke both English and Nauruan. Her reply to my question was positive and with a tinge of anger 'I am Nauruan.'

"At the same time there is a trend to keep foreigners from becoming residents. The husband of the native lady where we lived and operated our station, is not a native. The government has refused to allow him to stay on the island as a resident. He has only a temporary visa and must leave the country every six months. Then he can return on a new temporary visa. He was in Australia at the time of our visit, a fortunate situation for us, however, because we set up our station in his huge office which was not being used during his absence."

The Colvins packed up their gear, ready to head for the New Hebrides, and an entirely different learning experience. They had arranged "by radio" in advance, with written permission to operate there. They had rented an apartment owned by YJ8DE. With tickets in hand, they were ready to leave, but there was no place on Nauru to get a visa. The French Charge d'Affairs on Fiji had said no visa would be required for a 30-day stay in the New Hebrides.

With all seemingly in order, Iris and Lloyd and their hundred pounds of radio equipment were ready to fly to the New Hebrides. With C21DC and C21NI there with their wives to see them off, they took off, on Nauru Airlines (naturally) with gifts of beads, leis, and wreaths. "We felt as if we were leaving very close friends."

Approaching the island, they were told they could not take fresh flowers on to the island, so the airline hostesses got the leis and wreaths. YJ8DE, Reece, and his wife, Jean, met them on the ground. Given the choice of two entry lines for Customs the Colvins chose the British (the other line was French). While they displayed inoculation records, passports, and so on, shot records etc. Reece tagged along and let them know that the utilities had been turned on in the apartment and everything was ready. Their gear was coming out of the plane. So smooth....

Iris wrote that "The first hint of trouble occurred when they asked about a Visa for New Hebrides. We told them that the French consul in Fiji had assured us that we did not need a visa here (but we were in the British line). Next they asked for a ticket out of New Hebrides. I showed them other open tickets which we had purchased from various other places but unfortunately none of them was from New Hebrides. We told them that we wanted to buy a ticket from New Hebrides to Wallis Island, but this could not be done in Nauru because no one knew the amount to charge for the ticket. We showed them letters of credit that 'every other' country would accept as bond but nothing would satisfy them and they informed us that we must leave on the plane on which we had arrived. Lloyd was extremely angry and said he refused to leave."

Not to put thoughts in Lt. Col. Lloyd Colvin's head, but it should be remembered that shortly after the United States entered World War 2 in 1941, American military

forces established bases in the New Hebrides, probably saving that island chain by invasion by the Japanese. Not to mention that the U.S. military, in which Lloyd served for two decades, had saved Britain from Germany.

When the local police appeared on the scene, "we changed our minds about leaving, but we felt even more angry," Iris wrote. The police insisted they buy tickets to New Caledonia, and the Colvins agreed to do so, but the police would accept neither their letters of credit nor two kinds of travelers checks. They dug out all the cash they had, as a last resort, both U.S. and Australian, but didn't have enough of either to pay for the tickets to New Caledonia! "They refused to let us pay for the tickets with part of each currency because they could not calculate the exchange rate and the situation was rapidly becoming impossible." By now the airport officials, the pilots, and passengers already aboard the flight for New Caledonia were pretty irate. "They finally found someone who could sell us tickets and with much confusion, they put our baggage back on the plane and we reboarded, with many disgusted stares and comments from the other passengers," Iris said. "The hostesses, however, were very pleasant and looked pretty in their hair, redone with the flowers from our leis.

"Thus, before we hardly knew what had happened, we were up and away, headed for New Caledonia."

At least the Colvins managed to get their radio gear off of New Hebrides on the same plane. They landed in New Caledonia and said "the taking of radio gear in and out of New Caledonia was easy and our licenses to operate were obtained rapidly" (maybe it was just relative). "All in all, it was a great place for a DXpedition. We had a good time, and band conditions, although erratic, were good."

FK8KG was on the air 31 March to 28 April 1976, and produced 7,500 contacts on 80 through 10 meters, and DXCC #18,029, 28 August 1978. The first QSO was with JA1AB, the last with FG7AQ. WAC on 40-meter CW

The Colvins' digs on Fiji Island.

came in two hours and five minutes: F9YZ, K6HMO, ZL2AMP, PY1ARS/4, EA8BF, and JA1BZU.

The Colvins were "amazed" to find Noumea to be "the largest and most modern town that we have visited in the Pacific Ocean areas (excluding Honolulu). It is a 'little Paris' with many beautiful homes, apartment buildings, modern department stores, traffic jams, etc. Some of the local city buses have true hi-fi stereo sound on them for the enjoyment of the passengers. The only problem with Noumea is that everyone speaks French. We know a little French but could have enjoyed the city more if our French was better."

Iris and Lloyd didn't hold a grudge against the New Hebrides and they flew back there for an operation that lasted from 4 to 30 May 1976. The apartment they had reserved a month before wasn't available but Reece, YJ8DE, had found another one for them. This time they had license in hand, too - YJ8KG.

Once ensconced on the island, Iris and Lloyd heard some news that helped explain their less-than-friendly reception a month before. It seemed that some four years before, an American named Peacock (the Colvins were told) had bought $12 million in property in New Hebrides and then sold it to various foreign buyers. "Evidently," the Colvins said, "he didn't know or had neglected to take out a subdivision permit. Either that or else the New Hebrides government did not like the idea of so much land being sold to foreign buyers or else the thought of the huge profits going to an American speculator.

"In any case the sale of the properties were declared illegal. Peacock had to return all of the money to the investors - and was thus thrown into complete bankruptcy. Peacock appealed to the government to issue a subdivision permit. This failed. He even appealed to the United Nations to intervene in his behalf. This also brought no results. He then tried to sue the government of New Hebrides. At present (1976), he is collaborating with the rebels who advocate a change of government policy. He is also operating a clandestine radio station on the Island of Santos. What is even more to the point, he is an American and the illegal station is using amateur radio equipment."

Tuvalu

Poor Lloyd and Iris on vacation in the Pacific to get away from their own construction business at home, and now this! They must have been amused by it all.

As part of the deal to get into New Hebrides and operate, the Colvins gladly declared all of their radio equipment, piece by piece, and promised to take it out with them. They also declared that they would "not visit the island of Santos under any circumstances. We also declared that we would not do anything that might incite a revolution." It might have been easier to operate from, as Lloyd would have called it 10 years before, "Red China."

For the record, YJ8KG made 6,500 contacts (DXCC #18,072, 13 September 1978), the first contact with JA3BXF and the last with G2BOZ.

At any rate, "once inside the country there were no problems and everyone was extremely friendly." They got an apartment that had just been vacated by Robbie, YJ8AN, whom the Colvins had known from their visit to the Gilbert Islands 10 years before.

Iris wrote years later that "When Robbie was staying in the apartment (the one we would have occupied if we had made it on our first try) he simply ran a feed line to antennas located on the roof. We had a lower floor apartment and decided to erect our antennas on the top of a cliff a few feet back of the apt. building. This necessitated transporting the antennas up the hill. I was slipping and sliding and Lloyd was getting a terrific kick over the comedy of the situation. He asked me if I didn't see how funny it was. I said no, I did not. Not at the time anyway. Now as I look back it really was rather humorous."

The Colvins had settled on using an ordinary, coax-fed dipole for 80 meters, going outside to lengthen or shorten it to switch phone/CW. Here in Vila, New Hebrides, Iris did a lot of the climbing of the cliff to handle this chore. "It not only became easier but I began to enjoy it," she said. Otherwise, it was a pretty good setup, with a good view of the harbor and, from the cliff, of the town.

Iris and Lloyd learned that Reece was a big man on the island, a business owner and the person who raised two large sunken ships, artifacts of which were displayed in a the museum in New Caledonia. His car carried license plate #1.

In club talks beforehand the Colvins had often been asked about New Hebrides by people who had been there during the war. Iris said "WB6MIM asked me to 'find out what happened to 'Vic' whom all the Americans knew in New Caledonia. I was skeptical about finding any information, because Noumea is now so grown and probably completely changed. I did find out about Vic Brial. He passed away about six years ago but he is still well known around the area. The Brial family are well known wine merchants in the islands.

"His brother Benjamin is still alive and owns the hotel on Wallis Island. This information actually helped us tremendously. The officials were not going to give us a visa for Wallis until we had definite reservations for a place to say. We sent a telegram to the hotel and mentioned Benjamin Brial's name. The hotel did not wire back but telephoned that we did have definite reservations. The authorities then had no further excuse, and issued the visa for Wallis."

Unfortunately, operating permission for Wallis never materialized, so they went on to American Samoa, then to Western Samoa, where they were unable to get licenses, then to Hawaii, and home, ending their 1976 Pacific tour.

That summer was another round of tending to business, including exploring the possibility of expanding into mortgage lending and brokering. They offered the assets of Drake Builders as backing.

Iris applied to the FCC for a two-letter call of her choice, under the new rule for Extra Class holders with more than 25 years experience. Her first choice was W6QL, a call formerly held by Jim Wells, a famous DXer who had died. Iris got the call, so, as the YASME press release announced, it was now "Lloyd, W6 King George and Iris, W6 Queen Lady.

Iris and Lloyd began a winter Caribbean swing in October 1976 in the U.S. Virgin Islands, making 5,400 contacts plus an 881,000-point multi-single appearance in the CQ Worldwide Phone contest. They operated 8-31 October 1976 (first contact FG7XJ, last LU3FFG) and notched DXCC #17,943, dated 28 July 1978.

Stopping next at the British Virgin Islands, the Colvins operated 15 to 28 November 1976 from Beef Island. ZP5XU was the first contact of 9,000 QSOs; KP4CM the last. They worked 124 DXCC countries, a new record for them from a stop, and made 2,000 QSOs and 1.015 million points in the CQ WW CW contest as a multi-single station. Their VP2VDJ DXCC, #17,854, is dated 15 June 1978.

In their short report to the newsletters they mentioned that "Since 1954, YASME DXpeditions have been held 67 times in countries all over the world. Each of these efforts were real DXpeditions with QSOs world-wide with thou-

sands of hams. No other DX organization has such a record," and they said that they had just received word that they both had been picked for the CQ DX Hall of Fame, "following in the footsteps of such famous DXers as Gus Browning, Danny Weil, Dick Spenceley, and other world known DXers."

YASME Foundation board members at the end of 1976 were Don Wallace, W6AM, president; Danny Weil, vice president; Bob Vallio, W6RGG, secretary/treasurer; Rubin Hughes, WA6AHF, QSL manager; and directors Tom Taormina, WA5LES; Dick Spenceley, KV4AA, Dick McKercher, W0MLY; Rusty Epps, W6OAT;. "Nob" Itoh, JA1KSO; Marty Laine, OH2BH; Lloyd Colvin, W6KG, Iris Colvin, W6QL; and Dave Duff, VK2EO.

39· The Caribbean and Central America, 1976-79

We still intend to visit your country and are most anxiously hoping that you will permit us to bring our small radio amateur station with us and operate it temporarily while we are there. - Colvin followup letter to the Yemen Republic

Lloyd at the Fresno DX convention, 1977

Anguilla was a different Caribbean experience, the island being a United Kingdom dependency. Iris and Lloyd were surprised that there was no town or city on the island, "at least in the normal sense" and no central electric power system. Most people had their own electric generators; the Colvins' generator broke down just before they were scheduled to leave, cutting short their VP2EEQ operation.

From 9 December 1976 to 2 January 1977 Iris and Lloyd racked up some 8,000 VP2EEQ QSOs (first DJ9KA, last K2MJM), in 123 countries, with "heavy concentration" on 3.5 MHz. Their DXCC is #17,778, 18 May 1978.

In their note to newsletters they said that the stores, banks and other establishments were widely separated around the island, and they found the people "very friendly and helpful. A radio amateur license is obtainable immediately upon payment of $25 U.S. The country is relatively flat, which is a welcome change from most of the nearby countries which are mountainous, making it difficult to find a good radio site."

8/1 8th December, 1976

Mrs. Iris Colvin,
5211 Sacramento Avenue,
Richmond,
California 94804,
U.S.A.

AMATEUR RADIO LICENCE.

This is to certify that a Licence has been granted to Mrs. Iris Colvin of California, U.S.A. to operate a Ham Radio on the Island of Anguilla for twelve (12) months from this date.

The Call Letters to be used under this Licence are: VP2 EEQ.

It is understood that this Licence cannot be used for transmissions other than from Anguilla.

May we wish you every success in operating your Station here, and hope that you will derive much pleasure from it.

With best wishes.

Yours sincerely,

A. N. Hodge
Ag. Telecommunication Officer

Iris and Lloyd ended 1976 with more than 70,000 QSOs, from nine different DX countries. Adding that KV4AA, a YASME director and former president of the YASME Foundation, finished the year with over 35,000 QSOs, all using his U.S. Bicentennial special call sign AJ3AA, they said it was "most likely that every serious DXer in the world made at least one QSO with a YASME station in 1976. We hope that 1977 will be as good."

1977 opened for the Colvins on St. Maarten, the Netherlands Windward Islands, where they made 7,500 QSOs (first WA2ZQB, last DM4VUG) from 8 to 31 January 1977. They rolled up a PJ8KG DXCC, #17,859, dated 16 June 1978 and worked 121 countries overall. While there their transmitter's power supply capacitors burned out. The Colvins got word back to former YASME publicity director Frank Campbell, W5IGJ, who airmailed replacements to them.

Next stop was Antigua, operating as W6QL/VP2A 4 to 27 February 1977. Another QSO record for the duo was set, 10,000 (first VP2MAR, last JA7CUM), including operation in both the CW and 'phone weekends of the ARRL International DX Contest. Antigua/Barbuda was the first island Danny Weil landed on during the cruise of Yasme I. Iris and Lloyd worked 126 countries on the way to DXCC 17,840, 12 June 1978.

In the ARRL Phone Contest the Colvins' 4,000 QSOs in 48 hours was a two-day record for them. Iris said that "We remember world-famous contest operator Jim Neiger, W6BHY, telling us that in one contest he made 4 contacts a minute for an hour. At the time we did not see how that could be done, but we did almost the same thing in the contest just concluded." They also worked all continents in 30 minutes on 14 mc SSB on 21 February 1977: UF6VAG, VK4AK, YV4YC, IK7RNH, XE1PDF, and ZS6DN.

Last stop on this Caribbean swing was the DXers' paradise of Montserrat, where Lloyd and Iris rented a villa and set another personal record, with 11,700 VP2MAQ contacts, including, again, operation in both the ARRL CW and Phone DX Competitions (in those days the Competition was two weekends each, CW and Phone, and there was a special category for "One-Weekend Expeditions"). First QSO was WA2DXX, last was WB9RMC. Staying for the CQ WPX SSB Contest, the Colvins won a trophy. Their VP2MAQ DXCC certificate is #17,903, issued 3 July 1978.

The 1976/1977 Caribbean tour marked a change in the Colvins' travels, to a more structured approach. Their strict itinerary ran like a finely-tuned engine, with all the needed paperwork in hand before they left California. The other major change was in handling their growing celebrity, especially their appearances at club meetings both at home and abroad. On earlier trips they (mostly Iris) had taken copious notes in order to make "speeches" later. But more and more they were learning to just hang out and wing it, to speak off the cuff and show their slides. Neither Lloyd nor Iris leaned toward staying home and turning out long articles for the magazines; their time was better spent on the road.

VP2MAQ, Montserrat

Anguilla

In April 1977 Iris and Lloyd again rode the famous "Fresno Bus" from the Bay Area to Fresno for the joint DX convention of the Northern and Southern California DX Clubs. The "bus brochure's" thumbnail sketch for Lloyd said "W6KG - Lloyd, King George, is back from his fabulous DXpeditioning in the Caribbean, and will be on the Sunday morning program with Iris and some of his slides. It's just great having you back, Lloyd!"

And, for Iris, "W6QL - Queen Lady of the Lloyd and Iris DXpeditions. During the DX contest Iris was right in there, picking out calls in those MASSIVE pile-ups, and she would stay right there with it, hour after hour. Good show!" At the convention, Iris and Lloyd were presented several awards, mostly for their contest performances from the Caribbean that winter.

Also in April, both Colvins spoke at the Dayton Hamvention DX Forum, and found time to complete courses in "Real Estate Principles" and "Legal Aspects of Real Estate" They were deeply involved in catching up in business matters but still found time to make several presentations at local clubs, of their recent Pacific and Caribbean operations.

Iris and Lloyd would not take off again until the fall of 1978. Meanwhile, looking for new countries to conquer, inquiries/applications were made to Clipperton (France), China, Albania, Burma, Kamaran Islands, Comoros, French Somaliland, Somali Republic, Ethiopia, Yemen Arab Republic, Aden (PDR Yemen), Cameroon, Congo Republic, Mali, Chad, Gabon, Mozambique, Tanzania, Bahrain, Oman, United Arab Emirates, and Qatar.

Some 75,000 QSLs awaited the Colvins in the spring of 1977; most had already been answered by YASME volunteers and the task of filing them had begun. Their QSL collection now numbered about 250,000 cards. Both

Lloyd and Iris were on the air from their Richmond home but limited to working new DXCC countries. They kept their ears pealed to the two-meter Northern California DX Club net.

"The best part of our world travels," they wrote in the spring of 1977, "has been meeting people and making friends. The amateur radio operating was great, but takes second place to the experience of meeting people and making friends world-wide."

The persistent Colvins followed up in April through September 1978 with inquiries once again to Cameroon, Tanzania, Voltaic Republic (Burkina Faso), Albania, Ethiopia, Chad, Somalia, Mozambique, Comoros, and Gabon (rejected) and the Yemen Arab Republic. They refined their approach, with a new letter of introduction. Here's what they wrote to the Yemen Republic in April 1978

> We are writing to ask your help and consideration.
>
> Yemen Republic is one of the few countries in which there are not many persons actively participating on the air in the very interesting and worthwhile technical endeavor of amateur radio. More than 1,000,000 people throughout the world are licensed radio amateurs. They have trained themselves, on their own time, to understand the basic theory and operation of radio equipment and communication and have operated their small radio stations under on-the-air conditions after being licensed by their governments.
>
> It is an activity that is almost universally recognized as good for the people participating; good for the governments involved; and of great help to all concerned in the event of disasters, floods, earthquakes, or any other circumstances where

Montserrat

Lloyd and his beam at the U.S.V.I., W6KG/AJ3

> communication, to supplement or temporarily replace regular communication, is needed. It is also a valuable tool to help persons teach themselves the theory, operation, reception and transmission of radio.
>
> Amateur radio has been an active interest of ours for more than thirty years, We have traveled in many foreign countries, operating our small amateur radio station and showing interested personnel how it works and its advantages.
>
> We are American tourists who desire to travel to Yemen with our small radio and operate there for a short time. We would like to stay and operate for one month, but a period as short as one week would be satisfactory. We would stay in one of your hotels or similar building and are agreeable to any requirements that you may desire concerning our stay, operation and travel.
>
> Our equipment will weigh less than 150 pounds and we are financially able to pay for our travel both ways and while in the country of Yemen. We are ready to make the trip any time. We both hold the highest class of radio amateur licenses issued in this country. We are members of and represent several of the largest radio amateur clubs and societies in the U.S.A. We do not represent any political or official agency of the U.S. Government.
>
> Please consider our request and advise us by return mail if it is acceptable and can be accomplished.

A follow-up note on 25 May 1978 said "We still intend to visit your country and are most anxiously hoping that you will permit us to bring our small radio amateur station with us and operate it temporarily while we are there. We have traveled in Asia before and are fully award of conditions that we will encounter there."

Unfortunately, the director general of the Yemen Republic department of posts and telegraph replied that "It is regretted to inform that your request cannot be complied with due to unavoidable circumstances."

Off Again, 1978

The Caribbean is littered with islands, hence DXCC countries, and Iris and Lloyd Colvin set out for the area in October 1978, aiming to "finish it off" and make more DXCCs of their own. They opened from Guantanamo Bay 25 October from the resident Naval Amateur Radio Club station, KG4AN, using their own call sign, KG4KG. They were sponsored by a family on base, with arrangements made by Pat Benson, formerly KG4OO, and through 9 November 1978 they rolled up 1.7 million points in the CQ Phone DX Contest, won the ARRL Sweepstakes Section certificate on CW, and all together tallied 7,500 QSOs (first WA2TEG, last CO2OT). Their KG4KG DXCC certificate is #20,016, dated 28 April 1980 (122 countries total).

This was the first time the Colvins had lived in a totally military community in some years and they found it, "in general, very enjoyable." They had been assigned both KG4QL and KG4KG (calls are assigned by the Naval authorities, not the FCC) and chose to use the latter because it "sounded so good." The station's rotator and transmitter both were out of commission so they used their own equipment, and it took every inch of the coax they carried with them to place their triband beam on a nearby hill.

Iris and Lloyd enjoyed the ability to operate in non-U.S. 'phone bands. On the other hand, the FCC had begun issuing "peculiar" call signs in the Continental U.S., and many hams thought KG4KG was in the U.S. Fourth District.

St. Maarten

Following their Guantanamo operation, the Colvins traveled to Jamaica and broke new ground with their first operations on 160 meters. In the late 1970s, 160 was still an "off-band" for many reasons. One was that many hams were still using gear that didn't include the band; the Colvins themselves had traveled for years with a Collins S-Line that did not cover 160. When they switched to a new Japanese transceiver, that changed. Another problem is getting up an effective antenna for 160; the leap from 3.5 to 1.8 MHz is a big one in terms of wire and real estate. In 1978, it was not unknown for a highly-competitive contest operator, using an S-Line, to borrow someone's Kenwood TS-820 for the weekend, just to be able to get on 160 for a few more band multipliers.

Another consideration was that it was always the Colvins' goal to work the "average ham" and give him a brand-new country, as opposed to hand out "band countries" to the Big Guns. In a few years, as the new so-called "WARC" (World Administrative Radio Conference) bands became available, at 10.1, 18, and 24 MHz, Iris and Lloyd did not operate them, either. The average DXer was still working toward 5-Band DXCC, so operation on the traditional 80-10 meter bands made the most sense. Not to mention, WARC band operation would have required more antennas, perhaps an antenna tuner, etc.

The W6QL/6Y5 operation ran from 11 November to 7 December 1978, encompassing both the CQWW CW Contest (1.17 million points multi-single) and the ARRL 160 Meter Contest. Overall, Iris and Lloyd logged 8,600 QSOs (4Z4TT first, HR1KAS last) and worked 129 DXCC countries - DXCC #20,020, 28 April 1980. With a makeshift antenna for 160 Meters they worked 10 countries and 35 states in the U.S. The traditional 20 Meter band yielded a 24-minute All Continents on 28 November: JA6BVU, LU7XP, AA4BA, UK2RAX, FR7ZL, and ZL1CH. An "unusually fine QTH" in Port Royal, 13 miles out on a causeway in Kingston Bay; water on two sides and a clear shot to everywhere was a big help.

While in Jamaica the Colvins bumped into a team of 10 hams headed to Navassa Island for a major operation from that DXCC country. The team stayed briefly at the Colvins' hotel and warmed up by operating their station, before departing for Navassa on a fishing boat that docked right at the hotel. While they were making 22,000 QSOs at Navassa, W6QL/6Y5 racked up 5000 QSOs.

Jon Lindsay, W6OIG, was a member of the Navassa team and remembered

> While waiting for the boat to arrive, we checked into a hotel and subsequently found Iris and Lloyd there. I'd never met them before. They had all their gear set up around the hotel room. Lloyd was particularly proud of his antenna assembly(s). Each

Hy-Gain antenna was stuffed into a large metal tube for shipping. Of course, each beam was broken down into its singular components and each tube contained all that was needed to hoist the antenna, include antenna, mast and guy wires. He took me up on the top of the hotel to see the arrangement. He took these with him on many of his trips, rather than use suboptimal wire antennas.

Of note was the fact that Lloyd and Iris were staying in a fairly plush hotel. Lloyd said they always tried to stay in such a hotel, one that would allow him to set up his antenna on the roof and lead a fairly comfortable life. No outbacks, cots or tents for Lloyd. He probably would not do what we were about to do on Navassa Island.

I recall Lloyd as a distinguinshed individual, in keeping with his military experience. Iris was more earthy and I thought of her as the "real" ham. She could talk on the radio day and night. Lloyd was perhaps more interested in the hardware. But watching them "work" left you with the feeling that they had this routine down precisely! And they both really seemed to enjoy this lifestyle and seemed to know the special place they occupied in the ham world.

COUNTRY	DATE	LETTER
Clipperton Is	9 Dec 77	Direction des Telecommu Immeuble P.T.T. Bercy
People's Republic of China	9 Dec 77	Mr. Li Lin-Chuan, Direc Ministry of Posts an
People's Republic of China	10 Dec 77	His Excellency Chou Jun
People's Republic of China	9 Dec 77	Mr. Wang Lien-lung, Cha Peking University, P replied 5 May 78
Albania	10 Dec 77 2nd 12 Sept 78	Chief of Post & Telegra
Burma	10 Dec 77 2nd 25 May 78	Chief of Post & Telegra Replied 14 June 78
Kamaran	10 Dec 77 2nd 25 May 78 - 3rd 12 Sep 78	Chief of Post & Telegrap
Comoro Islands	15 Mar 78 2nd - 12 Sept 78	Post & Telegraph Office,
French Somaliland	10 Apr 78	Post & Telegraph Office,
Somali Republic	10 Apr 78 2nd 12 Sept 78	Chief of Post & Telegrap
Ethiopia	10 Apr 78 2nd 12 Sept 78	Chief of Post & Telegrap
Yemen ARAB Republic	10 Apr 78 2nd 12 Sept 78	Chief of Post & Telegrap
Yemen (Peoples Democratic Republic of Yemen	10 Apr 78	Chief of Post & Telegrap
Cameroon	25 May 78	Chief of Post & Telegraph
Congo Republic	25 May 78 2nd 12 Sept 78	Chief of Post & Telegraph answered 28 July 78

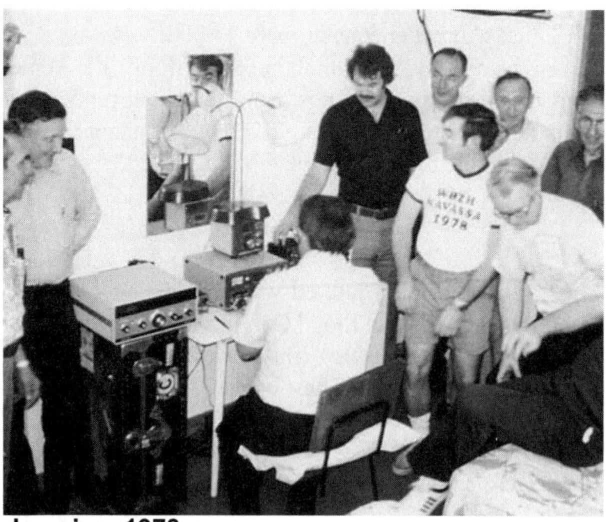

Jamaica, 1978

Three days after closing out their Jamaica log, Lloyd and Iris made the first of 9,000 QSOs in 126 countries from Bodentown, the Cayman Islands, as ZF2CI (ZF1 calls were for residents, ZF2 for visitors). Their operation lasted until 6 January 1979 (first QSO WA2JZL, last W2LT). Their DXCC #20,019 was issued 28 April 1980.

Their operating location near the center of Grand Cayman Island was at a "high spot" - 50 feet above sea level! They noted the large number of banks on the island, and tasted turtle steak for the first time, finding it "delicious." With frequent direct flights from Miami and other U.S. embarcation points, and prompt amateur radio licensing, the Caymans were, and are, one of the most popular DXpedition spots.

Reporting before their departure, Iris and Lloyd said "We have been finding it increasingly more difficult to take our radio gear with us on the same plane that we take. From Cayman the only place we can fly where an airline will guarantee our radio can fly with us is to Costa Rica. Accordingly, our next stop will be in Costa Rica in one of the more rare provinces, probably TI5."

Getting to Costa Rica may have been expeditious, but finding a place to operate in the Alajuela Province wasn't. The province had just a few local hams, who operated SSB only, so the lure of the W6KG/TI5 call sign was great.

It took the Colvins nearly a week of continuous searching to find a good place to operate from, and they finally came on the air on 13 January 1979. Operating through 3 February, they ran up 6,000 contacts in 123 countries, about half-and-half CW and SSB, including operating in the CQ WW 160 Contest (10,000 points). First contact was W1GLJ and the final log entry was DM3NWG.

They noted that more and more the widespread use of televisions, record players, and radios in other countries was making it hard to find a good place to play ham radio.

The best places were those away from cities, but that often meant unreliable or non-existent commercial power. "Another great problem," they said, "is doing anything when you are not fluent in the language of the country, and we speak very little Spanish."

In their search for an operating location Lloyd and Iris hired a young man off the street who spoke both English and Spanish. He turned out to be a refugee from neighboring Nicaragua, which at the time was involved in a civil war. "Our interpreter was not too much help because, like us, he was a stranger in the country of Costa Rica."

The W6KG/TI5 DXCC is #20,017, dated 28 April 1980.

One young American amateur remembered Iris especially fondly from this time, saying

> The first year I met them (either 1979 or 1980) I told Iris and Lloyd about the woman I was in love with who was still in Turkey [where he had visited], waiting for me to finish college. Every year Iris would ask me about [her], which always amazed me that she would remember me and my lady despite all that they had going on in their lives.
>
> It touched me deeply when I saw them that second year and Iris asked me about [my friend]. She kept track of my progress in college (79-82) and asked when we worked on the air. I have their QSLs from J6LOO and HR0QL. [Iris} was always the one who talked to me. She deserves all the accolades we can give her and Lloyd! I enjoyed about 4-5 Hamventions with them and our fellow DXers and think of them very fondly.

The Colvins next traveled to Tegucigalpa, Honduras, a country that had seen a change of government after a coup just the year before. But the political climate was relatively calm; Honduras's main problem in the 1970s and '80s was instability in neighboring countries.

The Colvins wrote that "While we were in Jamaica as W6QL/6Y5, Iris worked an amateur, HR1OL, in

Military transport plane, Guantanamo Bay

Guantanamo Bay

Honduras. In fact, it was our only contact with Honduras from Jamaica. When we landed at the airport in Honduras we asked if anyone could help us find HR1OL. It turned out that he is an ex-general in the Honduras army and ex-president of the country of Honduras! (Iris seems to have made all our most important QSOs lately.)"

They met with the local licensing official and learned that he had been active on the air during the Danny Weil era and had donated to the early YASME cause. This official met with HR1OL that evening and, the next day, Iris and Lloyd were issued the "special event" call sign HR0QL (highly unusual).

Iris enjoyed telling club meetings back in the States that winter that "On the ham bands one never knows what important persons they might communicate with. Before we arrived in Honduras we had contacted HR1OL, who said his name was Oswald. When we arrived at the airport we asked for him and the response was immediate: "I will take you to him." Everyone rushed around to be as helpful as possible. When we arrived at the magnificent office of HR1OL we learned that he was the former president of Honduras as well as the president of some airline companies. It may have been due to his influence that we received the special HR0QL call sign."

Operating from "an excellent site high in the hills," HR0QL bagged 10,000 QSOs in 129 countries (this total was their second-best; their record of 14,000 QSOs from VP2MAQ was going to be hard to beat!). Operation in the ARRL Phone DX Contest contributed 4,000 contacts to this total, and a 23-minute Worked All Continents was logged on 20 meter CW on 21 February 1979: VU2BK, VK4JU, F0CZG, 6W8EX, WA2WHE, and LU4KV. First contact of the operation was K1VHS; last was W5OB. The HR0QL DXCC is #20,015, 28 April 1980.

A recap of this operation was a big hit at a club meeting, as Don Calbick, W7GB, describes:

> Besides many DX QSOs with W6KG and W6QL, the only time I had contact with the Colvins was at an ARRL National Convention in Seattle in the early 80's. Lloyd and Iris were to give a

DXpedition travelog presentation and as I took my seat I saw a large bulletin board with a list of many Colvin DXpeditions. Nearby was a stack of slide carousel boxes and in the center aisle was the projector.

Iris got up on stage and asked the gathering of DXers which DXpedition they'd like to see. Someone suggested Honduras, so she went over to the stack and Lloyd handed her that carousel. What followed was the most enjoyable DX presentation I've seen. The entire program was spontaneous and to add flavor, they also had tape recordings of the operation.

Lloyd and Iris had booked into a hotel room from which they planned to operate /HR. Unfortunately, there was a very high noise level, making operating impossible. They explained this to the hotel manager and he said "no problem, we've got a motel on top of a nearby hill."

So they transported their gear and set up a station in the motel on top of the hill. According to the Colvins, the noise level was low, conditions were great and as it turned out, the DXpedition was a success. At this point, after much embellishment, the Colvins revealed their "secret." This motel was not a motel at all but a brothel! Of course all the DXers present just about rolled in the aisles.

They then played a tape of [HR0QL] operating CW and working several stations, and in the background we could hear music and girls squealing. This drew even more laughter! As far as I know, this is the only DXpedition done from a brothel!

The neat part of this program was the eloquence of presentation and the Colvins' personal touch. I never got to meet the Colvins personally but after this program, I felt as though I had known them for many years.

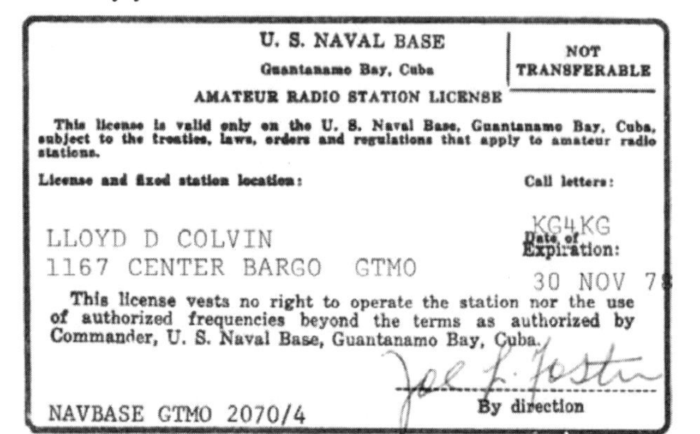
U. S. NAVAL BASE
Guantanamo Bay, Cuba
NOT TRANSFERABLE
AMATEUR RADIO STATION LICENSE

This license is valid only on the U. S. Naval Base, Guantanamo Bay, Cuba, subject to the treaties, laws, orders and regulations that apply to amateur radio stations.

License and fixed station location: Call letters: KG4KG

LLOYD D COLVIN
1167 CENTER BARGO GTMO

Date of Expiration: 30 NOV 7

This license vests no right to operate the station nor the use of authorized frequencies beyond the terms as authorized by Commander, U. S. Naval Base, Guantanamo Bay, Cuba.

By direction

NAVBASE GTMO 2070/4

Caymans Airport

Next stop, and last for this trip, was Belize, still a British colony (it gained independence in 1981). Their VP1KG operation from 9 March to 1 April 1979 included a 2.9-million point excursion in the CQ WPX SSB Contest. They turned in their highest contact total for this Caribbean swing, 12,500, and also worked the most countries, 136 (the VP1KG DXCC is #20,018, dated 28 April 1980). First QSO was with WA4LOF and last was old-timer G6CJ.

This autumn 1978 trip lasted just about six months. Operating as KG4KG, W6QL/6Y5, ZF2CI, W6KG/TI5, HR0QL and VP1KG, the Colvins made 50,000 QSOs in about four months on the air. The rest of the time away was spent "traveling, finding places to stay, and putting antennas and stations up and down."

Iris and Lloyd made "semi-formal" presentations to local radio amateur groups or clubs at all six stops. They said "The operations on the air became something like work, but the meeting of new people and making friends in new countries has been the most rewarding part of our world travel. We have now held 104 radio calls and visited in 135 different countries. Our 500,000 QSOs have resulted in a QSL collection of 250,000 QSLs - the largest alphabetically-filed QSL collection in the world."

The Colvins returned to California for the spring and summer with an amateur radio agenda including the Fresno Convention and the Dayton Hamvention. They also spoke in Missouri and around California at club meetings.

On 8 April 1979 they joined YASME Foundation directors W6RGG, N6SF (W6OAT) and OH2BH at Fresno. A request from George Collins, VE3FXT, for Foundation support was not approved (the Foundation actually had no funds to expend in those days). A motion by Lloyd was approved, and it read "that all officers and directors of the foundation actively pursue a course of supporting valid DXpeditions and making the YASME Foundation more visible to amateurs of the world."

Iris and Lloyd also caught up with business matters of Drake Builders (including an IRS audit). At this time, their daughter, Joy Gilcrease, and her husband, Richard Gilcrease, were handling business while Lloyd and Iris were traveling.

The Colvins wrote to friends that they expected to leave sometime in October 1979 and be gone for up to a full year. They observed that despite the "increased emphasis" on air travel, their ventures were no easier than, say, 10 years before. In the past they had had little trouble taking all their travel gear and radio equipment on the same plane they flew on; now, in 1979, increasingly airlines were requiring that excess baggage be sent as air freight, "which means that it will go on a different plane than the one that we take, and, also, may take up to several weeks to reach its destination."

By now Lloyd and Iris were hauling a solid-state transceiver (a Yaesu FT-901DM), a power supply, and the Heathkit SB-230 solid-state linear amplifier, along with a Hy-Gain TH-3 triband yagi broken down and stored in a PVC tube, several hundred feet of ½-inch coax, as well as numerous accessories. In 2002 terms, this would be about the same as traveling the world with a desktop personal computer, monitor, printer and scanner (not to mention antennas).

They said, too, that "We are sure that, on our future travels as in the past, the most enjoyable part will be meeting people and making friends in the countries we visit. Sometimes the actual radio operating becomes work without pay but the visiting of a new country and learning the customs, social activities, interests and government of the people is always an experience that we look forward to."

In July 1979, the West Coast DX Bulletin (Hugh Cassidy, WA6AUD) ceased publication after 11 years. In August 1979 this writer founded The DX Bulletin and wrote it until March 1986, a total of 325 issues. After that, Chod Harris, WB2CHO/VP2ML took over and published more than 500 more issues of TDXB. Both Cassidy and

Cayman Islands

Harris are now members of the CQ DX Hall of Fame. The Long Island DX Bulletin also was being published in 1979, and was widely read, and QRZ DX was founded in the fall of 1979.

Off to the Caribbean Again, 1979

Off to the Caribbean again, the first operation of fall 1979 was from Fort Nelson, Grenada, as J3ABV. From 13 October to 7 November 1979 the Colvins racked up 8,800 QSOs on 160 to 10 meters (first QSO KA1AWH, last VP2SAZ). Grenada had become independent from the U.K. in 1974 and early in 1979 the island's first prime minister, Eric Gairy, was overthrown in a coup. The island's government would lean toward communism until an invasion in 1983 by the U.S. and a contingent from the Organization of East Caribbean States.

Iris and Lloyd said in their later report on the visit that "The country of Grenada recently went through a revolution and the formation of a new government. This may, or may not, have had a bearing on the fact that we had to post a large, refundable, cash bond at customs before we could bring our radio gear into Grenada." It's somewhat amazing that they got their gear in, and got licenses, at all.

Despite the hurdles, it was a great operation. J3ABV made DXCC in "general operating" as well as working more than 100 DXCC countries during the CQ WW Phone Contest at the end of October. In all they worked 152 countries, a new record for them. Their DXCC was #21,886, issued 2 July 1981.

If at times Lloyd and Iris were finding their amateur radio operations to be "work," they were working harder and harder. Their next stop was at St. Vincent, from 15 November to 7 December 1979. They had been assigned call sign VP2SAX and anticipated the CQ WW CW Contest at the end of November. Lo and behold, two hot-shot contesters from Connecticut arrived on St. Vincent for the contest, and they had managed to get the call sign VP2SX! Needless to say, if anybody around the world failed to work St. Vincent that weekend, they just were not on the air.

The Colvins logged 9,000 contacts from VP2SAX, in 141 countries (DXCC #21,890, 2 July 1981. They also worked the ARRL 160 Meter Contest. First QSO was K4JI; last was N5AF.

Saint Vincent had received full independence on 27 October 1979, as "Saint Vincent and the Grenadines." A separatist movement in the Grenadines had resulted in a brief uprising (1979) on Union Island. Union Island, it will be recalled, is the island off which Danny Weil lost YASME II.The Colvins were happy that both in Grenada and St. Vincent they had accommodations near or at the top of high mountains, "with clear shots to everywhere. It is not always that we can find such ideal spots from which to operate. Most of the Caribbean countries are very mountainous, and it seems as if most of the places available for rent have a big mountain between them and the north, which is the direction for QSOs with North America and Europe."

St. Lucia became independent within the Commonwealth of Nations (British) in February 1979 and it was the next stop for the Colvins. It had enjoyed self-government as a member of the West Indies Associated States since 1967. They operated J6LOO there from 12 December 1979 to 4 January 1980, from a cottage by the sea near Fort Vieux. This was the first time in their records that they admitted to being concerned about what the salt spray might do to their equipment.

They rang up 9,000 QSOs, the first with K9ECE, the very Chicago fellow who had hosted Dave Tremayne, ZL1AV, in the early 1960s, the last with PY7PO, with radio amateurs in 130 countries on 10 through 160 meters, including, on 160, all 10 USA call districts for the first time from the Caribbean. Their DXCC was #21,887, 2 July 1981.

Grenada

Club meeting, Dominica

Two hams already on the island, J6LHV, a resident local, and J6LIM (VE2EWS) from Canada, came to visit with broken transceivers and the Colvins "with hard work and some luck repaired their rigs and got both amateurs back on the air."

The Colvins left for Martinique but found that their licenses had not arrived from France, so they went on to Dominica. Their J7DBB operation there from 15 January to 6 February 1980 netted another 9,000 contacts and 134 DXCC countries (DXCC #21,888, 2 July 1981). First QSO was with EA3OG, last in log was W5UFF. Dominica had obtained full independence in late 1978.

From 11 February to 2 March 1980 it was VP2KAH, within sight of Montserrat and Antigua. The Colvins had visited St. Kitts years before but at that time only residents were allowed amateur radio operation. In 1980 visitors were usually limited to signing their home call sign/VP2K but the Colvins were granted a call sign due to their having been there before and being "almost natives." Operation in both the ARRL Phone and CW International DX contest weekends swelled the overall QSO total to 12,000, in 137 countries, from a Conapee Beach location.

The Colvins were "delighted and amazed" on a visit to VP2KC, owned by Kit Carson, from where a big multi-multi contest operation, with 27 operators, the year before had racked up a giant score in the CQ WW Contest. They said "The location, amount of equipment, and number and type of antennas make VP2KC possibly the largest and most expensive amateur radio station in the world. On all of our trips, we have never seen a more elaborate station, anywhere. However, while making our 12,000 QSOs during our month here, almost no QSOs were made at VP2KC. We have noticed that many such large contest stations remain almost silent until the next big contest comes around."

Iris and Lloyd made more than a million points in each of the ARRL contest weekends. Overall, first QSO was WD4HWE, last was WA4SSU.

The final stop on this Caribbean tour was the island of Hispaniola and the Dominican Republic, where a military coup attempt had failed in 1978 and a 1979 hurricane had left more than 200,000 people homeless and caused an estimated $1 billion in damage. From 8 to 30 March 1980 they tallied 9,600 QSOs, including a rousing 4.4 million points in the CQ WPX SSB Contest and 2,900 points in the CQ 160 Contest. Their HI6XQL DXCC was #21,885, issued on 2 July 1981. YV5ANF was first in the log and WB5SYT made it in as the last QSO.

Vlad, UA6JD, said:

> I remember one QSO (among hundreds of QSOs with the Colvins). I was monitoring for HI6XQL on 7 MHz CW, and worked them a few times on other bands but not 7. When I tell this situation to a friend of mine, Ray,YS1RRD, during our regular QSO, we heard "break break" - it was Lloyd, HI6XQL, who says if UA6JAD (my call sign then) needs him on 7 he can QSY immediately. We QSY and I add HI6XQL on 7 MHz for my 5BDXCC. Since then I have a lot of contacts with Lloyd/Iris, but more remember this QSO when he first calling me and help. Real gentleman and good HAM spirit amateur.

This Caribbean swing mirrored the previous one: six months away, about four months on the air, and six DXCC countries operated from. They made 55,000 contacts and were happy to report that all their gear worked without a single failure. They were already announcing another trip in the fall and asked the press to spread the word that they would like to operate from Desecheo and the Kamaran Islands, if they could get the necessary permission.

Shortly after this, in April, the YASME Foundation, at its annual board of directors meeting, held at the Dayton Hamvention on 20 April 1980, established a "YASME Award" a certificate for presenting QSLs verifying contact with the holders of 30 different YASME DXpedition calls, including any calls held by YASME officers or directors, past or present. Yasme director Dick McKercher, W0MLY, became custodian, and would issue hundreds of the certificates until his death.

St. Lucia

Hurricane emergency station, Dominica

Iris and Lloyd once again spoke and showed slides of their travels at the 1980 Hamvention and generally made themselves available through the long weekend. Coincidentally, the local newspaper during the Hamvention weekend carried an Associated Press story (20 April 1980) called "Bad Days for the Caribbean," by AP writer Marc D. Cherney. The Colvins clipped and saved it.

> On the palm-fringed islands of the British Caribbean, descendants of slaves have long lived out their lives in poverty. The 1960s and 1970s brought most of them independence - but little letup in the poverty. Now there are the beginnings of violent political change. The prospect of instability in a 2,000-mile chain of new nations so close to U.S. shores has American diplomats worried.
>
> Two decades ago, the only independent nations in the Caribbean were Cuba, the Dominican Republic and Haiti. Then, after attempts to build a federation of the West Indies foundered on economic rivalries and insular nationalism, the British began spinning off one island after another as independent nations.
>
> Today, they include Jamaica, Dominica, St. Lucia, St. Vincent, Grenada, Barbados, Trinidad and Tobago, and Guyana. St. Kitts-Nevis, now literally self-governing, could follow soon, and so could Antigua.
>
> The area, indeed, is one of high [il]literacy rates and high unemployment. The annual per capita

income ranges from St. Vincent's $330 to Barbados' $1820 - in contrast to about $8,000 in the United States. The smaller islands generally have one or two-crop agricultural economies and an uncertain tourist trade to rely on. And among the larger islands, Jamaica's efforts to exploit mineral resources are foundering in an international climate of exploding oil costs and general inflation.

There has always been talk of these islands working together, but insular nationalism remains strong and U.S. diplomats are beginning to wonder if their traditional approach - channeling all long-term aid through such regional groups as the Caribbean Development Bank - is enough. Diplomatic sources say they are considering some direct aid to individual countries until the prospects for regional cooperation are better.

That is so because the British departure has left the Caribbean with great needs - and a vacuum of influence which Cuba has shown interest in filling. That in turn, has brought new expressions of interest from the United States, and from such other countries as France and oil-rich Venezuela. U.S. diplomats say they welcome the other Western aid.

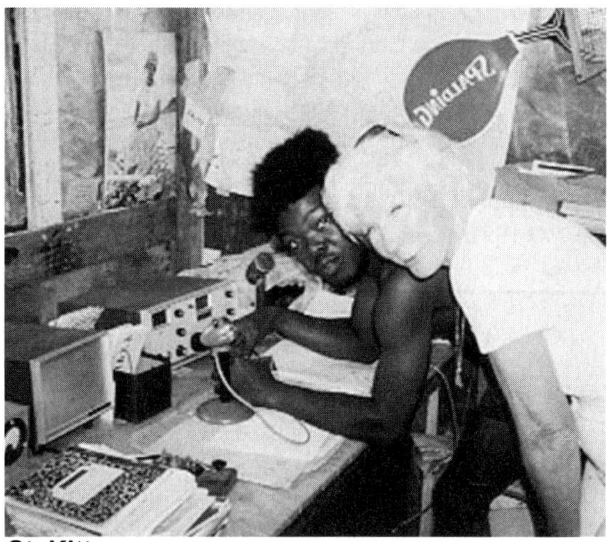

St. Kitts

Cherney outlined the problems that the islands faced, one by one, including Jamaica, St. Kitts-Nevis, Antigua, Dominica, St. Lucia, St. Vincent, and Grenada.

Iris and Lloyd read the article with great fascination, thinking, "We've been there."

40· The Mediterranean and Caribbean, 1980-81

[Visiting] the historic and biblical towns and areas of Israel is an experience that everyone should have sometime if possible." - Iris and Lloyd Colvin

After the usual spring and summer of tending to matters on the California home front, the Colvins set out again on 1 October 1980, flying directly to Athens, Greece. Before leaving they tallied YASME DXpedition operations, both theirs and those of others, at about 125.

It took the pair ten days to get operating permits from Greek authorities, and then only with the help of SV1JG and SV0AA. Ordinarily up to two months were required at that time to get operating permits. For some reason, Iris and Lloyd chose not to operate from Greece proper (not a rare DXCC country by any stretch of the imagination - in later years, however, they would have found a way to get on long enough to work 100 countries and add another DXCC certificate to their growing stack). Rather, they left immediately for the Grecian island of Crete.

W6KG/SV9 came on the air from Knossis Beach, Crete, on 12 October 1980, ending on 26 October. They finished the operation with the weekend of the CQ WW Phone contest, operating, as always, multi-operator, single transmitter, for 1.5 million points. Overall, they made 9,500 contacts, equally divided between CW and 'phone, in 142 countries and "werc on all bands permitted in Greece" (160 meters not authorized in Greece). Their DXCC was #23,379, 13 July 1982. First contact was OH6LD and last was CE4EM.

Their only comment about Crete was "The tourist business is a major industry in Greece, and, as a result, all of the islands have great numbers of hotels and everyone in the hotels and shops speaks English." (A definite plus after some of their Caribbean experiences and South American experiences to come.)

From Crete it was a short hop to Rhodes, in the Dodecanese group. Here they used Iris's call sign,

Club meeting, Crete

Rhodes

W6QL/SV5, to the tune of 6,500 contacts, from 29 October to 13 November 1980. Iris's DXCC there was #23,381, dated 13 July 1982. First contact was DF9RT, last one OK3KEE.

From Rhodes the Colvins flew to Israel, where they operated from the town of Hod Hasharan -- for the first time ever -- from a place different from where they slept. They set up their station in a farmhouse unoccupied for five years and slept in a house about a thousand feet away, they said. It was Lloyd's turn to use his call sign: from 21 November to 6 December 1980. W6KG/4X notched 5,000 contacts, including a 1.33 million-point CQ WW CW Contest performance. All told, they worked 120 DXCC countries (DXCC #23,380, 13 July 1982), the first QSO being OE7HHW, the last 4X4FU.

"We were surprised to learn," they said, "upon our arrival in Israel, that all hams there are prohibited from working any Arab country," but this was probably just their way of adding a little suspense! Their licenses listed 20 nearby countries with which contacts were not allowed (and this is generally still the case more than two decades later). In most cases, it is the other country, not Israel, that does the prohibiting. This is a pity for amateurs and especially for DXers and contesters. A contest operator in Israel has many juicy multipliers close by but is forbidden to work them. On a later swing through several Mideast countries Iris and Lloyd would face the situation in reverse but, in several cases, they kept a "special log" for working Israeli stations.

The Colvins also noted that since 1967 the USSR had forbidden all Soviet hams to communicate with Israeli stations; but, they said, "the USSR hams have interpreted this to mean that it is OK for them to work foreign hams, like ourselves, in Israel."

They also reported being able to visit "most of the historic and biblical towns and areas of Israel. It is an experience that everyone should have sometime if possible."

From Israel the Colvins flew home to California, then turned right around and headed back to the Caribbean, for a Christmas operation from French St. Martin. FG0FOL/FS was on the air from 19 December 1980 to 15 January 1981. From the Coralita Beach Hotel they "for all practical purposes, had this country all for ourselves as far as Ham Radio is concerned." 10 through 80 Meters, half CW and half SSB, yielded 9,000 QSOs with amateurs in 149 countries, their best since leaving the USA on 1 October of 1979. Their DXCC #23,374 was issued 13 July 1982, the first FG0FOL/FS contact was with N5BUP, the last with 4Z4DX.

They lamented their "very long" call sign slowing things down a bit, but "the rare call more than offset that problem. French St. Martin also was the most expensive of any country they could recall visiting. "All French countries tend to be expensive, plus the fact that the period from Christmas through New Year is the very peak of tourist activity in this part of the world.

"We found band conditions good, and we could generate pileups of stations wanting to work us at any time, on some band, during any 24-hour period," they said.

Iris and Lloyd ran into Mort Bardfield, W1UQ/PJ7UQ, and Claire Bardfield, K1YL/PJ7YL, who said:

Guadeloupe

Club meeting, Greece

We were fortunate to meet Lloyd and Iris on St. Martin in December of 1980. Claire and I had previously listened to the Colvin's presentation in Dayton, especially because we were also a husband-wife extra-class combo!

We owned a home in Simpson Bay and the Colvins were staying at the Coralita Hotel, on the French side of the Island, so it was a chance to meet them in person. At that time, four local hams were licensed on French St. Martin, Alain FG5AR; Terry FG5DYM/FS7 (W4GSM); Mort FG5UQ/FS7 (W1UQ); and Claire (FG5DYL/FS7 (K1YL).

We were not much for DXpeditions but had just gone to Nairobi with K1MM and K1MEM on a safari radio vacation so we had an interest in meeting the Colvins to find out their "secrets" of solving problems, shlepping equipment and getting licenses. They had lunch at our apartment with two toddlers at the table (who are now in their twenties, N1DDD and W1RES). The Colvins didn't have a car, and we obliged, rather than have them hitchhike around the island (which was perfectly safe!).

At the suggestion of the government radio bureau director, Jose, PJ2MI, I had just organized a radio club as a prerequsite for a club-licensed 2-meter repeater (still operating on 146.76). Our very first club meeting was to be held in a Chinese restaurant next to the government phone company "Landsradio." (The majority of the local Antillean hams worked for this local authority).

I asked Lloyd and Iris if they would take time off

from operating and come to the meeting, [although] I had apprehensions because I didn't think any of our members had heard of them.

Naturally, they were asked to give a talk, and let it suffice to say that the locals were a bit amazed at their travels. Their talk was the hit of the meeting and was impressive and unforgettable. Luckily, as I remember, three locals (old timers) had actually either worked or heard of the Colvins, and they were Erwin PJ7EF, Orlando PJ7RO and Vinny PJ7VL. Some of the younger hams could relate to the many countries because they worked in the long-distance department! I remember that the Chinese restaurant owner also listened carefully from a nearby table.

Naturally, the Colvins were welcome back, but their travels never returned them to the island. At any rate, I got a lot of credit for holding a heck of a meeting, but it was a a tough act to follow.

French licenses in hand, it was then off to Guadeloupe, 155 miles away, where they rented a cliff-side house, "a very good QTH," in Anza la Borde. Iris's sister Clara (wife of W7PSD) visited them there for a week. Operating from there 18 January to 12 February 1981 as FG0FOK they logged 7,600 QSOs with amateurs in 150 countries on 10 thru 80 meters, the first with WA8OVC and the final two-way with DF5RH. DXCC #23,373 was dated 13 July 1982.

Their call sign caused some tooth-gnashing. "Just about every fourth station [we] worked would not include the ending K with the call. For example we never signed our call just once, we always gave the call two or more times so it would be clear that the K was part of the call. This helped, but not enough. Once a ham thought the call was FG0FO, it was almost impossible to get him to change. Even after sending our call correctly at ten words a minute most hams would come back with the same wrong call. We do not intend to use a call ending in K ever again if it can be avoided. At our next stop, which will be Martinique, we will use the call FM0FOL."

Guadeloupe

Martinique

15 February to 8 March 1981 took Iris and Lloyd to Martinique, where they stayed in a house owned by the mother of FM7AD, at Ajoupa Bouillon. Working multi/single in both weekends of the ARRL International DX Competition they racked up nearly 5 million contest points. Overall, FM0FOL logged some 12,000 contacts, JR6RRD/3 the first and K4IM the last. Their DXCC certificate is #23,375, 13 July 1982.

Martinique finished up this October 1980 to March 1981 trip, once again with operation from six DXCC countries and another 55,000 Colvin QSOs. On their way home to California Iris and Lloyd visited and spoke at the National ARRL Convention in Orlando, Florida, then went to Dallas, Texas for a combined meeting of the Richardson Wireless Club and the North Texas Contest Club. April included an appearance in Cleveland, Ohio, and then on to the Hamvention, where they were honored by being the principal speakers at the Hamvention banquet. The week after that was, of course, another sojourn to the joint California DX Convention, which was now being held in Visalia (and is known to this day among hams as, simply, "Visalia").

In their usual end-of-trip report, the Colvins said they planned to take off again in the fall and, they said "If any of you can find a way for us to obtain permission to operate in a very rare spot such as Kamaran, China, Burma, or Albania, please let us know and we will go there and promise to work you from there."

The Kamaran Islands were still a DXCC country in 1981. They, along with China, Burma, and Albania, were needed by most DXers, none of them having had any amateur radio activity in many years. But the timing was wrong, and the Colvins were the "wrong" nationality. China would finally open amateur radio in the mid-1980s, in general only to Chinese citizens. Burma would see activity by visitors from Japan. And it was the early 1990s before a concerted effort would enable a major operation from Albania, involving both European and American amateurs.

In 1981 it seemed to many people that if anybody could get operating permission from some of the "closed countries," it might well be the Colvins. They certainly had the credentials as seasoned travelers and they were a little old to be suspected of being spies or subversives (Lloyd was 66, Iris 67). Here is what Lloyd and Iris wrote to Burma (they wrote similar letters to other countries around this time):

> We have been radio amateurs for most of our lives and hold the highest class (Extra Class) license issued in the United States. We are now retired and wish to visit Burma as tourists. We are retired (age 67) and represent no company or government. We have sufficient money to travel anywhere and pay for all other expenses.

Martinique

Martinique

> While we are in Burma, we would very much like to operate our small radio amateur station. We would like permission to bring our small radio station with us, and we would take it with us when we leave Burma.
>
> A couple of weeks ago there was radio operation from Burma by a Japanese group of radio amateur operators of which Mr. "Jin" Hiroto Sasaki was a member. They operated on 14 megacycle phone and CW and used the radio call XZ5A. They started operation on 22 May 1981 and were on the air for a little more than a week.
>
> We are writing this letter to ask you to please permit us to operate in the same manner and with the same restrictions and requirements as was granted to the Japanese Radio Amateurs. We are prepared to leave for Burma at any time and will anxiously await your reply. We hope that your answer is favorable and that you will allow us to operate our amateur radio station In Burma.

The fall of 1981 saw the Colvins finishing up the Caribbean, beginning with an operation from Hastings, Christ Church, Barbados. It lasted from 9 October to 1 November 1981. "We are deep into the YASME DXpedition again," they wrote, "after a breathing spell in the U.S.A. for nearly six months, during which time we talked at a number of radio clubs." Although visiting amateurs were not ordinarily granted call signs, Iris wrangled 8P6QL. They warmed up for this trip with 9,000 contacts in 150 countries, just two shy of their all-time high of 152 from Grenada, J3ABV, two years previous.

They stayed in a three-story hotel arranged by Woody, 8P6CC (VP6WR in the old days). "We had met Woody in Barbados 25 years ago, and we are surprised and happy to find that he has changed very little during the last quarter of a century," they later wrote.

"Our hotel has a sloping tile roof, and, during antenna erection, Lloyd stepped on a weak spot in the roof and fell through. Fortunately, the only serious damage was to our relationship with the Manager of the Hotel!!"

The 8P6QL DXCC was #24,292, dated 18 May 1983. KA9IQQ was first in the log and W8DNC was last. On the air for 48 hours in the CQ WW Phone Contest, they made 3,000 contacts. Another record for them from 8P6QL was a five-minute Worked All Continents, on 15 meters on 22 October, "with no pre-arranged schedules": F6HEW, ZS1LD, KA1CY, 4X4FU, VK4SS, and PY1DFJ.

The Colvins had confirmed for themselves that radio conditions seem most interesting close to the equator and their next stop, Trinidad, took them closer. The political situation there was unsteady following a rash of fire-bombing, arsons, and political shootings in 1980. They operated from Port of Spain from 11 November to 2 December 1981, making 9,000 contacts (first, K5ME; last, S83H). Their multi-single work in the CQ WW CW contest resulted in an impressive 1.7 million points. (9Y4KG, DXCC 24,293, 18 May 1983.)

Iris and Lloyd found Trinidad a "very difficult country to get your equipment in and out again," with complicated, time-consuming and costly regulations. It took eight days to get their equipment released from customs, even though their license was ready upon arrival thanks to the Trinidad Radio Club and their President, 9Y4NP, " but we did not realize that there was still the problem of posting a bond, obtaining revenue stamps, and engaging the services of a customs broker, etc."

They noted that Trinidad is an oil-rich country. "All the roads seem jammed full of automobiles almost continuously. One reason for this is that gasoline in Trinidad is possibly the cheapest in the world - only 40 cents per gallon (U.S.)"

The situation was even worse at their next stop, Guyana. They arrived less than four years after the horri-

Barbados

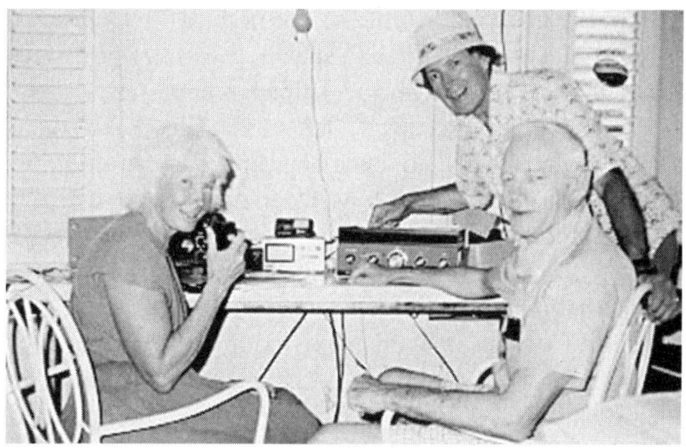

John Thompson, W1BIH/PJ9JT, visits on Curacao

ble Jonestown mass suicides and murders, and the country's economy and general conditions had been deteriorating for years. It took the Colvins three weeks to get their operating permission and to extricate their radio gear from Customs. "We were reported to be the first DXpedition to ever enter Guyana just to operate amateur radio. Guyana [has] a socialistic government, and our application to operate amateur had to be approved by a number of government sections. When we finally obtained our licenses, they were signed by the vice president of Guyana."

Operating from 23 December 1981 to 11 January 1982 from La Bonne Intention, W6QL/8R1 rang up 9,000 QSOs with hams in 144 countries, on 10 through 160 meters, with especially good results on 40. The W6QL/8R1 DXCC is #24,294, 18 May 1983; first QSO was WB8ZJW and last was WB1BVQ.

They noticed that the daily newspapers "often carried articles critical of capitalistic countries such as the U.S.A. For example, on the day of our arrival, the principal Guyana newspaper published the names of all United States CIA agents operating in Guyana. Articles like that, plus memories of the Jonestown disaster, did not make it any easier for us to obtain fast permission to operate amateur radio in Guyana."

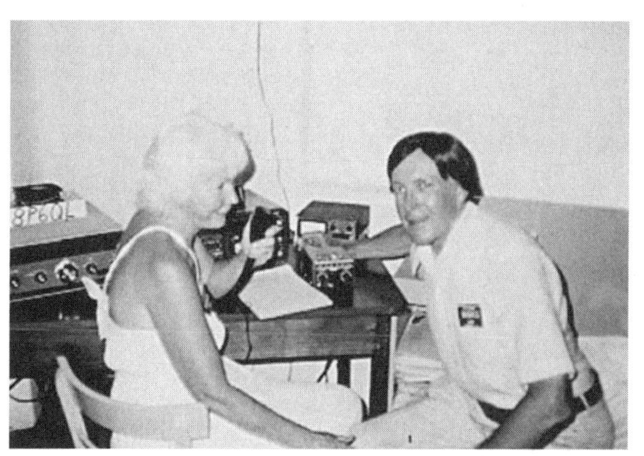

Barbados

On the good side, Guyana, formerly British Guyana, is the only entirely English-speaking country in South America, they said. "This fact helped us a lot in all ways."

Tony, G4UZN, received a letter from Lloyd and Iris in 1984, saying they had been the first to operate from Guyana on the new 10.1 MHz amateur band when it opened that January, making the first contact on 5 January. They also noted they later were the first to operate on that band from Abu Ail and the Galapagos.

The Colvins moved next door, to Paramaribo, Surinam, which gained independence from The Netherlands in 1975 (when some 40,000 people chose to retain their Dutch citizenship and moved to The Netherlands). Operating W6KG/PZ1 from 13 January to 1 February 1982, Iris and Lloyd racked up 9,000 contacts on 160 through 10 meters, and set a new record for themselves with 155 DXCC countries (DXCC #24,296, 18 May 1983). They made a welcome appearance in the CQ 160-Meter Contest.

They observed that the good results probably were because "PZ is fairly rare, January is a good month for DX, and, throughout the world, DX and DX operation seems to be at a peak level." First contact was with YV6CVK, and the last with HI8DAF.

Iris and Lloyd lived at and operated from the home of PZ1BU, a large estate on the Surinam River. This was the first time they had operated so close to another active ham, namely PZ1BU, who was on the air often at the same time as the Colvins, but they suffered only minor interference and then only when on the same band.

A short stop followed in French Guiana followed, the oldest of the overseas possessions of France and the only French territory now on the South American mainland, producing 5,000 FY0FOL QSOs, including the ARRL CW DX Competition. The 3- to 27 February 1982 operation began with G3FXA and ended with JI1QPU. Their DXCC is #24,291, 18 May 1983.

The Colvins now began island-hopping again, this time to Curacao, definitely a ham-friendly place. They were met right away by PJ2KI, PJ2WG and PJ2CZ, all members of the Radio Amateur Club in Curacao. They sped the Colvins through customs. Herman, PJ2CZ, had a small house ready for them at Westpoint and took them directly there. Jose (pronounced "Josee"), PJ2MI, who was in charge of amateur radio licensing then, had their radio licenses waiting for them. "We were very grateful that things were all set up for us because we had only a couple of days before the ARRL International DX Competition started," Iris remembered. The writer was there in 1974 and recalls the same sort of hospitality.

French Guiana

Operating W6QL/PJ2 from 4 to 21 March 1982, the Colvins racked up 9,000 QSOs with amateurs in 148 countries. They operated a little on the new 10 MHz, CW-only band that had come about after the 1979 WARC. First QSO was VE3GAU, last was P29AB, and their DXCC #24,295 was issued 18 May 1983.

Curacao ended yet another six-month, six-country swing for Iris and Lloyd Colvin, yielding 56,000 contacts,

Guyana

Guyana

six DXCC awards, and several contest certificates.

The YASME Foundation held its annual meeting on 24 April 1982, at "Visalia." A significant event was a plaque awarded to director Bob Vallio, W6RGG, for "19 years of dedicated service as Secretary/Treasurer and QSL Manager of the YASME FOUNDATION and for perpetuating the traditions of the YASME FOUNDATION." Iris and Lloyd Colvin's operations had taken YASME QSL chores into the stratosphere. YASME directors were W6QL, W6KG, VK2EO, W6OAT, JA1KSO, OH2BH, W0MLY, K5RC, W6AM, and VP2VB.

On 23 July 1982 Dick Spenceley, KV4AA, died of a heart attack, in St. Thomas. This milestone, the end of an era, in a way, sadly went mostly unreported in the amateur radio press until much later. Spenceley had been ill, and not active on his radio, for some time.

As Iris and Lloyd spent the spring and summer of 1982 planning their next six-month amateur radio expedition, they were more than ever focused on operating from new - for them - countries, sometimes from countries they had visited before but not operated from. Making DXCC from 100 different countries was now definitely a goal within their reach. At each stop, as they ploughed through the pile-ups of stations calling and filled pages of their log books, they also kept track of countries worked. By doing so, they did amateurs around the world a favor - operating from a particular spot on the globe, it is easy to be lured into "going with the propagation." For example, from the Caribbean, there's a "pipeline" to North America; from the Pacific there's a pipeline to Japan. Both have vast numbers of DXers and it's easy to sit there and fill page after page with contacts in the same area.

By chasing DX, Lloyd and Iris expanded their horizons, as they kept their ears tweaked for weak callers from areas out of the choice propagation.

The Colvins still wanted to activate some of the rarer countries, too, and beginning with this trip they took a new tack: they would apply for operating permission before leaving, then press for approval after they had arrived in the general area.

An example was the Somali Democratic Republic, where there were no licensed amateurs and had been no operations for years. In the early 1980s Somalia and Ethiopia (another very rare DXCC spot) were at war, fueled by support by both sides of the Cold War. The Colvins had been writing to Somalia for years. In early December, while they were in east Africa, they got word that a letter from Somalia had arrived at their home in California. Dated 5 December 1982, it was from Dr. Abdullah Osoble Siad, Somali Democratic Republic Minister of Posts and Telecommunication, and it said "In reference to your application of 17 November 82 we regret to inform you that according to our Law. No. 1 of 22 December 1975 Article 32 the Operation of Amateur service in our Country is forbidden therefore this Ministry is unable to give you the above mentioned licence.

"We have studied your references and noted your excellent experience in this aspects, however we can not make any exception for you."

The Colvins subsequently made a side trip to Somalia, and went to the office of Dr. Said. The following summer, in August 1983, the Colvins received a letter from Dr. Siad, saying:

> I'm very much pleased of having received your letter of 6 Aug 1983. I'm pleased, too, that you have visited many countries. I hope, as you desire, that you will visit other Countries. I remember very well the time you both spent with me in my office.
>
> I hope I have satisfied you as much as I could. I ask God to be always in your assistance. As I think I have told you, I have a son of name ... who studies first year of Agriculture in the Sam Houston State University. He is without Scholarship, so you can imagine how many difficulties he faces or I face in finding hard currency for his living and studies.
>
> I would be very much obliged to you, if you or you Foundation could suggest or find a way for settling the case of my son."

41 · East Africa and the Middle East, 1982

This will probably be one of the most difficult, most dangerous, and most expensive YASME DXpeditions ever. We will use the call G5ACI/AA. - Iris and Lloyd Colvin

The Colvins began their fall 1982 operations from Djibouti. Formerly called French Somaliland, the country had received its independence in 1977. Operating from Gabode as J20DU from 7 to 31 October 1982, the duo tallied 7,000 QSOs and tied their best effort for countries worked: 155 (the same number they worked from Surinam the previous winter). Their operation in the CQ Phone DX Contest netted a score of 1.8 million points, also a record for them. They easily added another DXCC certificate to the stack: #26,019, 29 August 1985. First contact was Y26DO, last was VE6WQ.

Before arriving in Djibouti the Colvins had made a stop in Doha, Qatar, where an American, Mike Smedal, had been very active for some time as A71AD. Using his address as a base of operations, Lloyd and Iris sent letters of inquiry to the Arab Republic of Yemen ("North" Yemen), to Ethiopia, and to the People's Democratic Republic of Yemen ("South" Yemen). All of these requests for amateur radio operating permission were either rejected or ignored.

But there was excitement in the air. In their report from Djibouti the Colvins said "NOW FOR THE BIG NEWS! We will operate from Abu Ail for 48 hours starting on or about 2 December 1982. This will probably be one of the most difficult, most dangerous, and most expensive YASME DXpeditions ever. We will use the call G5ACI/AA."

Iris and Lloyd cooked this up while they were in Djibouti. Abu Ail is basically a lighthouse on a spit of rock 350 feet high, off the coast of Djibouti, at the time an "international island" in the Red Sea. In 1982 it counted as a DXCC country, subsequently "deleted" from the DXCC list. They were invited to join with two Frenchmen, Christian Dumont, F0ECV, and Jean Michel

Mike Smedal, A71AD

Yemen

Gabouriaud, F6GBQ (recall that just a few years before Djibouti, and Abu Ail, had been French overseas territories). The two were J28DP and J28DL in Djibouti.

The operation took place 5 to 7 December 1982, "pretty much as planned." The first of 4,000 contacts (including 10 MHz) was across the water to J28DQ; the last, with PY5OC. The team squeaked by with 104 DXCC countries worked, allowing the Colvins to add the G5ACI/AA DXCC, #26,408, 16 April 1986, to their collection.

Everything had to be ferried to the island, first on the ship "Fahnous," then by dinghy: three transceivers, one amplifier, two generators, a tri-band beam, verticals and doublets, plus food, beds and gasoline. "We had very rough seas on the trip from Djibouti to Abu Ail," the Colvins wrote, but the return trip was smooth. "The loading, landing, and unloading of our small dinghy from the main ship, the 'Fahnous,' was difficult and dangerous, but was made successfully.

"Carrying all of our equipment almost straight up a very steep and high mountain to the lighthouse at the top was a difficult and tiring job. We first got on the air four hours after landing and operated continuously, using two stations simultaneously, for 41 hours."

Iris and Lloyd wrote an article about the operation that *73* published in July 1983. They said:

> All four of us had to sign agreements that we and our heirs and assigns accepted full responsibility for any casualty or mishap that might occur. We agreed not to interfere with the functions of the lighthouse in any way whatsoever. We agreed to take with us anything that we needed in the way of food, water, generators, gasoline, and anything else that we might use.

Arrangements were made for passage on the Fahnous (Arabian word meaning "lamp"), the supply ship that services the island monthly with food, water, and supplies. We agreed to pay for our passage and for the diversion of the ship from its regular route, putting us ashore on Abu Ail, continuing on to Jabal At Tair, and then returning 48 hours later to pick us up. We fully understood that the ship could not remain anchored near the island and that we must be ready and waiting to board quickly when the ship arrived.

We made a list of essentials, keeping in mind that time was limited and that everything must be carried up the steep cliff. The list included food and water for 48 hours, sleeping bags, gifts for the lighthouse keeper, and our radio equipment, consisting of two generators (one 1.5 kW and one 500 watts), gasoline, antennas, coax, a Yaesu 707, a Kenwood 520, a Yaesu 902DM transceiver, and a Heathkit 230 amplifier.

The good ship Fahnous left Djibouti with the four of us and our equipment aboard at about 3:00 pm on December 4, 1982. The seas were extremely high during the night but were somewhat calmer when we arrived at Abu Ail at 8:00 am the following morning. All of our equipment and ourselves had to be transferred from the ship to a small dinghy. The seas were still rough, and the dinghy was bouncing up and down alongside the ship as much as 6 feet with each wave. Both the loading of the dinghy and the landing ashore were dangerous and tricky. It would have been very easy to have lost our equipment or suffered injury ourselves. Fortunately, the only mishap was to the dinghy, which hit a reef during the landing, causing some damage to its side. After landing, we started the ascent up the cliff with our equipment, giving priority to the actual radio gear. The wind blew continuously, making the trek up even more difficult.

The Abu Ail group

Abu Ail

By about 10 am, G5ACI/AA was on the air, using the vertical antenna, the Yaesu 707, and the small generator. A little after noon, both stations were in operation. The four of us took turns operating for sessions of approximately two hours each. At the same time, work was continued to bring the rest of the supplies up the hill and to put up doublets for 40 meters and the new 30-meter band.

We encountered a number of minor setbacks and delays. The limited space at the top of the rock made it impossible to locate the two stations far enough apart for both to operate on some frequencies simultaneously without interference. The guys for the TH3 had to be located over the edge of the cliff, which slowed the process of erecting the beam. We also ran into some difficulties in getting either the Yaesu 707 or the Kenwood 520 on CW operation.

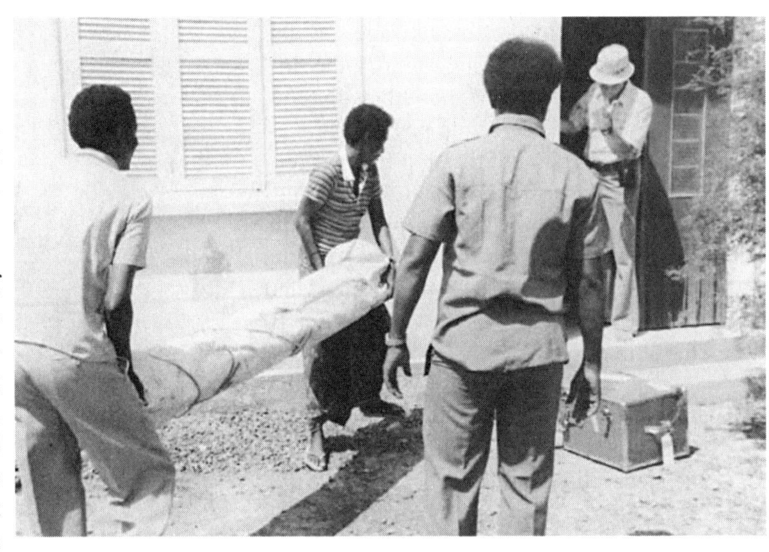
Djibouti

All obstacles were eventually overcome. We stayed on the air continuously, operating on 5 bands, both phone and CW. We made contact with over 4000 amateurs in 105 different countries.

At about 6:00 am on the morning of December 7th, we sighted the Fahnous approaching. We began dismantling the stations and pulling down the antennas. We bid farewell to the lighthouse keeper, who said that our visit was an exciting change from his usual routine. The descent was somewhat easier and faster than the ascent. The gasoline had been used up, and any extra food was left behind.

As we boarded the Fahnous, we felt tired and exhausted - but we also were very happy over the successful completion of Operation Abu Ail. A few observations and recommendations to anyone going on a similar DXpedition are listed:

1. Be sure that you obtain proper authorization and licensing, in writing, well in advance.

2. If possible, set up the complete operation, including all equipment (generators, etc.), at your home base before going on the DXpedition. We did this and discovered and corrected a number of time-consuming problems on the test run. These included how to solve problems of guying antennas, what interference we could expect to find, and what frequencies to use when simultaneously operating several stations close to each other. We found that we could operate a station on SSB and a station on CW simultaneously on the 10-meter band - but this could not be done on the other bands.

Kuwait

3. If you are going to climb a steep mountain carrying 50 to 80 pounds, figure out in advance the easiest way to do this. We found that the best thing was to reduce weight. Do not carry one pound of anything that is not required!

4. Try to think of everything in advance. We had small lights to use for nighttime operation, but we forgot to carry shades for the days. We had to waste some DXpedition time fabricating makeshift shades.

5. Don't forget to bring earphones - we almost did. With the noise from generators and two stations operating at once, the use of headphones was essential.

6. Remember to bring a soldering iron. Something will come up when it will be needed. We used ours on two occasions.

7. Figure out in advance how you are going to keep your logs. We were so anxious to get on the air that we got our log-keeping of the two stations a little mixed up at the start.

8. If you are going by sea, bring seasick pills and use them. We had the pills OK, but one of our operators thought that he wouldn't need them. He did.

9. If you have any equipment requiring batteries, bring along extras. Both of our keyers and one flashlight required battery changes.

We were very lucky as to the ability of our operators. All four operators carried their fair share of both the physical work and the operating. All operators were trained, experienced DXers, and the handling of the pileups was no major problem.

The YASME Foundation and everyone connected with it wish to thank the DX operators of the world for the general courtesy they showed in standing by until they could work the rare country of Abu Ail. "We gave a lot of hams their first contact with Abu Ail. It was not, however, the kind of operation to be done every weekend. We all came back very tired and worn-out, but also very happy. We hope that everyone who needed us for a new country made it.

Following the Abu Ail operation the Colvins traveled to Muscat in the Sultanate of Oman, which at the time had generally been supportive of Mideast peace initiatives and had entered into a security agreement with the United States in 1980. Despite that, Iris and Lloyd later said they felt "lucky" to have received permission to operate because "visitors, tourism and transit guests are in general not allowed." W6KG/A4 was on the air from the Sea View Hotel in Corniche from 24 December 1982 to 8 January 1983 10 through 80 meters; the new 30 meter band was not allowed there yet.

The Colvins generally had avoided operating from hotels, and the Sea View was full of television sets, but they had no problem with interference because all the local broadcast channels were UHF. Oman was the first of the oil-rich Middle East countries the Colvins visited and they found it "a booming frontier-land, with construction of new buildings, new roads, and general expansion everywhere. Much of the money for this comes from the new-found oil, but this fact is not mentioned or discussed much by the government or local newspapers," they said.

From the hotel Iris and Lloyd, with their triband beam and doublet antenna on the roof, logged 5000 QSOs with 127 countries, adding DXCC #26,023, 29 August 1985, to their collection. First QSO was VK6AFW, last was AB9V.

From Oman, the Colvins visited the People's Democratic Republic of Yemen, where the country's sole legal political party was the Yemen Socialist Party. In just a couple of days they took many more photos than usual,

Kuwait

undoubtedly because there was no amateur radio operation. On May 22, 1990, the PDR Yemen and the Yemen Arab Republic established a unified state.

W6KG/A7 was on the air from Doha, the capital of Qatar, from 14 January to 5 February 1983. At the time Qatar was similar to Oman - a newly rich oil country. Iris wrote of their time there that "Although the treatment of women here is more liberal than in the past, only approximately 5 percent of the people seen on the streets of the modern city of Doha or in the restaurants, are women. The biggest supermarket in town has one day set aside just for women and their families. It is not felt appropriate to have single men and single women go shopping together!"

The Colvins set up at a 14-story hotel: "For the first time ever, we operated from a skyscraper, with a perfectly clear view in all directions." In three weeks they logged 8,000 contacts in 135 countries on 10 through 40 meters, first in being EA8QO, last JA3DAY. Their W6KG/A7 DXCC is #26,024, 29 August 1985.

After this successful visit to Qatar the Colvins wrote "Visiting personnel of all kinds, and especially radio amateurs, are not encouraged to visit in the Arabian countries. We are extremely grateful to Mike [Smedal], A71AD, who has helped us enter and operate in several Arab countries."

Onward to Saudi Arabia, where U.S. personnel ran HZ1AB at the air force base at Dhahran. There were some 30 licensed operators there at the time. Iris and Lloyd were given permission to operate the station during the ARRL CW DX Competition on 19 and 20 February 1983, with four of the resident operators joining the fun. They made 1,500 QSOs with the U.S. and Canada (that's the nature of the ARRL contest). As a result, no DXCC for the Colvins from this country.

Iris and Lloyd observed that stores sometimes had no one working in them, and sometimes closed stores had no protection against theft (except the draconian Saudi laws).

They also visited two of the active locals, HZ1TA and HZ1TC in the Capital City of Riyadh,

Next door was the Colvin's next destination, Kuwait. They managed to get the special call sign 9K2QL and operated 28 February to 16 March 1983, including the ARRL Phone DX Competition.

They started out at the Marriott Hotel, housed in a docked ship much like the Queen Mary in Long Beach, California. But the portholes couldn't be opened and there was no way to get the coax line into their stateroom. So, they packed up and moved to one of the nearby chalets, convincing the manager that "our 3-element beam would be a desirable addition to the swimming pool and garden area of his 5-star hotel!"

6,500 QSOs resulted with 125 countries, the first UK6ACQ and the last with UA3HI (DXCC #26,022, 29 August 1985, and another contest award.

Last stop, Jordan, where King Hussein and the Royal Jordanian Radio Amateur Society support and encourage amateur radio. King Hussein was a longtime, active ham with the call sign JY1 (what else?). The society had "a beautiful club house and a first-class amateur station," the Colvins said, and qualified visiting radio amateurs were allowed to apply for permission to operate the club station.

Iris and Lloyd worked JY8KG from 17 March to 4 April 1983 before heading home. They logged 8,500 contacts. In the CQ WPX SSB contest they teamed up with JY4TJ and JY5MB, for 4.01 million pts. The JY8KG DXCC is #26,020, 29 August 1985. They worked 131 countries, and their first QSO was with OH2KT, the last, with UR2PRC.

After finishing this six-month swing through east Africa and the Middle East, Lloyd and Iris had visited all the Arab countries except Iraq. They felt that "the top communication officials in the Arab countries have an understanding of amateur radio and are favorable toward it. When amateur radio operation is not permitted it is usually [because of] the security officials. If, and when, fighting and political tensions diminish, we can expect more amateur radio operation."

Jordan

Jordan

This tour netted another 50,000 contacts for the Colvins, and yet more work for the untiring YASME Foundation volunteers in Castro Valley.

While they were home in Richmond preparing for the next trip, Iris and Lloyd got a letter from the recent president and DX columnist of the Radio Society of Great Britain, John Allaway, G3FKM:

> (12 September 1983) Dear Iris and Lloyd,
>
> Thank you so much for your letter dated August 20th. It was so nice to receive the latest news of your activities -- and particularly good to know that you still remember your visit to the UK and the time when we met. Sadly, both G2LB and G4IIJ have become Silent Keys since that time.
>
> Your adventures really are remarkable! I don't know how you both do it. You certainly have done an enormous amount of work on behalf of amateur radio and I am sure that you have found, as I have, that much pleasure is to be had from putting effort into the hobby - not only from taking from it.
>
> Thank you for your very kind invitation to visit you when in California. This I really would love to do! One of these days I will be dropping you a note to see if it is convenient. I am making a lightning

> visit to the USA soon, for the ARRL National Convention in Houston, and then returning home via Newington, but I'll only be over there for eight days and nowhere near W6. I wonder if you will be there? I don't think so if you are off to "finish off" South America.....
>
> May I take this opportunity to say a big "thank you" on behalf of so many of us for all the work you do on our behalf? Your efforts really are appreciated, and on a personal level I would like to thank you both for sending me all the bulletins/letters on your activities. I know that our members like to know about you via my column.

Iris and Lloyd made their annual appearances at Dayton, "Visalia," and elsewhere that spring, dug into business and family matters, then got ready to "finish off" South American countries in the fall. There were just seven left on that continent. They reported to friends that they were in good health. At this point they had visited in 154 countries (as hams count them).

"You may remember, [Iris and Lloyd wrote to many friends in September 1983], that we used to tell people that we hoped to visit all of the countries of the world. We are very much afraid that it will be impossible to visit all of the radio-amateur countries of the world. For example, it is not likely that we will visit Spratly Island. We do, however, have a new idea; perhaps we can visit all the countries of the world that have representation in the United Nations. We wrote to the U.S. Ambassador to the United Nations and found that the U.N. [has] representation from 157 countries.

We have tentatively decided that each year we will try to visit one continent and finish visiting the remaining countries there that we have not been to. We will also try to operate our amateur radio station in all such countries if we can get permission to do so.

"We like to stay in the U.S.A. each year from April through September. If you ever come to the U.S.A. during these months, please try to visit us here in California. We will always have a place for you to stay with us."

Iris and Lloyd sent an application for amateur radio permission to Uruguay but it was rejected in the summer of 1983. In the 1970s Uruguay was wracked by political strife, with thousands imprisoned and tortured. In 1981 a General Gregorio Alvarez was installed by the military regime in control, and the economy deteriorated. A demonstration in November 1983 in Montevideo drew 400,000 people; a general strike followed in 1984.

Iris and Lloyd also wrote to Argentina but no reply was received. They would visit Argentina in 1984 but would never operate from there (or from Uruguay).

42· South America, 1983-1984

If ever we had illusions of high-on-the-hog living for YASME DX trekkers, they were wiped clean that evening.
- Barry Boothe, W9UCW

In October the Colvins attended the ARRL National Convention in Houston, then flew directly to Bogotá, Colombia and opened up as W6QL/HK3, operating from 12 to 30 October 1983.

The Colvins received their operating permits almost immediately with help from the newest YASME Foundation director, Fred Laun, K3ZO, who was HK3NBB and the Liga Colombiana de Radioaficionados (LCRA). Iris and Lloyd had met the club's president, HK4BHC, in the U.S.A. In their report from Colombia, the Colvins said that it seemed like lots of the people involved in amateur radio licensing in Colombia were women.

"For example, the lady in charge of license applications from the Club to the Government is HK3AVH. The Chief of the Government licensing section is also a lady. Above her is the Vice Minister of Communications, another lady who is the daughter of the Colombian Consul General in San Francisco, whom we had met in the USA; and who kindly sent a letter to his daughter asking her to help us obtain our Colombian radio calls promptly. Another young lady amateur, HK3FKL, who speaks excellent English and Spanish, helped us and acted as an interpreter on several occasions."

W6QL/HK3 netted 6,000 contacts in 133 countries, in two weeks, the first being HK0TU, the last W4PRX. The Colvins rolled up a score of 1.8 million points in the CQ WW SSB Contest, and their DXCC is #26,031, 29 August 1985.

Next stop to "finish off South America" was San Andres Island off Colombia, where the most famous and active ham was Francisco, HK0BKX. Iris and Lloyd operated from his home, using their equipment and his antennas. Despite Francisco's activity over the years the Colvins found the pile-ups "tremendous. We had

San Andres

Galapagos

planned," they said, "for over a decade, to visit Francisco and operate there. It was great to finally make it."

From 4 to 20 November 1983 they logged just over 8,000 QSOs as W6KG/HK0, with amateurs in 126 countries. Their DXCC was #26,028, 29 August 1985; first contact was G4UZB, last HC2FN.

San Andres, a small island that caters to the tourist trade from South America, at least was warm compared to the cold of Madrid, Colombia, high in the Andes.

"San Andres reminds us," Lloyd and Iris wrote, "of some of the Caribbean Islands, but most of the people in the stores, as well as the visitors, speak Spanish. The local electric power system frequently stops for a few hours at a time. Fortunately, there was no major contest during our stay, so the power interruptions did not cause too much trouble. Most of the local stores have power generators for standby. When the power fails they bring their generators out on the sidewalk and let them run. During power outages, it looks unusual to see all the generators in front of the stores!"

Barry Boothe, W9UCW, had traveled to San Andres many times, and tells this story:

> This story has to do with the frugality that Lloyd and Iris displayed. Certainly, YASME funds were limited and had to be preserved for DX work.
>
> In the '80s, as Lloyd and Iris were putting together a trip to several spots in the Caribbean, they contacted me and my wife, Joyce, WB9NUL. One of

their destinations was to be San Andres Island, HK0, and perhaps Providencia. They knew we were very close friends with Francisco, HK0BKX, Justo, HK0COP, and others. They figured that after nearly 20 trips, we could provide details and introductions. We did our best to help. They were very appreciative and asked if we could get together at the Dayton Hamvention to chat about the islands. We agreed.

Bolivia

By Friday evening at Dayton we had not yet connected. As we entered a hospitality suite at our downtown hotel, Lloyd immedately suggested that we find a quiet place away from all the DX'ers. He said they had discovered "a great little place" nearby where we could have privacy, and "The food is really nice". They asked if they could take us there for dinner.

We had not eaten and had dressed up a bit, so we gladly agreed. Joyce, who was wearing a spaghetti strap evening dress with a high hemline asked if she should get her coat from our room. They said, "No, it's just around the corner." It was one of those cool April times and the blustery temperature outside was in the upper 30's. We all piled in an elevator and headed out. We walked and then jogged (I was miserable and Joyce was freezing!) almost four blocks before Lloyd said "I think it's just around this next corner." As we turned, we beheld the "great little place" where we could enjoy a peaceful, private dinner with our friends - It was called "Arby's"! [a fast-food restaurant].

Galapagos

Feeling a bit dumbstruck and certainly numb from the cold, we quickly entered. Iris told Joyce, "Let's go sit down and let the men get the food."

As Lloyd and I moved to the front of the line, he suggested that if I would cover the sandwiches, he would take care of the drinks and fries. That's how it worked. Conversation was difficult, for several reasons, and then there was the trip back to the hotel. My idea of calling a cab was rejected out-of-hand. If ever we had illusions of high-on-the-hog living for YASME DX trekkers, they were wiped clean that evening. Later, after we thawed, Lloyd asked if we could join them on San Andres and help with logging and other peripheral jobs. We pictured ourselves running out for food. We declined.

My buddy, Francisco (Pacho) had a few more stories about their tightfistedness after Lloyd and Iris left San Andres. Pacho said he was going to stay in the gas station business and leave the DXpeditions to more frugal types. Nevertheless, it was reassuring to know [the Colvins] guarded YASME [funds].

Before they left San Andres, the Colvins wrote letters seeking operating permission in Chile, San Felix Island, and Easter Island, anticipating arrival in Santiago, Chile, on or about 25 Jan 1984.

Meanwhile it was back to the continent, and Quito, Ecuador, another country high in the Andes, and the home of famous shortwave broadcast station HCJB as well as the place where the cubical quad antenna design was hatched.

Ecuador was suffering a decade of economic stagnation, exacerbated by heavy floods in 1983. Iris and Lloyd noted two earthquakes while they were there, "strong enough to rock the operating table. A devastating, major temblor occurred there in 1987.

Operating W6QL/HC1 24 November to 6 December 1983, filled log books with 5,000 contacts in 120 countries; first in W9IT, last entry JA2TCW. Their DXCC was #26,030, 29 August 1985. In the CQ WW CW contest they racked up 1.469 million points.

As was often the case, the Colvins' brief reports didn't say exactly where they operated from. In the case of Quito they did mention being able to see "several snow-capped volcanoes. We had our pictures taken with one foot in the Northern Hemisphere and the other foot in the Southern Hemisphere. After attending their first bull fight they said it was "something everyone should see at least once."

And, they said, "Both here and in Colombia, the major amateur radio clubs hold meetings every week, which is something that is not often done in the U.S.A. The Quito Amateur Radio Club is a very active organization with one of the finest buildings and antenna sites anywhere."

From the continent to the ocean once again, the Colvins flew to the Galapagos Islands for an operation 10 to 21 December 1983. They had arranged to meet Bud, HC8GI, the most famous ham resident there, active for years. Just three days after Lloyd and Iris arrived, Bud died. They had located their station at a small villa just a few hundred feet from HC8GI.

Everyone is familiar with the Galapagos's claim to fame, its national park status, and the visits of Charles Darwin. Iris and Lloyd toured the island and said "We saw a pair of giant, 100 year old tortoises making love. Many humans would like to enjoy that at 100 years old."

W6KG/HC8 racked up 6,000 QSOs with amateurs in 120 countries, putting DXCC #26,027, 29 August 1985, on the stack. First QSO was with W6YA, last with WA0GUD. There had been very little operation on CW from the Galapagos up until then, so they concentrated on Code.

Easter

Peru

1983 was World Communications Year and a few countries issued amateur radio call signs reflecting the event; the most famous for hams was, far and away, an operation from the extremely rare Laccadives Islands, that the government of India assigned VU7WCY. The Colvin's last South American stop of the year was in Peru, and they were awarded the call sign 4T4WCY. A call sign helps generate interest as hams scramble to learn what country holds the "4T" call sign block and prefix hunters check in to work another one.

Peru was politically in some turmoil in late 1983 as left-wing "Shining Path" guerillas gained influence and the U.S. government urged, with money, the Peruvian government to fight the growing of the coca plant, the Peruvians' top source of income.

They wrote "We spent New Years Eve here. The people celebrate the new year by building thousands of small bonfires. At exactly midnight, all of the electricity in this city of some 6 million people went out, and, from the high building where we are, we viewed the magnificent sight of the dark city with hundreds of small fires burning throughout the entire city and surrounding hills. It was not until 3 January 1984 that we found out that the magnificent sight was not entirely planned; terrorists had placed dynamite around many of the high voltage transmission line-towers leading into Lima, and blew up nearly 20 such towers at exactly midnight."

OA4OS helped the Colvins get their special call sign and OA4BI provided a penthouse apartment on a high building from which to operate. From 29 December 1983 to 13 January 1984 they made 7,000 contacts in 133 DXCC countries, the first contact HC2FN and the last YU7BPQ. Their 4T4WCY DXCC was #26,021, dated 29 August 1985.

Bolivia has never been particularly available on CW and the Colvins found their W6KG/CP6 call sign from there in good demand. They operated from 19 January to 2 February 1984, from Santa Cruz. The first contact was CX8DT and the last HB9CPR.

W6KG/CP6 produced some 4000 QSOs with hams in 128 countries, and DXCC #26,026, 29 August 1985.

At the time the Colvins were there, Bolivia operated under an "installed" president, who faced economic crises caused by Bolivia's crushing foreign debt. The Colvins reported that "For Americans, with U.S. dollars, Bolivia is the most economical South American country (costs about one-half U.S.) that we have visited.

Lloyd and Iris said that the Bolivian hams did a "magnificent job of making us welcome." They spoke at the LaPaz Amateur Radio Club in LaPaz, the National Radio Club in LaPaz, the Cochabamba Radio club in Cochabamba, and the Santa Cruz Radio Club in Santa Cruz. "Several dinners were held in our honor, special awards were given us, articles and photographs about us were in three Bolivian newspapers, and we appeared on TV in the capitol city of LaPaz."

LaPaz is the highest large city in the world, over 12,000 feet high. Iris and Lloyd went boating on the highest large lake in the world, Lake Titicaca.

The locals also were helpful to the Colvins at their next stop, Asuncion, Paraguay. An American there, Doug Wooley, ZP5XDW (now ZP6CW), was very active. Iris and Lloyd mentioned more than a dozen other local amateurs who pitched in with licensing and setting up their antennas, saying "We have never had that much help on any of our other stops." Operating from 9 to 22 February 1984, as W6QL/ZP5, they made 5,500 QSOs with 127 countries (DXCC #26,032, 29 August 1985.) First QSO was ZP5XDW, last was IK1AOI.

Paraguay had been relatively calm in the late 1970s and early '80s. The Colvins found Asuncion, city of nearly 2 million people, to seem to have a relatively high standard of living. The Colvins in their travels had often seen new high-rise buildings going up and it seemed one of these would be a good place to set up their radio station. They had sought such permission a number of times, and in Asuncion they were successful. ZP5PX, an architect, was friends with the architect of an eight-story apartment building being built next to a two-story hotel. The Colvins lived in the hotel and placed their antenna on the new building under construction next door.

Bolivia

Peru

From Paraguay, Lloyd and Iris flew to Easter Island, 2,300 miles from the coast of South America. They got there just before the start of the ARRL Phone DX Contest weekend, checked into their rented lodging, and rushed to install their station and beam antenna and doublets (they had their operating permissions, from Chile, in hand). With the station set up, antenna connected, all seemed ready to go as the contest starting gun approached. But "we turned on the receiver - it would not receive. Everything sounded mushy and the signals drifted. "

The 220-volt AC incoming line was only delivering 170 volts. So Iris and Lloyd jumped into a car and rushed to the home of Father Dave Reddy, CE0AE, Easter Island's most famous ham, who lent them a voltage regulator unit. That did the trick, getting their incoming voltage up to 200. They got on the air and stayed there for the next 48 hours.

From 2 to 16 March 1984 W6KG/CE0 made nearly 8,000 QSOs with radio amateurs in 127 countries. First was CE0AE, last was ZL2AAG, and their DXCC was #26,025, 29 August 1985.

The Colvins sent out their report on 16 March, saying "We would like to operate next on Juan Fernandez, but are having trouble obtaining transportation." Eventually, they got seats on a plane and got to Juan Fernandez about a week later. They described Juan Fernandez as "a rocky island with very high steep mountains extending nearly straight up from the sea. The only air strip is on top of one of these steep mountains, and only a very small plane can land there. After landing, there is a long walk to the sea where all equipment and persons must make a 2-hour trip in a small boat on the open sea to the small village, where the only people on the island live. It is located on a small circular cove surrounded by steep mountains.

The Colvins operated from within that inlet, with a reasonably good radio view to Europe and the U.S., but with steep mountains blocking the other directions. "From a radio standpoint, it was the worst location that we have ever operated from," they said. Electric power to the small village where they stayed was on just eight hours a day and the rest of the time "we tried our best to operate using a small generator that never seemed to work properly. In addition, there was no gasoline readily available to buy on the island, and we had to rely upon obtaining a little here and there from various inhabitants."

Despite these handicaps, the Colvins logged 5,000 contacts in 120 countries, from 22 March to 1 April 1984. They teamed up with Chilean amateur CE3ACA in the CQ WPX SSB contest for a 1.389 million-point score, too. The W6QL/CE0 DXCC was # 26,029, 29 August 198; first QSO was W9WHM, last was UN1JCC.

Celso, CE3ACA, had accompanied Iris and Lloyd to Juan Fernandez and while there acted as interpreter, radio repairman, and generator repairman. He did about a third of the operating, too. "We would never have had a successful operation without him," the Colvins recorded. Since there was no ship arrival or departure scheduled during their stay, and their small aircraft was strictly weight-limited, they left some of their radio gear behind in Chile.

Juan Fernandez was the last stop on this swing through South America, and after stopovers in Uruguay and Argentina Lloyd and Iris returned to California. From September 1983 to April 1984 they had tallied 55,000 QSOs from nine DXCC countries. "All in all, it was a great experience and we met many old friends and made some new ones," they said.

In April 1984, around the time of the Visalia and Dayton conventions, where the Colvins appeared, they also mailed inquiries for operating permission to these countries: Cameroon, Sudan, Chad, Guinea, Uganda, Central African Republic, Annobon Island (via Ecuadorial Guinea), Angola, Cape Verde, Zimbabwe, Sao Tome, Rwanda, Zaire, Burundi, Burundi, Madagascar (Malagasy Republic), Gabon, Mauritius, Uruguay, and Transkei (the short-lived, semi-independent South African area that was recognized by no other country and never did receive DXCC status). While this may seem like a long list, every time the Colvins returned from putting another half-dozen countries on the air, their list of possibilities grew that much shorter!

Uruguay

Also in 1984, Iris and Lloyd both were inducted as members of the Radio Club of America, an organization founded in New York in 1909 which, in succeeding years, recognized individuals of importance in radio. According to its charter, the RCA is "Organized for the interchange of knowledge of the radio art, the promotion of good fellowship among the members thereof, and the advancement of public interest in radio."

43· Foundation Business, Then Off to Africa, 1985

Don Wallace's accomplishments in ham radio and DX are almost beyond belief.
- Lloyd Colvin

Don Wallace, a friend of Don's, and Iris and Lloyd Colvin, 1984

The YASME board of directors, at their annual meeting in April 1984, counted the following as present: W6AM, W6RGG, K5RC, W6KG, W6QL, W0MLY, and W6OAT (Rusty Epps had given back the N6SF call sign he had held for a few years, in favor of his former W6OAT call.)

Don Wallace, W6AM, died on 25 May 1985. He was active right up to the very last. "On 24 May 1985," Lloyd and Iris wrote to friends, "he played in a golf tournament, was on the air from W6AM, and then, around 5 PM, decided to go to the Virginia Country Club, of which he was an active member, and play some gin-rummy. While playing cards, he suffered a stroke and was rushed to the hospital where he was kept alive by artificial means until he passed away the following day."

Lloyd and Iris flew to southern California for the funeral, and for a reception afterward at the "famous Wallace Antenna Farm" in Palos Verdes Peninsula.

At the service, Lloyd Colvin gave a eulogy to Wallace:

> I became a licensed radio amateur 56 years ago. A year earlier, 57 years ago, I bought my first copy of QST magazine. It cost only 25 cents. One of the main articles was about the accomplishments of Don Wallace, 6AM. At the beginning of the article was a large picture of an alarm clock with the hands set at 6AM. Underneath was a caption reading "To work DX, work 6AM at 6AM in the morning.
>
> Don Wallace's accomplishments in ham radio and DX are almost beyond belief. He entered ham radio in 1912 as "6OC". During the following 74 years, Don devoted most of his energy to amateur radio and DX. A few of his accomplishments I will list briefly:
>
> He was a leader in the use of high frequencies in the early days of radio when spark radio was used by all communicators. He was a Navy radio man in the early days of World War I. He was radio operator on the USS George Washington taking President Wilson to the Peace Conferences in Europe near the end of World War I. He was in charge of the vital communications between the United States and the President of the US while aboard.
>
> Don wrote several articles in QST and other publications about early long-distance communications on the higher (VHF) bands. He was issued a license and built an experimental radio station, 9XAX, in 1924. For his early-day accomplishments in amateur radio, Don was awarded the Hoover Cup by the Department of Commerce in 1923. In 1923, the Board of Directors of the ARRL honored Don for the "best all-around homemade amateur radio station.
>
> Don wrote and published one of the first amateur radio handbooks. In 1927, Don was elected Section Communication Manager for the ARRL in Los Angeles, a position that he held for three years. From 1927 to 1932, just about every other issue of QST had articles or notes about some radio activity, especially DX, by W6AM. From that time until now, Don has been one of the principal leaders of DX throughout the world.
>
> For many, many years, Don has occupied the No. 1 spot on the DX Honor Roll published by QST. If you count pre-World War 2 QSOs, Don has worked nearly every radio-amateur-country in the world, including countries that have been "deleted" through the years.

Many hams throughout the world know of Don's fantastic rhombic antenna farm, located here. For the last several years, Don has held open house, once a year, here at PalosVerdes Estates. Each year more and more hams have come from all over the world to visit one of the world's largest and most interesting ham stations and to hear a technical explanation by Don as to how it was built and how it operates.

Don traveled in nearly 100 countries, operating amateur radio in most of the countries. He has been an international ambassador for amateur radio goodwill in all of his foreign travels. Don was one of the first hams in the world to operate a complete one-KW amateur radio station in his automobile. He has enjoyed working hams all over the world as he traveled in his car, driving, and at the same time, working stations at 25 WPM by CW.

In 1975, Don accepted the position of President of YASME, an international DX organization. He held that position until now. In 1978, Don was elected to the exclusive CQ magazine DX HALL OF FAME. In 1984, just last year, Don was elected to the Quarter Century (Wireless Association) HALL OF FAME.

These are all concrete accomplishments of Don Wallace - but there is more that is more difficult to put into words. He was a tall man, a handsome man, a man who [fit in] anywhere. He was a fair man, a man who did not bear grudges. A man who could tell jokes. A person who spoke well of his fellow man. An intelligent man, who remained active right up to his passing.

Swaziland

Don Wallace, W6AM, W6AM! We are all calling you, Don. All of us in this group today are holding our very last QSO with you. We have enjoyed knowing you and talking to you through the years. It is with great regret and with great feeling that we say 73 and 88 to you for the last time.

Swaziland

For this, Wallace's son William Wallace, said "The Wallace family and I wish to thank you very much for your very eloquent talk on your experiences and thoughts of the life of our father Don C. Wallace.

"We will miss him but he will be remembered. Thanks to you and the rest whose lives he has touched. Your generous gift to come and speak at the Radio Ranch will not be forgotten."

On 9 September 1985, YASME Foundation secretary Bob Vallio, W6RGG, put out his now semi-annual announcement: "YASME DX-PEDITION GOES FORTH AGAIN."

Lloyd and Iris Colvin were taking off for South Africa and planned to be operating as W6KG/ZS early in October, through the CQ WW Phone DX Contest, and then travel to various nearby countries, operating half SSB and half CW in each country, for a period of about three weeks each. Although they do not guarantee it, Iris and Lloyd hope to visit and operate in Z53, A2, 7P, 3D6, S8, ZE, 7Q, 9J and possibly CN8; returning to the U.S. in Apr. 86."

These announcements were by now regular events and anxiously awaited by DXers, to find out just where the traveling Colvins might appear from next.

South Africa was a nice spot from which to begin another six months on the road, with many active local amateurs, comfortable accommodations, and painless licensing. Aided by their good friends in the Johannesburg branch of the South Africa Radio League, the Colvins were invited to operate from their "splendid" club station, and they did so, using their own equipment with the existing antennas.

Operating W6KG/ZS from 4 to 26 October 1985, Iris and Lloyd logged 6,500 contacts in 143 countries (DXCC #27,138, 28 May 1987). First QSO was with UA3LDC, last with G3UJE.

"To some extent," they said, "our DXpedition operation in Johannesburg was a little like someone going to the U.S.A. on a DXpedition; however, much of the groundwork for entering and operating in nearby DX countries was best done from Johannesburg. " One of the local hams, Wolfy Matz, ZS6BYM, recounted meeting the Colvins at the radio store operated by Julius, ZS6AF. Obviously the Colvins made quite an impression:

> Len, ZS6BYE, a very good friend, and I were having coffee with Julius, ZS6AF, at his emporium. "We're having a top-class meeting at the clubhouse

South Africa

> this afternoon," he informed us. "Lloyd and Iris Colvin are here from America, and they'll be giving a lecture about their travels." I looked at Julius blankly.
>
> "Don't say you've never heard of them," he said in a surprised tone. "They're only the most famous DXpeditioners in the world." "Len and I would love to go," I said, "but we're going cycling on our bicycles this afternoon."
>
> Julius rolled up his sleeves, and said in a threatening tone of voice "Do you think you can ride with a broken kneecap...?"
>
> "No," I answered, and that's why Len and I cycled to the Johannesburg clubhouse, ZS6TJ, looking quite ludicrous in our cycling gear. But I'm glad we came. Of all the lectures I have ever attended, this was the most unusual, and perhaps the most interesting.
>
> Lloyd, W6KG, and his wife Iris, W6QL, spend their time traveling around the world. Lloyd received his ham radio license in 1929, and Iris received hers during the war. They decided to make ham radio contacts from as many places in the world as they could manage.
>
> In 1965, the YASME Foundation, set up about 10 years previously [more like five years - ed] agreed to promote [the Colvins'] operations from rare DX areas, as well as assisting them in every way possible as far as publicity and QSLs were concerned. However, the Colvins had to pay their own way. At this [date] (1985) they have traveled to 166 countries, held 122 different call signs, and made some one million QSOs.
>
> They hail from Richmond, California (I often wonder why they hail; why don't they snow, or drizzle, from California?). They are semi-retired from their construction company in the United States, but their interests are looked after by their married daughter, who, incidentally, is also a ham.
>
> Whilst in Johannesburg, they have made ZS6TJ, the Johannesburg Branch of the SARL, their headquarters, and spend a lot of their time operating the club's equipment. In three weeks they have made 6,500 QSOs with 143 countries. They normally spend about a month in each place, meeting hams, making DX contacts, and sightseeing. I'm sure their experiences can fill a book. In fact, their lectures encompass, not the technical aspect of ham-

ming, but rather their adventures and experiences in each place they visit.

Iris is blonde and vivacious, and Lloyd a dapper, cheerful personality. They are both lively, articulate and wonderful people, full of good humour and patiently ready to answer questions about their lives and experiences. I chatted to Iris and found she and Lloyd are in their early seventies. In answer to a question I put to her, Iris replied "We've very impressed with the hams we have met in South Africa. In fact, ZS6TJ is one of the few working clubs with radio equipment available to a visitor that we have come across in our travels. Most other clubs are merely of a social nature."

"What equipment do you take when you travel?" I asked.

"We have a special bag, and carry a TH3 antenna as well as a radio rig for 40, 80, and 160 meters."

The way Lloyd and Iris conduct their lectures is quite novel. They have a banner hanging from a wall printed with the names of all the 166 countries they have visited. Lloyd or Iris ask a member of the audience to pick a country from the list, and then speaking alternatively, recount their experiences or impression, and sometimes give a few startling unknown facts about that country.

I was completely enthralled, and most impressed by their prodigious memories, as they recalled their feelings and impressions. One of their more hilarious escapades happened in Japan [this would have been in the late 1940s]. In a room in their hotel, they had set up their equipment. However, Iris complained that when her lips accidentally touched the mic, she suffered a minor shock. Lloyd offered to attach an earth to the offending piece of equipment, and accordingly dropped a wire out of the window, which he planned to attach to a pipe on the ground.

However, the wire became snagged on an obstruction on the window of the room directly below them. Lloyd went to the room below, and knocked on the door. It was opened by a luscious and most beautiful girl wearing the scantiest of clothing. Embarrassed, Lloyd explained the situation, but she obviously did not understand what he was saying. He was at that time, in his colonel's uniform, and this probably was the reason she allowed him into the room.

He unsnagged the wire, and while oggling this most beautiful creature he inadvertently opened the door of a walk-in closet (cupboard in our language) and stepped inside. Realizing his mistake, he came out of the cupboard, and looked straight into the eyes of the beautiful lady's husband, who had just arrived and seen him emerge from the cupboard. He did not try to explain, but made a bee-line out of the room. Lloyd avers that the velocity of a bullet coming out of the muzzle of a gun was slow compared to the speed at which he made his exit.

The next stop for the Colvins was Namibia (Southwest Africa), where they stayed at the Continental Hotel in Windhoek, while operating, as ZS3/W6QL, from the club house and station of the South [West] Africa Branch of the South African Radio League. They used both their own antennas and radio equipment, with help setting up from, especially, "another husband and wife team, ZS3KB-ZS3MB, the club president ZS3HB (also chief justice of the supreme court) and the club secretary/treasurer ZS3TW."

Lesotho

At this time, many countries were coming around to the more logical way of assigning call signs to visitors, with the country prefix first, followed by the visitor's own, home call sign. Hence, ZS3/W6QL. Iris said this generated a lot of confusion: "Many amateurs answered us incorrectly. Some called us W6QL/ZS3, others ZS3/W6, others just ZS3, or just W6QL, and still other such strange calls as ZS3QL, or ZS3A, etc etc. If you worked us, please check your log and make sure that the call is correct and send your QSL showing the correct call."

The Namibia operation, from 29 October to 26 November 1985, coincided with the CQ WW CW contest, and Lloyd and Iris racked up a 1.7 million-point score. Overall, they made nearly 10,000 contacts in 146 countries, and garnered DXCC #27,140, 28 May 1987. First QSO was ZS1OZ, last was N4MM.

Namibia

Next stop, the Kingdom of Lesotho, and a short plane ride from Windhoek takes them there. Lesotho is a tiny country completely surrounded by South Africa, formerly known as Basutoland, and became an independent country in 1966. A little more than a year after the Colvins' visit a military coup shuffled power within the kingdom.

Iris and Lloyd operated from the Maseru Casino Hotel, a first for them, and, in another first, set up their entire station and antennas twice in the same country. The hotel manager allowed them to put their antenna on an unused portion of the hotel some 600 feet from the main casino lounge and the occupied hotel rooms. He gave them an unused room in the vacant portion for an operating room, while they had another "elegant sleeping room near the casino, swimming pool, restaurant, etc. It appeared to be a perfect setup. If something is too good to be true...."

After everything was installed, there was no radio or TV interference observed. Looking good! But when they turned on their transceiver; all they could hear was "a very loud hum and roar." They suspected that the trouble was caused by a 33,000-volt transmission line some 1,000 feet away. This was positively confirmed by borrowing a portable short-wave receiver from a good friend, Gunter, 7P8CI. Then, using that same receiver, they found a spot at the main entrance to the hotel grounds that was free of interference. Lloyd and Iris moved antennas and operating position, not to mention their personal items, some 1,200 feet to a small room in the quiet area. "The double installation took us two days of hard work, and, in addition, we both got serious sun burns."

Undeterred, 7P8KG logged 5,000 contacts from 28 November to 21 December 1985, with hams in 137 countries. DXCC was #27,135, 28 May 1987; first QSO K8MFO, last AA4CM.

It's another short hop to Swaziland, fully independent since 1968, where the Colvins opened up as 3D6QL on 21 December 85. Through 10 January 1986 they worked 6,500 contacts in 130 countries, operating from the Highlands Inn, Piggs Peak.

Iris and Lloyd put together a "small, multination, two-vehicle convoy" in Lesotho and drove to Swaziland. They were joined by 7P8CM (English), 7P8DF (German), the girlfriend of 7P8DF, a native of Lesotho; and were assisted in the operation by 7P8CI (Austrian).

The others were going to Swaziland to conduct the first major amateur radio satellite operation from there. The satellite operation was very successful with some 400 QSOs with amateurs in 40 different countries.

The Colvins recalled "When we reached the border of Swaziland, we were all granted entry without question except the German, 7P8DF. Germany had no [instant] reciprocal visa agreement with Swaziland. It took a lot of time and effort to finally get 7P8DF into Swaziland. No one showed any interest or asked any questions about our large amount of radio gear but they were most reluctant to issue a visa to the German."

The 3D6QL DXCC was #27,134, 28 May 1987. First QSO was 9H4G; last, WF4I.

Iris and Lloyd said that all their spare time in Swaziland was spent trying to get visas to enter Mozambique, but thus far, their efforts have resulted in failure. They ran an advertisement in the Times of Swaziland newspaper in late December that read "HELP. Two tourists want to visit Mozambique. If you can help, please call Mr. or Mrs. Colvin, Room 6, Highlands Inn, Pigg's Peak, Phone 71144.

Filling in a semi-circle around South Africa, the Colvins trekked to Botswana next and set up in a motel which they said was in the general area where a raid was made a few months before by South African Army troops and some 15 people were killed. Locals Mac, A24KM, and Gerald, A22TJ, helped out in station setup. From 15 January to 2 February 1986, A25/W6KG completed 5500 QSOs. (DXCC #27,139, 28 May 1987.) First contact was with OK1MNW, and the last with ON5JY.

Completing the semi-circle, the penultimate stop on this African trip is Zimbabwe, formerly Rhodesia and independent only since 1980. An operating period of 10 February to 4 March 1986 includes both the Phone and

CW ARRL DX Contest weekends, for which the Colvins win awards for both. Operating from the Tesseskane Hotel in Harare, they run up 8,000 contacts in 135 countries. The total contacts second-best for this trip.

Molly, Z21JE, an active amateur and mainstay of the local branch of the Zimbabwe Amateur Radio Society, met the Colvins upon arrival and "spent a lot of her time helping us in all sorts of ways." Molly's husband was not a ham, but her brother, Dudley, was Z22JE.

Their hotel near downtown Harare, the capitol city, had a stairway leading to the flat, concrete roof which made it easy to put up the TH3 beam and a "longwire" for the low frequency bands. The hotel had "a TV room, three radio channels in each room, an intercom system, several public address systems, a busy nightclub with all kinds of electronic equipment, plus individual radios and TVs in the hotel.

"The ease of putting up our antenna was great, but when we were ready to operate, we crossed our fingers and held our breath. There was NO interference from our station to any of the above listed TVs etc, etc."

Lloyd secured the metal push-up mast that supports the TH-3 and electronically ground it to a large metal water tower on the roof. "We suspect (but cannot prove) that such an antenna installation may have helped eliminate TVI and BCI problems. In any case, we were very thankful to cause no interference."

Iris noted that Zimbabwe had sent them, in advance of their arrival, a "License to possess a Radio Transmitting Station. This was a great help in clearing customs. A permit to operate our equipment was issued after we were in the country."

The W6QL/Z2 DXCC was #27,139, 28 May 1987. First QSO was VS6TQ, last was KA3DBN.

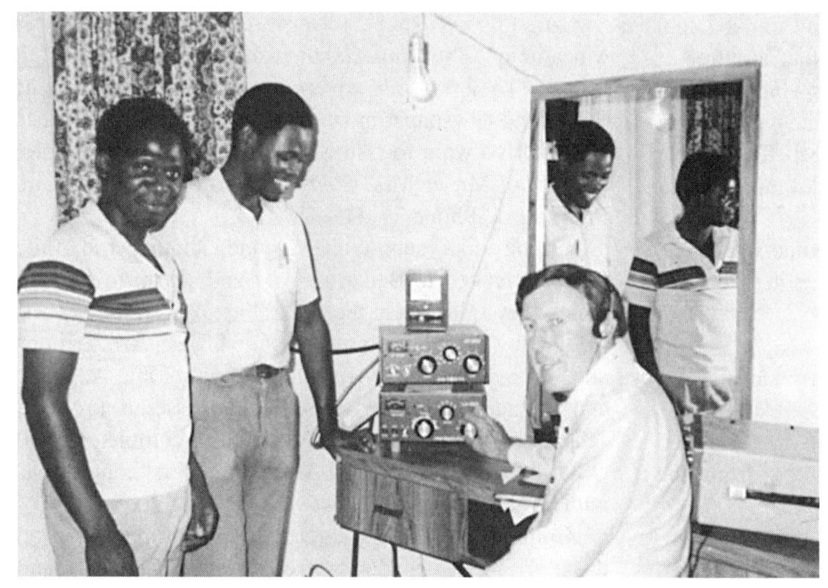

Botswana

The next operation, and last for this trip, was from Zambia (formerly Northern Rhodesia, independent since 1964), as 9J2LC, where "two wonderful radio amateurs, 9J2FC and 9J2KL, helped us obtain licenses and a great location," at the Andrews Motel, in Lusaka. In just 15 days on the air, ending 24 March 1986, Iris and Lloyd logged 8,000 contacts in 137 countries. Their DXCC was # 27,136, 28 May 1987; first QSO G3VJE, last K4CEB.

Their report to the press at the end of the 1986 Africa trip said "We would like a QSL card for every one of the 50,500 total QSOs that we had. We promise that ALL QSLs received will be answered. An SASE and contribution is appreciated - but NOT required. We want to thank all of the radio amateurs of the World for their generally fine DX operation during a time that is supposed to be a period of low sunspot activity and decreased DX opening!"

The Colvins offered to appear at radio clubs and conventions during the spring and summer, mentioning having appeared either on TV or Radio in nearly 100 countries, including, in January 1986 on National Radio in Botswana. They were the subject of a two-hour, prime-time interview and travel discussion, they said.

Their brief "radio resume" at this time also said that after 20 years of world travel, Iris and Lloyd in early 1986 had been at one time or another members of nearly 100 radio clubs, including honorary and associate memberships in many foreign clubs, as well as long-time life members of the American Radio Relay League. Iris had served as manager for the Sixth District ARRL QSL Bureau, and Lloyd was a founding director of the ARRL Foundation, they said.

The Colvins once again made their annual Visalia and Dayton appearances in April 1986.

Also in April 1986, the YASME board met. After Don Wallace's death the previous May, Lloyd Colvin became the foundation's president. Danny Weil was vice president, Bob Vallio was secretary, and "Mac" McHenry, W6BSY, joined the team as treasurer. Directors were K5RC, W0MLY, W6OAT, K3ZO, W6QL, JA1KSO, OH2BH, N7NG, and Heather Pike, VK2HD. Allen B. Harbach continued as the foundation's "resident agent in Florida, where the YASME Foundation is chartered.

At this meeting, YASME directors discussed how the foundation might be able to encourage amateur radio in countries where it was dormant or non-existent. Director Fred Laun, K3ZO, with considerable experience as an employee of the United States Information Agency, later presented a paper, on May

2, 1986, to the other board members:

As I mentioned during our meeting at Visalia, on April 20, I have had an idea germinating in my mind for some time about the way to encourage amateur radio's establishment as an institution in countries where it is currently non-existent, or where a very few foreigners and no nationals of the country in question are licensed. The idea would be to attempt to establish amateur radio as a long-range institution, and not, primarily at least, to put the country on the air immediately .

At least initially, I would anticipate very little cost to YASME, expenses being limited to those connected with correspondence. If the correspondence bears fruit, there might be travel expenses at a later date. And now here are the details of my proposal:

During my rather extensive overseas experience, and during conversations with other amateurs who have served for long periods overseas, or, in some cases, with residents of countries whose authorities have not been favorable to amateur radio, I have been disappointed to find that some countries view amateur radio either with suspicion or regard it as unimportant compared with the very serious problems their countries face. Naturally, being an enthusiastic ham, I am convinced that such countries are missing the boat, and that they would likely encourage the establishment and growth of amateur radio if they became aware of its very real advantages.

There are also countries which permit amateur radio, but where, for one reason or another, only foreigners are licensed. Some of this may simply be a carryover from colonial days, where amateur radio was a pastime to be engaged in only by the citizens of the mother country, and locals were discouraged from participation. In a few places, this attitude created a mentality on the part of the local residents whereby they never even gave a thought to getting involved in amateur radio "It was only a toy "for the colonials."

Reasons often given or implied for official reluctance to permit amateur radio are, in order of importance;

1) Security concerns. Amateurs might use amateur radio to provide communication for a revolution or coup d'etat, or to send messages to a foreign correspondent or government hostile to the government of the country in question.

Zambia

2) Fear of Technology. To some of the world's leaders, science is beyond comprehension in many respects. To permit something like amateur radio to flourish is to set in motion a mechanism which may lead to unforeseeable consequences of a mysterious nature which these leaders fear might very well push things out of their control.

To begin with, they feel that they cannot adequately monitor such communications with the deficient monitoring systems under their control.

3) Fear of direct people-to-people communication over the heads of governments. Such communication might open the eyes of the subjects of a cruel dictatorship that people in other lands live better lives than they do. (The transistor revolution and the world information explosion are more important factors in this than amateur radio, however, and they will not abate.)

4) "Given a country's poverty, amateur radio is a luxury which cannot be afforded."

5) Import restrictions, put into effect to prevent precious foreign exchange reserves from flowing out of the country, make it difficult for amateurs to acquire amateur radio equipment.

The IARU is the logical organization to sell the non-ham countries on the benefits of amateur radio, but unfortunately the IARU is one more example of "government by committee," where even one negative expression can shoot down an idea. While the IARU contributed greatly to the establishment of amateur radio in China and Turkey, there is no systematic IARU program for contacting all of the world's nations with a pro-amateur radio program, even though resolutions are regularly passed at regional IARU meetings encouraging the growth of amateur radio in the developing world.

South Africa

Other programs, such as the ARRL program to send low-powered CW transceivers to developing countries, assumed a wish by would-be amateurs in those countries to learn CW before they could enjoy the fruits of the amateur radio service. If I may say so, this is an unfortunate example of the cultural bias built into some of our international amateur radio programs, and explains why success of some such programs has been indifferent. What will be more impressive, I believe, will be a demonstration of instant access to more advanced communications techniques than CW.

I have been associated with the Department of State Amateur Radio Club for some years, and that organization does have the advantage of having members posted in almost every country where the United States has diplomatic relations, including in some where amateur radio is not permitted. These amateurs, I am certain, would be willing to provide us information on the state of amateur radio in their countries of assignment, but neither they nor the DOSARC are in a position to push for improvement of amateur radio conditions in these places.

YASME may not be the only organization to be in a good position to help advance the cause of amateur radio in the ways I will describe, but I think it will do nicely.

Since YASME has been closely associated with the idea of sending DXpeditions to rare countries for all the years of its existence, I think I should take some time to explain here why the program I propose is different and, I believe, unique.

I do not propose that we send DXpeditions [to countries with which] we will be corresponding; it may be that after some back-and-forth communications there will be an opportunity to demonstrate amateur radio in a few countries, but the opportunity should result from an invitation by the government in question, and the main purpose should be to establish amateur radio as a robust and permanent institution in that country, rather than the short-term completion of DX contacts.

I would propose that the following be the order of procedure toward implementing

1) Determination of which countries are to be target countries.

2) Collection of as much information relevant to amateur radio in those countries as possible.

3) Correspondence with the authorities in those countries, making the following points:

a) Expressing surprise that the country has not chosen to take advantage of a radio service which obviously provides so many benefits to the nations which engage in it (free transfer of technology, self-training of communications technicians and operators; development of national industry; emergency communications, etc.)

b) request that the government state what its position on amateur radio is.

c) offer to assist in establishing amateur radio according to the national processes of the country in question, by providing whatever information is requested.

Following receipt of the reply, and depending on its content, YASME could be prepared to send an emissary, with or without equipment, to further discuss or demonstrate amateur radio. Useful models for action might be the Finnish efforts in the Sudan or the Yugoslavs' work with the Iraqis.

For obvious reasons, it may be necessary to have someone other than a U.S. citizen write the letters to some countries, and for the letter to be mailed in some country other than the U.S.

At some point, the situation in one or another country might develop to the point where it would be advantageous to provide equipment for a number of budding amateurs in that country. In this connection, I have already asked friends of mine in industry to advise when it is possible to generate a complete SSB signal in a single IC chip. This could revolutionize the home construction of gear by overseas amateurs where import of finished equipment is prohibited or too expensive. One of us could carry a number of chips in our coat pockets.

This has been a general outline of my idea. Hopefully it will give rise to your own interesting variations, which could be discussed at the next annual meeting of YASME. In the meantime, I will be delighted to entertain correspondence with any and all of you in the further refinement of this idea.

In 1986 Lloyd got himself listed in Who's Who in California, published by the Who's Who Historical Society:

COLVIN, LLOYD DAYTON, construction co. president; b. Apr. 24, 1915, Spokane, Wash.; s. George R. and Edna M. (Teeter) C: m. Iris Atterbury, Aug. 11, 1939; 1 dau Joy b. 1940; edn: BS, Univ. of Calif. 1938; grad. AUS Command and Gen. Staff Coll., Ft. Leavenworth, Ks. 1947; grad. Univ. of Heidelberg, Ger. 1955. Career: served to col. (ret) US Army Signal Corps, 23 years; supvr. Constrn. Projects in Alaska, N.J., N.C., Calif. And Europe; currently pres. Drake Builders, Richmond, Calif.; coauthor (w. Iris Colvin) best selling Book of the Month Club book, How We Started Out Building Our Own Home In Our Spare Time and Went On To Make A Million Dollars in the Construction Business (Prentice Hall); rec: amateur radio. Address: Drake Builders, 5200 Panama Ave Richmond 94804.

After the Colvins' winter 1985/spring 1986 trip they visited a number of clubs and reported they had now traveled to 217 different countries, held more than 200 different calls, and had made some 1,250,000 QSOs, resulting in the largest alphabetically-filed collection of QSLs in the world. The QSL collection had approximately 650,000 QSLs; it weighed more than 5,000 pounds, and, on the basis of spending one-and-a-half minutes to mark a QSL off the log, answer it, and file it, it would take one person working 40 hours a week nearly 5 years just to file the cards, they calculated.

Iris and Lloyd said they had worked DXCC from approximately 125 different calls, and held the largest number of DXCC certificates of any couple in the world (this was just about right, when including the multiple DXCCs they held under some of their U.S. call signs). Both also held 5-band DXCC certificates; both had been on the ARRL DX Honor Roll for the last decade; and they both held amateur Extra Class operating licenses. They said their travels and operating on DXpeditions had been fun, but that the greatest pleasure was in meeting people all over the world, making friends with them, and sharing a unique hobby.

Zambia

In the spring of 1986 the Colvins made new inquiries to India, Burma, Madagascar (Malagasy Republic), Rodriguez, Chagos Islands, Zaire, and Mauritius. Only Zaire seems to have replied, in the negative.

On 11 August 1986 Ed Peck, K6AN (formerly W6LDD) died. He was 67, a native of Oakland,

California, and the husband of Mary Rachel Peck. Ed Peck was an early YASME Foundation director and assisted the foundation for many years as its lawyer.

The foundation' by-laws and its legal status were to a large extent drafted and obtained by Peck, and in later years these documents were used by the Northern California DX Foundation and the International DX Foundation in obtaining their special IRS status.

The Colvins dedicated their fall 1986 expedition to the memory of Don Wallace, W6AM, who had died in May 1985 and had traveled to nearly 100 countries himself in his lifetime.

The Colvins wrote to friends they had made recently in South America that "The friends that we visit during our various DXpedition travels has always been the most enjoyable part of our trips.

'We usually show slides and give talks here in the States, which serves as a constant reminder of the good times we have had and the great new acquaintances we have made.

“Sometimes, however, it seems that time passes so quickly, we neglect to tell our far-away friends about the marvelous memories that we have of them and their country."

A new "YASME Supreme Award" was announced, a trophy for holding QSLs from 60 different YASME calls signs. As with all YASME awards, there was no charge for the YASME Supreme Awards.

At the end of the year, 215 of the traditional YASME awards had been issued and foundation awards manager W0MLY had issued 44 of the new YASME Supreme plaques.

In early October Lloyd and Iris arrived on Mauritius, having already sent requests for operating permission there several times. The American Charge' d'Affaires in Port Louis, Robert C. Perry, wrote to the country's permanent secretary, Mr. Bihnod Bacha:

> Two U.S. Citizens, Lloyd Colvin and Iris Colvin have traveled some 12,000 miles from San Francisco, California, to Mauritius in hopes that they can operate their small amateur radio station in Mauritius for a short time. Together, they have been radio amateurs for 100 years.
>
> They are both over 70 years of age and are devoting their remaining days to traveling around the world with a small radio set. They hope to visit and operate their little radio set in all U.N. countries of the world. They are close to accomplishing this goal and may never get a chance again to visit Mauritius.
>
> Anything you and the Ministry of External Communications can do to help them reach their goal will be greatly appreciated.

44. Indian Ocean and Mexico

Who are you sending messages to? - Comoros Islands police

We never squander our money - we use it all for DXing. - Lloyd Colvin

Comoros Islands

Sadly, neither the formal letter of recommendation, nor the on-going efforts of Jacky, 3B8CF, managed to break loose the necessary permits "We tried everything that we could think of, but they are not issuing amateur licenses to foreign amateurs at the present time," they said. So the Colvins proceeded to nearby Reunion Island. Operating 22 October to 16 November 1986 from Plaine de Caifre FR/W6QL logged 8,000 QSOs in 148 countries, on 10 to 160 meters.

With help from Guy, FR5ZG, and Herik, FR5DX, on the island, they also ran up a half-million point score in the CQ WW Phone DX Contest. First QSO from Reunion was ZS5WX, last was OK3YCA, and their FR/W6QL DXCC was #27,850, 23 May 1988.

The Colvins' next stop, the island of Mayotte, was controlled by France although France has never designated it as one of its departements, or administrative districts and has postponed, over the years, many referendums on the island's future. The United Nations and the Organization of African Unity, however, recognized Mayotte as part of the country of Comoros. Iris and Lloyd arrived 21 November 1986 with their French licenses, staying in a small hotel atop a hill overlooking La Tortue Bigatu. There were no embassies or consulates there, so they could do very little to arrange for their next stop.

Comoros

A requirement on many of the islands and countries in the Indian Ocean, was that you cannot buy a one-way airline ticket to anyplace, Iris and Lloyd lamented. "You must buy a roundtrip ticket," they said, "and if you only use one half the ticket you are entitled to a refund sometime in some amount. We are collecting lots of unused airline tickets and sure hope it is true we will get refunds. So far we [have] had very little success in getting any money back on unused tickets."

Operating FH/W6KG through 9 December 1986 they made 8,000 QSOs in 144 countries and earned DXCC #27,849, dated 23 May 1988. Operating in the CQ WW CW DX Contest, Iris and Lloyd, both now over 70 years old, made the excellent score of 1.557 million points, with more than 2,000 contacts. "The pile ups on the air are enormous when you are the only station on the air in a rare DX location," they said.

First QSO was UZ6PWA and the last was W7CMO. The Colvins completed their operation and began station tear-down still not exactly sure where they would go next but "We will be on from another country very soon."

After the delay, the Comoros Islands were chosen as the next YASME DXpedition operation for the Colvins. They stayed on Grande Comore, at the Coelacanthe Hotel. Both the two licensed amateurs there at the time; Bill, D68WB, and Alain, D68AM helped. D68WB was an American doctor who had been here on the Comoros for 10 years. He was associated with a missionary group, all of whom assisted us in many ways, Iris and Lloyd reported.

"The procedure to obtain a license includes a meeting

and approval of the Commander in Chief of the Armed Forces of the Comoros Islands. It cost 160 American Dollars for our licenses, which is the highest that we have ever paid," they said. "This is not very much, however, when compared to other prices here; for example, two Coca Colas cost six American dollars."

Operating 13 to 30 December 1986, D68QL logged 9,000 QSOs in 152 countries, for DXCC #27,847, 19 May 1988. First QSO was DL8RBR, last was N4OI.

This is their "best record" so far on this trip. Their tribander was in an "excellent location" on the shore of the Indian Ocean, with two guy ropes secured to a rock actually in the Indian Ocean. The Colvins were visited twice by local police, asking "Who are you sending messages to?"

Kenya

Due to the very hot and humid weather, "for the first time ever, we really enjoyed having air-conditioning available (usually we disconnect the air conditioning unit and connect our radio to that electric circuit)."

On 9 January 1987 the Colvins arrived in the Seychelles Islands, where an army mutiny had been thwarted in 1982. Iris and Lloyd visited during a window of opportunity, because in the late 1980s there were several more coup attempts. They arrived at the airport late in the afternoon and looked over a list of ten hotels to choose from. They asked for one on the way to town from the airport, and got reservations at the Eureka Guest House. "When we arrived at the Hotel, we were surprised to see a triband beam on a 60-foot tower. It turned out that this is the former QTH of VQ9R, who departed the island about 10 years ago."

S79KG used VQ9R's beam the first night, but "the transmission line was damaged, the rotator would not turn, and some of the rubber ends of the traps were missing. With great reluctance, we installed our own triband beam on our own telescopic mast just ten feet behind the one owned by VQ9R. We often wonder how much interaction there was between the two beams. In any case, we worked out great.

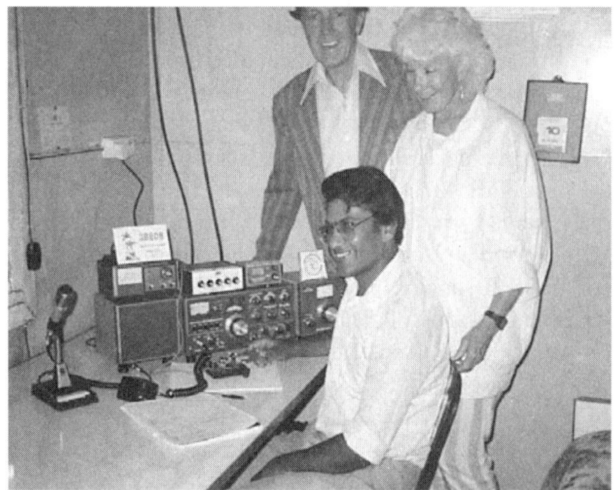
Mauritius

"The Seychelles were really on the air in a big way this January, because Ian Shepherd, G4LJF, was active as S79LJ during most of January 1987, on his own DXpedition."

Operating 9 to 27 January 1987 Lloyd and Iris made 9,000 QSOs in 130 different countries, for DXCC #27,251, 23 May 1988. First QSO was A92EN, last was YU1UV.

Iris and Lloyd left immediately for the Maldives Islands, an independent republic since 1968. The population, concentrated around the capital city of Male, was about 55,000.

On 4 February 1987 as they were leaving the Maldive Telecommunications office, Iris, who was 72 years old, slipped off their steps, fell, and broke her leg just below the hip joint. There was neither a qualified physician nor facility there to do the extensive surgery necessary.

Lloyd wrote that "With considerable difficulty and pain, Iris was transported on a stretcher by small boat to the island where planes land, and from there by airplane (we had to buy tickets for four seats) to the country of Sri Lanka, where Iris had a four-hour operation. She now has a pound of metal in her leg and has started the rather long time required to get back on her feet and walking again."

Later, their good humor intact, Lloyd and Iris would note that dictionaries define "Mal" as "evil, harmful, hurt," and defines "dive" as "fall or plunge"

"The radio amateurs in Sri Lanka, 8Q7CH in the Maldive Islands, and the radio amateurs world-wide have been very helpful and kind to Iris. Their support, gifts, and messages have gone a long way in making Iris determined to get well as soon as possible," they said at the time.

Joe Reisert, W1JR, said "When Iris had a bad fall, I wrote her in a hospital (I think in Maldives or that area) and she sent me a personal note from there thanking me for my note, along with other info. They were true human beings."

After Iris's surgery she and Lloyd flew right back to the Maldives from Sri Lanka, Iris on crutches, and they were on the air on 18 February. Operating through 3 March, they made 4,000 contacts in 120 countries, resulting in DXCC #27,853, 23 May 1988. Their first contact was UA4PNL, the last JL1OYU.

After five months of island-hopping, the Colvins finished their Indian Ocean trip back on the east coast of Africa, in Kenya. With the help of missionaries Lynn Raburn, 5Z4DU, and his wife, Brenda, 5Z4BF, Iris and Lloyd were able to both live and operate from the Brackenhurst Baptist International Conference Centre, 7800 feet above sea level in the mountains of the Tigoni area, 27 miles from the capital city of Nairobi.

5Z4KG made 7,000 QSOs with 135 countries from 10 to 25 March 1987. Their DXCC was #27,852, 23 May 1988. First QSO was W8PHZ, and the final entry was W7RO.

This trip resulted in yet another 45,000 contacts in Colvin log books. Their 152 DXCC countries from one spot, the Comoros, was just three less than their all-time best of 155. Now it was back to California, the usual springtime conventions and club meetings, and writing yet more letters, to Bangladesh, Bhutan, Burma, Mozambique, Rwanda, Yemen, Lebanon, Cameroon, Syria, Tunisia, Libya, C.A.R., Chad, Sudan, Niger, Zaire, and Thailand. For a number of these countries, the

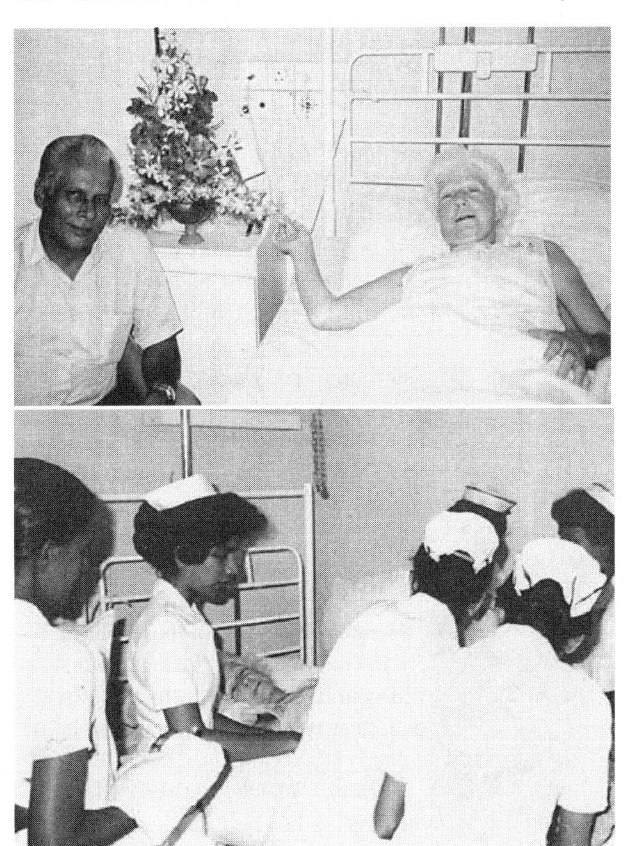

Sri Lanka

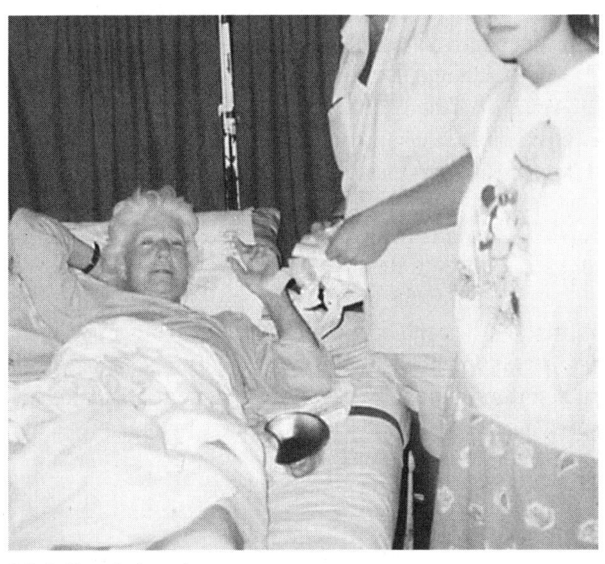

Maldive Islands

Colvins were prepared to travel there at a moment's notice, either from their home in California or from where they happened to be in the world.

Iris and Lloyd joined the Traveler's Century Club in 1986, a group promoting worldwide travel. Requirements for full membership was having visited 100 countries, which Lloyd and Iris easily met and surpassed. The TCC "Country List" is interesting in comparison to the ARRL DXCC Country List. In order for a TCC member to count a visit to a country it is necessary only to land in an airplane there, not even to disembark. Iris and Lloyd, on the other hand, not only disembarked, they lugged radios, antennas, battled bureaucracies, stayed weeks, and, well, you get the idea.

The Provisional Military Government of Socialist Ethiopia, Ethiopian Telecommunications Authority - Head Office, wrote to the Colvins on 16 June 1987 "concerning your intention to operate amateur radio station while you visit here in the country."

Goshu Abebe, Radio Division Manager, said "We appreciate your endeavours and devotion to radio amateurship; however, we regret to inform you that at present, all operations of radio amateur station in the country has been suspended indefinitely and thus permissions are no longer granted to anyone."

A letter of 6 June 1987 from the Yemen Telecommunications Corporation in Aden was a disappointment, informing the Colvins that "We wish to advise you that legislation governing the use of amateur radio in P.D.R. Yemen has not yet been enacted. Hence, we regret, we are not in a position to grant you authorization for operating a Amateur Radio station during your forthcoming visit to P.D.R. Yemen."

The Colvins heard from The People's Republic of Bangladesh in a letter of 20 July 1987, from Kazi Agdus

Salam, Divisional Engineer (Wireless & Frequency), as to "Consideration for using Amateur Radio licence.

"With reference to your letter dated 21.5.87 on the subject quoted above and the undersigned is directed to inform you that no foreign Nationals are allowed to use Amateur Radio as per existing Government Rules. It is regretted to inform that Bangladesh Telegraph and Telephone Board is unable to consider this case. This is for your kind information please."

And, finally, the Burma P&T said "I am directed to inform you that the operation of Radio amateur activities has been suspended in Burma since 10th January 1964 until further notice. Therefore it is regretted that we are not in a position to oblige your request."

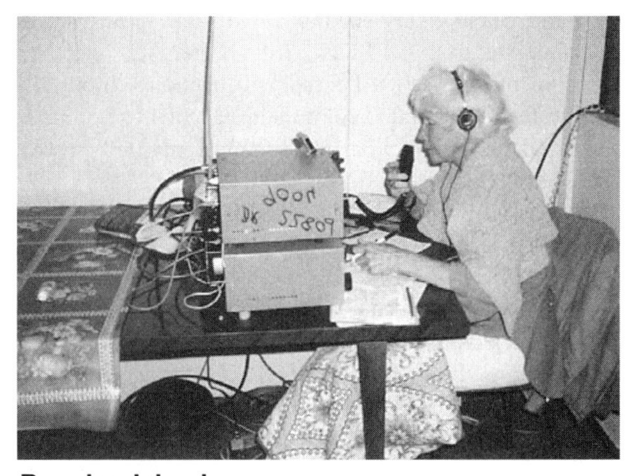

Reunion Island

Off to Mexico

In October 1987 Iris and Lloyd did something a little different, hopping into their van and driving to Mexico for a six-week stay; after all there was a DXCC to be made from a country right next door! Leon Fletcher, N6HYK, visited the Colvins at their Richmond home about this time and later wrote about the Colvins in 73 magazine:

"Are you flying or driving to Mexico?"

"Driving! Flying costs too much!" Lloyd's tone was friendly but firm. "And we never fly first class. We never squander our money - we use it all for DXing."

Lloyd offered to drive them to lunch, in his very old, very dented, very decrepit light tan Volkswagen van. As it sputtered along, Fletcher asked, "What will you be driving to Mexico?"

Mexico

"This!" Lloyd said.

"Unfortunately" Fletcher said ,"I paused, and in that split second he seemed to read my mind. He added, "We've driven this thing for years with no problems. It will make the trip. We've not worried."

"For lunch, Fletcher said, "he drove us to a modest coffee shop nearby. Obviously they'd eaten there often - they didn't look at the menu, ordered quickly, ate fast, talked little. I tried to pick up the check, but Lloyd beat me to it.

"The speed of the lunch told me that they wanted to get back to their home to return to preparing for their trip to Mexicali. As soon as we were back at their place we said our farewells. As I drove away, I could see Lloyd already out on the roof of their building, where they stored their antennas; he was starting to pack gear."

Before leaving Fletcher had asked about their future plans, and Lloyd said "We've been corresponding with government officials in a number of countries for some time, trying to get permission to operate, especially in those places in the top twenty of the most wanted-countries. We'll go to the first one that sends us authorization. "

"Clearly," Fletcher said, "Lloyd means it when he takes Iris's hand-as he does often-and says,

'Hamming is everything in our lives.' Iris nods in agreement."

Iris and Lloyd operated XE2GKG from Villa Xochtle, Mexicali, from 20 October to 28 November 1987, to the tune of 8,000 contacts on 160 through 10 meters, 122 countries, including about 1,500 QSOs each in the Phone and CW weekends of the CQ WW DX Contest. First QSO was with VE6CS, last with NL7G.

While there the Colvins experienced two major 6.2 strength earthquakes. "It was scary but the only damage to the motel in which we stayed was that many items fell off shelves. Fortunately, none of our radio equipment was damaged."

They again experienced the Spanish-English language barrier (Lloyd spoke passable French). XE2CN, Francisco, and his family, helped in many ways, they said.

Iris and Lloyd went home to re-pack, and sent a letter to Mr. Chiran S. Thapa, Secretary, Royal Palace, Katmandu, Nepal. They briefly described amateur radio and its one million adherents around the world:

> They must pass a test for proficiency and are then issued a license to operate their equipment on special frequencies set aside for their use by international agreement. Their communications can not be used for any kind of business or commercial use. These radio amateurs have a splendid history of helping in times of disaster, such as earthquakes, high winds, fires and other catastrophes.
>
> Your country of Nepal has two active amateurs who are doing a great job of spreading good-will and favorable information about Nepal to other radio amateurs worldwide on special occasions, a few foreign radio amateurs have been permitted, for short periods of time, to visit Nepal, bring their small radio equipment and operate on the radio amateur frequencies.
>
> The two radio amateurs presently active in Nepal are Father Marshall D. Moran of the St. Xavier's Godavari School, call-sign 9NlMM, and Krishna B. Khatry, Ministry of Communications, Panchayat Plaza, Katmandu, call-sign 9N1MC. Both of these radio amateurs have communicated with us often, while we have been traveling around the world, and both are aware of our great desire to visit Nepal and operate our amateur radio for a short time while there.
>
> We have made an unusual record in amateur radio; the two of us together have been licensed radio operators for over a hundred years, traveled in 181 countries and have communicated with more radio amateurs than any other two radio amateurs in the world. We have never caused any trouble in all of our world travels. We believe that in a small way we have helped to spread good-will and understanding world-wide.
>
> We are enclosing our application for permission to operate our amateur radio station while in Nepal. Will you please be so kind as to try to help us in securing favorable consideration of our request? Both Father Moran and Krisna Khatry have indicated to us that amateur radio is not widely used nor understood in Nepal. We promise you that our operation there will in no way cause complications of any kind, and that our operation will only serve to further the good-will for which Nepal is already famous.
>
> Please try to pass the enclosed application on to your communications officer with an indication of your approval.

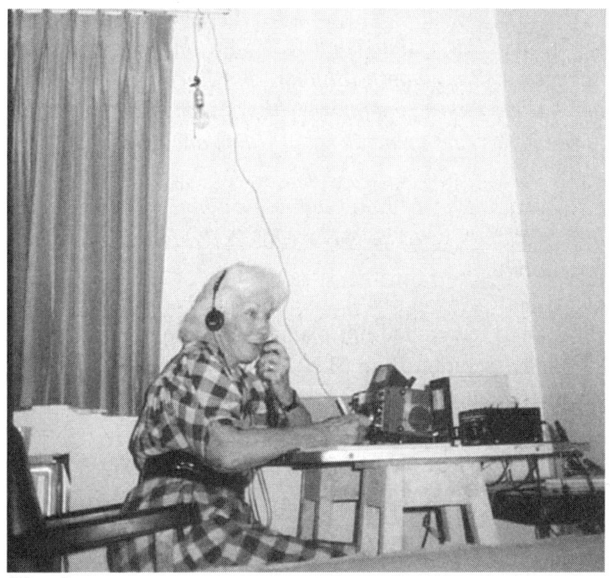

Mexico

It worked! - The Colvins were granted permission to operate if they would be there "on the occasion of the birthday of His Majesty the King of Nepal, which is on 29 December 1987. Amateur Licences are issued on special occasions only. Please, therefore, come here during the last week of December" said Krishna B. Khatry, Chief Engineer.

45. The Himalayas and Grecian Isles

We tried everything that we could think of to obtain licenses in Bangladesh, Bhutan, and Burma. We did our best to leave these countries with a favorable impression of ourselves and amateur radio. - Iris and Lloyd Colvin

I said to Lloyd "Don't antagonize him," but evidently that is exactly what he was trying to do, and it worked. - Iris Colvin

Iris and Lloyd had not planned an itinerary following their operation from Mexico. They'd sent inquiries to many countries in Asia and Africa, none of which had generated a positive response. The welcome from Nepal was a no-brainer; get on an airplane for Nepal! After that, Lloyd and Iris would wing it.

Their Licence No. 23/1987 gave the Colvins permission to operate 9N5QL from 27 December 1987 to 5 January 1988, as Nepal celebrated the king's birthday. The station would be in Katmandu, transmission power: 200 watts; on frequencies and designation of emission: 14000 to 14350 KHZ, 21000 to 21450 KHz, 28000 to 29700 KHZ; C.W. and S.S.B. In addition, they would be allowed 7000 to 7100 KHZ, 3700 to 3720 KHZ, and 3515 to 3530 KHZ only between 1300 and 0100 UTC "on the conditions that special care will be taken to avoid interference to any service; but if any interference is noticed, these frequencies will be withdrawn."

While the Colvins were there, a group of Japanese hams also operated, as 9N5YDY. They helped Iris and Lloyd get their station set up, as did Father Moran, 9N1MM, and Chief Radio Engineer Krishna Khatry, 9N1MC, the only two permanent amateur licensees in Nepal

Iris and Lloyd "had one of our all-time lucky breaks. We arrived past midnight and asked a taxi cab driver to

Nepal

Nepal

take us to a hotel. He did not know that we were radio amateurs and neither did the hotel managers. When we awoke in the morning, we found that we were in a room on the top, fourth floor, all by itself, and on the roof above, an additional floor with a sundeck about 10 ft x 10 ft with regular stairs leading to it. It was a magnificent place for our beam. We could have searched for a month and not found a better QTH located on a high place in the center of Katmandu in the Himalayan Mountains."

They netted 3,500 QS0s, half phone and half CW, on 10 through 40 meters, with hams in 121 countries. First QSO was UZ4WWF, last YU7GMN. The 9N5QL DXCC was #28,173, 20 October 1988.

While in Nepal the Colvins tried to get through to the right people in Bhutan, a very-needed country among DXers. Nothing was heard in reply but they went there anyway, in January 1988. It was hopeless. From there, in late January, they flew to Rangoon, Burma, where they had been sending letters for years and as recently as the previous summer. They spent five days in Rangoon, and visited the U.S. ambassador there, who sent a letter on their behalf. Lloyd and Iris also tried to find the "proper government officer" in person, without success.

So, having left California for Nepal in somewhat of a rush, and with no plans ahead, they flew home. On 31 January 1988 they wrote to U Ne Win, Chairman, Burma Socialist Programme Party, in Rangoon, saying that they'd been there for five days and "On receipt of indication of

Bangladesh

approval of our request, we will return to Burma immediately."

Iris and Lloyd additionally delivered a primer on amateur radio, and it's worth reading:

> The following are additional points that we wish to submit in connection with our application to operate our amateur radio station in your country for a short time:
>
> 1. The communication department of every country in the world can authorize qualified persons to operate amateur radio on frequencies set aside world-wide for such use and which are defined in detail by the ITU.
>
> 2. Out of some 300 countries in the world there are only a very few who have no active amateur radio stations on the air. ALL countries of the world, including Burma, have at some time had legally authorized amateur radio operation.
>
> 3. There are some two million radio amateur operators in the world today, and, to some extent, we represent these two million radio amateurs, all of whom are anxiously awaiting to hear and hopefully communicate with Burma.
>
> 4. Throughout the world, radio amateurs have many times been of great service in times of emergency. The two of us participated in a disaster in Japan. For 12 hours we provided the only communication from a town severely damaged by earthquake and where nearly a thousand persons were killed.
>
> 5. One of the factors holding back the permission for radio amateur operation is sometimes the possibility of interference to existing radio operations such as police, military, or other government operation. We have had wide experience in assignment and use of all radio frequencies.
>
> The possibility of such interference can be eliminated in Burma if the following is done:
>
> a) At first, do not authorize amateur radio operation on the frequencies 3500 to 4000 KC and 1800 to 2000KC. (Most of operation within Burma is between 1800KC and 7000KC)
>
> b) Authorize operation in the 7000 to 7300KC band of only 7000KC to 7100KC CW and SSB.
>
> c) Authorize CW and SSB operation in the bands 14000 to 14350KHZ, 21000 to 21450KC, and 28000 to 29700KC. (These frequencies operate on long skip and are generally heard and used for distances more than 1500 miles).
>
> d) Enclosed is a license which was very recently issued to us by the country of Nepal. Our operation was a complete success and resulted in no interference or problems of any kind.
>
> 6. To further the use and understanding of amateur radio in Burma, and upon conclusion of at least a two-week operation by us in Burma, we will donate our ICOM 751A 100-watt transceiver to the Burma Communication T. & T. Section. It will receive on all HF frequencies and has extremely good calibration. The unit can be used in many ways including: a) An excellent unit for use in a government monitoring station; b) a 100-watt transceiver for use on all amateur radio bands.

Indonesia

This would allow someone in Burma to join the ranks of the other two million radio amateurs in the world.

Unfortunately, this was simply another case where amateur radio operating permission wasn't about to be allowed to "American capitalists."

Iris and Lloyd left in early February 1988 for Bangladesh, adding it to the list of countries they had visited, but unable to pry loose permission to operate. From there, they applied, successfully, for operating permission in Sri Lanka (where Iris had her broken leg repaired the previous year). They actually had permission in 1987 and Sri Lanka had been on the itinerary, but they had cancelled it so Iris could get home for further medical care.

So, the Colvins flew from Bangladesh to Sri Lanka, picked up their operating permits, and opened up on 28 February 1988 from Colombo, as W6KG/4S7. They stayed at the same place as when Iris was hospitalized, at the home of Paul, 4S7PVR, operating using their ICOM 751A transceiver and Paul's triband beam and said {Paul] "and his family were very kind to us. His wife is a good cook.

"A year ago, Iris was in the hospital with a broken leg for about two weeks. Many amateurs visited her in the hospital. It was a great pleasure to see them again on this return visit. This time Iris is walking."

The Colvins attended a meeting of the Radio Society of Sri Lanka and were "very happy" to be elected honorary members. They sent a note to the DX newsletters and columnists that they were next flying to Indonesia, where they hoped but "are not sure" of getting a license.

W6KG/4S7 made close to 5,000 QS0s with amateurs on 10 through 40 meters in 137 countries. Their DXCC was #28,172, 20 October 1988. First QSO was UV6UG, last was DL7UX.

The Colvins did get to operate from Indonesia, as YB0AQL, where they were "astonished to discover that Indonesia is the fifth largest country in the world, and that the capital city, Jakarta, has a population of nearly eight million people!"

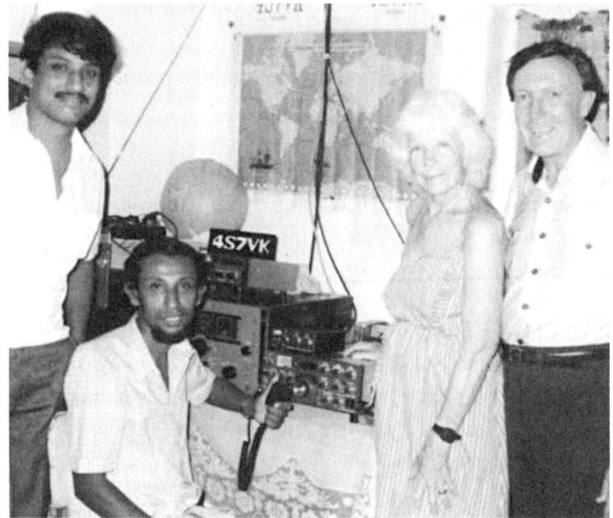

Sri Lanka

We stayed at the QTH of "Wan", YB0SY, who was a DXer with big monoband beams for all bands. Iris and Lloyd, as usual, used his antennas but their own equipment.

YB0AQL made nearly 8,000 QSOs including radio amateurs in 133 countries, including the CQ WW WPX Contest in March 1988 (multi-single, 2.7 million points). The log book runs 8 to 27 March 1988 and the YB0AQL DXCC was #28,170, 20 October 1988. First QSO YB3CN; last, I1AGC.

An Indonesian newsletter noted the Colvins' visit, saying "They [the Colvins] were surprised that there are so many amateurs, and that they are so well organized. The Colvins were favorably impressed with the ORARI [the Indonesian amateur radio society] and its many projects and activities. They have contacted Indonesian hams from all of their various world locations, and believe them to be great amateurs with good stations and excellent operating ability. They think that the Indonesian people are extremely pleasant, friendly, and helpful.

"Lloyd and Iris will be leaving the end of March for the United States, where they will attend two of the largest [conventions] in the world, one in Visalia, California, and the other in Dayton, Ohio. They have key spots on the convention programs and their slide presentation this year will include the Indonesian operation.

"The Colvins express their hope to again visit in these beautiful islands."

Returning to California after this successful operation, Lloyd and Iris wrote, as always, to summarize their latest six-month sojourn, which included Mexico (XE2GKG), Nepal (9N5QL), Bangladesh, Bhutan, Burma, Sri Lanka (W6KG/4S7), and Indonesia (YBOAQL).

"We stayed for several weeks," they said "in each of these countries, and operated ham radio in each country that would permit it. We tried everything that we could think of to obtain licenses in Bangladesh (S2), Bhutan (A5), and Burma (XZ). We were unable to obtain licenses in these 3 countries. We did our best to leave these countries with a favorable impression of ourselves and amateur radio. The best thing that we can report is that all three countries did not come right out and refuse us licenses. They all said that maybe, at some future date, they would comply with our request. We will try to keep after them and perhaps we can return some day and operate."

Back Stateside, Lloyd and Iris fired off inquiries to Albania, Laos, Vietnam, Egypt, Iraq, Libya, Tunisia, Iran, Sudan, Chad, Somalia, Niger, Yemen (Aden), Afghanistan, Sao Tome, Zaire, Mozambique, and the United Arab Emirates.

From Yemen, A. Mohsin, For the Acting Director General, Yemen Telecommunications Corporation, said "We would like to advise you that further to our letter of 24/6/87 addressed to you, we are still not in a position of

Nigeria

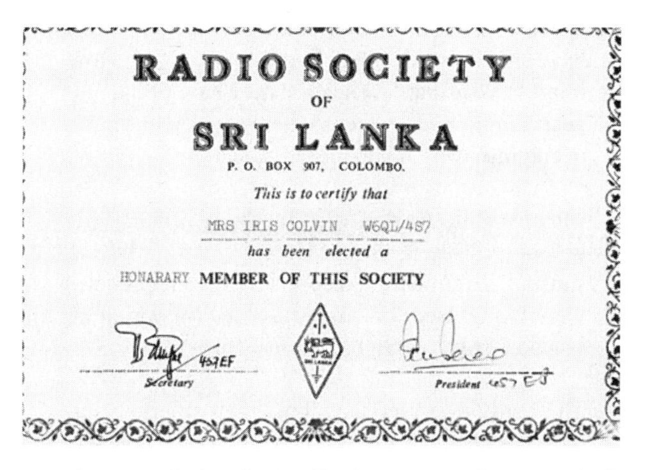

RADIO SOCIETY
OF
SRI LANKA
P. O. BOX 907, COLOMBO.
This is to certify that
MRS IRIS COLVIN W6QL/4S7
has been elected a
HONARARY MEMBER OF THIS SOCIETY

Secretary *President*

granting permission for Radio Amateur station. We shall advise you in future, when the legislations concerning use of Radio Amateur is approved in P.D.R. Yemen."

The director of telecommunications, United Arab Emirates, Ministry of Communications, Abu Dhabi, wrote "We would like to inform you that we do not issue amateur radio station licences for visitors to the country, but grant to holders of U.A.E. Visas only."

Also in June 1988, Woudneh Taddesse, Radio Regulatory Branch Chief, The People's Democratic Republic of Ethiopia, Telecommunications Authority - Head Office, advised "This is in reply to your letter of 31 May 1988, requesting for permission to operate your amateur radio station for certain period during your visit in this country. We regret to inform you that even though we appreciate your devotion in the amateur radio operation, we can not comply with your request at present since this kind of activity has been suspended for an indefinite period in the country."

Things were looking so dismal that Lloyd and Iris also enquired into operating from Canada's Sable Island - a DXCC "country" but probably not, from their point of view, a very exotic place!

Iris Atterbury and Lloyd Colvin were married 11 August 1939. To celebrate their 50th anniversary, a year early, a large ham gathering took place in Richmond. Their friend Armond Noble, N6WR, editor and publisher of Worldradio, put a photo of the Colvins on the cover of his September 1988 issue, and wrote this story [the careful reader will note a few, minor errors, but no matter]:

> Mr. and Mrs. DX celebrated their 50th wedding anniversary. Naturally, it was done in the presence of a large gathering of Amateur Radio operators. Guests found food, drink and the world's largest QSL card collection. If, over the past 60 years. you've never worked a Colvin, your rig must have been in the pawnshop!
>
> Lloyd and Iris (when you say Lloyd and Iris, calls and last names seem superfluous, for there is no

other, Lloyd and Iris they could be confused with) have made well over 1 million QSOs. How impressive is that? Well, since they have about 100 years of amateur license between them, that's averaging over 10,000 QSO's a year!

Lloyd was licensed in 1929 (at the age of 12). Iris got her ticket in 1945. He missed the DXing sunspot maximum of the '40s as the frequencies were being used for other purposes (in which he played a role as a Signal Corps officer).

The really awesome fact about their stunning statistics is that the vast majority of those contacts was made from overseas. In what they have accomplished, there is no one in second place or even third place. They have worked DXCC from well over 100 countries. Right, over 100 contacted from each of 100+ countries they have used as the base of operations. In total, they've operated from 140 countries. It would be possible to apply for your own DXCC award by just submitting cards of Colvin contacts.

They have received over a half-million QSL cards and kept them neatly filed in metal file drawers; it's almost a ritual that visitors to their Richmond, CA home delve into the files to find their own cards. This combined 100 years of marriage and DXing all resulted from a blind date while both were attending the University of California in Berkeley.

Lloyd went into the Army in 1940 and spent 23 years in uniform. The DXpeditioning started in 1965, so the contacts have been made in a far more compressed time frame than the 10,000 QSOs a year over their careers would indicate.

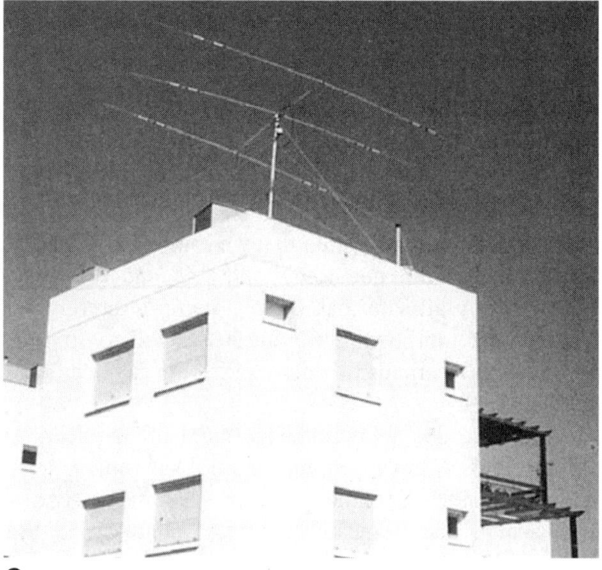

Cyprus

For the few who don't study their geography, we'd like to point out that to operate from as many countries as they have (and they've gone to others where operating permission was denied), one has to go far beyond where there are Hiltons. You are really in "don't drink the water" country. One is surrounded by diseases not even mentioned in the medical texts in this country.

Operating as they have done is a feat in itself, but just try [to] imagine all the planning and effort it has taken! In the near future, QST will be running a major article on the worldwide accomplishments of this intrepid duo.

And they're not sitting on their laurels. At the present time, work is progressing toward getting them to some other far-flung corner of the planet.

Malta

Here is the admirable fact about them. They have chosen, with the risks in the backwaters of the world (injury is not unknown to them), this avenue of Amateur Radio. Others have put sums into "super-stations" which, if put into banks instead, would have accrued enough interest to fund DXpeditions.

One group operates in comfort. The Colvins have chosen heat and bugs, giving out the contacts instead of just taking them in. There's a big difference. They have had their share of "Only one flight a week to there," and the plane looks like a rubber band is keeping it in the air.

You are in your comfortable shack working them. They are in a country on the equator where air conditioning is something the people see in ads when a year-old copy of Life magazine is passed around.

When some amateurs travel, they are concerned whether they get first or second seating on the luxury cruise line. The Colvins are just hoping the food has been cooked. Some get their excitement from the shoot-em-ups on TV; for the Colvins it's real bullets down the main street. "Well, dear, we timed it just right. Sure didn't want to miss this month's coup."

Malta

The next time we jot one of their calls into the logbook, an extra tad of appreciation to W6 King George and W6 Queen Lady [for the] contact may be in order. - N6WR.

Looking for more destinations

While others might at this point have rested on their laurels, not the Colvins. In September 1988 they sent inquiries to Rwanda, Ecuatorial Guinea, C.A.R., Angola, Burundi, and Gabon. Then, in mid-October, they flew to Cyprus, a week before the CQ WW SSB Contest. "We thought that we had plenty of time to be there before the contest started. We were wrong! We missed our flight at Frankfurt, Germany even though we were in the airport an hour before the flight, only to find that on international flights they now want all passengers to check in two hours before flight time. We had to wait 48 hours for the next flight."

When they got to Larnaca, high winds delayed them getting their antennas up for another 48 hours. 5B4TI, Mike and 5B4WW, Maryann, were a great help. "We got on the air just minutes before the contest. All went well then, and we had a good operation on all bands. We worked 100 countries during the contest."

Operating W6KG/5B4 26 October to 7 November 1988, Iris and Lloyd made 1.292 million points in the contest, and 4,000 QSOs overall, in 133 countries, snatching DXCC #29,813, 11 July 1990. First QSO was GW8AAT, last was LU3XPM.

One of the strange (to non-DXers) places on the DXCC Country List is the British Bases Area of Cyprus; the best, perhaps only, similar DXCC "entity" is Guantanamo Base, Cuba. "Separate administration" is the DXCC country criteria that applies. Iris and Lloyd operated from The British Base Area (call sign prefix ZC4) and said "We were extremely lucky to receive permission to operate in ZC4 land. Many radio amateurs from the USA and other counties have tried to obtain ZC4 licenses with no success. ZC4 licenses are given normally to British Military Personnel located in Cyprus. Alan, ZC4AB, and his wife, Diane, permitted us to sleep in their quarters while we actually operated from the Radio Club located in the British Episcopi Garrison, using our own radio equipment. Alan and Diane helped us in a number of ways for which we are most grateful."

From 11 to 27 November 1988 ZC4ZR made 7,500 contacts with 117 countries, including the CQ WW CW contest and a 1.3 million-point score. The first QSO was W1YY and the last was LU2YE. The DXCC certificate was #29,812, 11 July 1990.

Reporting later they said "during the time we operated as ZC4ZR and W6KG/5B4 on Cyprus, for a little over a month, the value of the U.S. dollar declined in value by 5% which directly resulted in an increase of 5% in our cost of living." (How do you get to be millionaires? By watching the pennies, that's how.)

In this same, brief report mailed to the DX press in the U.S. and some other countries, the Colvins closed with "Look for us from another rare spot soon." They never mailed a report until their on-the-air operating was finished and it was time to leave, but clearly they did not know where they would be next. This was a major, and recent, change in their modus operandi. At this point, they had made DXCC from more than 100 DXCC countries, so that milestone was behind them. From Cyprus, there were many European countries close by from which they had not operated, including Greece. But these seemed not to interest Iris and Lloyd, in part because they already had at least visited them (most, 35 years before when Lloyd was in the Army, assigned to Germany).

The Colvins left Cyprus and flew to Germany, then on to the island country of Malta. From the Grosvenor Hotel they put 9H3JM on the air for an operation 10 December 1988 to 6 January 1989. They got radio licenses on the spot and got their radio equipment out of Customs fairly easily, except for the triband beam antenna. The Board of Trade insisted that they obtain an import license for the antenna only.

Joe, 9H1GY, had recommended a good hotel in the middle of the island. The nearest TV receiving antennas were 200 feet away, so TVI wasn't a concern. Iris and Lloyd were "favorably impressed" with amateur radio in Malta, which had three active clubs. They were treated to Christmas and New Year's dinner meetings of all three clubs, with turnouts of about 60 people, including many family members.

9H1GY, 9H5L, 9H5CL, and 9H5BU all helped the Colvins install their antennas, in strong winds once again. While in Malta, they contacted 9H4G, who had the only YASME Supreme Award in Malta.

The 9H3JM operation netted 6,000 QSOs in 132 countries and DXCC #29,815, 11 July 1990. First contact was UV0CC, last was JO1NZT.

In their departure report from Malta, Lloyd and Iris said they were headed to Africa, but didn't know where. It turned out to be Nigeria, where they stayed in Lagos. Preparations there were underway for a transfer to a democratic government, from one being run by a military general. They had flown from Malta to Nigeria, flying first to the Nigerian capital of Lagos, then flying again to Sokoto, in northern Nigeria. From there they made a 10-hour car trip to Niamey, the capital of Niger, since there was no direct airline between Nigeria and Niger. They had to change cars at the border, after "a rough car trip. We carried all our 400 pounds of equipment and belongings." The Colvins had sent many letters to Niger and had a few names of authorities to talk to but were unable to bust loose operating permission.

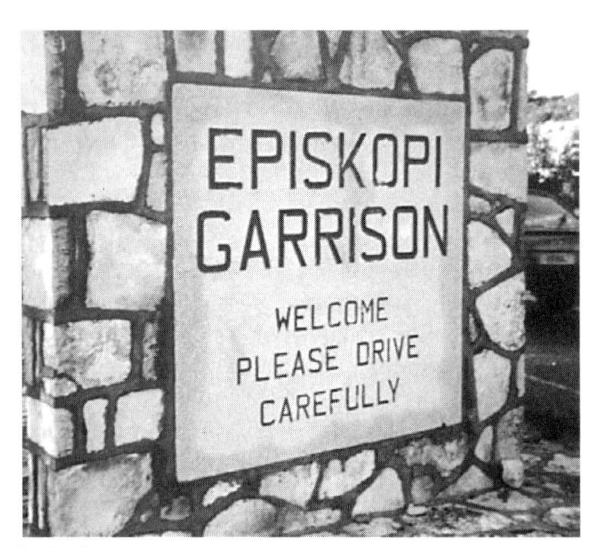

U.K. base, Cyprus

Nigeria

Iris told the story of the border crossing:

Nigeria was at war over Biafra. We had rented a car and driven down the west coast of Africa, well aware of the political tension and the fact that anyone crossing the border was probably an enemy. We had a room outside the city but had gone into town to meet some hams. Going back to our hotel we were stopped by a local man. We had all of our ham equipment in the trunk - absolute proof that we were foreign spies.

When the young sergeant came over to inspect and demanded that we open the trunk and then open all the boxes, Lloyd said "If you want to see, you are dressed in fatigues, you open them." Actually, he was sharply dressed (even if in fatigues). I said to Lloyd "Don't antagonize him," but evidently that is exactly what he was trying to do, and it worked.

The sergeant became angry and ordered us to follow him into police headquarters. When we came to the big iron gates of the department the guards let the sergeant through and slammed the gates shut so that we could not get in. The sergeant was angrier than ever and when we finally reached the police captain the sergeant was beside himself with rage and said "I order you to arrest these people." Then the police captain became angry and said "I don't take orders from you."

The captain asked us to explain the problem and to see the "personal effects" in our car. We opened a couple of the suitcases, which satisfied him. He said "you are going to run into more difficulty with road blocks before you reach your hotel. I will give you an escort." So we whipped through about 10 more roadblocks, following a motorcycle escort.

Back in Lagos, Nigeria, the Colvins stayed and operated from the home of Cal, KH6HSS/5N0, and W6QL/5N0 made 3,000 QSOs with 124 countries, on 10 through 40 meters. Their DXCC was #29,814, 11 July 1990; first QSO was LY2WR, last was 5N9MBT.

This was the last overseas operation for this trip, and the Colvins returned to the U.S. in time for the April 1989 meeting of the board of directors of the YASME Foundation. Danny Weil was still on the board but was absent, as usual, and the board began efforts to encourage him to "become more active in amateur radio." 353 YASME Awards had been issued, along with 104 YASME Supreme awards

Hal Sears died 31 May 1989. He had traded the call sign K5JLQ for W5NC since his involvement with the YASME Foundation in the early 1960s. At the time of his death he still lived in Houston, and was a Quarter Century Wireless Association director. According to a brief obituary that appeared in Worldradio, he was first licensed in 1921 as 9CPV, and "became a commercial marine operator for Tropical Radio Telegraph in 1925. His career shifted to aeronautical radio in 1930 and he because associated with Western Air Express (later TWA) along with Herbert Hoover Jr., W6ZH.

“In the late '30s W5NC became involved with geophysical exploration and in the early '40s established his own instrument company, which became world renowned. Hal first served on the QCWA Board of Directors in 1969 and was elected vice president for the 1977-79 term of office," Worldradio said.

46. The Americans are Coming!

Your amateur radio operation from the USSR was a special permit which was granted to you because of your vast experience in world-wide DXpeditions and great contribution to friendship and mutual understanding between our two nations. - Vasily Bondarenko, chief, Central Radio Club, USSR

Baku, Azerbaijan

Where to next for the Colvins? The winter 1988/89 trip had been catch-as-catch can, and Iris and Lloyd had logged only four new amateur radio countries on that trip. As it turned out, they went from that itinerary-less trip to an ambitious, hectic itinerary of every republic of the Soviet Union.

A friend had been to the USSR in early 1988, and as soon as he got back he suggested to Lloyd and Iris that they would enjoy a trip to the USSR, and, with good luck, might even be permitted to operate ham radio there.

The Colvins immediately contacted the Russart Travel Bureau in San Francisco, and asked if a visit could be arranged to all 15 republics of Russia. The answer was yes and a $1,500 deposit for such a trip was made in November 1988 and Lloyd and Iris immediately wrote to the chairman of the U.S.S.R.'s Central Radio Club, in Moscow:

> The YASME Foundation is a world-wide amateur radio club. It is a non-profit organization and none of its officers or members are paid any money and have no connection with any government or political organization. The President is Lloyd Colvin, W6KG, and Iris Colvin, W6QL, is a YASME Director.
>
> We are planning to visit all 15 USSR Republics during May and June 1989. We have made applications for visas and travel permissions through INTOURIST in San Francisco, California. Please send us the addresses for the Amateur Radio Clubs in each of the fifteen republics, and, if possible, please advise them that we are coming and will contact them when we arrive there.
>
> Together, we have been radio amateurs for 104 years. We have traveled to 189 countries and want to add the USSR to our list of countries visited. From our home amateur radio station in San Francisco, California, we have contacted every country in the world that is counted as a separate country. We have had more than 1,000,000 amateur radio QSOs and hold the largest collection of QSLs in the world, numbering approximately 550,000 QSL cards. These are filed alphabetically and occupy two rooms, reaching from the floor almost to the ceiling.
>
> Hoping to hear from you soon, because we will be leaving for USSR in May 1989. - Lloyd Colvin, W6KG, and Iris Colvin, W6QL.

Yerevan, Armenia

There is no record of a written response from the CRC before their departure but they may have got it by telephone. Lloyd later wrote that they had special permission to operate ham radio in all 15 republics with the requirement that they operate from club stations, using the club station call sign followed by their own stateside call, so they knew they would not have to take radio equipment and antennas on this trip! Lloyd later said "But we made up for that by dragging along a lot of heavy [winter] clothing we had absolutely no need for."

After attending the Dayton Hamvention in late April 1989, they flew to Chicago and then took an inaugural flight of American Airlines, non-stop Chicago to Stockholm, Sweden. At O'Hare Airport, passengers got a surprise welcome - American Airlines officials and diplomatic representatives of both the U.S. and Sweden made speeches; a band played, and cocktails and snacks were served. Some of the hostesses wore formal dresses, and the men "looked elegant" in coats and ties. The captain of the airline introduced a special international pilot who would accompany him on the flight.

Tallinn, Estonia at UR1RWX

The flight left Chicago right on time and arrived in Sweden also exactly on schedule. From there, the Colvins took another American Airlines flight to Moscow.

"Entering Russia," Lloyd said, "was less complicated than expected." No one asked to see passports or visas until after they showed up at their 22-story hotel in Moscow. "The hotel facilities, transportation, restaurants, and people were found to be much more like the USA than expected. All the people near Moscow and Leningrad looked like similar groups of people in America. Only when people started to talk in Russian were you aware that you were in Russia."

In Glasnost 1989 the Colvins were struck by the streets and highways crowded with vehicles. News of their visit was passed to hams in the next country ("republic") they would visit. They described friendly welcomes at each stop, where hams awaited their arrival at railway stations or airports and welcomed them with banners and gifts of flowers for Iris. The hams at every stop providing

Minsk, Belarus

transportation to the club stations and enthusiastically showing off their own country. "We were continually being hosted at parties where much food and vodka was consumed," they wrote.

The weather was cold only in Moscow and Leningrad and even then it was never freezing. Otherwise it was "sunny, pleasant and mild."

The complex schedule of travel, most by air, and hotel schedules went exactly as planned, with representatives always present to provide transportation, accommodations, and assistance. The Colvins complimented INTOURIST on a job well done.

The club stations were mostly in factories, schools, or electrical or other manufacturing plants. All of the radio equipment at the club stations was home built and many of the stations had very large and elaborate antenna systems. Most stations ran from 200 to 1000 watts. One thing was noticed as being different from the USA - Most of the stations visited did not have an SWR meter in the antenna circuit.

Their INTOURIST-arranged schedule read:

"Air:Minsk, Kiev, Kishinev, Tbilisi, Baku, Ashkhaba, Dushanbe, Samerkand, Bukhara, Tashkent, Frunze/, oscow, Leningrad, Tallin.

"Hard train: Tallin-Riga-Vilnius-Minsk. Sitting train: Moscow-Leningrad." Here's the itinerary and results:

2 to 4 May, Hotel Belgrade, Moscow, Russia, U3WRW/W6KG, /W6QL; UZ3AWA/W6KG, /W6QL; 30 QSOs. First QSO RB5WL, last IK2FOD.

4 to 7 May, Hotel Karelia, Leningrad, Russia, UZ1AWA/W6KG, /W6QL, 280 QSOs; first DL5XAS, last RT4UA

7 to10 May, Hotel Viru, Tallin, Estonia,

UR1RWW/W6KG, /W6QL, 1010 QSOs, first WA4DRU, last UA0FZ.

11 to 12 May, Hotel Latvia, Riga, Latvia, UQ1GXX/W6KG, /W6QL, 360 QSOs, first QSO YU4PH, last UB5MV.

12 to 15 May, Hotel Letuva, Vilnius, Lithuania, UP1BWR/W6KG, /W6QL, 640 QSOs, first VE3XK, last RA3APS.

15 to 17 May, Hotel Planeta, Minsk, Byelorussia (Belarus), UC1AWB/W6KG, \W6QL, 390 QSOs, first I5BZ, last I1GMF.

17 to 20 May, Hotel Lybed, Kiev, Ukraine, UT4UXX/W6KG, /W6QL; 240 QSOs, first QSO UZ4SWF, last LU5DWD.

19 May, RT0U/W6KG, /W6QL Ukraine, 200 QSOs, first QSO K2SHZ, last K9LJN.

Диплом
Киргизия
W6QL
ВЫДАЕТСЯ ЗА УСТАНОВЛЕНИЕ ДВУХСТОРОННИХ РАДИОСВЯЗЕЙ С РАДИОСТАНЦИЯМИ КИРГИЗИИ
НАЧАЛЬНИК ФРУНЗЕНСКОГО РАДИОКЛУБА
№ 1 YL-USA
ПРЕДСЕДАТЕЛЬ СПОРТИВНОЙ КОМИССИИ UM8MDX
„6" VI 1989 г.

Kirghiz

20 to 22 May, Hotel Intourist, Kishinev, Moldavia (Moldova), UO4OWA/W6KG, /W6QL, 640 QSOs, first QSO UZ9XWV, last F1JNE.

22 to 26 May, Hotel Iveria, Tbilsi, Georgia, UF7FWO/W6KG, /W6QL, 25 QSOs, first QSO G4TRM, last UB5EDG.

26 to 29 May, Hotel Azerbajdzan, Baku, Azerbaijan, UD7DWB/W6KG, /W6QL, 910 QSOs, first UA3PTW, last OZ1FRR.

29 to 30 May, Hotel Ashkabad, Ashkhabad, Turkmenistan, UH9AWE/W6KG, /W6QL, 30 QSOs, first OK3TMM, last UL7RDI.

30 May to 1 Jun, Hotel Tadzikistan, Dushanbe, Tadzhikistan, UJ9JWA/W6KG, /W6QL, 160 QSOs, first QSO 5B4TI, last UA6LQY.

1 to 3 June, Hotel Samerkand, Samarkand, Kazakhstan, UL8NWC/W6KG, /W6QL, 40 QSOs, first RH8AD, last LX2KQ.

3 to 4 June, Hotel Bukharo, Bukhara, Uzbekistan

Georgia

4 to 6 June, Hotel Uzbekistan, Tashkent, Uzbekistan, UI9AWD/W6KG, /W6QL, 1100 QSOs, first QSO UO7OB, last NK6A.

6 to 7 June, Hotel Ala-Too, Frunze, Kirghiz, UM9MWA/W6KG, /W6QL, 160 QSOs, first QSO DA1CT, last WA6AHF.

7 to 8 June, Hotel Belgrade, Moscow, UZ3AWA, 160 QSOs, first QSO K3ZO, last Y21FC.

The word "whirlwind" comes to mind. Iris and Lloyd keep notes on the people they met, and the list reads like a Who's Who of famous hams in the Soviet Union. As soon as they were back home in California they wrote letters to the more than a hundred hams on the list to say:

> This letter is to tell you how much we enjoyed meeting you and having such a marvelous visit in Lithuania. It was great operating ham radio in the club stations (the TV tower of course was the most unusual). We will always remember the nice trip on Ted's boat and the visit to the castle on the lake. We are happy to have seen the new club station that you expect to build; it is going to be a great location.
>
> It was a great pleasure meeting all of you. We will always remember the good times we had together. We are happy that we had the opportunity of becoming acquainted with you, and regret that we did not have longer. Perhaps you can one day visit the USA. We will be most happy to see you and show you some of our country.
>
> Since returning to the USA, we have given talks and shown slides of our Russian Experience.

Everyone here in America is especially interested in hearing all about the Soviet Union. They are all hoping that we will soon have greater communication and friendship between the two countries.

Also in June 1989 they wrote to Nick Kazansky, vice president of the Radio Sports Federation of the USSR:

We arrived in Moscow 2 May 1989 and told you in person that we intended to visit all 15 Republics of the USSR in May and June. We had previously informed you of such trip by letter dated 27 November, 1988. We are very happy to inform you that we have just concluded the trip and did visit all 15 Republics and operated the club amateur radio stations in all but one of the 15 republics. The missing republic was Armenia where we were unable to locate a club radio station during the one day we were there.

We made a total of approximately 9,000 QSOs with radio amateurs in 118 countries on our USSR tour. Without exception we found the USSR radio amateurs to be kind, cooperative and glad to meet us and help us. About half of our radio amateur contacts were made using radio phone and half were on continuous wave (CW).

The short wave equipment worked well and antennas were in general excellent. This is something the USSR amateurs can be very proud of because they assemble and make their own transceivers and antennas, while most of the radio amateurs outside the USSR find it fairly easy to buy radio gear and antennas already assembled and sold by commercial manufacturers.

We thank you, and all radio amateurs of the USSR,

Moldova, RO5OC

Tadzik. UJ8JX, UJ8JCQ, UJ8JS (and his family)

Vilnius, Lithuania

Riga, Latvia

for the way we were welcomed and allowed to operate while in the USSR. We will never forget you and are honored to call you our friends.

A letter from Vasily Bondarenko, chief, Central Radio Club of the USSR, to the Colvins said "We are pleased that your tour to all 15 Republics of the USSR was successful and you were able to operate club stations in these Republics.

"Your amateur radio operation from the USSR was a special permit which was granted to you because of your vast experience in world-wide DXpeditions and great contribution to friendship and mutual understanding between our two nations."

Ashkhabad, Turkmenistan

Patriots of the Fatherland

On 28 May 1989, as they neared the end of their epic USSR visit, Lloyd and Iris were subjects of an article in the soviet newspaper Patriots of the Fatherland, a features-oriented publication that circulated throughout the Soviet Union. The article was entitled "YASME DXpedition Around the World. This is a translation from the original:

In our land are sojourning the well-known travelers, husband and wife, Lloyd, W6KG, and Iris, W6QL, Colvin, wealthy radio amateurs who visit and operate amateur radio from rare countries, and have received 600,000 QSLs (confirmation of contact). They have courteously agreed to reply to our several inquiries for readers of our column.

Question: The YASME ROUND THE WORLD DX PEDITION - Please give us a few words about yourself.

W6QL: I would like to say that, together, we have been licensed for 104 years. I obtained my first radio amateur license in 1945. Since then, I have operated from many DX stations.

W6KG: I have been interested in amateur radio and radio communications since the year 1929. I had never seen an amateur radio station when I built my first transmitter and receiver. I learned about radio amateurs, how they meet, what they do, and how they operate on the air from articles in the magazine "Boy's Life"! which is published in the USA by the Boy Scouts. Fortunately, this simple transceiver worked. I sent in an application and soon received my amateur radio operator's license, and started operating on the air. At first I could not send nor receive Morse Code (CW) very well. I could make a contact, copy the call sign and report, and that was all. I would tell him that there was much QRM (noise/interference) on the bands and then run upstairs calling out "Mama, Mama, I have just talked to someone in Colorado." After much practice, I learned to send and receive the code better and have more and more satisfactory QSOs.

Leningrad, at UZ1AWA

Q: Is this visit private, or under the auspices of YASME?

W6QL: We are sponsored by the YASME FOUNDATION and they are very helpful, but we pay for our tour out of our own funds. We want to visit all the countries of the world, if that is possible, and operate amateur radio from these countries, therefore, we are both tourists and radio hams.

Q: What is YASME?

W6KG: YASME began with a young Englishman, Danny Weil, 29 years old. He owned and built by hand a small yacht and set out to sail around the world. He sailed to the American Virgin Islands where he met the well-known radio amateur, Dick Spenceley, KV4AA, who told him that he should become an amateur radio operator, because, as he sailed around the world, many amateurs would like to communicate with him, and also, be willing to help with his travel expenses. Danny Weil did just this. He studied Morse Code, passed his amateur radio exam, and, in one month's time was on the air making contacts at 25 wpm. Danny was dedicated to his effort and had the extraordinary ability to work extremely hard for long hours. This is an amazing accomplishment that few people can match. YASME was the name of Danny's yacht, from which The YASME Foundation derived its name. We call ourselves THE YASME DXPEDITION.

Q: Give us your impression of meeting Soviet radio amateurs.

Kiev, Ukraine

W6QL: We were surprised. Although we have had many communications with Soviet amateurs, we did not know what to expect when we arrived here. We were surprised to find the people here very much like the people in the USA. We were surprised at the hospitality of the people, and even more surprised at the possibility of operating amateur radio from here. The first QSOs were made from the club stations U3WRW/W6KG and U3WRW/W6QL. We also made many contacts from the Leningrad club station UZ1AWA; UR1RWW and UR1RWX, Tallinn; UQ1GXX, Riga; UP1BWR, UPlBWW/W6KG, UP1BYL/W6QL, Villnius; UC1AWB, Minsk, UT4UXX, Kiev. I operated with great satisfaction from all of these stations. Their equipment was home-made but very good. Soviet hams are very considerate, their on-the-air operation is excellent, and many of them speak English. I especially like the club stations in schools and the youth programs.

Q: What are your wishes for Soviet hams?

W6KG: I wish that Soviet radio amateurs had resources to travel to other countries and exchange equipment and ideas. Iris and I have visited 198 countries. We have always left on friendly terms and believe that travel brings many more benefits than harm. At present, we are traveling to all 15

A Soviet flyer promoting Amateur Radio from 1989

republics of the Soviet Union.

Q: Are mutual Soviet-American radio expeditions possible?

W6QL: In the future, we expect that Soviet and American radio amateurs will make ever greater allied expeditions. For the present, we are happy to have had a successful operation from almost all of the Soviet Republics, which we believe to be the first such operation by Americans. I hope that in the future more Soviet amateurs will visit the USA and obtain permission to operate.

Q: Besides DX, what are you most enthusiastic about?

W6KG: We collect and exchange QSLs and letters from radio amateurs. Next comes photography, at which we are really amateurs, but we show slides of our travels and operations at many clubs and conventions.

Moscow, at UZ3AWA (RW3AH)photo).

Tashkent, Uzbekistan, UI9AWD club station

Q: What are your plans?

W6QL: I have already mentioned that earlier this year we traveled to the country of Niger, and, although we had a long correspondence with them, permission to operate was not granted when we arrived in the country. A month ago, however, we received a letter from Niger stating that a license would now be issued. We plan to re-visit Niger toward the end of this year. We will also try to gain permission and operate from some of the neighboring countries, such as Zaire. Our plans for the future include visiting all the countries we have not yet visited, and operate wherever we can obtain licenses.

W6KG: I hope that in the future, tourist exchanges between all countries will become better.

After their return, Iris summarized her feelings in a story she entitled "Impressions of USSR after visiting all 15 republics." With the fall of the Berlin Wall and Glasnost, Iris had caught the fever:

> After being told that the people had no cars, food or luxuries of life, my first impression was of great surprise that everything is very much the same as in the US. There is plenty of food and clothing. The Intourist Hotels at which we stayed were almost on a par with our middle class hotels. There was no shortage of any of the essentials.
>
> As I studied more closely the differences between the two countries, my second impression was that in this Communist/Socialist society, there was no trade with the outside world, there was a lack of personal enterprise even within the country itself, and, in general, an absence of the rights and free-

doms that we are fortunate to enjoy in the US.

The Club Stations, as well as all personal radio stations, use hand-made equipment that the operator has made himself. Some of the equipment is very good and some not so good. It certainly cannot compare with the excellence of commercially made equipment. In the early days of radio, hams in the US built their own equipment.

Even today, hobbyists and students experiment with and learn about construction, but they can buy computer designed and manufactured radio sets, with diodes, shielding against TVI etc, with which the home-built sets cannot compete.

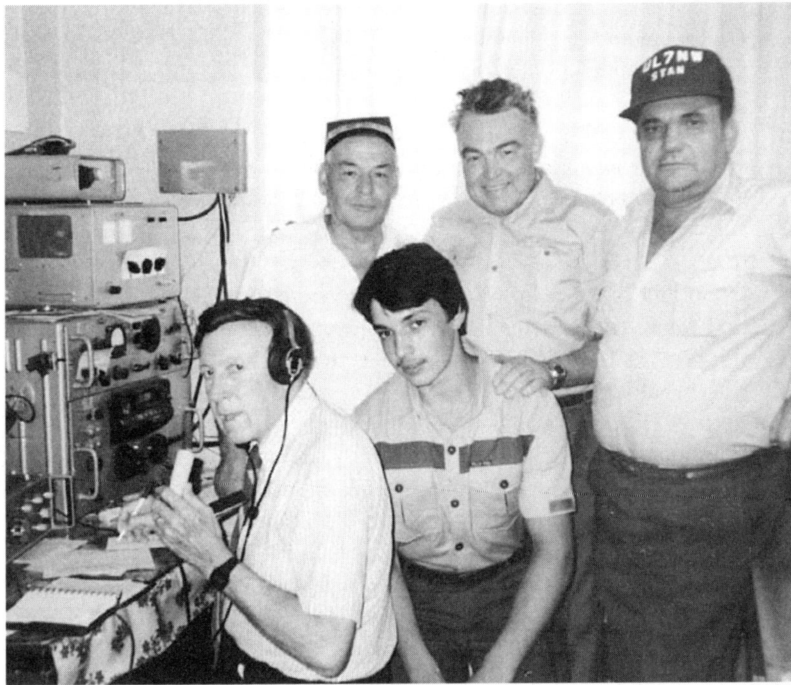

Chimkent, Kazakh (UL7HV, UL7NA, UL7NEA, UL7NW).

M.S. Gorbachev, Secretariat Communist Party of the Union of Soviet Socialist Republics, is making great strides. Friendship is a key theme in the many, many slogans, bill boards, notices, statues etc. seen everywhere in the USSR.

I noticed a great many programs for education and improvement of the younger generation. We must all remember that personal freedom is to be prized, and its survival depends upon an acceptance of the personal obligations and responsibilities that go with it. We must face judgment for our own lives.

Also in June 1989, Iris and Lloyd sent follow-up inquiries to Sudan, Chad, Guinea, and Cameroon, saying "No answer has been received as of this date. We intend to arrive in your country very soon now.

"Please help us and let us operate our small amateur radio in your country. We have now traveled in 207 countries. There are some 2 million amateur radio operators in the world. There have been only very limited such operation in your country and we feel that permission for us to operate in your country for a short time will bring much favorable publicity in all parts of the world."

They sent new inquires to Uganda, Central African Republic, Guinea Bissau, and Gabon. Business matters on the home front were now entirely in the hands of their daughter Joy and her husband Richard. Lloyd, at age 74, and Iris, at 75, deserved a breather after their USSR trip, which could have exhausted folks half their ages. But, come November, it was back to Africa.

47· The Honors and Awards Begin

Our one-month stay here was the most expensive of all of our DXpeditions. - Lloyd and Iris Colvin, on Bahrain.

You will make us very happy if you will please arrange some way to comply with our request. - Colvin letter to various African countries

The long-anticipated operation from Niger finally became a reality from 3 to 26 November 1989. The Colvins stayed at the Sabka Lahya Hotel and made 8,000 QSOs in 161 countries, as good as their efforts of 20 years before. Their 5U7QL DXCC was #30,575, 24 April 1991; first QSO was YU3NY, last was K8MFO.

They quickly traveled to Burkina Faso and came on the air from the OK Inn, in Ouagadougou, on 26 November 1989. Operating through 18 December they logged a fabulous 14,000 contacts as XT2KG; Lloyd got on briefly for 82,000 points in the CQ WW CW Contest.

Niger

In their report following the operation they called it a "magnificent operation," and darned right it was. "There was never a time," they wrote, "when there were no DX stations to be heard either on 10, 15, 20 or 40 meters.

"We have now traveled in 208 countries and worked DXpeditions from more than half of them, covering a period of 60 years, from 1929 thru 1989. Our most successful operation before coming here was 11 years ago in the country of Montserrat using the call VP2MAQ, where we worked nearly 14,000 amateur radio stations in 149 countries.

"We beat that record here, working nearly 14,000 stations located in 161 countries in a 3-week period, using the call XT2KG. Operation here, and elsewhere, supports the following observations: Band conditions peak on an 11-year cycle, operation is better near the equator than at either polar region, and it helps to be in a rare country with an unusual call sign."

The first XT2KG QSO was with N2DT and the last was KK7Y. Their DXCC was #30,557, 24 April 1991.

Leon Fletcher, in reviewing personal notes of the Colvins, observed that the Colvins were positive thinkers, rarely dwelling on the many problems and obstacles they faced in their travels. Customs officials, language differences, protocols, and telecommunications officials probably lead the list, with Iris and Lloyd (and anyone else, for that matter) weak from travel "road food."

"But they seem never to have complained, or, more importantly, given up," Fletcher said.

After more than two weeks in Niger, the Colvins took a bus (!) to Ouagadougou. Notes in their log book there records:

Niger

Wednesday, November 22: Downtown to negotiate baggage that had been turned into (driver) who had to turn it over to [unreadable word] - much difficulty.

Thursday, November 23 (Thanksgiving Day): More difficulty with customs. Got letter authorizing release of baggage.

Friday, November 24: Letter from Ambassador David Chen to Head

Commissioner; received license - rushed to Duane - Got baggage late - Evening assembled beam.

Saturday, November 25: Put up beam - cut down tops of trees - fixed 40 meter switch.

Sunday, November 26: Worked on beam 4 o'clock - Lloyd worked contest until midnight.

Monday, November 27: XT2KG (their call in Burkina Faso); Final blew up--used only transceiver; beam won't rotate properly.

The Colvins had passed through Bahrain in 1983 but couldn't get permission to operate. It was different in 1990, though, and they came on the air as A92QL, from 8 to 21 January 1990, from the Bait al Hoora Hotel, a 7-story building overlooking the Arabian Gulf. "Everyone has helped us to get a special license that is not in the normal sequence of assignment. The President of the Bureau of Wireless Licensing, Abdulla S. Al Thawadi, [as well as] A92BE, A92BW, and A92EV have been especially helpful. We had good propagation and no TVI."

A92QL logged 5,000 QSOs with amateurs in 126 countries. A92QL's DXCC was #30,576, 24 April 1991, and first QSO was UB4UFZ, last was W8CZN.

They observed that the value of the US dollar continued to decline. "Our one-month stay here was the most expensive of all of our DXpeditions."

After this, Iris and Lloyd flew to not only a new country for them but a new continent, as well - Australia. They said later that this was "not a planned stop." The YASME Foundation at the time had, on its board of directors, Heather Pike, VK2HD, who greased the skids by putting the word out around Sydney and arranged for the Colvins to stay with Harry Mead, VK2BJL, and his wife Mary, at a "perfect location, on a hill." Afterward, Iris and Lloyd reported "We used Harry's antennas and our own equipment; Mary did the cooking. Both Harry and Mary put up with our operation at odd hours, day and night, and fed us with our favorite foods."

From 25 January to 7 February 1990, operating as VK2GDD, they made 4,000 QSOs, on 10 thru 40 meters, with 142 different countries, in just two weeks. Their VK2GDD DXCC was #30,805, 22 August 1991; first QSO was ZK1DD, last was VE1NH.

They got their operating permits immediately, and Harry provided tips on band openings, undoubtedly helping in that 146-country total.

Iris and Lloyd were asked by The Wireless Institute of Australia to speak over their Sydney radio station and also at a special meeting of the Wireless Institute of Australia. The radio interview was conducted by Stephen Pall, VK2PS.

Australia

Bahrain

New Zealand

Tahiti

Switching to the Southern Hemisphere seems to have had a good effect, because the Colvins next flew to New Zealand for a 6,000 QSO performance (148 countries) as ZL0AKH. They operated from 8 to 28 February 1990 from Pakuranga, Auckland, earned a certificate in the ARRL CW Contest, and later were awarded DXCC #30,692, dated 7 February 1991. First QSO was VK9NS, last was IK4CIE.

Operation was from the stations of ZL1AMO, ZL1BQD, and ZL1BMU, using their antennas and the Colvin's equipment. "Other hams joined in making our operation successful and very enjoyable.

"We received our licenses before we arrived in New Zealand. Not many countries will issue licenses before arrival in the country, but it's a big help to the visiting radio amateur.

"We operated in Australia previously and found that both countries use identical electric plugs. We wish that all countries did. We were surprised to learn that New Zealand is only about 1,000 miles from the South Pole, making it one of the most southerly countries from which we have operated."

On a trip that began in Africa, then in the Middle East, then in Australia, Iris and Lloyd finished up on Tahiti, homeward bound. FO0XXL operation was 3 to 27 March 1990. This was the second stop on the current trip on which no "report" exists, which is surprising. These were scattered among 8,000 contacts (first WB8OHO, last DL9BW). Their DXCC is #30,694, 7 Feb 1991.) Operating multi-single, Iris and Lloyd's work in the ARRL DX Phone contest was Continental (Oceania) Leader.

Back home, to accollades

While they were in Australia, Lloyd was awarded a plaque for 60 years of ARRL membership, in January 1990.

In February, 1990, the ARRL gave the Colvins a "To Whom it May Concern" letter to assist them in gaining entrée to other countries, saying:

"Lloyd and Iris Colvin are lifelong Amateur Radio operators and Charter Life Members of the American Radio Relay League. A particular interest of theirs is operating Amateur Radio stations from locations throughout the world, always in strict accordance with the regulations of their host country. This they have succeeded in doing so from more than 180 countries, in the process making hundreds of thousands of radio contacts with other amateur stations. Any assistance you can provide to the Colvins will be appreciated by the worldwide Amateur Radio community."

The ARRL Board of Directors at its July 1990 meeting recognized the Colvins, unanimously voting "that the Board of Directors confer on Iris and Lloyd Colvin, W6QL and W6KG, a special award as Amateur Radio Ambassadors of the decade 1980-1990 for their tireless efforts to foster international goodwill by means of Amateur Radio."

In April 1990 QST published an article about the Colvins, entitled "Radio Roulette in the Soviet Union." It was the first article about the YASME Foundation, or the Colvins, ever published in QST. No feature article had been published before in QST about either the Colvins or Danny Weil.

May 1990 saw Iris and Lloyd asking again, of the Malagasy Republic, Malawi, Uganda, Walvis Bay, Burundi, and Rwanda. They had a new form letter to send, to the "Telecommunications Office, Radio Licensing Department":

> We hereby apply for permission to operate our amateur radio station while we are visiting in your country later this year or the first of next year, for a period of time not exceeding 30 days. We have been active radio amateurs since the first days of radio. We both hold the highest type of amateur radio license issued in the United States, which is the "Extra Class" license.
>
> Since our retirement from the construction business, we have been traveling all over the world, carrying our small amateur radio station, which weighs approximately 150 pounds, with us and operating it in many countries. We have visited 209 countries, a list of which is enclosed. We call our travels "The YASME DXpedition". The word "YASME" was the name of the boat in which a radio amateur, named Danny Weil, went forth in 1954 on a continuous world-wide trip devoted primarily to communication with other radio amateurs throughout the world. He was the first person in the world to make such a journey, and we are more or less following in his foot steps.
>
> We represent no government or company. We do represent the YASME FOUNDATION, which is an informal association of leading radio amateurs around the world. None of the YASME Foundation members or officers receive any pay, and our travels are paid for entirely from our own funds, which are sufficient to meet all requirements.
>
> We are in good health and have had wide experience in visiting various parts of the world. We have always been very careful to avoid any "Problems" in the countries visited. We have worked hard to give a good impression of ourselves and amateur radio. Our operation will bring much favorable publicity to your country among the more than one million radio amateurs in the world.
>
> We will comply in full with the radio rules and regulations of your country. If desired, we will be happy, at the end of our stay there, to submit a copy of our radio log, showing all contacts made, frequency of operation, time of contact, etc.
>
> Anxiously awaiting your reply and hoping that it is favorable, we are, sincerely....

What more can you do? Well, include a list of countries visited (more than 200), copies of FCC licenses, copies of French licenses, copies of British licenses, and an extract from QST magazine, April 1990 [an article written by the author of this book].

In July 1990, Lloyd and Iris sent a follow-up letter:

> We are leaving the United States soon for a 6-month tour of East Africa. We can be reached, however, at the address and phone number at the top of this page [their California home]. Please give this letter your utmost consideration.
>
> We want to visit your country and, if at all possible, we want to operate our hobby of amateur radio while there. We would like to have permission to operate for no more than 3 weeks; we will comply with any requirements you want. Some of the restrictions that you might impose are:
>
> a) Operate only in a place that you designate;
>
> b) Pay for someone to monitor our operation;
>
> c) Submit copies of our radio logs, showing who was contacted, when and where;
>
> d) Any other requirements that would make our hobby operable in your country.
>
> We are trying to visit all of the countries of the world and operate amateur radio in them. We are both past 75 years of age, but we are in excellent health. We require no doctor, no medication, and no assistance. This may be our very last chance to enjoy our hobby in your country, and we can guarantee that our operation there will bring only good will and favorable publicity to you and your country.
>
> Although there has been some radio amateur operation during the last 8 years, it has been somewhat limited. We will bring with us all of the equipment that we need, which consists of a radio receiver and transmitter and an antenna. The equipment weighs approximately 150 pounds.
>
> You will make us very happy if you will please arrange some way to comply with our request.

They enclosed copies of their US (FCC) radio licenses, a list of countries visited, and a copy of the complete text of the April 1990 QST article.

THE WINNIPEG AMATEUR RADIO CLUB INC. PRESENTS A

NATIONAL C R R L CONVENTION

" AMATEUR RADIO COMMUNICATIONS . . . AT THE CROSSROADS"

AUGUST 18–19–20 1989
WINNIPEG MANITOBA
CANADA

Satellite Communications

DX EME

CRRL Forum

Communications Canada

Repeaters

Packet Radio

Ladies/Childrens Programs

Indoor/Outdoor Flea Market

Vendor Booths

Antique Radio Display

Wouff Hong Ceremony

General Interest Displays

Many Prizes And More!

W6KG – W6QL

Featuring As Keynote Speakers . .

IRIS AND LLOYD COLVIN

WORLD FAMOUS GLOBE-TROTTING DXERS

INTERNATIONAL INN

ADJACENT TO WINNIPEG INTERNATIONAL AIRPORT

CANADIAN AIRLINES INTERNATIONAL IS THE OFFICIAL CARRIER TO THIS CONVENTION. DELEGATES AND THEIR SPOUSES CAN SAVE 15% OR MORE OFF REGULAR ECONOMY FARES IN CANADA.

CALL YOUR TRAVEL AGENT INDICATING THAT YOU ARE A DELEGATE TO THE CONVENTION (#2055) CALL TOLL FREE: IN CANADA 1-800-268-4704 . . IN U.S.A. 1-800-426-7000 . . IN TORONTO 675-8256

For More Information Write To: WINNIPEG AMATEUR RADIO CLUB INC.
CONVENTION 89 COMMITTEE P. O. BOX 352 WINNIPEG MANITOBA CANADA R3C 2H6

48· Africa and Southeast Asia

I know from experience that you can get at least five cups [of tea] per bag if you do it right" - Lloyd Colvin.

We now have 221 countries visited during our lifetime, and have DXCC certificates from more than half of them. - Iris and Lloyd Colvin

The expected YASME press release came out on 15 August 1990, announcing that the Colvins would once again depart for parts somewhat unknown. First stop would be Dar es Salaam, Tanzania. "They hope to use the call 5H0KG there. The well-known DXer, Tom Warren, 5H3TW, has a place for the Colvins to stay and operate. After that, Lloyd and Iris hope to visit many of the remaining countries where they have not bcen before, and to operate in such countries if permission to operate can be obtained. They plan to visit both Mozambique and Madagascar, but, as of now, do NOT have permission to operate."

In a way, it was getting easier, because there were fewer countries to choose from! The Colvins left early, compared to previous years' trips, arriving in Tanzania in mid-August 1990. Tom Warren, K3TW/5H3TW lived there, working for the US Department of State, and Lloyd and Iris stayed there, saying they had now visited Tom in three continents: North America, Asia, and Africa.

Tom told a story. "I went to Tanzania with a case of tea bags, enough for several lifetimes. After Lloyd had enjoyed a cup of tea here, he slipped the bag aside. I said something like 'I have thousands of tea bags,' and Lloyd said 'I know from experience that you can get at least five cups per bag if you do it right.'"

The Colvins called their Dar es Salaam operation, as 5H0QL, "very satisfactory." From 22 August to 7 September 1990 they made 7,000 QSOs and worked 139 countries, half CW and half SSB. The inevitable 5H0QL DXCC was #30,804, 22 August 1991. First QSO was JR2PAP, the last I2FGT.

They said, from Tanzania, that "We go now to Malawi,

Mozambique

Malawi

where it seems possible that we can get permission to operate. We tried to get such permission several times in the past, with no luck. However, several licenses have been issued in the last few months. So wish us luck, and we hope to see you all on th air from Malawi."

Operation from Malawi did take place, from the Capital City Motel in Lilongwe, where "the food and accommodations were excellent, and there was a great high place to erect our antenna." The Colvins got their licenses in the city of Blantyre, 350 kilometers from Lilongwe and, although the licenses were issued promptly and the authorities were "most friendly and cooperative," the two licenses cost $115 each, the most Iris and Lloyd had ever paid for amateur radio permits.

The 12 to 29 September 1990 7Q7KG operation tallied 7,000 QSOs in 155 countries, and DXCC # 30,693, 7 Feb 1991. First QSO was 4X6DW, last was WB4TGB.

Their next operation was from Mozambique, "one of the rarest DXpeditions that we have ever made." Iris and Lloyd had been sending inquiries to Mozambique for years, but were rebuffed, as were many others seeking amateur radio permits there.

In Malawi, they met a woman in the office of the US ambassador to Malawi. She and her husband had previously worked in the U.S. embassy in Mozambique. She suggested the best place to stay in Maputo, the Cardoso Hotel, and provided a list of names of people who might be helpful in Mozambique.

One of those people she mentioned knew someone else "who often helped business people in their efforts to do business" in Mozambique. She turned out to be the daughter of an ex-president of Mozambique, and spent "many days" contacting officials of the government on behalf of the Colvins.

The U.S. ambassador, Melissa F. Wells, was getting ready to leave the country but found time to write "a nice letter" on the Colvin's behalf to the Minister of Transportation and Communications.

Going through another friend of a friend, they finally got to see the vice minister of communications, Mr. Rui Jorge Comes Lousa, and licenses were issued. Iris and Lloyd also had already got permission from the hotel to install and operate their radio station, providing they had licenses from the government.

C9QL came on the air 16 October 1990. Operating through 5 November, the Colvins ran up more than 5,000 contacts, including 345,000 points in the CQ WW Phone DX Contest (DXCC # 30,802, 22 August 1991. 131 countries were worked; WA3TIH was the first QSO, YB5QZ the last.

Operating 40 through 10 meters, both phone and CW, they said "We could easily have worked twice this number of QSOs if the bands were open all 24 hours a day. They were not. We were lucky when we could hear any signals at all on any of the four bands for as much as 12 hours a day."

Earlier in the year Walvis Bay, a port city in Northwestern Namibia, had been announced as a new DXCC country, and contacts back to 1977 would count, but few DXers had a contact or a QSL (it would be deleted from the DXCC list in 1994). Control of Walvis Bay had bounced around for centuries; in 1992 it was under South African jurisdiction. Walvis Bay was a "new one" for DXCC and the Colvins headed there.

The Colvins stayed at the home of Capt. John Smith, ZS9S, using his antennas with their equipment. Smith and Ian, ZS9A, were the only active permanent residents of Walvis Bay. Five years before, Ian had been the Postal Chief and in charge of issuing ZS3 licenses to radio amateurs in the area now known as Namibia (then Southwest Africa). During their stay, the Colvins had a surprise visit by two YASME directors, Martti Laine, OH2BH, and Wayne Mills, N7NG, on their way to operate from Penguin Island.

Mozambique

Tanzania

Burundi

Tanzania

Walvis Bay

From 17 November to 11 December 1990 ZS9/W6KG made 8,000 QSOs in 152 countries, and in the CQ WW CW DX Contest racked up 1.1 million points. The ZS9/W6KG DXCC was #30,801, 22 August 1991. First QSO was K5OVC; last K1FWF.

Iris and Lloyd had reported "weak signals" while in Mozambique, and discovered when they got to Walvis Bay that their IC-751A transceiver's front end had failed. In addition, while passing through Namibia on 15 November, a thief stole their big bag containing some 220 feet of RG8 coaxial cable, the cable they carried to feed their tribander. While in Namibia two hams, Rudy Wischers, V51W, and Derek Moore, V51DM, both of ElectroCom Company there, found new coax for them, and repaired the transceiver, "at below their cost."

Over the next two months, Lloyd and Iris visited Algeria and Zaire (where a letter on their behalf from the American embassy failed to get them operating permission). They also made side trips, to test the waters, in Congo, Rwanda, and Uganda.

The Colvins' final operation on this trip was as 9U5QL, Burundi, from 29 January to 16 February 1991, from the Club du lac Tanganyika, in Bujumbura. They made 9,500 contacts from this rare country, the first with 9H1NB and the last with WD5GJS, and garnered DXCC #30,691.

Macao

Iris and Lloyd were happy with this six-month Africa swing, saying that "All of our operations on this trip were rare countries." Indeed they were especially in demand from Mozambique, Malawi, Walvis Bay, and Burundi. "All this," they said, "made us very much wanted by hams world-wide. The pile-ups were enormous but we enjoy them and have learned, on both fone and cw, to copy stations with many others calling on top of them." In total they made nearly 40,000 QSOs.

In the summer of 1991 the Colvins looked at the list of countries they had not yet operated from and sent letters to many of them, including Albania, Bangladesh, Lebanon, Iraq, Uganda, Yemen, Burma, Mongolia, Mauritius, Angola, Malaysia, Rwanda, Bhutan, Chagos Islands, and Iran. As usual, most did not bother to reply. One that did was Iran. The Minister of Posts, Telegraph, and Telephone, Directorate General of Telecommunications, Hossein Mahyar, said "I am sorry to inform you that in compliance with the current Radio Rules and Regulations of our country, operation of your amateur radio station could not be permitted at present. I hope to be of your assistance in the future.

In October 1991 H.F. Hatton, "Commander Royal Navy, British Representative, British Indian Ocean Territory, Diego Garcia" (Chagos) said

> Your letter of 13 September 1991 applying for an amateur radio license and permission to operate the same in the British Indian Ocean Territory was received by me today. For clarification, the Chagos Archipelago forms the BIOT of which Diego Garcia is the largest and only inhabited island. However, Diego Garcia has strictly limited access to military personnel and those employed by the Defence Departments of the 2 countries, Britain and the USA. Thus, the only non-Defence

Department personnel permitted to land in the BIOT are those who arrive using their own sailing vessels and who stay for limited periods in the deserted outer islands of the Archipelago.

As your correspondence does not show that this method of travel is normal for you, I would need to know in greater detail your intentions for arrival before I could accede to your request.

Southeast Asia

The Colvins flew to Thailand in November 1991 for an operation from Bangkok. John Vajo, HS0ZAA, custodian of HS0AC, club station of the Radio Amateur Society of Thailand, helped them both get their operating permits and permission to use the club station. Vikrom, HS1HB, the RAST president, also helped. Staying at the Sri Guest House 20 minutes away, they used their own equipment at the club station and made 1,500 contacts in 120 countries, signing HS0ZAP. Getting "home" licenses, they said, is much more difficult. First QSO was with DU1GW, last with P29JR. Lloyd and Iris attended the SEANET convention, in Chiangmai, Thailand.

After this, the Colvins visited Vietnam in November 1991, but were unable to obtain licenses.

On 1 January 1992 the Colvins opened up from Cambodia. They stayed at the Monorom Hotel in Phnom Penh and operated, using their own gear, from a station at PTT headquarters (XU8DX), but they did have their own call - XU8KG. Two directors at the PTT office, about a three-minute walk from their hotel, were very familiar with amateur radio and were of great help. In 10 days of operation they made 800 QSOs in 105 countries; first QSO was 9V1QG, last was XU8DX.

Lloyd wrote that "Iris stayed up all night long the last day of our license permit. She started at 80 countries and

Macao

Cambodia

got to 105 countries worked in the last few minutes of our operation."

Next stop, Laos, where the Colvins stayed at first the Ekalath Hotel, then at the Saysana Hotel, both in Vientiane. A Japanese amateur there was licensed as XW8KPL, and they signed XW1QL from his station. From 27 January to 2 February 1992, 1,600 QSOs and 115 DXCC countries resulted (first QSO JA7DAT, last YU4CA).

They said that "In both here in Laos and in Cambodia, we had difficulties with the authorities over the problem of visas. It has been difficult in these two countries, and in other countries nearby, to receive permission to stay more than one week, which is not long enough to set up station, obtain an operating permit, and work DXCC."

The U.S. Embassy in Laos provided a letter:

The Embassy of the United States of America presents its compliments to the Ministry of Foreign Affairs of the Lao People's Democratic Republic and has the honor to request the Ministry of Foreign Affairs and the Ministry of Communications issue an amateur radio operator's license to Mr. Lloyd Colvin and Mrs. Iris Colvin. These two retired American citizens are currently traveling throughout the world operating their

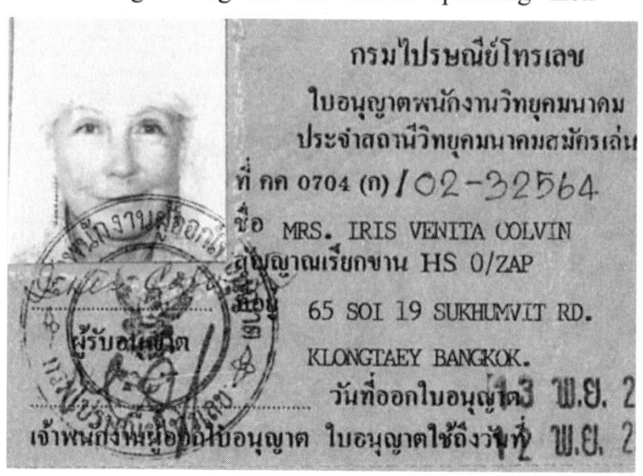

กรมไปรษณีย์โทรเลข
ใบอนุญาตพนักงานวิทยุคมนาคม
ประจำสถานีวิทยุคมนาคมสมัครเล่น
ที่ คค 0704 (ก)/02-32564
ชื่อ MRS. IRIS VENITA COLVIN
สัญญาณเรียกขาน HS 0/ZAP
65 SOI 19 SUKHUMVIT RD.
KLONGTAEY BANGKOK.
วันที่ออกใบอนุญาต 13 พ.ย. 2
เจ้าพนักงานผู้ออกใบอนุญาต ใบอนุญาตใช้ถึงวันที่ 12 พ.ย. 2

amateur radio. "They wish to operate their radio in Laos for a period of three weeks, after which they will exit Laos. The Embassy notes these two American citizens have successfully operated their amateur radio in over 220 countries. [Not accurate, of course, the fault of the embassy personnel.]

The Embassy further requests the Ministry of Foreign Affairs grant a visa extension to Mr. and Mrs. Colvin so they can remain in Laos a further three weeks to operate their radio.

The Embassy of the United States of America takes this opportunity to renew to the Ministry of Foreign Affairs of the Lao People's Demo-cratic Republic the assurances of its highest considera-tion.

Obviously, the letter didn't work.

The Colvins next operated as V85KGP from Brunei, 11 to 26 February 1992, from Bandar, Seng Kurong Village, making 1500 QSOs, and another DXCC with 130 countries worked. First QSO was VK6HQ, last was V63AO.

Lloyd and Iris found this very oil-rich country also very expensive, the lowest-priced hotel they could find was $80 a night, with the rest costing $200 and up. "This government is ruled by a sultan. He has two wives, which he must provide for equally. The majority of the popula-tion are Muslims. We hear the call-to-prayer from the many mosques five times a day, beginning at 4 a.m. The stores here do not keep late hours and the government has seen fit to order more stores to [stay] open for business and work longer daily hours.

"There are not many amateurs here, so we are much in demand and have had many QSOs. Our countries worked from here is the largest for this trip. We have worked more than 100 countries on each stop and hope to qualify for DXCC from each one of them. We now have 221 coun-tries visited during our lifetime, and have DXCC certificates from more than half of them."

Cambodia

Brunei

The Colvins had no trouble getting licenses in Macao - XX9TQL and XX9TKG, and they chose Iris's to use. They had been told beforehand that licensing would be difficult, but they had applied five months before and the licenses were ready when they arrived. They made 1200 QSOs, the first AH6JF and the last XU8DX, in 112 countries. Operation was from 11 to 24 March 1992, from the Kingsway Hotel and the Macao Moke Car Park.

"It is getting hard to find a place to operate without TV sets nearby," they said. "We finally solved this problem when we found an automobile sales and rental business only one story high. We rented space there to operate with no nearby TV sets. We worked more than 100 different countries, with no TVI complaints.

"We did have to walk from our hotel to the ham sta-tion, but it was worth it."

Macao wrapped up their six months in the Far East, with operations and DXCCs from five countries: Thailand, Cambodia, Laos, Brunei, and Macao.

The YASME Foundation met in April 1992 and dis-cussed "preparing a history" of the foundation. Two years later no progress on this had been made and Iris Colvin agreed to hold archival material until a further decision could be made. (Iris was elected president of the founda-tion, on 17 April 1994).

In December 1992, Iris and Lloyd flew to Aruba, a DXCC Country they had not operated from, but were unable to get reciprocal licenses. While there they also applied to Venezuela for operating permission but nothing materialized. After some two weeks on Aruba, they flew home to California. Unknown to them, Macao was to be their last DX operation.

In November 1992 Iris and Lloyd officially retired from their positions in the family business, Drake Builders. On January 24, 1993, the birthday of their son-in-law Richard Gilcrease, Lloyd gave him a desk plaque saying "Richard Gilcrease, President." Drake Builders continued as a family corporation. Lloyd and Iris were its president and secretary until they retired in November 1992, when Richard and Joy Gilcrease became president and secretary, respectively. The primary business is still real estate and it is also the entity behind the Gilcreases' publishing house in Galveston, Texas, VanJus Press.

49· Turkey, 1993, the Last Colvin Expedition

Lloyd and Iris Colvin were the most spectacular and most enduring team of DXers ever to show on the bands. They were the friends of every journeyman DXer and the passing of Lloyd will leave a void that possibly never can be filled.. - Hugh Cassidy

In Turkey

In late November 1993 Iris and Lloyd Colvin flew to Istanbul, Turkey, where they had permission to operate. They took their radio equipment and antennas with them; no itinerary after Turkey had been announced. Iris and Lloyd took photos during their first few days in Istanbul, and they had one of their special log books prepared for use.

Iris made about 80 contacts beginning 15 November 1993, as TA1/W6QL, and ending 1 December. The first contact was UA1CBV, and the last, I1TBE.

About two o'clock in the morning on 11 December, the Colvins' daughter Joy, in California, got a telephone call that Lloyd was in the hospital with pneumonia and might not recover. He had left the hotel on a cold rainy night to get something to eat. He was found incoherent and disheveled, probably from a stroke, and, since he was American, he was taken to the American hospital. Fortunately, Lloyd and Iris were registered with the American consulate, and Iris was notified by them the next day.

Joy began to pack and realized she did not have an active passport. She called the State Department in Washington, at about seven in the morning on the east coast. "Amazingly, I reached someone with a sympathetic ear. She told me to go to the San Francisco passport office with appropriate photos and documents, and mention her name - they would be able to issue a passport as soon as they opened at 9 AM. I packed, got passport photos made at an all-night copy service, picked up my passport, and was on a plane from SF airport at 10 AM.

"The hospital staff was very kind," Joy said, "and there was a lady there who spoke English. When I got there, Lloyd was only barely conscious, his eyes seeming to roll unseeing from side to side. On my first visit, as I spoke to him, he made a tremendous effort, arrested that side-to-side eye movement, and with great difficulty focused directly on me. When I said, "I love you, Daddy," a single tear rolled down his cheek. Iris and I visited daily for a week following, but he never really was conscious again." Lloyd Colvin died on the morning of 13 December 1993.

Lloyd with TA2CII, left adn TA2DS

Joy remembered that Richard Dailey, the American consul general in Istanbul, "helped us tremendously both before and after Lloyd's death, with the transportation of the body back to the States. I am forever grateful."

Before leaving, Iris donated their amateur radio equipment and antennas to the amateur radio club in Istanbul.

Lloyd's desire was to be cremated and his ashes scattered at sea, beyond the Golden Gate Bridge. Joy said "Iris and I sent notification to some hundreds of hams whom Lloyd considered his friends, but there were only a dozen or so family and closest friends on the boat that sailed out into San Francisco Bay on a beautiful sunny December day to scatter his ashes."

Harvey McCoy, W2IYX wrote in The Long Island DX Bulletin, which he published, just a week later, that

> Lloyd Colvin, W6KG, succumbed to a fatal heart attack on 13 December '93 in Istanbul, Turkey. He and his XYL Iris, W6QL, were there on another of their DXpeditions, (which have taken them to nearly 200 Countries throughout the world. Your editor first met Lloyd in 1943, (when he was a US Army Signal Corps officer stationed in Alaska) and I was a civilian engineer in the Office of the Chief Signal Officer, US Army. I had just invented FS RTTY and Lloyd helped me in preparing the new format for RTTY transmissions on military radio circuits. Iris will be back home as you read this - and will appreciate your messages of condolence.

Of the cards and letters that arrived over the next months, as word spread, one, written to Iris from Virgilio Soria Valle, EA4CQT, in strained English, was typical:

Dear Iris: I've let passed a time from the death of Lloyd, now I think that you will be better.

The first thing that I want to tell you is that I am very sorry about the gone of Lloyd. You, both, "The Colvins" gave me a lot of countries, a lot of wonderful time, days, looking for you, reading about news from you: where you will go next?, How long time you will be there?, could I work them?, on both modes? You gave me a lot of illusion and happiness.

Now I remember when you where in 9Q, I worked you on SSB, but not on CW. One night you where on CW, and I was calling you, you didn't take me, and I called and called again in the "pile up", I thought that you had to hear me!, but the "pile up" was so big, and I was with 80 watts and an inverted vee, that it was impossible. But I worked you on SSB!, new one for me.

Dear Iris, I only can tell you that Lloyd now is in heaven, Sure!, and he has the best pile up's. He now has the rarest call signs from the rarest places, and he now gives the 59 or 599 to everybody, now everybody can have a QSO with Lloyd, with a prayer, that's all for a good contact!

Now I am finishing my University, and the radio is my Hobby, but I think that radio amateur is more than a hobby, I think that is a lifestyle. If you need anything from here, do not hesitate to ask for it. You have here a friend. I am proud to have the same illusion [sic] that Lloyd and you. God bless you Iris.

Lloyd's death coincided with the publication of ARRL's DXCC Yearbook in early February 1994. The author of an article that appeared there is also the author of this book:

Lloyd Colvin, W6KG: End of a DX Era

At year's end the DX world was saddened to learn of the death of Lloyd Colvin, W6KG, at age 78. Fittingly, the end came (on December 13) in Istanbul, Turkey. It was appropriate that Lloyd, who spent at least half of every year for the past three decades away from his Richmond, California home, would depart this earth from a DX location.

Lloyd was born April 24, 1915, in Spokane, Washington. His wife of 55 years and DXpedition partner Iris Colvin, W6QL, survives him. At the time of his death both Lloyd and Iris had received permission to operate from Turkey. Lloyd did not operate but Iris did, briefly, as TA1/W6QL. Lloyd suffered an apparent stroke, was hospitalized, and died shortly thereafter.

TA1/W6QL's beam in Turkey

Lloyd was first licensed in 1929. His early interest in radio led to a career in the U.S. Army Signal Corps. He served 23 years, retiring as a lieutenant colonel in 1961. "I got into the Signal Corps," he told 73 magazine in 1981, "through ham radio. In fact, almost everything we've done in our lives has been directly connected with ham radio. I guess I found my military life interesting because much of it was similar to running ham stations."

In the 1950s and '60s Lloyd was a general contractor and president of Drake Builders, building houses, apartments and hospitals in the San Francisco Bay area.

In 1965 the Colvins began traveling the world, and in the nearly three decades that followed they visited 223 countries, operating from more than half of them. Their reputations preceded them, helping them obtain operating permission when most others couldn't (although a few stone walls stood, in countries such as Burma and Bhutan).

One of the Colvins' most ambitious trips was in 1989, when they operated from 14 of the 15 then-republics of the Soviet Union (they couldn't find a club station in Armenia in the one day they were there). Their last major trip was to Southeast Asia in 1992. Over the years the Colvins made more than a million contacts and had more than half a million QSLs on file. Lloyd Colvin was an ARRL Charter Life Member (as is Iris), a life member of the Northern California DX Club, and an honorary member of countless DX clubs and groups. Lloyd was an ARRL member continuously since 1930, and was a strong and loyal supporter of the League and of its DXCC program.

The Northern California DX Club, of which the Colvins were long-time members (and Iris, the first female member), recorded Lloyd's death, and the following story also was quickly published in Worldradio:

> Lloyd Dayton Colvin, W6KG, who with his wife Iris, W6QL, visited and operated from hundreds of DX locations all over the world, is a Silent Key. He was 78. After initially rallying, Lloyd succumbed to the effects of a stroke suffered in early December during a visit to Istanbul, Turkey with his wife. His death there was attributed to cardiac arrest.
>
> First licensed in 1929 at age 12, Lloyd earned an electrical engineering degree from U.C. Berkeley. After graduation, he embarked upon a 23-year army career, retiring in 1961 as a lieutenant colonel. His service years had provided the opportunity to operate from many foreign countries, establishing the penchant for DX globetrotting that marked his later years. Issued WPX #1, he was featured on the cover of CQ in 1957.
> Lloyd and Iris married in 1938, and Iris obtained her license in 1945, as W6DOD. By the mid-sixties, their successful real estate investments and contracting business in Alaska and California had ensured financial independence for the Colvins. DX history was about to be made.

REPUBLIC OF TURKEY
MINISTRY OF TRANSPORTATION
GENERAL DIRECTORATE OF RADIOCOMMUNICATIONS

GEÇİCİ AMATÖR TELSİZCİLİK BELGESİ
(Temporary Amateur Radio License)

ADI ve SOYADI (Name and surname) : LLOYD DAYTON COLVIN
MESLEĞİ (Job/Title) : RETIRED
DOĞUM YERİ VE TARİHİ (Place and Date of birth): WASHINGTON 1915
TÜRKİYEDEKİ ADRESİ ve KALMA SÜRESİ (Address and period of stay in Turkey) : GRAND LORD HOTEL -İSTANBUL (Üç ay)
PASAPORT NUMARASI (Passport Number) : 050922277
İSTASYON ADRESİ (Station adress) : GRAND LORD HOTEL İSTANBUL
ÇAĞRI ADI (Call sign) : TA4/V6KG
BELGE SINIFI (Class of license) : EXTRA CLASS (TÜRKİYE'de A)
CEPT LİSANSI (Cept license) : ---
VERİLİŞ TARİHİ (Effective Date) : 08.11.1993
BİTİŞ TARİHİ (Expiration Date): 06.02.994
AMATÖR TELSİZ CİHAZLARI (Amateur Radio Equipment) : İCOM 751 A S.N.: 02445

Bu geçiçi belge, yabancılara Türkiye de Amatör Telsizcilik Yönetmeliği esaslarina göre yukarıdaki özelliklere sahip bir Amatör Radyo istasyonu kurma ve işletme müsaadesi verir.

This temporary license permits the holder (Foreigner) to set up and operate an Amateur Radio Station in Turkey as specified above in accordance with the Radio Amateur Requlation in force in Turkey.

BELGE SAHİBİNİN İMZASI (signature of the Holder)

İŞLETME VE DIŞ İLİŞK. DAİ.BŞK. (Headi of Operation and International Relations Dept.)

B. Gürbüz [illegible]BULUT
TELSİZ UZMANI

ONAY (Approval)
[illegible] Kemal [illegible]MAZ
Bölge Müdürü

The YASME Foundation had been established in 1961 to aid the worldwide sailboat DXpeditioning of Danny Weil, VP2VB, a young Englishman. By 1964, Weil had married, lost the last of five boats and abandoned his odyssey. The foundation had also sponsored several other DXpeditioners. Although it now had no funds, YASME could provide licensing and QSL assistance to DXers who could pay their own way - as the Colvins could. Thus YASME and the Colvins began a 30-year association.

The Colvins' YASME saga began in the fall of 1965 on Saipan; the last completed trip, to Southeast Asia, was in 1992. They traveled to 221 DXCC countries, operating in about two thirds of them and generated over one million QSOs. YASME volunteers answered over 750,000 QSLs, which are cataloged and stored in the Colvin home - the world's largest QSL collection.

Lloyd and Iris frequently operated from countries, such as Abu Ail in 1982 and Burundi in 1991, which had seen no amateur operation for many years. In 1989, glasnost brought the opportunity to tour all 15 Soviet republics, and to operate in all but Armenia.

Both Colvins were Life Members of the Northern California DX Club, and honorary members of many other DX organizations. On average, they were abroad for six months each year, and were rarely home for Christmas. When at home, they frequently entertained traveling DXers they had met on their sojourns; their annual Fourth of July party was well attended by local and visiting DXers.

In addition to Iris, Lloyd is survived by a daughter, Joy Gilcrease, formerly W4ZEW, and granddaughters Justine and Vanessa Gilcrease. His ashes have been scattered at sea.

Hugh Cassidy, WA6AUD, also writing in the NCDXC newsletter The DXer, said

> As far as I am concerned, Lloyd and Iris Colvin were the most spectacular and most enduring team of DXers ever to show on the bands. They were the friends of every journeyman DXer and the passing of Lloyd will leave a void that possibly never can be filled. They were a DXing phenomenon which lasted so long that it often was taken for granted, something that would always naturally occur. It

wasn't and it won't. Of all the qualities that should be remembered and admired, it was their always-openness with any DXer, and the total countries worked was not their criterion. If you were a DXer, you were a friend and the Colvins had friends. Always.

Members of the club may have their favorite story about the Colvins, some of which have been heard before, and some of which bear repeating. One which I have often retold is how Lloyd fell off the roof of their three-story home over by the edge of the bay in Richmond. This happened about twenty years ago. Years back, the Colvins had a home up atop the Berkeley Hills on Grizzly Boulevard. They sold that home and traveled the world DXing for some years. Returning to the Bay Area, they purchased property adjacent to present Interstate 580 and built a three story home. Most club members are familiar with the site: the poles for their low band antennas are often noted.

One morning upon arising, and when Iris was fixing breakfast, Lloyd said he was going up to the flat rooftop of the structure to check the antennas. Up there, he carefully studied the various arrays, edging backwards to get better angles for viewing or to check certain things. He kept moving back until his heel struck the low parapet around the rooftop, and Lloyd toppled over the side of the building. He hit the ground three stories down, fortunately hitting softened ground.

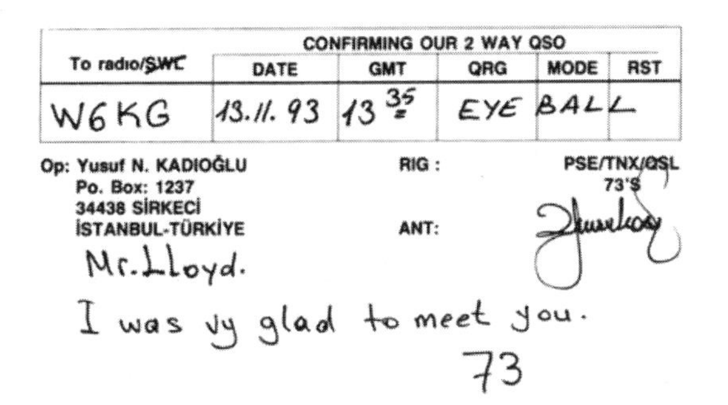

To radio/SWL	CONFIRMING OUR 2 WAY QSO				
	DATE	GMT	QRG	MODE	RST
W6KG	13.11.93	13 35	EYE	BALL	

Op: Yusuf N. KADIOĞLU
Po. Box: 1237
34438 SİRKECİ
İSTANBUL-TÜRKİYE

RIG :

ANT:

PSE/TNX/QSL
73's

Mr. Lloyd.
I was vy glad to meet you.
73

After a moment or two, maybe even three, of running through the peripheral check-list. Lloyd got to his feet and climbed back up the stairs to the kitchen on the upper floor. There Iris, still busily engaged, was astonished to see Lloyd come up the stairway. 'Lloyd! I thought you were up on the roof,' Iris said in surprise. 'I was,' was Lloyd's reply, 'I was.'

Some may have some reservations about this story. Some years ago, when writing for CQ, I wanted to use this item in a DX column and checked with Lloyd and Iris before sending off the copy, They confirmed that this is the way it happened. Absolutely!

It is a sad task to tell such things and to realize that they must be put in the past tense. But there are other factors. One eventually learns that one lives as long as one is remembered, and is dead when forgotten. For DXers, Lloyd Colvin will live a long time - a very long time.

Garry Shapiro, NI6T, was editor of The DXer at time. He wrote:

In their 30 years of DX travel, Lloyd and Iris Colvin had countless adventures, some of which have become folkloric. For the benefit of younger members, here are a few, as told by Rubin Hughes, WA6AHF.

His friends say Lloyd was loathe to spend money on fancy accommodations, and scrupulously avoided hotels. Once, in a nameless South American country [it was in Central America], Lloyd and Iris arranged to stay a week at a 'private boarding house' with very reasonable rates, where setting up the ham rig would present no problems. Their first night proved sleepless, with a constant clamor, loud male voices, and a lot of coming and going. The next morning, they found that no men were present - all the other residents of the building were women. They quickly arrived at the correct assessment: they had moved into a brothel. Undaunted, they stayed the rest of the week!

On another occasion, in another third world QTH, Lloyd and Iris were in the street. Suddenly, a man tore Iris's watch from her wrist and took off running. Lloyd took off in hot pursuit of the retreating figure, yelling for assistance as he chased the man into ever narrowing streets and alleys. Of course,

> no one spoke English, and few would have been inclined to help, in any case. Undaunted, Lloyd cornered his man in an apartment house, shook him violently, took the watch, and returned, breathless, to present the regained prize to Iris. Iris looked at the recovered valuable and said, "Lloyd, that isn't my watch!"

The writer John Troster, W6ISQ, said

> I have seen Lloyd and Iris's slide shows of their wonderful DXpeditions at conventions for years. Most of us go someplace and return to our local radio clubs to deliver an evening program on that one trip. Not the Colvins. They would tell us about their latest seven DXpeditions. And they always made it sound so easy. Lloyd would go a couple of rounds with the immigration and customs people, then he would charm the hotel people into declaring it an honor for the Colvins to erect their equipment on the roof. It was almost like a repetitive social event with them.
>
> But do you remember those photographs of Lloyd? No matter who the pictures were taken with, whether it be host amateurs, the head of the local Posts and Telegraph, the busboy who put up the antenna, or all by himself at the rig... Lloyd always had on his coat and tie. It might be the West Coast of Africa, a Pacific Island, South America, wherever, he always sported his ubiquitous coat and tie. Always the gentleman, that's how I'll remember Lloyd.

At the time of Lloyd's death, the ARRL Pacific Division director was Brad Wyatt, K6WR, who said

> My memories of Lloyd Colvin, W6KG, are, of course, intertwined with the activities with Iris and go back many years -- from meetings at the NCDXC, the DXpeditions before YASME and since, the Fresno/Visalia International DX Convention and, of course, the July 4 parties at the Richmond QTH.
>
> Personally, Lloyd and Iris provided me with a number of new countries throughout the years -- so many I cannot now recall. It was always fun to hear their voices and the keying from their travels. I shared with many others the retelling of the stories of the DXpeditions, with slides and without, at the International DX Conventions. It is hard to cite specific examples as there were so many presentations over the years.
>
> Probably the most concrete memories are tied with the more recent July 4 parties at the Richmond QTH. How many of us remember the searching of the QSL card files looking for how many QSLs they had received from us, the circular stairways to Heaven and Hell (with the Devil down below in costume), the refreshments and the food and, above all, the great fellowship with Lloyd and Iris and all those who came by that day?
>
> Lloyd Colvin, W6KG, was unique, one of those rare individuals who, all too infrequently, touches our lives with enormous impact. Fortunately for us, we all were here and lived through this time and were all lucky that he appeared in the Amateur Radio world!

An informal mailing list had popped up in the 1970s, called the "Extra Class Couples." In a newsletter in the spring of 1994 the editor, Chris Baldo, AI6S, said

> For those who may have forgotten the past history of ECC, a review would be in order. At a dinner where there were hams gathered, two of the couples held Extra [Class amateur] licenses. The conversation got on the subject of Extra Class Couples when one of the Extras, Lloyd Colvin, W6KG said 'I wonder how many Extra Class Couples there are.' After a few moments another Extra, Betty Baldo, KB6P, added 'I'll find out.' Then the quest began. At last count there were about 100 couples. The proposed future count should have many more.
>
> The two amateurs mentioned rightfully deserve to be considered the Founders of ECC. They were the catalysts for the group's beginning. The sad note is that they both became Silent Keys just recently, W6KG in December and KB6P in July. We as a group and individually extend to their families and friends our deepest sympathy."

Lloyd Colvin was an ARRL benefactor. He had taken out a life insurance policy, with the proceeds to go to the League. According to the ARRL the date of the original arrangement was "apparently in the mid-'60s." The League was asked, for the purposes of this book, but was unable to pin down a date. Lloyd Colvin's life insurance policy, according to the League, was on himself, and was in the amount of $100,000, with the ARRL as the beneficiary. Lloyd paid the annual premium, $3,028. The League said that in the late 1980s "we determined that for the same premium a policy with a value of about $154,000 could be obtained from a different company. Lloyd was agreeable to the change."

After Lloyd's death in December 1993, the League wrote,

"COLVIN AWARD TO BENEFIT AMATEUR RADIO DXING"

While he never made a public fuss about it, Lloyd Colvin, W6KG, gave an insurance policy on his own life to the ARRL and made an annual contribution to the League to pay the premium. The proceeds of the policy were to be used in accordance with an agreement between Lloyd and the ARRL.

Upon Lloyd's death late last year (1993), the proceeds of the policy, more than $150,000 - became available to the League.

It was Lloyd's intent that the income generated by this endowment, to be called the Colvin Award, would be used to further the aspect of Amateur Radio that he held most dear: the strengthening of international friendship through DXing. While the details of how the Colvin Award will be administered are still being worked out, the ARRL will proudly honor Lloyd's memory by fulfilling his intent.

During his lifetime, W6KG made great personal contributions to international friendship through Amateur Radio. Through the Colvin Award, those contributions will continue in perpetuity.

The League says that following Lloyd's death "the proceeds from the policy were placed in a permanently restricted fund. In other words, we can't touch the principal. The investment income goes into a separate fund from which the Colvin Award grants are made. Consistent with Lloyd's wishes and with the policies of the ARRL Board, the Colvin Award is conferred in the form of grants in support of Amateur Radio projects that promote international goodwill in the field of DX."

As of late-2002, grants had been made to the 1996 World Radiosport Team Championship, Heard Island DXpedition; (1997) China Radio Sport Association (BS7H DXpedition); (1998) St. Brandon DXpedition; US Amateur Radio Direction Finding Team (Hungary); (1999) Clipperton DXpedition, Campbell Island DXpedition; (2000) WRTC-2000; Agalega DXpedition (2001); Tromelin DXpedition, Kingman Reef/Palmyra DXpedition; (2001) St. Peter & St. Paul Rocks DXpedition; Albuquerque ARDF Championships; and (2002) Ducie Island DXpedition; WRTC-2002; and the Baker Island DXpedition.

Iris Colvin, and the Foundation, Carry On

At the time of his death Lloyd was president of the YASME Foundation. At their annual meeting in April 1994 Iris was named to that position.

The same weekend, at the 1994 International DX Convention in Visalia, California, the Lifetime Achievement Award (the "Spirit of DX Award"), was presented to Iris by the Southern California DX Club. This annual award was to the person who "has contributed much to the betterment of DX in Amateur Radio," the SCDXC said. "All DXers will recognize Iris as part of the famous YASME DXpeditions."

Also at the convention, a "new ARRL Award in honor of Lloyd Colvin, W6KG, which will benefit Amateur Radio DXing" was announced. This was the gift Lloyd left to the ARRL in the form of a life insurance policy.

In April 1995 Iris joined the Quarter Century Wireless Association's Old Old Timer's Club, #3359 (she was first licensed in 1945). In August 1995, Iris asked the FCC to assign the call sign W6KG to the YASME Foundation. The request was granted.

After Lloyd's death, Iris was not active on the radio but did attend the annual Visalia convention.

After Iris died in 1998, the house in Richmond, California was sold but the big pole and antennas are still there, in 2003, and can be seen from Interstate 580. Lloyd had said in 1981 " There are some amazing statistics about our QSL collection. For one, the weight. We're a little worried about the structure of our home which, incidentally, we built. It was designed to carry a good load, but the QSLs we have amassed weigh over a ton!" (The house survived.)

The YASME Foundation directors met on 21 April 1996 and carried a motion that President Iris Colvin and Director Ruben L. Hughes, WA6AHF, "shepherd the project to complete the biography of Mr. and Mrs. Colvin, and the History of the Foundation."

In early 1998, the foundation received a bequest from the Colvin Family Trust. The foundation continued as a "Public Charity" until 2000 when it was reclassified as a Private Foundation.

This book was commissioned by the YASME Foundation.

Appendix 1: CQ DX Hall of Fame

1. Gus Browning, W4BPD Nov. 1967
2. John M. Cummings, W2CTN Mar. 1968
3. Stewart S. Perry, W1BB Aug. 1968
4. **Richard C. Spenceley, KV4AA Mar. 1969**
5. **Danny Weil, VP2VB Sept. 1969**
6. H. Dale Strieter, W4DQS May 1970
7. Stuart Meyer, W2GHK Oct. 1970
8. **Martti Laine, OH2BH Jan. 1972**
9. Ted Thorpe, ZL2AWJ Aug. 1972
10. Chuck Swain, K7LMU Aug. 1972
11. C.J. (Joe) Hiller, W4OPM Mar. 1973
12. Ernst Krenkel, RAEM Apr. 1974
13. Frank Anzalone, W1WY June 1976
14. **Lloyd & Iris Colvin, W6KG & W6QL Nov. 1976**
15. Geoff Watts, Editor & Publisher Jun. 1977
16. **Don C. Wallace, W6AM 1978**
17. Joe Arcure, W3HNK Dec. 1979
18. Hugh Cassidy, WA6AUD Apr. 1980
19. Eric Sjolund, SMØAGD Apr. 1981
20. Franz Langner, DJ9ZB 1982
21. Dr. Sanford Hutson, K5YY Jan. 1983
22. Rodney Newkirk, W9BRD Feb. 1984
23. Ronald Wright, ZL1AMO Apr. 1985
24. Herb Becker, W6QD Apr. 1985
25. Jim Smith, P29JS/VK9NS Apr. 1986
26. Kan Mizoguchi, JA1BK Apr. 1987
27. John Troster, W6ISQ Apr. 1988
28. Charlie Mellen, W1FH Apr. 1994
29. Carl Henson, WB4ZNH Apr. 1995
30. **Rusty Epps, W6OAT May 1996**
31. **Robert Vallio, W6RGG May 1997**
32. Robert Ferrero, W6RJ May 1997
33. Frank Schwab, W8OK May 1997
34. Robert White, W1CW May 1998
35. Anthony W. DePrato, WA4JQS May 1998
36. **Wayne Mills, N7NG May 1999**
37. Chod Harris, WB2CHO/VP2ML May 2000
38. John Kanode, N4MM May 2000
39. Robert Allphin, K4UEE May 2001
40. Robert Eshleman, W4DR May 2001
41. Lee Bergren, WØAR May 2002

***YASME personages**

Appendix 2: Danny Weil Call Signs

G7DW/MM 1954-1955
KZ5WD, Canal Zone, 1955/56 (no logs)
FO8AN, Tahiti, early 1956 (no logs)
VR1B April 1956
VK9TW, Nauru, July 1956
VR4AA, Honiara, Solomon Islands, August 1956
VK9TW, Papua, September 1956

(YASME I sunk 24 October 1956)

YV0AA/YV0AB, Aves Island, July 7-14, 1958
VP2VB, Brit Virgin Islands, July 14-28, 1958
VP2KF, St. Kitts, August 28-September 4, 1958
VP2AY, Antigua, September 8-October 14, 1958
VP2MX, Montserrat, October 22-27, 1958
VP2KFA, Anguilla, November 1-8, 1958
VP2DW, Dominica, December 1958
VP2LW January 4-11, 1959
VP2SW January 15-24, 1959

(YASME II sinks late January 1959)

VP2GDW March 1959 ("Less YASME")
VP4TW April 1959 ("Less YASME")
VP7VB, Bahamas, September 1959

HK0AA, Bajo Nuevo, June 11-17 1960
VP5VB, Jamaica, May 17-June 4, 1960
KZ5WD July 1960
HC2VB September 1960
HC8VB October 1960
FO8AN/Clipperton, November 1960

FO8AN, Nuku Hiva, Marquesas, January/February 1962
FO8AN, Tahiti, February-July 1962
ZK1BY, Suvorov Island, Manihiki,
October 17-28, 1962
ZM6AW, Rep. of Western Samoa,
November/December 1962
FW8DW, Wallis Island, January 1963
VR2EO, Fiji, March 1963

Appendix 3: Colvin Operations/Call Signs

3D2KG January 1976
3D6QL December 1985
W6KG/4S7 February 1988
W6KG/4X November 1980
4T4WCY December 1988
5H0QL August 1990
5L2KG August 1967
W6KG/5B4 October 1988
W6QL/5N0 February 1989
5T5KG March 1967
5U7QL November 1989
5V1KG November 1967
5Z4KG March 1987
6W8CD February 1967
W6QL/6Y5 November 1978
7P8KG December 1985
7Q7KG September 1990
8P6QL October 1981
8Q7QL February 1987
W6QL/8R1 December 1981
9G1KG October 1967
9H3JM November 1988
9J2LC March 1986
9K2QL February 1983
9L1KG July 1967
9N5QL December 1987
9U5QL January 1991
9Y4KG November 1981
W6KG/A25 January 1986
W6KG/A4 December 1982
W6KG/A7 January 1983
A92QL January 1990
W6KG/AJ3 October 1976
C21NI March 1976
C9QL October 1990
W6KG/CE0 March 1984
W6QL/CE0 March 1984
W6KG/CP6 January 1984
CT2YA October 1966
CT3AU September 1966
D68QL December 1986
DL4ZB 1951-1957
DL4ZBD 1951-1957
DL4ZC 1951-1957
F/W6KG 1951-1957
FA8JD 1944-1945
FG0FOK January 1981
FG0FOL/FS December 1980
FH/W6KG November 1986
FK0KG March 1976
FM0FOL February 1981
FO0XX 1985-1986
FO0XX/MM 1985-1986
FO0XXL March 1990
FR/W6QL October 1986
FY0FOL February 1982
G5ACI/AA December 1982
G5ACI/MM December 1982
G0/W6KG 1966
GC5ACH/W6KG July 1966
GC5ACI/WB6QEP June 1966
GD5ACH/W6KG May 1966
GD5ACI/WB6QEP May 1966
W6QL/HC1 November 1983
W6KG/HC8 December 1983
HI6XQL March 1980
W6QL/HK3 October 1983
W6KG/HK0 November 1983
HR0QL February 1979
HS0ZAP Nov/Dec 1991
HZ1AB Feb (19-20) 1983
J2AHI 1946-1947 by Lloyd and
J2USA 1946-1947
J20DU October 1982
J3ABV October 1979
J6LOO December 1979
J7DBB January 1980
JA2KG 1947-1949
JA2US 1947-1949
JY8KG March 1983
K2CC 1950-1951
K4WAB 1957-1959
K6WAP 1957-1961
K7KG 1950-51/1965-66
KG6SZ/KC6 October 1965
KC6SZ October 1965
KG4KG October 1978
W6BWS/KG6 1946
W6KG/KG6 1946
KG6SZ September 1965
KL7DTB 1956-1962
KL7KG 1956-1962
KL7USA 1967-1970
KX6SZ Oct 1965/Jan 1966
W6QL/PJ2 March 1982
PJ8KG January 1977
W6KG/PZ1 January 1982
RT0U/W6KG May 1989
RT0U/W6QL May 1989
S79KG January 1987
W6KG/SV5 October 1980
W6QL/SV5 October 1980
W6KG/SV9 October 1980
TA1/W6QL December 1993 .
TU2CA September 1967
TY2KG November 1967
W6KG/TI5 January 1979
U3WRW/W6KG May 1989
U3WRW/W6QL May 1989
UC1AWB May 1989
UD7DWB May 1989
UF7FWO May 1989
UH9AWE May 1989
UO4OWA May 1989
UP1BWR May 1989
UP1BWW May 1989
UP1BYL May 1989
UQ1GXX May 1989
UR1RWW May 1989
UR1RWX May 1989
UT4UXX May 1989
UZ1AWA May 1989
UZ3AWA May 1989
UI9AWD Jun 89
UJ9JWA Jun 89
UL8NWC Jun 89
UM9MWA Jun 89
V85KGP Feb/Mar 1992
VK2GDD January 1990
VP1KG March 1979
VP2EEQ December 1977
VP2KAH February 1980
VP2MAQ March 1977
VP2SAX November 1979
VP2VDJ November 1976
VR1Z Feb 1966/Dec 1975
VR8B January 1976
W6QL/VP2A February 1977
W2USA 1951-1952
W4KE 1951-1953
W4ZEW 1951-1953
W6AHI Lloyd and Iris
W6ANS 1933-1938
W6DOD by W6QL
W6IPF 1933-1938
W6KFD 1933-1938
W6KG YASME official
W6QL YASME official
W7KG 1929-1951
W7YA 1929-1951
WA6DFR 1958-1962
WW6ITU May 1975
XE2GKG October 1987
XT2KG November 1989
XU8KG January 1992
XW8QL Jan/Feb 1992
XX9TQL March 1992
YB0AQL March 1988
YJ8KG May 1976
W6QL/Z2 February 1986
ZB2AX August 1966
ZC4ZR November 1988
ZD3I May 1967
ZF2CI December 1979
ZL0AKH February 1990
W6QL/ZP5 February 1984
W6KG/ZS October 1985
ZS3/W6QL October 1985
ZS9/W6KG Nov/Dec 1999

Appendix 4: Colvins' List of Countries Visited

Appendix: Colvin Countries-Visited List, as maintained by them

United States
Canada
Formosa
China
Cuba
Morocco
Portuguese Guinea
Macao
Portugal
Azores
Madeira Islands
Germany
East Germany
West Germany
Philippine Islands
Spain
Balearic Islands
Canary Islands
Rio de Oro
Cueta and Melilla
Republic of Ireland
Liberia
Iran
Ethiopia
France
Corsica
French Oceania
French Guiana
England
Guernsey
Jersey
Isle of Man
Northern Ireland
Scotland
Wales
Hungary
Switzerland
Liechtenstein
Haiti
Dominican Republic
Colombia
Korea
Panama
Honduras
Thailand
Vatican City
Saudi Arabia
Italy
Sardinia
Japan
Okinawa
Bonin Islands
East Caroline Islands
West Carolines
Tuvalu
Guam
Mariana Islands
Johnston Island
French West Africa
Gold Coast
Midway Island
Puerto Rico
American Samoa
U.S. Virgin Islands
Marshall Islands
Canal Zone
Norway
Luxembourg
Austria
Finland
Belgium
Denmark
Netherlands
Brazil
Suriname
Sweden
Egypt
Crete
Dodecanese
Greece
Turkey
Guatemala
Costa Rica
Ivory Coast
Dahomey Republic
Mali Republic
British Virgin Islands
Bahamas
Gilbert Islands
Fiji Islands
Singapore
Hong Kong
India
Mexico
Nicaragua
El Salvador
Venezuela
Albania
Gibraltar
Gambia
Monaco
Tunisia
Ceylon
Libya
Nigeria
Mauritania
Togo Republic
Senegal Republic
Jamaica
Algeria
Barbados
Guyana
San Marino
Ghana
Sierra Lconc
West Malaysia
East Malaysia
Trinidad
Ebon Atol
Tangiers
Republic of Nauru
New Hebrides
New Caledonia
Australia
Andorra
Western Samoa
Sint Maarten
Saint Martin
Montserrat
Anguilla
Saint Kitts
Montserrat
Guantanamo Bay
Cayman Islands
Belize
Grenada
St. Vincent
St. Lucia
Dominica
Martinique
Guadeloupe
Curacao
Israel
Somali
North Yemen
South Yemen
Djibouti
Oman
United Arab Emirates
Qatar
Bahrain
Abu Ail
Kuwait
Jordan
San Andres
Ecuador
Galapagos Islands
Peru
Bolivia
Paraguay
Chile
Easter Island
Juan Fernandez
Uruguay
Argentina
Syria
Kenya
South Africa
South West Africa
Lesotho
Swaziland
Botswana
Zimbabwe
Zambia
Lebanon
Mauritius
Reunion
Mayotte
Comoros
Seychelles
Maldive Islands
Nepal
Bangladesh
Bhutan
Burma
Indonesia
Taipei
Cyprus
Cyprus (British)
Malta
Niger
Aruba
Russian SFSR
Armenian SSR
Azerbaijan SSR
Byelorussian SSR
Estonian SSR
Georgian SSR
Kazakh SSR
Latvian SSR
Lithuanian SSR
Moldavian SSR
Tadzhik SSR
Turkmen SSR
Ukrainian SSR
Uzbek SSR
Burkina Faso
New Zealand
Tanzania
Malawi
Mozambique
Walvis Bay
Congo
Zaire
Burundi
Rwanda
Vietnam
Cambodia
Laos
Brunei
Cameroon
Saipan

[Aruba]
[Turkey]

*Notes: Aruba and Turkey added by the author.

Montserrat appeared twice on the Colvins' list.

The Colvins tallied 223, through Saipan, including Montserrat twice.